Plant Pathology: An Integrated Study

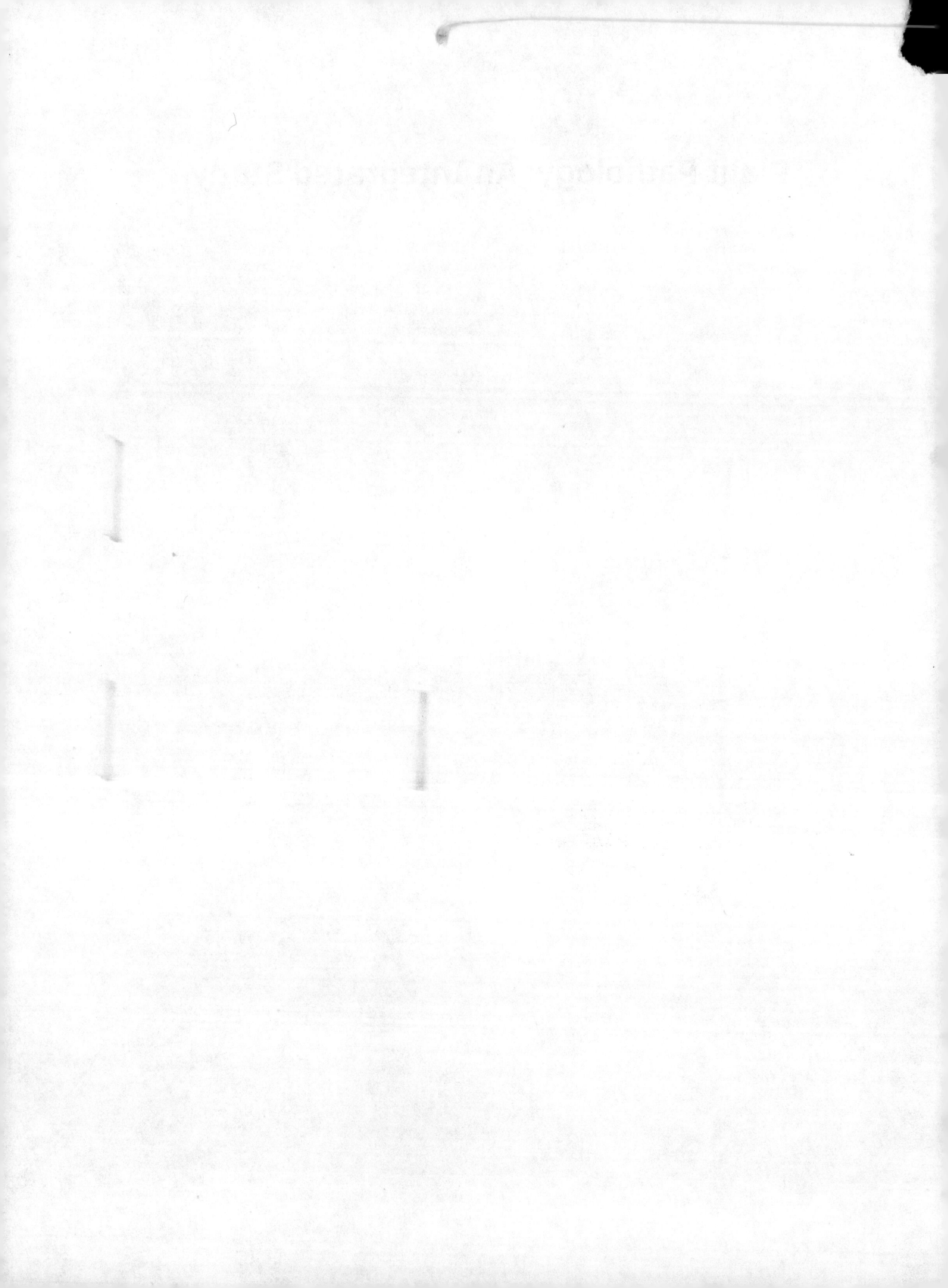

Plant Pathology:
An Integrated Study

Edited by Chris Frost

Syrawood Publishing House
New York

Published by Syrawood Publishing House,
750 Third Avenue, 9th Floor,
New York, NY 10017, USA
www.syrawoodpublishinghouse.com

Plant Pathology: An Integrated Study
Edited by Chris Frost

International Standard Book Number: 978-1-68286-671-9 (Hardback)

Cataloging-in-Publication Data

Plant pathology : an integrated study / edited by Chris Frost.
p. cm.
Includes bibliographical references and index.
ISBN 978-1-68286-671-9
1. Plant diseases. 2. Diseased plants. 3. Plants--Wounds and injuries. I. Frost, Chris.
SB601 .P53 2019
632--dc23

TABLE OF CONTENTS

PREFACE

This book is designed to provide thorough information in the area of plant pathology. It is the study of the plant structure, the nature and effect of diseases they are susceptible to. Diseases in plants are often caused by either pathogens or abnormal physiological conditions. Plant pathology has relevance and applications across a number of other scientific fields such as crop production and management, forest science, etc. It also covers aspects of disease resistance and management in plants, causative pathogens, etc. This book unfolds the innovative aspects of plant pathology which will be crucial to the progress of the field in the future. From theories to research to practical applications, case studies related to all contemporary topics of relevance to this field have been included herein. This book also consists of researches that have transformed this discipline and aided its advancement. It is appropriate for students seeking detailed information in this area as well as for experts.

This book has been the outcome of endless efforts put in by authors and researchers on various issues and topics within the field. The book is a comprehensive collection of significant researches that are addressed in a variety of chapters. It will surely enhance the knowledge of the field among readers across the globe.

It gives us an immense pleasure to thank our researchers and authors for their efforts to submit their piece of writing before the deadlines. Finally in the end, I would like to thank my family and colleagues who have been a great source of inspiration and support.

Editor

1

AFLP Fingerprinting for Identification of Infra-Species Groups of *Rhizoctonia solani* and *Waitea circinata*

Bimal S. Amaradasa[1]*, Dilip Lakshman[2] and Keenan Amundsen[3]

[1]*Department of Plant Pathology, University of Nebraska-Lincoln, Lincoln, NE 68583, USA*
[2]*Floral and Nursery Plants Research Unit and the Sustainable Agricultural Systems Lab, Beltsville Agricultural Research Center-West, Beltsville, MD 20705, USA*
[3]*Department of Agronomy and Horticulture, University of Nebraska-Lincoln, Lincoln, NE 68583 USA*

Abstract

Patch diseases caused by *Thanatephorus cucumeris* (Frank) Donk and *Waitea circinata* Warcup and Talbot varieties (anamorphs: *Rhizoctonia* species) pose a serious threat to successful maintenance of several important turfgrass species. Reliance on field symptoms to identify *Rhizoctonia* causal agents can be difficult and misleading. Different *Rhizoctonia* species and Anastomosis Groups (AGs) vary in sensitivity to commonly applied fungicides and they also have different temperature ranges conducive for causing disease. Thus correct identification of the causal pathogen is important to predict disease progression and make future disease management decisions. Grouping *Rhizoctonia* species by anastomosis reactions is difficult and time consuming. Identification of *Rhizoctonia* isolates by sequencing Internal Transcribed Spacer (ITS) region can be cost prohibitive. Some *Rhizoctonia* isolates are difficult to sequence due to polymorphism of the ITS region. Amplified Fragment Length Polymorphism (AFLP) is a reliable and cost effective fingerprinting method for investigating genetic diversity of many organisms. No detailed analyses have been done to determine the suitability of AFLP for inferring infra-species level of *Rhizoctonia* isolates. The objective of the present study was to develop AFLP fingerprinting to identify infra-species level of unknown *R. solani* Kühn and *W. circinata* isolates. Seventy-nine previously characterized *R. solani* (n=55) and *W. circinata* (n=24) isolates were analyzed with AFLP markers generated by four primer pairs. Unweighted Pair Group Method with Arithmetic Mean (UPGMA) correctly grouped *R. solani* and *W.circinata* isolates according to their AG, AG subgroup or *W.circinata* variety. Principle component analysis (PCA) corroborated UPGMA clusters. To our knowledge this is the first time AFLP analysis has been tested as a method to decipher the AG, AG subgroup or *W.circinata* variety across a wide range of *Rhizoctonia* isolates.

Keywords: AFLP; *Rhizoctonia solani*; *Waitea circinata*; anastomosis groups; turfgrasses

Introduction

Patch diseases caused by multiple *Rhizoctonia* species pose a serious threat to growth and maintenance of several important turfgrass species in the southern and transition zones of the USA [1,2]. The transition zone refers to the central part of the country where climatic conditions are not favorable for either cool-season or warm-season turfgrasses. However, both turfgrass types are routinely grown and managed in this region. The form-genus *Rhizoctonia* includes uninucleate, binucleate, and multinucleate species, and of these, multinucleate *Thanatephorus cucumeris* (Frank) Donk (=*R. solani* Kühn) and *Waitea circinata* Warcup and Talbot varieties *agrostis*, *zeae*, *oryzae*, *circinata*, and *prodigus*; and binucleate *Ceratobasidium cereale* Murray and Burpee (=*R. cerealis* Van der Hoeven, AG-D) have been reported from diseased turf lawns and golf greens [1-4]. *Rhizoctonia solani* is a genetically diverse species consisting of many Anastomosis Groups (AGs) [5,6]. Six AGs have been reported to cause blight in turfgrass with AG 1(–IA and –IB), AG 2 (–2IIIB and –2LP), and AG 4 being more common on infected turfgrasses than other AGs [7,8].

Reliance on field symptoms to identify *Rhizoctonia* causal agents can be difficult and misleading. In general, *Rhizoctonia* affected turfgrasses show circular areas of blighted brown colour leaves. Microscopically, all *Rhizoctonia* species look more or less similar, i.e. nonsporulating mycelia with 90 degree branches having dolipore septa [8]. However, *R. solani* isolates can be distinguished from *W. circinata* varieties by colony morphology on Potato Dextrose Agar (PDA). *Rhizoctonia solani* produces brown to dark brown sclerotia on PDA whereas sclerotia of *W. circinata* are orange to salmon in the formative stages and darkens to brown as cultures age [1,2,8]. Also, sclerotia of *W. circinata* are frequently submerged in the media unlike *R. solani* which are formed on the agar surface. Although, colony morphology on PDA can differentiate *R. solani* from *W. circinata*, colony features are not reliable to distinguish AGs within these species. It is common to isolate multiple *Rhizoctonia* species and AGs from infected turfgrasses. Different *Rhizoctonia* species and AGs vary in sensitivity to commonly applied fungicides [9-12] and they also have different temperature ranges conducive for causing disease [8]. Therefore, correct identification of the causal pathogen and its AG is important to predict the disease progression and make future disease management decisions. For plant breeders, knowledge of the main causal pathogens at different locations is important for selecting appropriate turfgrass germplasm with resistance to *Rhizoctonia* blight.

The classical method of grouping isolates of *Rhizoctonia* is based on anastomosis with tester strains. However this method is sometimes difficult to interpret and may take excessive amounts of time when grouping many isolates. Some isolates which are known as Bridging

***Corresponding author:** Bimal S. Amaradasa, Department of Plant Pathology, University of Nebraska-Lincoln, Lincoln, NE 68583, USA
E-mail: bamaradasa2@unl.edu

Isolates (BI) can anastomose with more than one AG leading to further confusion [13].

It is also important to note that anastomosis reactions cannot be used to distinguish subgroups within an AG because subgroups anastomose with each other [14].

Although analysis of Internal Transcribed Spacer (ITS) region is a well-tested method for identifying *Rhizoctonia* species, it can be cost prohibitive for investigating a large number of samples. Some *Rhizoctonia* isolates are difficult to sequence due to polymorphisms in the ITS region [15-17]. These isolates may require cloning before sequencing, which adds more time and cost to the analysis. ITS sequence polymorphism may make it difficult to group *Rhizoctonia* isolates to their AG subgroups [18].

PCR based fingerprinting method Amplified Fragment Length Polymorphism (AFLP) is a cost effective alternative for assessing genetic diversity of many organisms. AFLP is a multilocus marker and does not need prior knowledge of sequence information to generate polymorphic markers. Technology advancements have made AFLP technique a relatively cheap, easy, fast and reliable method to generate hundreds of informative genetic markers [19]. No detailed studies have been done to ascertain whether AFLP on *Rhizoctonia* should be performed at genus level, species level or infra-species level (i.e. at AG and AG subgroup level or *W. circinata* variety). The objective of this study was to determine if AFLP is capable of revealing AG, AG subgroup or variety of unknown *R. solani* and *W. circinata* isolates commonly occurring on cool-season turfgrasses in comparison to the conventional ITS sequence analysis. Since no detailed analyses have been done to determine whether AFLP is suitable for grouping multiple species of *Rhizoctonia* together or whether this method is appropriate for deciphering the genetic diversity of isolates within a single *Rhizoctonia* species, we analyzed *R. solani* and *W. circinata* isolates together as well as separately.

Materials and Methods

Isolates used in this study and ITS sequence analysis

Rhizoctonia isolates (n=71) used in this study were collected from lawns and golf courses of Virginia and Maryland during summer months of 2007 to 2009 (Table 1). A previous study had identified these isolates to species, AG or AG subgroup level using colony morphology on PDA, anastomosis reactions, and ITS sequence analysis as described by Amaradasa et al. [20]. The present study also included eight tester strains consisting of *R. solani* (AG 1-IB, 2-2IIIB and 5) and *W.circinata* (var. *zeae* and *circinata*) (Table 2). Accordingly, there were 55 *R. solani*

Isolate	Origin	Host	Management Type	Species acronym	Anastomosis group	GenBank Accession no‡.
ANP 202B	Annapolis, MD	Tall fescue	Lawn	Rs	AG 2-2IIIB	JX631193
ANP 205A	Annapolis, MD	Tall fescue	Lawn	Rs	AG 2-2IIIB	JX631194
ANP 205B2	Annapolis, MD	Tall fescue	Lawn	Rs	AG 2-2IIIB	JX631195
ANP 309A	Annapolis, MD	Tall fescue	Lawn	Rs	AG 2-2IIIB	JX631196
ANP 301B	Annapolis, MD	Tall fescue	Lawn	Rs	AG 1-IB	JX631170
ANP 306B	Annapolis, MD	Tall fescue	Lawn	Rs	AG 1-IB	JX631171
ANP 109B	Annapolis, MD	Tall fescue	Lawn	UWC	WAG	JX631224
ANP 304	Annapolis, MD	Tall fescue	Lawn	UWC	WAG	JX631225
BELT 114	Beltsville, MD	Tall fescue	Lawn	Rs	AG 2-2IIIB	JX631189
BELT 150	Beltsville, MD	Tall fescue	Lawn	Rs	AG 2-2IIIB	JX631190
BELT 262	Beltsville, MD	Tall fescue	Lawn	Rs	AG 2-2IIIB	JX631191
BELT 267	Beltsville, MD	Tall fescue	Lawn	Rs	AG 2-2IIIB	
BELT 26	Beltsville, MD	Tall fescue	Lawn	Rs	AG 1-IB	JX631156
BELT 2	Beltsville, MD	Tall fescue	Lawn	Rs	AG 1-IB	JX631157
BELT 5	Beltsville, MD	Tall fescue	Lawn	Wcz	WAG-Z	JX631239
BELT 159	Beltsville, MD	Tall fescue	Lawn	Wcz	WAG-Z	JX631237
BELT 228	Beltsville, MD	Tall fescue	Lawn	UWC	WAG	JX631221
BLBG 6	Blacksburg, VA	CBG/ABG	Golf green	Rs	AG 2-2IIIB	JX631186
BLBG 13	Blacksburg, VA	CBG/ABG	Golf green	Rs	AG 2-2IIIB	JX631185
BLBG 20C	Blacksburg, VA	CBG/ABG	Golf green	Rs	AG 2-2IIIB	JX631180
BLBG 22C	Blacksburg, VA	CBG/ABG	Golf green	Rs	AG 2-2IIIB	JX631181
BLBG 32C	Blacksburg, VA	CBG/ABG	Golf green	Rs	AG 2-2IIIB	JX631182
BLBG 320	Blacksburg, VA	Tall fescue	Lawn	Rs	AG 1-IB	JX631162
BLBG 510	Blacksburg, VA	Tall fescue	Lawn	Rs	AG 1-IB	JX631165
BLBG 430	Blacksburg, VA	Tall fescue	Lawn	Rs	AG 1-IB	JX631164
BLBG 350	Blacksburg, VA	Tall fescue	Lawn	Rs	AG 1-IB	JX631163
BLBG 211	Blacksburg, VA	CBG/ABG	Golf green	Wcc	WAG	JX631228
BLBG 216	Blacksburg, VA	CBG/ABG	Golf green	Wcc	WAG	JX631229
BLBG 202	Blacksburg, VA	CBG/ABG	Golf green	Wcc	WAG	JX631230
BLBG 8	Blacksburg, VA	CBG/ABG	Golf green	Wcc	WAG	JX631227
HDN 102	Herndon, VA	CBG/ABG	Golf green	Rs	AG 2-2IIIB	JX631201
HDN 208By	Herndon, VA	Tall fescue	Golf rough	Rs	AG 2-2IIIB	JX631202

HDN 225	Herndon, VA	Tall fescue	Golf rough	Rs	AG 2-2IIIB	JX631203
HDN 111A	Herndon, VA	Tall fescue	Golf rough	Rs	AG 1-IB	JX631166
HDN 122A	Herndon, VA	Tall fescue	Golf rough	Rs	AG 1-IB	JX631167
HDN 302	Herndon, VA	Tall fescue	Golf rough	Rs	AG 1-IB	JX631168
HDN 115A	Herndon, VA	Tall fescue	Golf rough	Wcz	WAG-Z	JX631235
HDN 211	Herndon, VA	Tall fescue	Golf rough	Wcz	WAG-Z	JX631236
HDN 222A	Herndon, VA	Tall fescue	Golf rough	UWC	WAG	JX631226
LB 312	Leesburg, VA	Tall fescue	Lawn	Rs	AG 2-2IIIB	JX631192
LB 317	Leesburg, VA	Tall fescue	Lawn	Rs	AG 2-2IIIB	JX631183
LB 325	Leesburg, VA	Tall fescue	Lawn	Rs	AG 2-2IIIB	JX631184
LB 4114	Leesburg, VA	Tall fescue	Golf rough	Rs	AG 2-2IIIB	-
LB 4118B	Leesburg, VA	Tall fescue	Golf rough	Rs	AG 2-2IIIB	JX631197
LB 4303	Leesburg, VA	Tall fescue	Golf rough	Rs	AG 2-2IIIB	JX631198
LB 4316	Leesburg, VA	CBG/ABG	Golf green	Rs	AG 2-2IIIB	JX631199
LB 4319	Leesburg, VA	Tall fescue	Golf rough	Rs	AG 2-2IIIB	JX631200
LB 123	Leesburg, VA	Tall fescue	Lawn	Rs	AG 1-IB	JX631158
LB 124	Leesburg, VA	Tall fescue	Lawn	Rs	AG 1-IB	JX631159
LB 127	Leesburg, VA	Tall fescue	Lawn	Rs	AG 1-IB	JX631160
LB 234	Leesburg, VA	Tall fescue	Lawn	Rs	AG 1-IB	JX631161
LB 4217	Leesburg, VA	Tall fescue	Golf rough	Rs	AG 1-IB	JX631169
LB 204	Leesburg, VA	Tall fescue	Lawn	Rs	AG 5	JX631204
LB 319	Leesburg, VA	Tall fescue	Lawn	Wcz	WAG-Z	JX631233
LB 228	Leesburg, VA	Tall fescue	Lawn	Wcz	WAG-Z	JX631234
LB 4116	Leesburg, VA	Tall fescue	Golf rough	Wcz	WAG-Z	JX631238
BSF 69	Richmond, VA	Tall fescue	Lawn	Rs	AG 2-2IIIB	JX631176
BSF 50	Richmond, VA	Tall fescue	Lawn	Rs	AG 2-2IIIB	JX631175
BSF 42	Richmond, VA	Tall fescue	Lawn	Rs	AG 2-2IIIB	JX631174
BSF 90	Richmond, VA	Tall fescue	Lawn	Rs	AG 2-2IIIB	JX631177
BSF 207	Richmond, VA	Tall fescue	Lawn	Rs	AG 2-2IIIB	JX631178
BSF 209	Richmond, VA	Tall fescue	Lawn	Rs	AG 2-2IIIB	JX631179
BSF 214	Richmond, VA	Tall fescue	Lawn	Rs	AG 2-2IIIB	JX631188
BSF 127	Richmond, VA	Tall fescue	Lawn	Rs	AG 2-2IIIB	JX631187
BSF 13	Richmond, VA	Tall fescue	Lawn	UWC	WAG	JX631223
PW 326	Woodbridge, VA	Tall fescue	Lawn	Rs	AG 1-IB	JX631173
PW 353	Woodbridge, VA	Tall fescue	Lawn	Rs	AG 1-IB	JX631172
PW 220	Woodbridge, VA	Tall fescue	Lawn	Wcz	WAG-Z	JX631232
PW 119	Woodbridge, VA	Tall fescue	Lawn	Wcz	WAG-Z	JX631231
VABCH 8	Virginia Beach, VA	Tall fescue	Lawn	Wcz	WAG-Z	-
VABCH 10	Virginia Beach, VA	Tall fescue	Lawn	Wcz	WAG-Z	-

*ABG: Annual Bluegrass; CBG: Creeping Bentgrass; Rs: *R. solani*; Wcz: *W. circinata* var. *zeae*; Wcc: *W. circinata* var. *circinata*; UWC: Unidentified *W. circinata* species.
‡ITS sequence.

Table 1: Geographic origin, host, management type, species, and anastomosis group of isolates used in this study*.

Isolate	Species Acronym*	AG	Host‡	Location	Donor†
EDHGED 2-1	Wcc	Not assigned	ABG	California, USA	FW
BSCCST 17-1-1	Wcc	Not assigned	ABG	California, USA	FW
AVGCAV	Wcz	WAG-Z	ABG	California, USA	FW
M008	Wcz	WAG-Z	Rice	Japan	MC
Rh102/T	Rs	AG 5	Unknown	Unknown	LB
Rh 63/T	Rs	AG 5	wheat crown	California, USA	LB
Rh146	Rs	AG 2-2IIIB	Bentgrass	Georgia, USA	LB
BM2	Rs	AG 1-IB	Unknown	Unknown	BM

*Rs: *R. solani*; Wcc: *W. circinata* var. *circinata*; Wcz: *W. circinata* var. *zeae*
‡ABG: Annual Bluegrass
†FW: Frank Wong, University of California Riverside, USA (currently Bayer Crop Science, USA). LB: Lee Burpee, University of Georgia, USA. MC: Marc Cubeta, North Carolina State University, USA.

Table 2: Rhizoctonia and Waitea tester isolates used in this study.

isolates including tester strains, which consisted of 33 AG 2-2IIIB, 19 AG 1-IB, and three AG 5. A total of 24 isolates represented *W.circinata* with 13, six and five isolates belonging to *W.circinata* var. *zeae* (Wcz), var. *circinata* (Wcc) and an unknown *W.circinata* group (UWC), respectively. In addition to these isolates, ITS phylogram included Genbank deposited ITS sequences of *W.circinata* varieties *agrostis* and *prodigus*.

Generation of AFLP markers

DNA was purified using the QIAGEN DNeasy plant mini kit (QIAGEN Inc., Valencia, CA) according to the manufacturer's instructions. AFLP analysis explained below was based on the method described by Ceresini et al. [21] and Vos et al. [22]. All reaction plates contained the isolate BELT 267 in duplicate to ascertain reproducibility of AFLP fragments. DNA samples were digested with restriction enzymes *Eco*RI (New England BioLabs, Beverly, MA) and *Mse*I (New England BioLabs). Thereafter, digested products were ligated with *Eco*RI Double Stranded (ds) adapter (EA1: 5′-CTCGTAGACTGCGTACC-3′ and EA2 3′-CATCTGACGCATGGTTAA-5′) and *Mse*I ds adapter (MA1: 5′-GACGATGAGTCCTGAG-3′ and MA2: 3′-TACTCAGGACTCAT-5-3′). Both digestion and ligation reactions were done in one step by preparing a reaction mixture of 20 µl having 2 U of each restriction enzyme, 1.2 U of T4 DNA ligase (New England BioLabs), 0.1 µM of *Eco*RI adapter, 1 µM of *Mse*I adapter and 100 ng of DNA template. The reaction mixture also included 1× *Eco*RI buffer (50 mM NaCl, 100 mM Tris-HCl, 10 mM $MgCl_2$, 0.025% Triton X-100), 1× *Mse*I buffer (50 mM NaCl, 10 mM Tris-HCl, 10 mM $MgCl_2$, 1 mM DTT), 1× T4 ligase buffer (50 mM Tris-HCl, 10 mM $MgCl_2$, 10 mM DTT, 1 mM ATP), and 2µg of bovine serum albumin. The reaction mixture was incubated overnight at room temperature to complete digestion and ligation reactions and thereafter, diluted ten-fold by adding sterile TE (Tris-EDTA) buffer and stored at -20°C for later use. The first amplification (pre-amplification) was carried out with one selective nucleotide for each primer: *Eco*RI primer + *A* (5′-GACTGCGTACCAATTCA-3′) and *Mse*I primer + *C* (5′-GATGAGTCCTGAGTAAC-3′). Each sample of 25 µl included 5µl of digestion and ligation reaction from the previous step, 0.5 µM each of *Eco*RI and *Mse*I primers, 1× Taq polymerase reaction buffer (10 mM Tris-HCl, 50 mM KCl, 1.5 mM $MgCl_2$), 0.2 mM dNTP and 1 U of Taq polymerase (New England BioLabs). The PCR was performed in a thermocycler (MJ Research PTC-200, Global Medical Instrumentation, Ramsey, MN) with initial denaturation at 94°C for 2 min followed by 20 cycles of 30 s at 94°C, 1 min at 56°C and 1 min at 72°C. PCR products were diluted ten-fold with TE buffer and used as the template DNA for selective amplification using four *Eco*RI and *Mse*I primer pairs with three selective nucleotides (*Eco*RI primer + *ACA* and *Mse*I primer + *CAA*, *Eco*RI primer + *AAA* and *Mse*I primer + *CTA*, *Eco*RI primer + *AAC* and *Mse*I primer + *CAC*, *Eco*RI primer + *AGT* and *Mse*I primer + *CTG*). *Eco*RI and *Mse*I primers were synthesized by Eurofins MWG Operon (Huntsville, AL) and IDT (Coralville, IA), respectively. All *EcoRI* primers were end labeled with fluorescence dye 6-FAM™ at the 5′ end. Each selective PCR mixture of 20 µl included 4 µl of diluted preselective reaction, 0.5 µM each of *Eco*RI and *Mse*I primers, 1× standard Taq polymerase reaction buffer, 0.2 mM dNTP and 1 U of Taq polymerase. The PCR reaction was performed for 36 cycles with the following cycle profile. Cycle 1 with 30 s DNA denaturation step at 94°C, 30 s annealing step at 65°C, and 1 min extension step at 72°C. The same conditions were used in cycle 2-12 as in cycle 1, but included a progressive drop in the annealing temperature of 0.7°C in each cycle. Cycles 13-36 included 30 s at 94°C, 30 s at 56°C and 1 min at 72°C. A final extension of 5 min at 72°C completed the reaction. Presence of AFLP banding profiles were confirmed by electrophoresing 5µl of PCR samples in 1.7% agarose gel for one hour at 100 V and visualizing ethidium bromide stained gels under UV light.

AFLP data capture and analysis

We used the size standard GeneScan™ 500LIZ® (Applied Biosystems, Foster City, CA) with capillary electrophoresis system ABI 3730 (Applied Biosystems) to capture AFLP fragments. The size standard was added to each PCR sample for automated data analysis and is essential for precise DNA fragment size comparisons between electrophoresis runs. GeneScan 500LIZ has a DNA fragment sizing range of 35-500 bp with 16 single-stranded labeled DNA fragments. Samples amplified with primers were analyzed by loading a denatured cocktail containing 0.5 µl of PCR sample, 9 µl Hi-Di formamide (Applied Biosystems, Warrington) and 0.5 µl of 500LIZ. The GeneMapper software V4.1 (Applied Biosystems) was used to extract and analyze raw data files obtained from the ABI 3730. AFLP products and size standard fragments can be distinguished since FAM dye-labeled AFLP products are associated with blue signals/peaks while LIZ dye-labeled size fragments generate orange signals. 500LIZ electropherograms had clear, tall peaks without any missing or extra ones and GeneMapper software correctly detected them. The *Analysis Range* settings of the software were changed 50 through 500 to limit the allele calling analysis within that range. Thereafter, samples were analyzed using the *Advanced* peak detector algorithm. DNA fragments smaller than 50 bp were not scored to avoid artefacts of primer-dimer formation. The *Advanced* mode uses the defined size standard values to select peaks of the size standard on an electropherogram. This is achieved by ratio matching where the software uses relative distances between neighboring peaks to correctly define sizes. A Size Standard Curve is generated for each sample and this is used as a reference to accurately compare and capture the size of AFLP amplicons among sample runs. AFLP amplicons of *R. solani* and *W. circinata* isolates were analyzed separately as well as together to determine the suitability of this method in each situation. Initially, we created a bin set by setting the minimum peak intensity to 100 relative fluorescent units (rfu) for peaks generated by FAM labeled AFLP products. A bin set is a set of allele definitions specific to a set of samples with a set of analysis conditions. GeneMapper software is capable of scoring alleles directly from a new set of samples without any bin set or using a previously generated bin set. Better results were obtained by first generating a bin set using a low peak amplitude threshold (ex. 100 rfu) which captures most of the AFLP fragments and then applying that bin set to analyze the same sample set with higher Peak Amplitude Threshold settings in order to filter weak signals and background noise. Once peaks were scored, a binary table was generated of ones and zeros relating to presence absence of alleles. We tested binary tables generated with peak capture thresholds of 500, 1000, 1500, and 2500 rfu. Each scoring table of zeros and ones generated by four AFLP primers was imported to NTSYS version 2.2 [23] and converted to a different similarity indices using *Qualitative data* tab of *Dis/similarity* module. The compared indices included Dice, SM, Phi, O, and Y [24-27]. These similarity values were used in MEGA 5 [28] software to construct an UPGMA [25] tree in order to cluster isolates according to their genetic distances. Cophenetic goodness-of-fit tests were also performed as described in NTSYS to ascertain how well the distance matrices are represented by UPGMA dendrograms. For this, COPH module was used to produce a cophenetic value matrix [29] for each UPGMA dendrogram and compared to the relevant distance matrix using the MXCOMP program to compute the correlation between the two matrices. Cophenetic correlation of > 0.9 is a very good fit while 0.8 to 0.9 is a good fit [23]. The genetic distances generated by different similarity indices were used to compute eigenvalues and eigenvectors in NTSYS using *Eigen* function in the *Ordination* module. NTSYS was then used to perform Principal Component Analysis (PCA) [30] by plotting the first three eigenvectors for each similarity index tested.

Genetic variability among subpopulations of *R. solani* and *W. circinata* as shown by UPGMA analysis was determined by analysis of molecular variance (AMOVA) [31] in GenAlEx version 6.5 [32]. Since AFLP markers generate a binary matrix without any information of intra-individual variation (heterozygosity), AMOVA was performed by calculating PhiPT (Φ_{PT}) which is an analogue of Wright's F_{ST}. Normally, F_{ST} and its analogues are greater than zero but rarely exceed 0.5. An F_{ST} value of 0.05 or less is generally considered as reasonably low and may be interpreted to mean that structuring between subpopulations is weak [33,34].

Results

All four AFLP selective primers produced a large number of polymorphic alleles for each isolate. For instance, the average number of alleles scored per *R. solani* isolate per selective primer was 64. The two BELT 267 samples in each plate gave similar fingerprinting patterns indicating high reproducibility of the AFLP technique. Binary tables generated with a peak capture threshold of 1000 rfu gave better results than other rfu values in terms of grouping isolates to their correct AGs. None of the UPGMA dendrograms constructed with different similarity matrices could correctly group all *R. solani* and *W. circinata* isolates when analyzed together (results not shown). A few isolates of AG 2-2IIIB grouped with *W. circinata* var. *zeae* (Wcz) group while *W. circinata* var. *circinata* (Wcc) cluster consisted of a few Wcz isolates. Therefore, we did not proceed with the analysis of combined AFLP data for *R. solani* and *W. circinata* isolates.

AFLP analysis of *Rhizoctonia solani* isolates

The AFLP primer pairs *Eco*RI-*AAC* and *Mse*I-*CAC*, *Eco*RI-*AGT* and *Mse*I-*CTG*, *Eco*RI-*AAA* and *Mse*I-*CTA*, and *Eco*RI-*ACA* and *Mse*I-*CAA* produced 230, 234, 265, and 213 alleles, respectively for the isolates analyzed. Allele 7 produced by *Eco*RI-*AAA* and *Mse*I-*CTA* primer pair was monomorphic across all *R. solani* isolates. AG 1-IB isolates had a total of three monomorphic alleles while AG 2-2IIIB resulted in two. There were no clones resembling isolates with same DNA fingerprinting pattern among AG 1-IB or AG 2-2IIIB. All the alleles (942 in total) produced by four primer pairs were pooled together to make a single binary matrix for calculating genetic similarity of isolates. We compared the dendrograms produced by different similarity indices to the results of ITS sequence analysis in a previous study having all but BELT 267 and LB 4114 isolates (Supplemental Figure 1) [20]. Though all the similarity indices tested largely grouped *R. solani* isolates to their correct AG, the Dice coefficient based UPGMA tree generated the highest cophenetic correlation value of 0.8968 (Figure 1) and corresponded very well with ITS analysis (Supplemental Figure 1). The tree consisted of three clusters that represented the correct AG or AG subgroup (i.e. AG 1-IB, AG 2-2IIIB or AG 5) of each *R. solani* isolate studied (Figure 1). The PCA for *R. solani* isolates clearly separated them into AG or AG subgroup along dimension 2 (Figure 2). This difference was tested using AMOVA, which showed significant difference among AG subgroups (p=0.0001) representing 14.2% of the total genetic variance (Table 3).

AFLP analysis of *Waitea circinata* isolates

The same selective primers mentioned above viz., *Eco*RI-*AAC* and *Mse*I-*CAC*, *Eco*RI-*AGT* and *Mse*I-*CTG*, *Eco*RI-*AAA* and *Mse*I-*CTA*, and *Eco*RI-*ACA* and *Mse*I-*CAA* produced, 97, 122, 91, and 175 alleles, respectively for the 24 *Waitea* isolates. All alleles were polymorphic across the *Waitea* isolates. However, there were one, five and 20 monomorphic alleles among Wzc, Wcc, and UWC isolates, respectively. Similar to *R. solani*, there were no clonal isolates among *W. circinata* isolates. Binary tables generated for each primer set were pooled to produce a single table of 485 alleles and used to calculate genetic similarity values. The Y coefficient gave the best UPGMA tree with the highest cophenetic correlation value of 0.8 (Figure 3). Although UPGMA tree largely corresponded with ITS phylogram (Figure 4), there were few differences. BELT 159 and BELT 5 with colony morphology similar to Wcz on PDA grouped separately in the ITS analysis (Figure 4), whereas AFLP dendrogram had them grouped together with the rest of Wcz isolates (Figure 3). Though the five UWC isolates formed a single large cluster in both ITS phylogram and AFLP

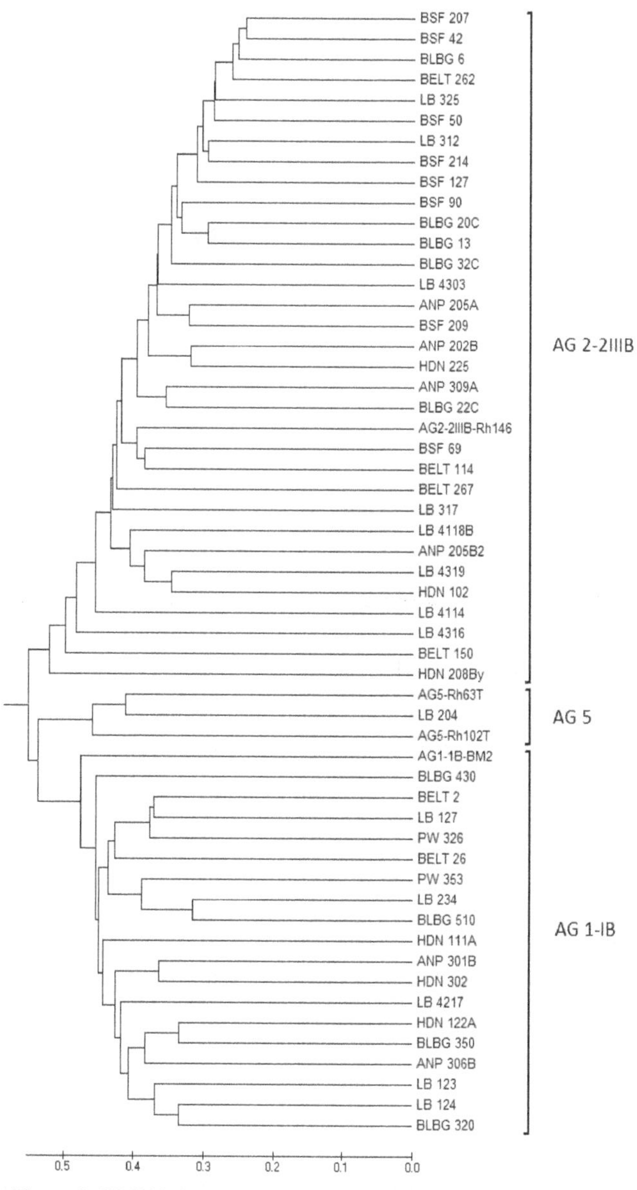

Figure 1: UPGMA dendrogram for *R. solani* isolates derived from Dice's genetic distance matrix. Isolates of *R. solani* anastomosis groups AG 1-IB, 2-2IIIB, and 5 grouped separately and are indicated in the tree.

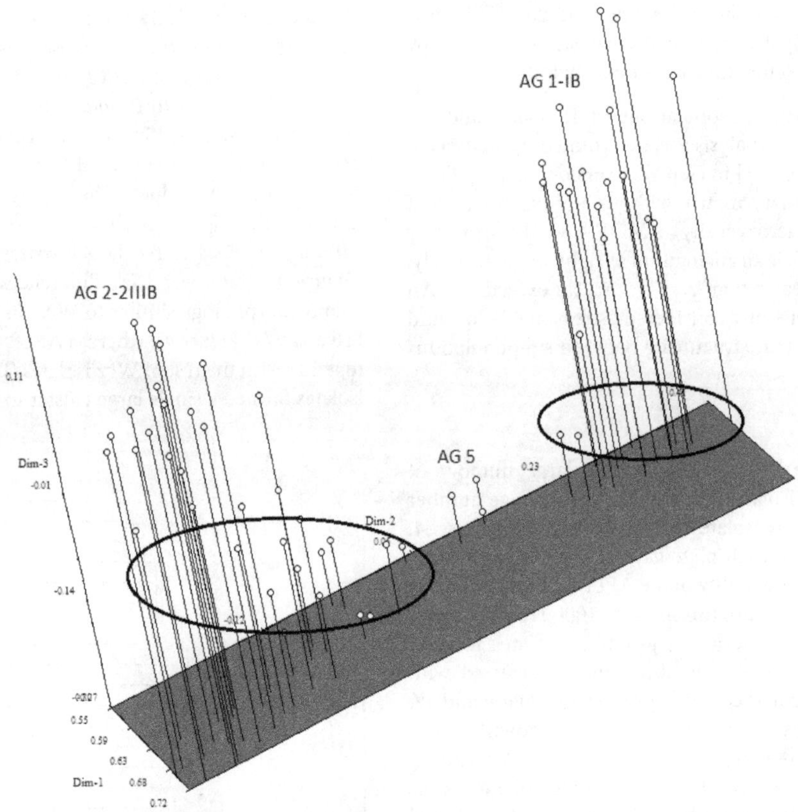

Figure 2: Placement of *R. solani* isolates on a three dimensional plot using PCA based on Dice's genetic distance matrix.

Group	Source of variation	df	SS	% Variation	Φ_{PT}	P value
R. solani[a]	Among populations	2	982	14	0.142	0.001
	Within populations	52	7605	86	0.858	0.001
	Total	54	8587			
W. circinata[b]	Among populations	2	348	12	0.120	0.0001
	Within populations	21	1840	86	0.880	0.0001
	Total	23	2188			

P value for randomization test for ΦPT is based on 999 permutations across the full data set.
R. solani group consisted of three subpopulations namely, AG 1-IB, AG 2-2IIIB, and AG 5.
W. circinata group consisted of varieties zeae, circinata and an unidentified subgroup.

Table 3: Analysis of molecular variance (AMOVA) of *R. solani* and *W. circinata* isolates based on four AFLP primer data.

dendrogram, they resolved differently thereafter. UWC isolate BSF 13 grouped close to *W. circinata* var. *prodigus* in the ITS tree (GenBank accessions HM597147, HM597146, and HQ850254), while UWC isolate ANP 109B grouped with *W. circinata* var. *agrostis* (AB213578 and AB13572) (Figure 4). The rest of the UWC isolates (ANP 304, BELT 228, and HDN222A) grouped in between (Figure 4). The AFLP dendrogram (Figure 3) agreed with the ITS tree clearly by grouping BSF 13 and ANP 109B in two sub-clusters but ANP 304 and BELT 228 did not cluster closely as in ITS phylogram. We did not have AFLP data of varieties *agrostis*, *prodigus* and other GenBank accessions used in the Figure 4 for comparison. Wcc isolates clustered similarly in both AFLP and ITS trees. PCA clearly separated *W. circinata* isolates to their subgroups Wzc, Wcc and UWC along dimension 2 and 3 (Figure 5), which was corroborated with an AMOVA that showed significant difference among *W. circinata* subgroups (p=0.0001), representing 12.0% of the total genetic variance (Table 3).

Discussion

The present study investigated the applicability of the AFLP technique for grouping *R. solani* and *W. circinata* isolates into their infra-species level. In this method, genomic DNA is digested with two restriction endonucleases and two double stranded oligonucleotide adapters are ligated to each fragment. These modified fragments are amplified by two primers recognizing the adapter sequences and adjacent restriction site/s using PCR [22,35]. The resulting banding patterns are highly reproducible and the proportion of the genome analyzed is larger than other DNA fingerprinting techniques such as RAPD [36]. When both *W. circinata* and *R. solani* isolates were analyzed together, AFLP markers did not result in an acceptable dendrogram. It is possible that co-migration of AFLP amplicons generated by genetically distant *Rhizoctonia* and *Waitea* isolates have caused this. High variability of AFLP fingerprinting profiles among distant taxa reduces similarities among them to level of chance [19].

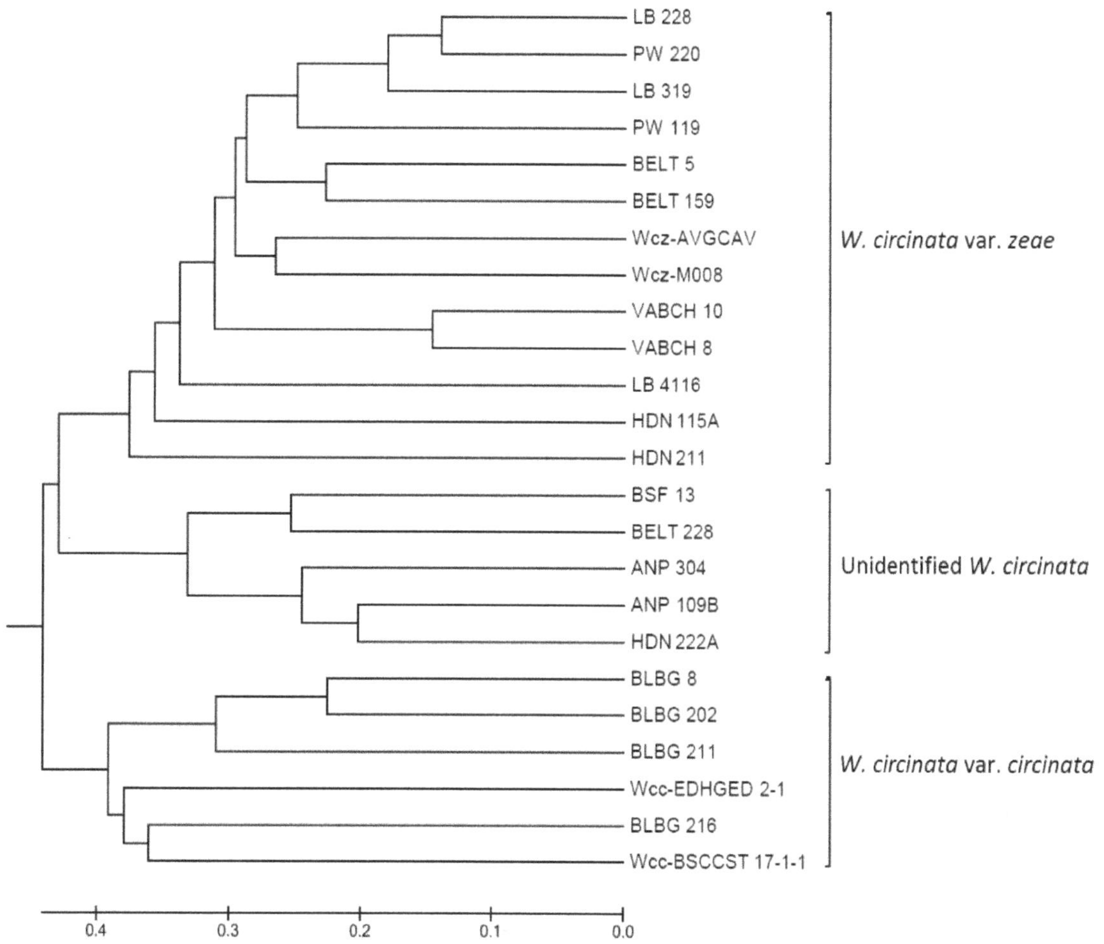

Figure 3: Dendrogram based on UPGMA clustering algorithm for *W. circinata* isolates derived from Dice's genetic distance matrix. *W. circinata* variety of each cluster is indicated in the tree.

Therefore, AFLP is not useful to make phylogenetic inferences among higher taxonomic levels but is more suitable for deriving relationships among closely related lineages.

The AFLP markers used in our analysis was able to accurately resolve *R. solani* isolates to AGs and AG subgroups. Clustering of isolates within the UPGMA dendrogram corresponded well with ITS pylogram and was also corroborated by the PCA scree plots. The AMOVA results also showed significant difference between these subgroups. Both PCA and AMOVA results are positive indicators of the confidence of UPGMA clusters.

ITS sequence analysis grouped Wcz isolates BELT 159 and BELT 5 separately from rest of the Wcz cluster. There was high sequence dissimilarity of 8.5% between above two isolates and other Wcz (sequences dissimilarity not shown). It is possible that ITS region polymorphism of these two isolates have contributed to the discrepancy. Previous studies have reported the ITS region polymorphism within *Rhizoctonia* isolates and how it can compromise accuracy of phylograms [17,18]. Contrary to the results of ITS sequence clustering, AFLP analysis grouped these two isolates within the Wcz cluster. Since AFLP generates multilocus markers, the effect of polymorphism on a single locus is negligible.

Contrary to our findings, previous reviews indicate AFLP and other DNA fingerprinting techniques are more efficacious for studying genetic variation of *Rhizoctonia* at the individual level rather than at the subgroup level within an AG or different AGs [6,37]. However, these hypotheses were based on few studies without proper investigation of analyzing a large number of *Rhizoctonia* and *Waitea* isolates belonging to different AGs. Ceresini et al. [21] employed the AFLP technique to evaluate genetic diversity of isolates within AG3 obtained from potato (PT) and tobacco (TB). AFLP analysis on 32 PT and 36 TB isolates placed them into two distinct groups based on their host. A similar genetic diversity study of *R. solani* AG4 isolates obtained from the rhizosphere of six vineyards in Mexico was reported by Meza-Moller et al. [38]. They analysed 41 *Rhizoctonia* isolates using AFLP markers, which revealed three main groups in the UPGMA dendrogram and six groups from principal component analysis. None of the above studies included different AGs. López-Olmos [39] grouped isolates of AG 2-3, AG BI, and AG 5 from common bean using AFLP. However, each AG was represented by only one or a few isolates, thus limiting applicability of AFLP fingerprinting in deriving AGs of unknown isolates. No peer reviewed documentation is available on performing AFLP on different *W. circinata* varieties. Therefore, we feel our analysis is unique since we used a large number of *R. solani* and *W. circinata* isolates to test applicability of AFLP in resolving isolates to AG and AG subgroup level.

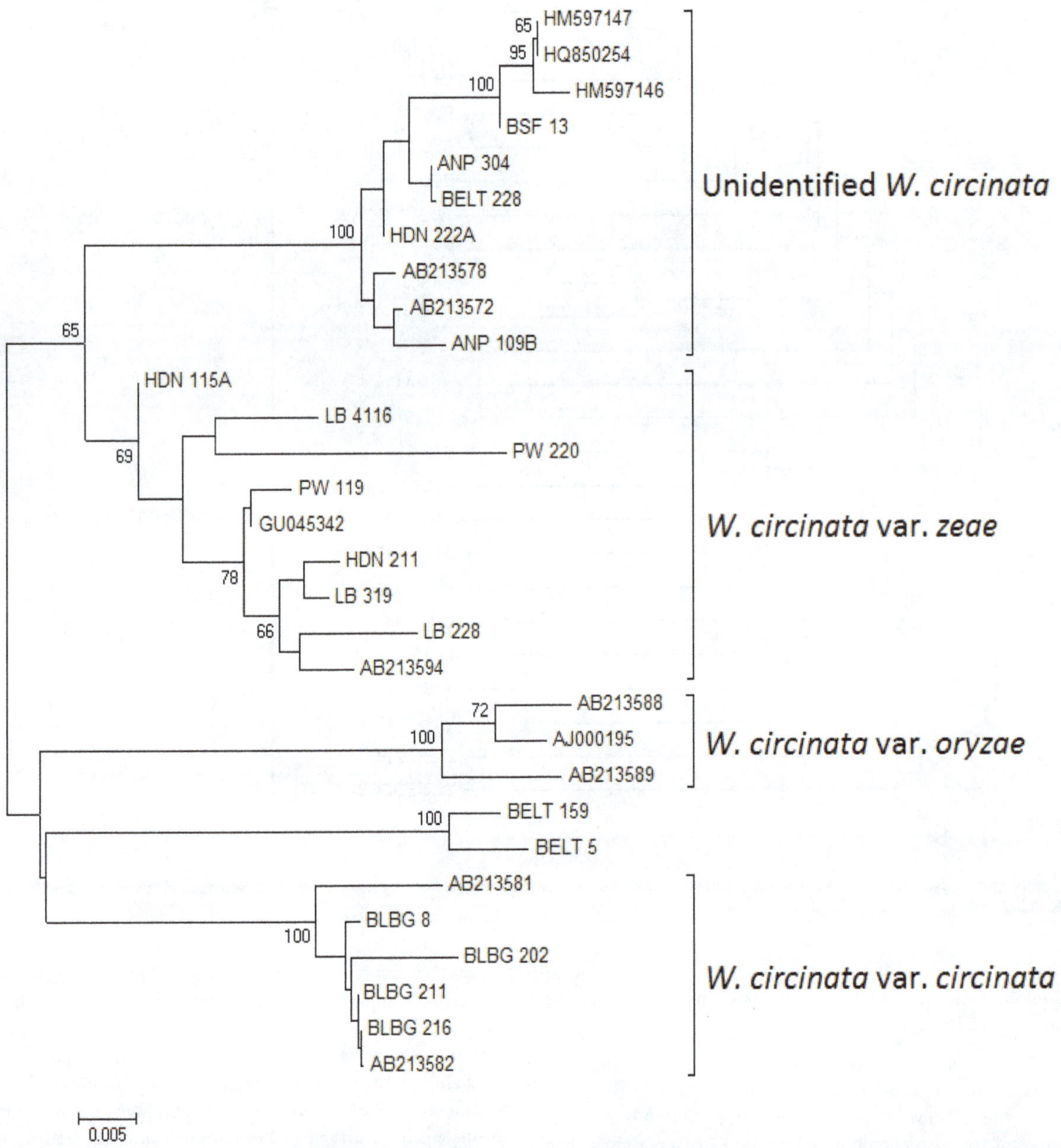

Figure 4: The Neighbour-joining tree of *W. circinata* isolates based on ITS sequence analysis. Bootstrap values 65% and above assessed by 500 replications are shown next to the branches. The tree is midpoint rooted. The clades of *W. circinata* varieties *circinata*, *oryzae*, *zeae* and unidentified *W. circinata* group (UWC) are indicated on the tree. Taxons starting with AB, AJ, GU, HM, and HQ are GenBank accessions. UWC isolate BSF 13 grouped closed to *W. circinata* variety *prodigus* (accessions HM597147, HQ850254, and HM597146) whereas UWC isolate ANP 109B is closely clustering with variety *agrostis* (AB213578 and AB213572). GU045342 and AB213594 are *W. circinata* var. *zeae* ITS sequences. AB213588, AJ000195 and AB213589 are *W. circinata* var. *oryzae* ITS sequences while AB213581 and AB213582 belong to *W. circinata* var. *circinata*.

We used the ABI 3730 electrophoresis system to capture AFLP amplicons since it is sensitive enough to differentiate fragments having one base pair difference. This aided in scoring a large number of polymorphic fragments for the four AFLP primers employed (942 and 485 fragments across *R. solani* and *W. circinata* isolates, respectively). However, size standard used with the ABI 3730 capillary gel electrophoresis system limited the longest fragment size that can be scored to 500 bp. Although, scanning gel images can record longer fragments, this method results in less number of total markers compared to capillary gel electrophoresis system and also requires additional labor. When GeneMapper is used it is important to optimize the peak capture amplitude to get the best results. A peak having low amplitude may be generated from background noise and not represent a true AFLP fragment. Therefore, it is necessary to test with different peak detection levels and choose the best for a particular data set. Best peak height depends on run conditions of the capillary electrophoresis system and AFLP samples. Our data set gave better results when the peak capture threshold was set to 1000 rfu. GeneMapper is also capable of analyzing multiplexed AFLP fragments. Multiplexing refers to the labeling of amplicons of different isolates with different fluorescent

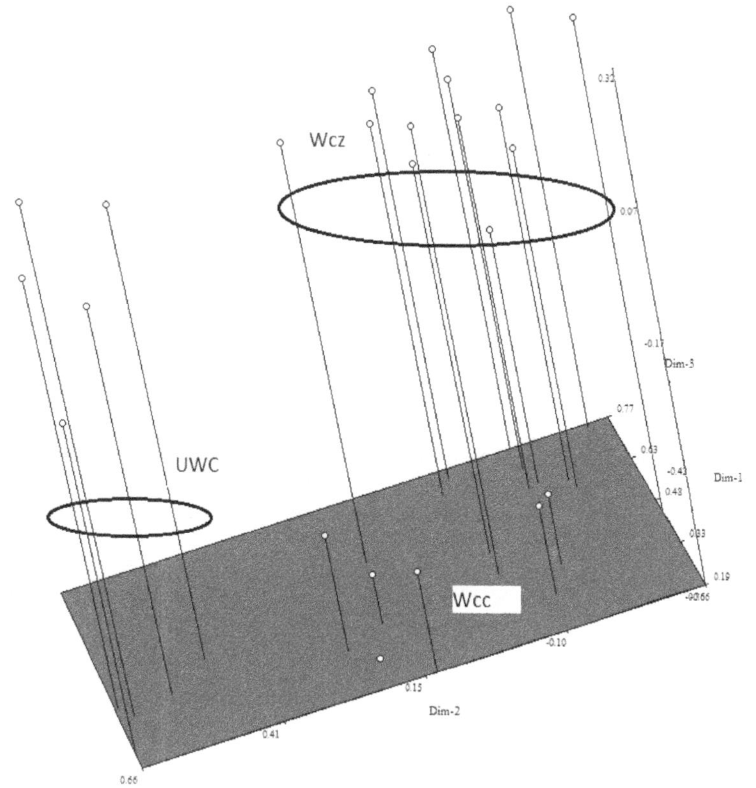

Figure 5: Three dimensional PCA scatter diagram of studied *Waitea circinata isolates* along first three principal components based on Y genetic distance matrix.

dyes and analyzing as a single sample. This dramatically reduces the cost of AFLP analysis via capillary electrophoresis.

The main disadvantage of AFLP lies on the dominance nature of its markers which make it unable to determine homologous alleles. Due to this, amid all of the benefits of the AFLP technique such as high reliability, ability to generate molecular markers from uncharacterized organisms, etc., it cannot replace co-dominant makers such as RFLP and microsatellites, completely. Nevertheless, our results showed AFLP can be used successfully to determine genetic structure of unknown *R. solani* and *W. circinata* isolates infecting cool-season turfgrasses by identifying to infra-species level. *Rhizoctonia* species infecting turfgrasses are difficult to identify using disease symptoms. Accurate identification of causal pathogens to AG or AG subgroup is important since they have differential sensitivity to fungicides and environmental conditions. Our analysis show AFLP is a good alternative for classical methods to characterize a large number of unknown *R. solani and W. circinata* isolates to infra-species level reliably and cost effectively.

Acknowledgements

This research was partially funded by the United States Golf Association.

References

1. Burpee L, Martin B (1992) Biology of Rhizoctonia species associated with turfgrass. Plant Disease 76: 112-117.
2. De La Cerda KA, Douhan GW, Wong FP (2007) Discovery and characterization of Waitea circinata var. circinata affecting annual bluegrass from the western United States. Plant Disease 91: 791-797.
3. Kammerer SJ, Burpee LL, Harmon PF (2011) Identification of a new Waitea circinata variety causing basal leaf blight of seashore paspalum. Plant Disease 95: 515-522.
4. Toda T, Hayakawa T, Mghalu JM, Yaguchi S, Hyakumachi M (2007) A new Rhizoctonia sp. closely related to Waitea circinata causes a new disease of creeping bentgrass. Journal of General Plant Pathology 73: 379-387.
5. Lübeck M, Poulsen H (2001) UP-PCR cross blot hybridization as a tool for identification of anastomosis groups in the Rhizoctonia solani complex. FEMS Microbiol Lett 201: 83-89.
6. Sharon M, Kuninaga S, Hyakumachi M, Sneh B (2006) The advancing identification and classification of Rhizoctonia spp. using molecular and biotechnological methods compared with the classical anastomosis grouping. Mycoscience 47: 299-316.
7. Zhang M, Dernoeden PH (1995) Facilitating anastomosis grouping of Rhizoctonia solani isolates from cool-season turfgrasses. HortScience 30: 1260-1262.
8. Smiley RW, Dernoeden PH, Clarke BB (2005) Compendium of Turfgrass Diseases. (3rdedn), American Phytopathological Society, St. Paul, MN.
9. Martin SB, Campbell CL, Lucas LT (1984) Response of Rhizoctonia blights of tall fescue to selected fungicides in the greenhouse. Phytopathology 74: 782-785.
10. Martin SB, Lucas LT, Campbell CL (1984) Comparative sensitivity of Rhizoctonia solani and Rhizoctonia-like fungi to selected fungicides in vitro. Phytopathology 74: 778-781.
11. Carling DE, Helm DJ, Leiner RH (1990) In vitro sensitivity of Rhizoctonia solani and other multinucleate and binucleate Rhizoctonia to selected fungicides. Plant Disease 74: 860-863.
12. Kataria HR, Hugelshofer U, Gisi U (1991) Sensitivity of Rhizoctonia species to different fungicides. Plant Pathology 40: 203-211.
13. Ogoshi A (1987) Ecology and pathogenicity of anastomosis and intraspecific groups of Rhizoctonia solani Kühn. Annual Review of Phytopathology, 25: 125-142.
14. Carling DE (1996) Grouping in Rhizoctonia solani by hyphal anastomosis reaction. In B Sneh, S Jabaji-Hare, S Neate, G Dijst eds Rhizoctonia species:

Taxonomy, Molecular Biology, Ecology, Pathology and Disease Control. Kluwer Academic Publishers, Dordrecht.

15. Boysen M, Borja M, del Moral C, Salazar O, Rubio V (1996) Identification at strain level of Rhizoctonia solani AG4 isolates by direct sequence of asymmetric PCR products of the ITS regions. Curr Genet 29: 174-181.

16. Strausbaugh CA, Eujayl IA, Panella LW, Hanson LE (2011) Virulence, distribution and diversity of Rhizoctonia solani from sugar beet in Idaho and Oregon. Canadian Journal of Plant Pathology 33: 210-226.

17. Grosch R, Schneider JH, Peth A, Waschke A, Franken P, et al. (2007) Development of a specific PCR assay for the detection of Rhizoctonia solani AG 1-IB using SCAR primers. J Appl Microbiol 102: 806-819.

18. Pannecoucque J, Höfte M (2009) Detection of rDNA ITS polymorphism in Rhizoctonia solani AG 2-1 isolates. Mycologia 101: 26-33.

19. Mueller UG, Wolfenbarger LL (1999) AFLP genotyping and fingerprinting. Trends Ecol Evol 14: 389-394.

20. Amaradasa BS, Horvath BJ, Lakshman DK, Warnke SE (2013) DNA fingerprinting and anastomosis grouping reveal similar genetic diversity in Rhizoctonia species infecting turfgrasses in the transition zone of USA. Mycologia 105: 1190-1201.

21. Ceresini PC, Shew HD, Vilgalys RJ, Cubeta MA (2002) Genetic diversity of Rhizoctonia solani AG-3 from potato and tobacco in North Carolina. Mycologia 94: 437-449.

22. Vos P, Hogers R, Bleeker M, Reijans M, van de Lee T, et al. (1995) AFLP: a new technique for DNA fingerprinting. Nucleic Acids Res 23: 4407-4414.

23. Rohlf FJ (2005) NTSYS-pc: numerical taxonomy and multivariate analysis system. Version 2.2. Setauket, New York: Exeter Software.

24. Dice LR (1945) Measures of the amount of ecologic association between species. Ecology 26: 297-302.

25. Sneath PHA, Sokal RR (1973) Numerical taxonomy. W.H. Freeman and Co., San Francisco, CA.

26. Sokal RR, Michener CD (1958) A statistical method for evaluating systematic relationships. University of Kansas Science Bulletin 38: 1409-1438.

27. Sokal RR, Sneath PHA (1963) Principles of numeric taxonomy. W.H. Freeman, San Francisco, pp. 359.

28. Tamura K, Peterson D, Peterson N, Stecher G, Nei M, et al. (2011) MEGA5: molecular evolutionary genetics analysis using maximum likelihood, evolutionary distance, and maximum parsimony methods. Mol Biol Evol 28: 2731-2739.

29. Rohlf FJ, Sokal RR (1981) Comparing numerical taxonomic studies. Systematic Zoology 30: 459-490.

30. Hotelling H (1933) Analysis of a complex of statistical variables into principal components. Journal of Educational Psychology 24: 417-520.

31. Excoffier L, Smouse PE, Quattro JM (1992) Analysis of molecular variance inferred from metric distances among DNA haplotypes: application to human mitochondrial DNA restriction data. Genetics 131: 479-491.

32. Peakall R, Smouse PE (2012) GenAlEx 6.5: genetic analysis in Excel. Population genetic software for teaching and research--an update. Bioinformatics 28: 2537-2539.

33. Wright S (1978) Evolution and the Genetics of Populations. Variability within and among natural populations. Vol 4. The University of Chicago Press, Chicago.

34. Balloux F, Lugon-Moulin N (2002) The estimation of population differentiation with microsatellite markers. Mol Ecol 11: 155-165.

35. Lübeck M (2004) Molecular characterization of Rhizoctonia solani. In DK Arora, GG Khachatourians eds Applied mycology and biotechnology Volume 4: Fungal genomics. Elsevier B.V., Amsterdam.

36. Majer D, Mithen R, Lewis BG, Vos P, Oliver RP (1996) The use of AFLP fingerprinting for the detection of genetic variation in fungi. Mycological Research 100: 1107-1111.

37. Cubeta MA, Vilgalys R (1997) Population Biology of the Rhizoctonia solani Complex. Phytopathology 87: 480-484.

38. Meza-Moller A, Esqueda M, Sanchez-Teyer F, Vargas-Rosales G, Gardea AA, et al. (2011) Genetic Variability in Rhizoctonia solani Isolated from Vitis vinifera based on Amplified Fragment Length Polymorphism. American Journal of Agricultural Biology 6: 317-323.

39. López-Olmos K, Hernandez-Delgado S, Mayek-Perez N (2005) AFLP fingerprinting for identification of anastomosis groups of Rhizoctonia solani isolates from common bean (Phaseolus vulgaris L.) in Mexico. Revista Mexicana de Fitopatologia 23: 147-151.

2

Environmental Factors for Germination of *Sclerotinia sclerotiorum* Sclerotia

Michael E Foley[1]*, Münevver Doğramacı[1], Mark West[2] and William R Underwood[1]

[1]*USDA-Agricultural Research Service, Sunflower and Plant Biology Research Unit, Fargo, ND 58102-2765, USA*
[2]*USDA-Agricultural Research Service, 2150 Centre Ave., Suite 300 Fort Collins, CO 80526, USA*

Abstract

Basal stalk rot of sunflower is an economically important and rather unique disease among crops that are susceptible to *Sclerotinia sclerotiorum*. This disease is the result of myceliogenic germination of sclerotia whereby the vegetative hyphae infect the sunflower below the soil level. In contrast, sunflower head rot and similar diseases of susceptible crops result from carpogenic germination to produce airborne ascospores that infect above ground senescent or wounded tissues. Research was initiated on several factors reported to affect sclerotia germination as a prelude to genomic investigations of myceliogenic and carpogenic germination. Specifically, the effects of inoculum development temperature, sclerotia development temperature, conditioning temperature, conditioning of hydrated and desiccated sclerotia, and the duration of sclerotia desiccation on germination strain Sun-87 sclerotia were reevaluated. As reported previously, we were not able to use conditioning temperature from -20°C to 30°C to differentiate myceliogenic and carpogenic germination for either hydrated or desiccated sclerotia. Besides conditioning temperature, inoculum production temperature, sclerotia formation period and temperature, and desiccation failed to distinguish the two forms of germination. The high level of variability for sclerotia germination between experiments indicates the critical nature of repeating all experiments aimed at understanding factors that influence sclerotia germination. Thus, other methods will be required to discover a reliable and non-confounded method that clearly differentiates myceliogenic and carpogenic germination of *S. sclerotiorum*.

Keywords: Ascospore; Carpogenic; Conditioning; Germination; Head rot; Myceliogenic; Sclerotinia; Stalk rot; Temperature

Introduction

Sclerotinia sclerotiorum is a necrotropic fungal plant pathogen with a broad crop and non-crop host range that produces hard, asexual resting structures called sclerotia [1,2]. Sclerotia are a means by which a quiescent state in the soil can be maintained in the absence of suitable host or conditions favorable for germination and active growth [3]. Myceliogenic germination of sclerotia follow by root invasion of hyphae leads to basal stalk rot disease of sunflower, whereas carpogenic germination leads to head rot [2]. Some information exists on conditions for carpogenic germination; less information exists related to myceliogenic germination.

Huang [4] reported a method that changes the germination behavior from carpogenic to myceliogenic based on the temperature during conditioning of mature sclerotia of strain Sun87. Carpogenic germination occurred exclusively for sclerotia conditioned for 4 weeks at 0.5°C to 25°C; whereas, only myceliogenic germination occurred if conditioned at -20°C to -10°C. Additionally, Huang and Kozub [5] and Huang et al. [6] examined the effects of inoculum (mycelial mat) production temperature and desiccation of sclerotia after development, respectively on germination. Huang's group reported that the favorable inoculum production temperature for carpogenic germination of Sun87 was 10°C for daughter sclerotia formed at 20°C, and that myceliogenic germination of Sun87 occurred most readily when the sclerotia were formed at 20°C to 25°C and were desiccated prior to germination.

We sought a reliable and non-confounded method to obtain exclusively myceliogenic or carpogenic germination to investigate the germination transcriptome regulating the two forms of sclerotial germination. Therefore, the objective of these experiments was to verify that the conditioning temperature treatments [4] would clearly distinguish carpogenic and myceliogenic germination.

Materials and Methods

Sclerotia for *Sclerotinia sclerotiorum* (Lib.) de Bary strains Sun87 (also known as 2148) originating from Canada [7] was provided by Dr. K.Y. Rashid, Agriculture and Agri-Food Canada. Strain 1980 originating from Nebraska [8] was provided by Dr. J. Rollins, University of Florida.

Effect of sclerotia formation period and temperature and conditioning temperature on germination of Sun87

Two experiments were conducted as described below.

Experiment 1: Sclerotia were formed by transferring plugs of mycelial mat to the center of 100 × 15 mm Petri dish containing full strength potato dextrose agar (PDA). These dishes were incubated in the dark at 10°C and 20°C for 4 and 8 weeks to facilitate growth of mycelia and formation of sclerotia (Figure 1). The sclerotia were then transferred to an open Petri dish in a laminar flow hood for a 1 day desiccation treatment or remained hydrated on the PDA. The desiccated and hydrated sclerotia were conditioned at -20°C, -10°C, 0.5°C, 10°C, or 30°C for 4 weeks as per Huang [4]. Thereafter, sclerotia were placed in glass 100 × 15 mm Petri dishes containing moist, sterile sand, sealed with parafilm, and carpogenic or myceliogenic germination (Figure 2) was determined after 4 weeks at 16°C in low continuous light. Each treatment had 6 biological replications with 20 sclerotia per replicate petri dish.

***Corresponding author:** Michael E Foley, USDA-Agricultural Research Service, Sunflower and Plant Biology Research Unit, Fargo, ND 58102-2765, USA
E-mail: michael.foley@ars.usda.gov

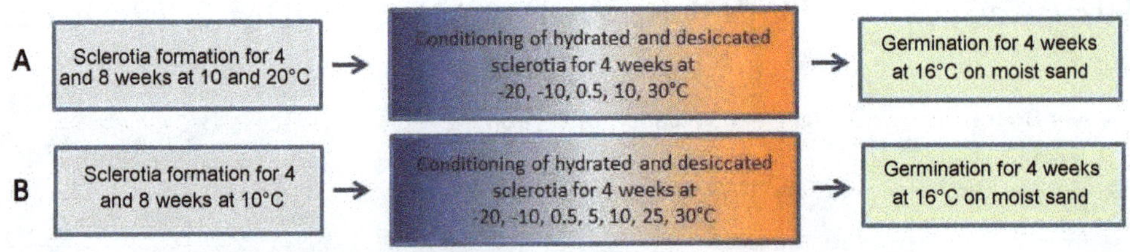

Figure 1: Experimental design to determine the effects of sclerotia formation period, development temperature, and conditioning temperature on germination of *Sclerotinia sclerotiorum* strain Sun87. (A) Experiment 1, (B) Experiment 2.

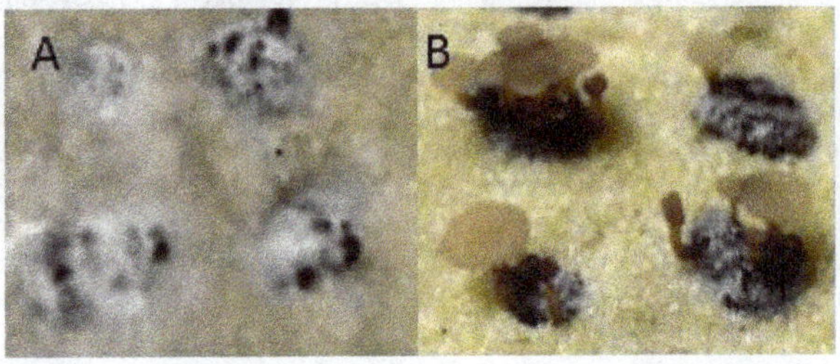

Figure 2: Myceliogenic (A) and carpogenic (B) germination on sand after 4 weeks incubation of *Sclerotinia sclerotiorum* strain Sun87 sclerotia.

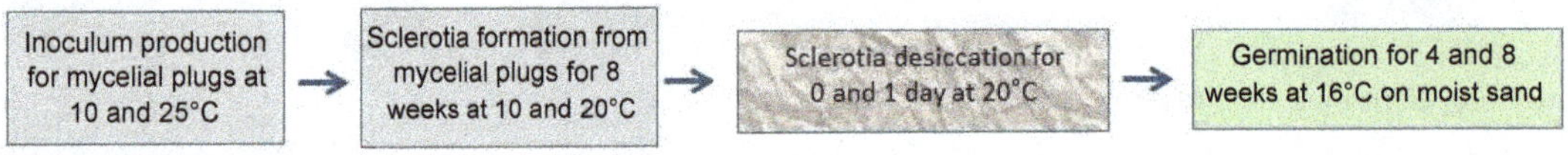

Figure 3: Experimental design to determine the effects of sclerotia inoculum production temperature, formation temperature, and desiccation time on germination of *Sclerotinia sclerotiorum* strain Sun87 and 1980 for experiments 3 and 4.

Experiment 2: Experiment 1 was repeated with slight modification; that is, elimination of the 20°C sclerotia formation temperature and addition of 5°C and 25°C conditioning temperature.

Effect of inoculum production temperature, sclerotia formation temperature, and duration of sclerotia desiccation on germination of Sun87 and 1980

Two experiments with identical designs were conducted as described below.

Experiment 3: Inoculum for procuring plugs of mycelia mat for subsequent sclerotia formation was produced at 10 and 25°C (8 h light/16 h dark) over a 7 to 10 day period in Petri dishes containing ½ strength PDA (Figure 3). Plugs were then transferred to the center of 100 × 15 mm Petri dish containing full strength PDA. These dishes were incubated in the dark at 10°C and 20°C for 8 weeks to facilitate growth of mycelia and formation of sclerotia. The sclerotia were then transferred to a Petri dish for desiccation treatments of 1 to 21 days or remained hydrated on the PDA; however, only 0 and 1 day data are presented. Thereafter, sclerotia were placed in glass 100 × 15 mm Petri dishes containing moist, sterile sand, sealed with parafilm, and germination was determined at 16°C after 4 and 8 weeks of incubation under continuous low light. Each treatment had 4 biological replicates with most replicate petri dishes having 20 sclerotia but some having as few as 7. In this experiment several petri dishes for Sun87 (the plugs of mycelial mat for sclerotia formation produced at 25°C) were contaminated; therefore, the data was excluded.

Experiment 4: Experimental design was identical to Experiment 3 (Figure 3).

Statistical analysis

Data collected from the experiments revolved around counting the number of sclerotia germinating out of a set number of sclerotia exposed to a treatment condition. Therefore, experiments were evaluated by fitting generalized linear mixed models in order to test the factors of interest in the experiment. These models fit binomial proportions to treatments with a logit link. Proportions of sclerotia that germinated were compared among treatment conditions using Likelihood Ratio, Chi-Square or Z statistics.

Results

Effects of sclerotia formation period, development temperature, and conditioning temperature on germination of Sun87

We observed 88% and 98% carpogenic germination in the first experiment when sclerotia were developed for 4 and 8 weeks respectively at the 10°C formation temperature followed by conditioning them in a hydrated state at 10°C (Figure 4). Comparatively, carpogenic germination at 8 weeks was 85% and 51% for sclerotia conditioned at 0.5°C and 30°C, respectively. Conversely, carpogenic germination for sclerotia formed for 4 and 8 weeks at the 20°C formation temperature and conditioned in a hydrated state at 10°C was observed to be significantly ($P<0.01$) lower at 17% and 23% respectively when compared to sclerotia formed at the 10°C temperature. However, when sclerotia were desiccated prior to germination, carpogenic germination was limited to between 0% and 21% for either formation period (4 or 8 weeks) or for either development temperature (10°C or 20°C) and for all conditioning temperatures (Figure 4). Moreover, no more than 6% myceliogenic germination was observed among treatment means in experiment 1 regardless of the sclerotia formation period and temperature, and conditioning temperature (Supplemental Figure 1a).

Because germination levels in experiment 1 for Sun87 sclerotia formed at 20°C were decreased when compared to those formed at 10°C (Figure 4), we eliminated the 20°C sclerotia formation temperature and added two additional conditioning temperatures (5°C and 25°C) to be evaluated in experiment 2. The conditioning temperatures in experiment 2 that were most efficacious for carpogenic germination were similar to those in experiment 1, i.e., 0.5 and 10°C. In both experiments for all conditioning temperatures and sclerotia formed at 10°C, germination was generally greater for sclerotia formed for 8 weeks relative to 4 weeks (Figures 4 and 5). Overall carpogenic germination ranged from 1% to 98% in experiment 1 and from 5% to 99% in experiment 2 for sclerotia formed at 4 and 8 weeks at 10°C and conditioned in a hydrated state between -20°C to 30°C (Figures 4 and 5). When sclerotia were desiccated and conditioned at -20°C to 30°C, carpogenic germination was limited (0% to 40%) regardless of sclerotia formation period (Figure 5); these results are similar to those from experiment 1. For myceliogenic germination, treatments in experiment 2 resulted in 0% to 34% germination (Supplemental Figure 1b).

Effect of inoculum production temperature, sclerotia formation temperature, and duration of sclerotia desiccation on germination of Sun87 and 1980

Observations taken from experiments 1 and 2 did not lead to a high certainty of treatment conditions for obtaining strictly myceliogenic or carpogenic germination using the range of treatments, including conditioning temperatures as reported by Huang [4]. Therefore, we further utilized sclerotinia strains Sun87 and 1980 to examine two additional factors that could possibly separate carpogenic and myceliogenic germination, but were not explored in experiments 1 and 2 (Figure 3). These factors were: 1) the temperature for inoculum production (the material used to obtain mycelial mats for sclerotia formation), and 2) the duration of sclerotia desiccation. Strain 1980 was specifically chosen for experiments 3 and 4 to evaluate its germination since its genome has been sequenced, making it ideal for further genomics studies [9].

Evaluation of Strain Sun87

In experiment 3, the 8 weeks germination evaluation revealed that hydrated Sun87 sclerotia had 84% and 61% carpogenic germination in response to the 10°C and 25°C inoculum production temperatures, respectively, when the sclerotia formation temperature was 10°C

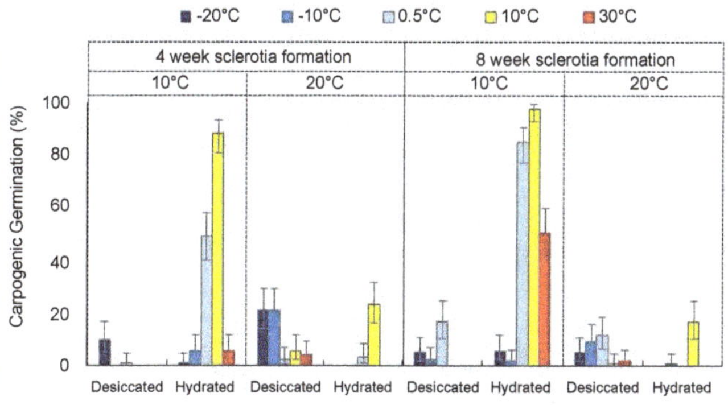

Figure 4: Carpogenic germination of *Sclerotinia sclerotiorum* strain Sun87 sclerotia in experiment 1. Sclerotia were formed for 4 and 8 weeks at 10°C and 20°C followed by conditioning under desiccated and hydrated conditions between -20°C to 30°C (see Figure 1A).

(Figure 6a). Germination was 85% and 79% for the same conditions in experiment 4 (Figure 6b). When the sclerotia formation temperature was 20°C (Figure 6a) carpogenic germination at 8 weeks was observed to be 66% and 10% in response to the 10°C and 25°C inoculum production temperatures, respectively, in experiment 3. However, there was no carpogenic germination in experiment 4 with these treatments (Figure 6b).

Eight week germination rates for sclerotia strain Sun87 under the

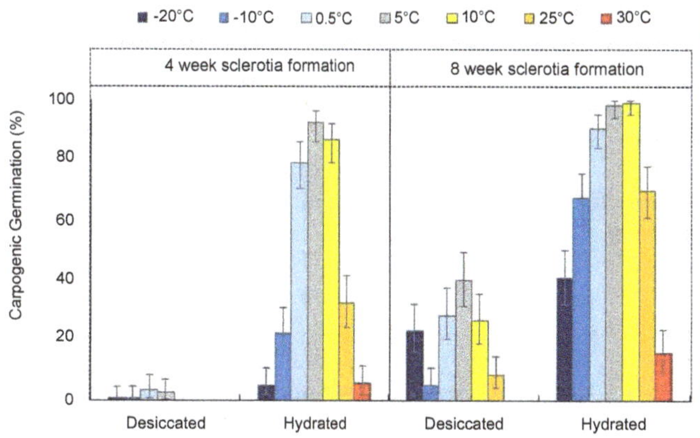

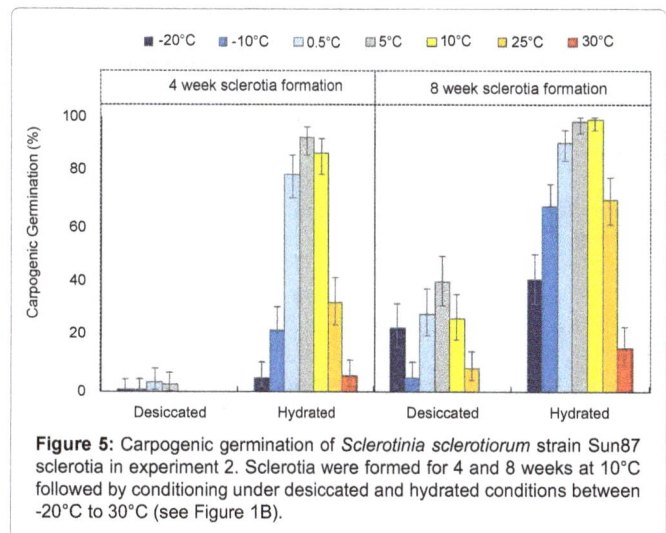

Figure 5: Carpogenic germination of *Sclerotinia sclerotiorum* strain Sun87 sclerotia in experiment 2. Sclerotia were formed for 4 and 8 weeks at 10°C followed by conditioning under desiccated and hydrated conditions between -20°C to 30°C (see Figure 1B).

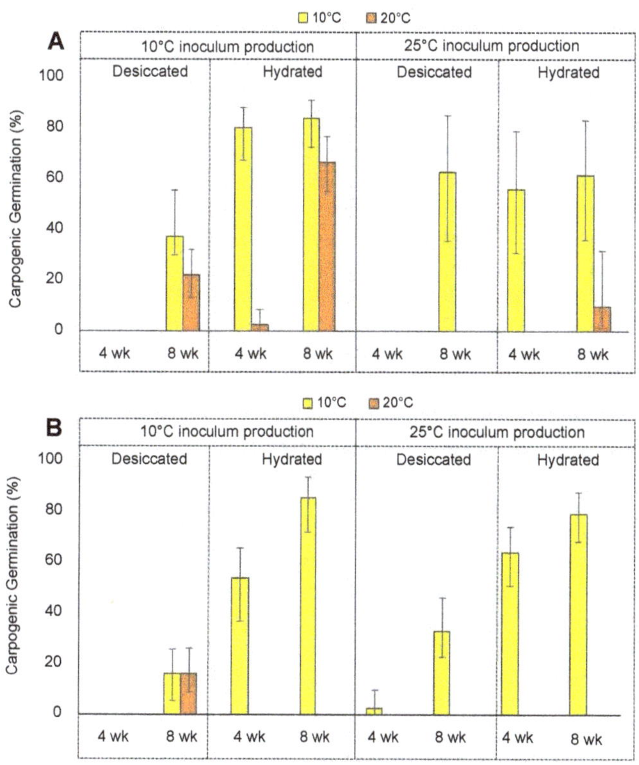

Figure 6: Carpogenic germination of *Sclerotinia sclerotiorum* strain Sun87 sclerotia in experiment 3A and experiment 4B. Inoculum production temperature for sclerotia was 10°C and 25°C, the sclerotia formation temperature was 10°C and 20°C, and germination was evaluated at 4 and 8 weeks. Prior to germination sclerotia were kept hydrated or desiccated for 1 day (see Figure 3).

desiccated study conditions showed consistency between experiments 3 and 4 although the levels of germination differed. Analysis of germination rates for both experiments showed a significant interaction between inoculum production and sclerotia formation temperatures with significance of $P<0.01$ for both. Evaluation of germination showed an increase from 37% carpogenic germination for the 10°C inoculum production temperature (N=4 reps, 64 sclerotia tested) to 63% carpogenic germination for the 25°C inoculum production temperature (N=1 rep, 16 sclerotia tested) when the sclerotia formation temperature was 10°C in experiment 3 (Figure 6a), although this was not found to be highly statistically significant compare to 25°C inoculum production temperature ($P=0.0755$, $Z=-1.436$). We qualify this result by noting that experiment 3 encountered contaminations for strain Sun87 in 3 of the 4 replicates when inoculum production occurred at 25°C, and do not consider this convincing evidence that increasing sclerotia formation temperature from 10°C to 20°C increases carpogenic germination. However for the same treatment conditions in experiment 4 (Figure 6b), the increase from 10°C to 25°C inoculum production temperature resulted in an increase in germination from 16% to 33% ($P<0.01$, $Z=-2.489$) for 10°C sclerotia formation temperature. For sclerotia formed at 20°C in experiment 3 (Figure 6a), the 8 weeks germination evaluation revealed 22% (N=4 reps, 79 sclerotia tested) and 0% (N=1 rep, 20 sclerotia tested) carpogenic germination in response to the 10°C and 25°C inoculum production temperatures, respectively. A similar decrease in germination was observed for carpogenic germination with 16% (N=4 reps, 80 sclerotia) for the 10°C inoculum production temperature and 0% (N=4 reps, 80 sclerotia) or the 25°C inoculum production temperatures in experiment 4 (Figure 6b). However, neither of these decreases were found to be statistically significant ($P>0.25$)

Evaluation of Strain 1980

Both experiments showed a statistically significant effect of sclerotia formation temperature on 8 week germination for hydrated sclerotia ($P<0.001$ for both experiments), but also statistically significant interaction between inoculum production temperature and sclerotia formation temperature on 8 weeks germination for hydrated sclerotia ($P=0.074$ for experiment 3 and $P=0.044$ for experiment 4). In experiment 3, the 8 weeks germination evaluation revealed that hydrated 1980 sclerotia increased ($P=0.073$) from 26% (N=4 reps, 66 sclerotia) to 35% (N=4 reps, 76 sclerotia) carpogenic germination in response to the 10°C and 25°C inoculum production temperatures, respectively, when sclerotia formation temperature was 10°C (Figure 7a). However the 8 weeks evaluation revealed no significant difference ($P=0.297$) for carpogenic germination in response to the 10°C and 25°C inoculum production temperatures, which was 71% (N=4 reps, 77 sclerotia) and 65% (N=4 reps, 80 sclerotia) respectively, when sclerotia formation temperature was 20°C (Figure 7a). For experiment 4 (Figure 7b), carpogenic germination was 23% (N=4 reps, 71 sclerotia) and 25% (N=4 reps, 70 sclerotia) after 8 weeks for the 10°C and 25°C inoculum production temperature, respectively, in combination with the 10°C sclerotia formation temperature and represented no significant difference. However, germination was 66% (N=4 reps, 80 sclerotia) and 48% (N=4 reps, 80 sclerotia) after 8 weeks for the 10°C and 25°C inoculum production temperature in combination with the 20°C sclerotia formation temperature (Figure 7b) representing a significant decrease ($P<0.01$). Significant increases were observed for carpogenic germination in both experiments when the sclerotia formation temperature was increased from 10°C to 20°C with germination increasing from 30% (N=8 reps, 142 sclerotia) to 68% (N=8 reps, 157 sclerotia) in experiment 3 and from 24% (N=8 reps, 141 sclerotia) to 57% (N=8 reps, 160 sclerotia) in experiment 4. For experiments 3 and 4, there was essentially no carpogenic germination after 8 weeks for desiccated 1980 sclerotia with any treatment combinations (Figures 7a and 7b).

Finally, for both Sun87 and 1980, there was no more than 5% myceliogenic germination (Supplemental Figures 2 and 3) for any combination of treatments. However, significant myceliogenic germination ($P<0.0001$), although very low, occurred only for the desiccated treatments. Additionally, no consistent differences in germination were observed between sclerotia desiccated beyond 1 day for either strain (data not presented).

Discussion

Our objective was to confirm previous reports outlining a reliable method to absolutely distinguish myceliogenic and carpogenic germination as a prelude to investigating the transcriptomes associated with the two forms of *S. sclerotiorum* germination. We procured strain Sun87 and repeated a study by Huang [4], which indicated different conditioning temperatures clearly distinguished carpogenic from myceliogenic germination. Our experiments were done in nearly the same manner as previously reported [4]. However, we conditioned sclerotia in our experiments in both the hydrated and desiccated states due to ambiguity in the methods, and a report by Huang et al. [6] indicating that desiccation of sclerotia was an important factor for myceliogenic germination. Huang [4] reported 100% myceliogenic germination for Sun87 sclerotia following conditioning at -20°C. However, we observed ≤5% myceliogenic germination of Sun87 sclerotia that were conditioned at -20°C in both the hydrated and

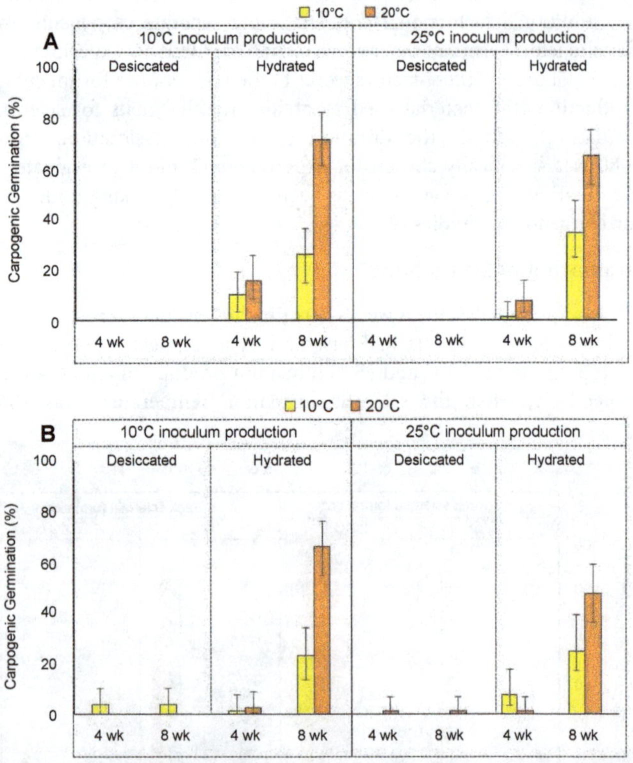

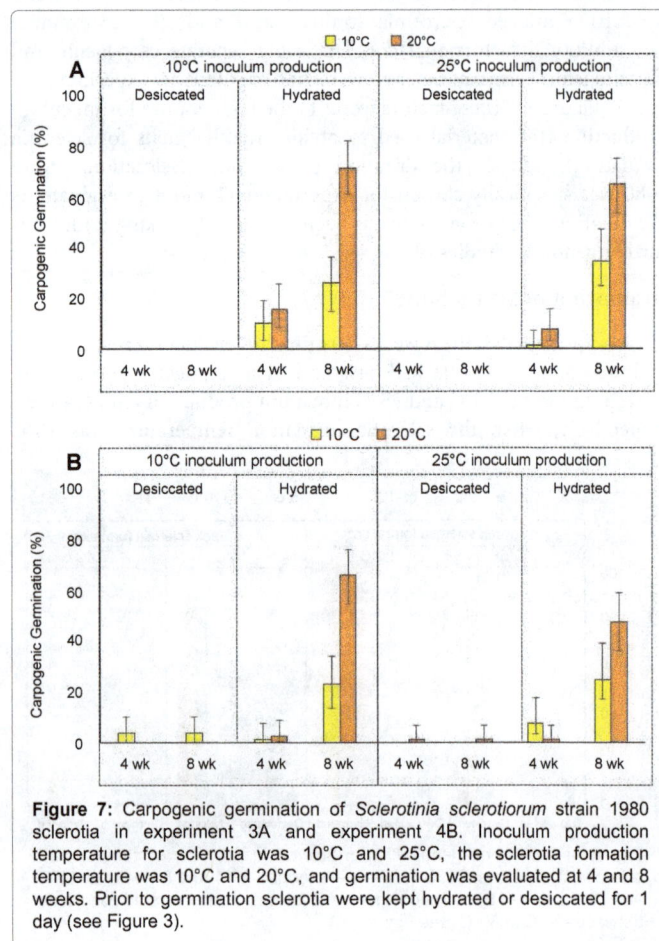

Figure 7: Carpogenic germination of *Sclerotinia sclerotiorum* strain 1980 sclerotia in experiment 3A and experiment 4B. Inoculum production temperature for sclerotia was 10°C and 25°C, the sclerotia formation temperature was 10°C and 20°C, and germination was evaluated at 4 and 8 weeks. Prior to germination sclerotia were kept hydrated or desiccated for 1 day (see Figure 3).

desiccated state (Supplemental Figure 1). We now assume that Huang [4] conditioned Sun87 sclerotia in a hydrated state because he obtained 90% to 100% carpogenic germination at conditioning temperatures of 0.5°C to 30°C for 8-week old cultures. Similarly, we obtained levels of carpogenic germination approaching 100% at conditioning temperatures of 0.5°C to 10°C for 8-week old cultures. Unfortunately, in our experiments distinct separation of the two forms of germination was not apparent. For example, in experiments 1 and 2, in a hydrated state, a conditioning temperature, such as -10°C that resulted in 2% to 34% myceliogenic germination, also resulted in 2% to 68% carpogenic germination. Beyond that, we obtain inconsistent results when we repeated the experiment 1 (i.e., experiment 2) to examine the effect of sclerotia formation period and temperature, and conditioning temperature on germination. Thus, the use of conditioning temperature alone to clearly differentiate myceliogenic and carpogenic germination for Sun87 sclerotia did not meet our objective.

We examined several other factors that might affect germination such as inoculum production temperature and duration of sclerotia desiccation. Huang and Kozub [5] examined inoculum production temperatures of 10°C, 25°C, and 30°C. They reported 84% carpogenic germination for Sun87 sclerotia whose inoculum was produced at 10°C and the sclerotia formed at 20°C; no more than 35% germination was observed for the 25°C and 30°C inoculum production temperatures for sclerotia formed at 20°C. In contrast, our results indicated that the 10°C and 25°C inoculum production temperatures for Sun87 resulted in 61% to 85% carpogenic germination, particularly when the sclerotia were formed at 10°C and they were kept hydrated (Figures 6a and 6b). Conversely, for strain Sun87 carpogenic germination was significantly lower ($P<0.0001$) if the sclerotia were produced at 20°C relative to 10°C when hydrated. However, for strain 1980 carpogenic germination was significantly higher ($P<0.0001$) if the sclerotia were produced at 20°C relative to 10°C (Figures 7a and 7b) when hydrated. It is known that carpogenic germination depends on the geographic origin of the strain and the temperature at which sclerotia are produced [7]. The different results from our two strains may be due to different geographic origins and thus temperature requirements. Ultimately, we were not able to verify that a low inoculum production temperature was the main factor affecting carpogenic germination of Sun87. Rather, a low temperature (10°C) during sclerotia formation seems to be the more important factor to enhance carpogenic germination of Sun87 and 1980.

Our first experiment indicated that drying sclerotia prior to conditioning had little effect on subsequent myceliogenic germination (Supplemental Figure 1). However, Huang et al. [6] reported that desiccation of strain S9 sclerotia at 25°C and near 0% relative humidity for 3 weeks significantly increased myceliogenic germination and hyphal growth at 25°C. We had data that air drying of Sun87 sclerotia for 1 day in a laminar flow hood facilitated complete desiccation (data not presented). Nevertheless, based on Huang et al. [6] we reasoned that prolonged desiccation may facilitate additional benefits for sclerotia germination similar to that which occurs when dormant seeds are subjected to afterripening under warm, dry conditions [10]. However, there was no benefit to a period of desiccation beyond 1 day for either strain 1980 or Sun87. In fact, desiccation still resulted in greater carpogenic than myceliogenic germination, even though the carpogenic germination levels were lower than if they were hydrated.

We did not identify a treatment that clearly distinguished myceliogenic from carpogenic germination for strain Sun87 based on inoculum production temperature, sclerotia formation period and temperature, conditioning temperatures, and desiccation. An important outcome of our experiments was the high level of variability for sclerotia germination indicating the critical nature of repeating experiments aimed at understanding factors that influence sclerotia germination, which may not have been done in past studies. In any event, other methods, such as injuring the sclerotia rind [11], manipulating the quantity and quality of light [12,13] and manipulating the type of germination using chemical treatments [14] are possible solutions to discovering a reliable and non-confounded method that clearly differentiate myceliogenic and carpogenic germination of *S. sclerotiorum*.

Acknowledgement

Ms. Cheryl Huckle and Ms. Michelle Gilley provided technical assistance.

References

1. Boland GJ, Hall R (1994) Index of plant hosts of *Sclerotinia sclerotiorum*. Can J Plant Path 16: 93-108.
2. Bolton MD, Thomma BP, Nelson BD (2006) *Sclerotinia sclerotiorum* (Lib.) De Bary: biology and molecular traits of a cosmopolitan pathogen. Mol Plant Pathol 7: 1-16.
3. Adams PB, Ayres WA (1979) Ecology of *Sclerotinia* species. Phytopath 69: 896-899.
4. Huang HC (1991) Induction of myceliogenic germination of sclerotia of *Sclerotinia sclerotiorum* by exposure to sub-freezing temperatures. Plant Path 40: 621-625.
5. Huang HC, Kozub GC (1993) Influence of inoculum production temperature on carpogenic germination of sclerotia of *Sclerotinia sclerotiorum*. Can J Microbiol 39: 548-550.
6. Huang HC, Chang C, Kozub GC (1998) Effect of temperature during sclerotial formation, sclerotial dryness, and relative humidity on myceliogenic germination of sclerotia of *Sclerotinia sclerotiorum*. Can J Bot 76: 494-499.
7. Huang HC, Kozub GC (1991) Temperature requirements for carpogenic germination of sclerotia of *Sclerotinia sclerotiorum* isolates of different geographic origin. Bot Bull Acad Sinica 32: 279-286.
8. Steadman, JR, Marcinkowska J, Rytledge S (1993) A semi-selective medium for isolation of *Sclerotinia sclerotiorum*. Can J Plant Path 16: 68-70.
9. Amselem J, Cuomo CA, Van Kan JAL, Viaud M, Benito EP, et al. (2011) Genomic analysis of the necrotrophic fungal pathogens *Sclerotinia sclerotiorum* and *Botrytis cinerea*. PLoS Genet 7: e1002230.
10. Foley ME (1994) Temperature and water status of seed affect after ripening in wild oat (*Avena fatua*). Weed Sci 42: 200-204.
11. Huang, HC (1985) Factor affecting myceliogenic germination of sclerotia of *Sclerotinia sclerotiorum*. Phytopath 75: 433-437.
12. Huang HC, Kokko EG (1989) Effect of temperature on melanization and myceliogenic germination of sclerotia of *Sclerotinia sclerotiorum*. Can J Bot 67: 1387-1394.
13. Sun P, Yang XB (2000) Light, temperature, moisture effects on apothecium production of *Sclerotinia sclerotiorum*. Plant Dis 84: 1287-1293.
14. Steadman JR, Nickerson KW (1975) Differential inhibition of sclerotial germination in *Whetzelinia sclerotiorum*. Mycopathologia 57: 165-170.

3

Comparative Study of Powdery Mildew (*Erysiphe polygoni*) Disease Severity and Its Effect on Yield and Yield Components of Field Pea (*Pisum sativum* L.) in the Southeastern Oromia, Ethiopia

Teshome E* and Tegegn A

Sinana Agricultural Research Center, Bale-Robe, Ethiopia

Abstract

Field pea or "dry pea" (*Pisum sativum* L.) is an annual cool-season food legume which grows worldwide and is the major pulse crop in the highlands of Bale next to Faba bean. The experiment was conducted for two consecutive cropping seasons; 2011/12 and 2012/13 at Sinana agricultural research center (SARC) on-station research site. The objective was to find out the effect of Powdery mildew disease on field pea yield and yield components. Local field pea cultivar was used with a fungicide Benomyl at a rate of 2.5 kg/ha and four fungicide application schemes (spraying every 7 days, 14 days, 21 days and no fungicide spray) arranged in randomized complete block design (RCBD) with 3 replications. Logistic model (ln [y/ (1-y)]) was employed to analyze the Field experiment data using SAS procedure. The association between disease parameters and yield and yield components were assessed using regression and correlation techniques. ANOVA has shown significant difference ($p \leq 0.05$) among treatments for disease severity. The highest diseases severity (41.98%) and Area Under Disease Progress Curve (AUDPC) (1458.33% days) and the lowest disease severity (13.89%) and AUDPC (471.15% days) were recorded from a plot with no fungicide treatment and plot sprayed every 7 days, respectively. Similarly, the highest disease progress rate (r) (0.044227 units-day^{-1}) and the lowest r (-0.006122 units-day^{-1}) were recorded from a plot with no fungicide treatment and plot sprayed every 7 days, respectively. Regarding the yield and yield related parameters; ANOVA has shown significant variations ($P \leq 0.05$) between treatments for number of pods per plant, seeds per plant, TKW and grain yield. The highest number of pod per plant (21.75), seed per plant (89.5), TKW (189.81 g) and grain yield (2945.6 kg/ha) were recorded from plots sprayed every 7 days; while the lowest were from non-sprayed plots. On the other hand, the higher grain yield loss of 21.09% and the lowest loss (8.53%) were recorded from plots without fungicide spray and plot received spray at 7 days interval, respectively. The linear regression between powdery mildew severity index and grain yield revealed significant difference ($P \leq 0.0001$) between treatments; and the estimated slope of the regression line obtained for Powdery mildew severity index was -34.16. Correlation analysis has shown that Powdery mildew disease severity have significantly strong negative correlation with grain yield ($r= -0.76120$, $P \leq 0.01$). Similarly, grain yield has significant strong negative correlation ($r= -0.76298$, $P \leq 0.0001$) with AUDPC.

Keywords: Field pea; Yield loss; Powdery mildew; Disease progress rate (r) and disease severity index

Introduction

Field pea or "dry pea" (*Pisum sativum* L.) is an annual cool-season food legume that grows worldwide [1]. In Ethiopia, it is among the major food legume crops produced ranking third in terms of area of production and yield next to Faba bean and chick pea [2]. It is important crop in providing quality vegetable protein in the diets of Ethiopians [3]. It also plays an important role in soil fertility restoration and controlling disease epidemics as a suitable rotation and break crop where cereal mono-cropping is predominant at areas like Bale and Arsi, Ethiopia. In Ethiopia, the area of field pea production and yield per unit area is increasing from time to time, according to Central Statistical Authority of Ethiopia [4], in 2009/2010, out of 1,489,308 ha of land covered by pulses the area occupied by field pea was 226,533 ha and the annual production was estimated at about 235,872.10t with the average annual productivity of 1.041 t/ha. Although cereal crops are the major crops cultivated in Bale highlands, food legumes are also one of the most important pulse crops produced by Bale farmers. Field pea, despite its importance, is very low in productivity which is far below its potential. This low productivity is mainly attributed to several yields limiting factors; among which, the inherent low yielding potential of the indigenous cultivars [5], diseases like Powdery mildew (*Erysiphe Polygoni*) and Ascochyta blight (*Mycosphaerella pinodes)* and some insect, pests are the major production constraints [5]. Powdery mildew caused by the obligate biotrophic fungus *Erysiphe polygoni* DC is an air-borne disease of worldwide distribution, being particularly important in climates with warm dry days and cool nights [6]. Even though it is severely damaging, the level of loss on field pea due to this disease is not known in Bale area. Therefore, this trial was initiated with the objective of quantifying the magnitude of loss caused by Powdery mildew on yield and yield components of Field pea.

Materials and Methods

Description of experimental site

The experiment was conducted for two years; in 2011/12 and 2012/13 at Sinana Agricultural Research Center (SARC) research site. The location represents the major Field pea production area of Bale highlands and is a hot spot for the development of Powdery mildew. The area is characterized by bimodal rain-fall pattern where the first rainy season occurs from March to June called "Ganna" (short season) and the second is from August to December which is called "Bona" (main

***Corresponding author:** Teshome E, Sinana Agricultural Research Center, P.O. Box 208, Bale-Robe, Ethiopia, E-mail: ermiastafa@gmail.com

season), the two seasons are locally termed in line with the time of crop harvest. SARC is situated at 07° 07' N latitude and 40° 10'E longitude with an elevation of 2400 m.a.s.l. The area receives 750 mm to 1000 mm high mean annual rain fall and have mean annual temperature of 9°C to 21°C. The area is dominantly characterized by a soil type which have a pellic vertisol character and is slightly acidic.

Treatments and design

The experiment was arranged in three replications of RCB Design. Local field pea cultivar was evaluated on plot size of 2 m × 1.2 m with between row, plot and replication spacing of 0.2 m, 1 m and 1.5 m, respectively. Powdery mildew disease development was initiated through natural infection and the disease infection gradient was created by spraying a fungicide Benomyl@2.5 kg/ha at a fixed spray interval of every 7, 14, and 21 days and a control plot receiving no fungicide spray was included for treatment comparison. A Fungicide was applied using knapsack sprayer with spray volume of 60.6 ml per 2.4 m^2 plot. Fungicide application was started immediately after the development of the first observable disease symptom. Seed rate, fertilizer rate, weeding and other all agronomic packages were done as per the recommendation for the crop. Disease scoring was conducted in a 1-9 disease scoring scale [7]. The disease data recorded based on scale mentioned above was converted to percentage severity index (PSI) according to Wheeler [8]:

$$PSI = \frac{\text{Sum of Numerical Ratings X100}}{\text{Number of Plants Scored X Maximum Score on Scale}}$$

Data management and statistical analysis

Variables for field experiment data under different treatments were analyzed using logistic model, ln [y/ (1-y)] [9] with the SAS Procedure [10]. The slop of the regression line estimated the disease progress rate in different treatments. AUDPC values were calculated for each treatment using the standard formula [9]. ANOVA was performed for disease severity index, AUDPC [9], and rate of disease progress (r) according to SAS procedure. LSD technique at the 5% probability level was used for treatments mean separation. Logistic model, [ln [(Y/1-Y)], (Vander Plank, [11]) was used for estimation of disease parameters from each treatment. These parameters were used in analysis of variance to compare the disease progress among the treatments.

$$\textbf{AUDPC} = \sum_{i-1}^{n-1} 0.5\left(x_{i+1} + x_i\right)\left(t_{i+1} - t_i\right)$$

Where, X_i= the PSI of disease at the ith assessment

t_i= is the time of the ith assessment in days from the first assessment date

n= total number of disease assessments

The association of Powdery mildew disease severity with grain yield was analyzed using linear regression analysis by plotting yield data against Diseases severity. Correlation between grain yield and yield related parameters with the disease parameters (AUDPC, Powdery mildew disease severity and r (disease progress rare)) were assessed and correlation coefficient values were computed to establish their relationships.

Yield loss estimation

The relative losses in yield and yield components were determined as a percentage of that of the protected plot. Losses were calculated separately for each of the treatments with different levels of disease severity, as:

$$RL\ (\%) = \frac{(Y_1 - Y_2)}{Y_1} \times 100$$

Where, RL% = percentage of relative loss (reduction of the parameters; i.e. yield, yield component),

Y1 = mean grain yield on the protected plots (plots with maximum protection)

Y2 = mean grain yield on unprotected plots (i.e. unsprayed plots or sprayed plots with varying level of disease).

Results and Discussions

The combined analysis of variance over years has shown that there was statistically significant difference ($p \leq 0.05$) between treatments for parameters such as Powdery mildew disease severity, AUDPC, Disease Progress Rate (r), Number of pods per plant, Number of seeds per plant, Thousand Kernel Weight (TKW) and Grain yield (Table 1). In contrast, for the parameters such as Plant height and Total biomass the difference was not statistically significant ($p \leq 0.05$). The highest Powdery mildew disease severity (41.98%) was recorded from a plot without fungicide treatment, while lowest disease severity of 13.89% was recorded from plot sprayed at 7 days interval (Table 1 and Figure 1). In general, both the disease severity and AUDPC has shown a linearly increasing trend as the spray interval is increasing (Table 1 and Figure 1). This finding is supported by different studies that fungicides have dramatically reduced Powdery mildew disease severity [12,13].

Similarly, the highest AUDPC of 1458.33% day was calculated from a plot with no fungicide treatment; while the lowest AUDPC (471.15% day) was calculated from a plot with a fungicide treatment at every 7 days. This result has supported with a finding of [13], when they found the highest AUDPC from control plot and the lowest from fully

Treatment	PmDS (%)	AUDPC (% days)	Disease Progress Rate (r)	Diseases Severity Reduction (%)
@ 7DI	13.89	471.15	-0.006122	66.91
@ 14DI	28.40	985.19	0.013149	32.35
@ 21DI	34.88	1205.55	0.020656	16.91
No spray	41.98	1458.33	0.044227	-
CV (%)	3.35	122.88	0.0082	
$LSD_{(p \leq 0.05)}$	9.35	9.90	37.96	

Note: DI=Days Interval of Spray; LSD=Least Significant Difference; CV=Coefficient of Variation; PmDS=Powdery Mildew Disease Severity

Table 1: Effect of Fungicide application on powdery mildew disease severity (%), AUDPC (%-days), disease progress rate (r) and percent disease severity reduction (%).

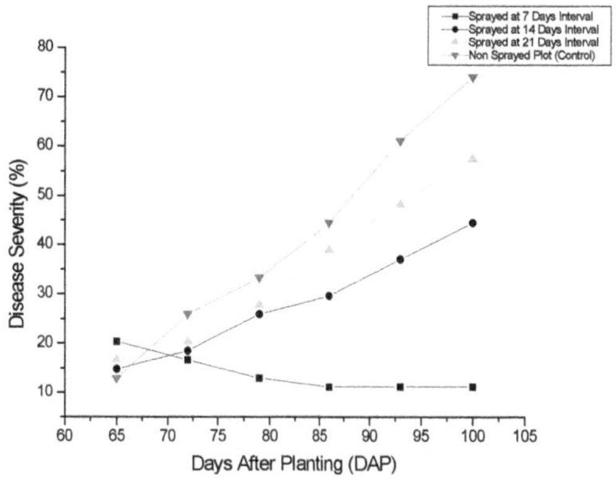

Figure 1: Influence of fungicide spray on powdery mildew disease severity (%).

controlled plot. Whereas, Disease progress rate (r) of -0.006122 units day^{-1}, 0.013149 units day^{-1}, 0.020656 units day^{-1} and 0.044227 units day^{-1} were calculated from plots sprayed every 7, 14, 21 days and No fungicide spray, respectively.

The highest percent disease severity reduction of 66.91% was obtained from plot received a fungicide application at a weekly interval; whereas, the lowest powdery mildew disease severity reduction (16.91%) was recorded from a plot treated with a fungicide at 21 days interval. This result supported the result of [14] that four sprays of Karathane (0.1%) at weekly interval gave effective control of powdery mildew. And the plot with a fungicide treatment at 14 days interval has reduced the powdery mildew disease severity by 32.25% (Table 1). Similarly, this result agrees with [12], they found that the highest disease severity reduction from fully controlled plot while the lowest disease reduction was from a plot with no fungicide spray.

With regard to yield related traits, the maximum number of pods per plant (21.75) was recorded from the plot with fungicide sprays at 7 days interval while the lowest (14.88) was from the plot with no fungicide spray.

This result is exactly in agreement with [13] result when they found the highest number of pods/plant from treated plot while the least number of pods/plant was recorded from Control plot. In case of seeds per plant, the maximum number (89.5) was recorded from the plot sprayed a 7 days interval and the lowest (57.23) was obtained from a plot with no fungicide treatment. The current result is supported by the finding of different scholars; the disease have the potential to reduce total yield biomass, number of pods per plant, number of seeds per pod, plant height and number of nodes [15]. Similarly, ANOVA for TKW and grain yield has shown statistically significant ($p \leq 0.05$) variations between treatments. The Maximum TKW (189.81 g) was recorded from plot which has received a fungicide treatment at 7 days interval where the smallest TKW of 175.23 g was recorded from unsprayed plot. With regard to grain yield, the maximum grain yield (2945.6 kg/ha) was obtained from a plot which has received a fungicide spray at 7 days interval where the smallest grain yield of 1873.5 kg/ha was recorded from a plot with no fungicide spray (Table 2). This result is supported by Shah et al. [16]. They found the maximum grain yield from a plot where the Powdery mildew was fully controlled and the minimum yield was from a plot with no treatment for the disease. Similarly, it was reported that the disease can cause 25% to 50% yield losses, reducing total yield biomass, number of pods per plant, number of seeds per pod, plant height and number of nodes and the disease also affects green pea quality [17].

Yield loss estimation

Losses in yield and yield related traits as a function of Powdery mildew disease infection was assessed as a comparison of the control

Treatment	#Pod/ plant	#Seed/ plant	Plant height (cm)	TKW (gm)	Grain yield (kg/ha)
7DI	21.75	89.50	121.86	189.81	2945.6
@ 14DI	19.16	80.01	118.53	187.87	2511.7
@ 21DI	17.89	75.33	116.01	183.07	2049.4
No spray	14.88	57.02	114.22	175.23	1873.5
$LSD_{(p<0.05)}$	3.32	24.81	NS	11.65	303.03
CV (%)	14.98	27.30	20.99	5.27	10.91

Note: DI=Days Interval of Spray; LSD=Least Significant Difference; CV=Coefficient of Variation; PmDS=Powdery Mildew Disease Severity

Table 2: Field pea yield and yield components as influenced by fungicide treatment against powdery mildew.

Treatment	#Pod/plant (%)	#Seed/plant (%)	Plant height (%)	TKW (%)	Grain yield (%)
7DI	-	-	-	-	-
14DI	11.91	10.60	2.73	1.02	8.53
21DI	17.75	15.83	4.80	3.55	17.61
No spray	31.59	36.29	6.27	7.68	21.09

Note-7DI- sprays at seven days interval; 14DI-sprays at fourteen days interval and 21DI-sprays at twenty one days interval

Table 3: Percent (%) losses in yield and yield related traits of field pea as a function of powdery mildew disease infection.

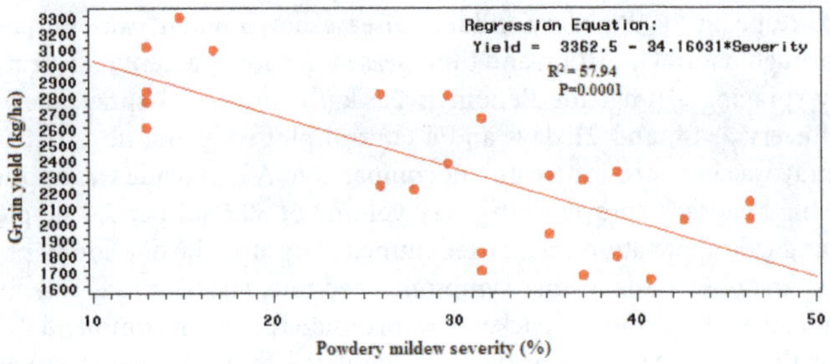

Figure 2: Estimated relationship between powdery mildew severity index and field pea grain yield loss at Sinana.

(Unsprayed) plot and the plot sprayed at 7 days interval. The plot sprayed at 7 days interval is considered as fully controlled plot and losses in yield and yield components is calculated based on this treatment. The highest loss in number of pod per plant (31.59%) was recorded from plot with no fungicide spray while the lowest loss (11.91%) was from plots treated with fungicides at 7 days interval. This result is similar with [18] findings who have recorded the losses in number of pods/plant from 100% infected crops were estimated to about 21% to 31%. Similarly, the highest loss in number of seeds per plant (36.29%) was from a plot without fungicide treatment; while the lowest (10.60%) was from plot treated with a fungicide at 14 days interval. In the same manner, the maximum loss in grain yield (21.09%) was obtained from plots without any fungicide treatment while the lowest loss of 8.53% was recorded from plot received a fungicide treatment at every 14 days interval; where loss of 17.61% was recorded from plot treated with a fungicide at an interval of 21 days (Table 3).

The result from this study supports the finding from different experiments; the disease can cause 25% to 50% yield losses [12,18-20], A finding from (Dixon, [20]) also supports our result, he found from his study that from a heavily infested plot with powdery mildew disease and with no any treatment; the pathogen has caused up to 50% yield losses and reduced pod quality significantly.

Simple linear regression model was employed to assess the relationship between Powdery mildew severity at a weekly interval as predictor variable and yield as a dependent variable. The linear regression between powdery mildew severity index and grain yield revealed there was significant difference ($P \leq 0.0001$) between treatments. The estimated slope of the regression line obtained for Powdery mildew severity index was -34.16. The estimate showed that for each unit increase in percent severity index of Powdery mildew, there was a Field pea grain yield loss of 34.16 kg/ha (Figure 2). Based on coefficient of determination (R^2) value, the equations explained about 57.94% of losses in grain yield was occurred due to Powdery mildew severity. F-statistics calculated showed very highly significance ($P \leq 0.0001$) of the over-all probability of the equation (Figure 2).

Similarly, pair wise Pearson correlation analysis was employed

	PMDS	AUDPC	r	#Pod/plant	#Seed/plant	Plant height	BMS (kg)	TKW (gm)	G. yield kg/ha
PMDS									
AUDPC	0.99776***								
r	0.91023***	0.90333***							
pod/plant	-0.58854**	-0.57610**	-0.66069***						
Seed/plant	-0.43624*	-0.44545*	-0.40659*	0.72195***					
Plant height	-0.26985^{NS}	-0.25347^{NS}	-0.14546^{NS}	-0.12064^{NS}	-0.23672^{NS}				
BMS (kg)	-0.24653^{NS}	-0.25570^{NS}	-0.39367^{NS}	0.47339*	0.46309*	-0.76925***			
TKW (gm)	-0.56522**	-0.55550**	-0.61705**	0.27480^{NS}	0.16206^{NS}	0.02598^{NS}	0.34566^{NS}		
G. yield kg/ha	-0.76120***	-0.76298***	-0.79800***	0.70565***	0.51163*	-0.33647^{NS}	0.73066***	0.61770**	

Note: PMDS- Powdery Mildew Disease Severity (%), AUDPC- Area Under Disease Progress Curve (%-days), r-Disease Progress Rate, #pod/plant-number of pod per plant, #Seed/plant-Number of Seed per Plant, BMS-Biomass Yield, TKW-Thousand Kernel Weight and G. yield -Grain yield

Table 4: Pair wise Pearson correlation coefficient among disease parameters, yield and yield related parameters of field pea.

to assess the relationship of disease parameters with yield and yield components. Powdery mildew disease severity has significant negative correlation with number of pods per plant (r= -0.58854, P ≤ 0.1). Similarly, number of seeds per plant and TKW (g) have significant negative correlation (r= -0.43624, P ≤ 0.05; r= -0.56522, P ≤ 0.01), respectively with powdery mildew disease severity. Likewise, Powdery mildew disease severity was found to be significantly strongly negatively correlated with field pea grain yield (r= -0.76120, P ≤ 0.01) (Table 4). On the same way, AUDPC have significant very strong positive correlation with Powdery mildew severity and disease progress rate (r) (r=0.99776, P<0.0001; r=0.91023, P<0.0001). Significant negative correlation was also found (r= -0.57610, P ≤ 0.01; r= -0.44545, P ≤ 0.05; and r= -0.55550, P ≤ 0.01) between AUDPC and number of pods per plant, number of seeds per plant and TKW (g), respectively. AUDPC have strong negative correlation (r= -0.76298, P ≤ 0.0001) with grain yield. Disease progress rate (r) has significant negative correlation with number of pods per plant (r= -0.66069, P ≤ 0.0001) and TKW (r= -0.61705, P ≤ 0.001) and have strongly significant positive correlation (r=0.91023, P ≤ 0.0001; r= 0.90333, P ≤ 0.0001) with Powdery mildew diseases severity and AUDPC. Likewise, disease progress rate (r) has strongly significant negative correlation (r= -0.76298, P ≤ 0.0001) with grain yield (Table 4). With regard to the association between grain yield and some yield related parameters; grain yield have significantly strong positive correlation (r=0.70565, P ≤ 0.0001; r=0.73066, P ≤ 0.0001) with number of pods per plant and Biomass yield, respectively. Similarly, grain yield has significant positive correlation (r=0.51163, P ≤ 0.001; r=0.61770, P ≤ 0.001) with number of seeds per pod and thousand kernel weight (TKW), respectively.

Conclusion

Field pea is the major pulse crop grown in the highlands of Bale next to Faba bean. However, some diseases like Powdery mildew (*Erysiphe polygoni*) and Ascochyta blight (*Mycosphaerella pinodes)* have put its productivity under question. This study will better contribute towards the management of field pea diseases, particularly of Powdery mildew, which is most important disease of field pea at highlands Bale, Ethiopia. The fungicide spray frequency has made a statistically significant difference on field pea productivity. The highest grain yield of 2945.6 kg/ha was recorded from plot sprayed at 7 days interval. Whereas the highest grains yield loss of 21.09% was recorded from plot without fungicide treatment. Therefore, for the management of field pea Powdery mildew disease, based on disease pressure and the prevailing environmental condition 2-3 times spray of a fungicide Benomyl at a rate of 2.5 kg/ha within 7-10 days interval after the disease development is recommended from the result of the current study.

Acknowledgement

Without the support of some individuals the successful completion of this experiment would have not been realized. Oromia agricultural Research Institute (OARI) is duly acknowledged for fully funding this work and Sinana Agricultural Research Center (SARC) also deserve great thanks for implementation of the activity. All pulse and oil crops research case team staff have played their unreserved role, we would like to say thank you all. Our special thanks would go to Mr. Dagne Kora for his unreserved assistance in data analysis and interpretation.

References

1. McKay K, Blaine S, Gregory E (2003) Field crop production. North Dakota State University Agriculture and University Extension Morrill Hall, Fargo. 58: 105-562.
2. CSA (2008) Central statistics agency of Ethiopia, Addis Ababa, Ethiopia. 53-85.
3. Kemal A (2002) An integrated approach to pest management in field pea, (*Pisum sativum* L.) with emphasis on Pea Aphid, *Acyrthosiphon pisum* (Harris). University of Free State, Bloemfontein, South Africa 313 p.
4. CSA (2010) Agricultural sample survey 2009/10. Report on area and production of crops private peasant holdings, Meher Season. Addis Ababa. Statistical Bulletin no. 446. Volume 1.
5. Asfaw T, Tesfaye G, Beyene D (1994) Cool season food legumes of Ethiopia. Proceedings of the first national cool season food legume review conference, 16-20 December 1993. ICARDA, Aleppo-Syria, 1994, 197.
6. Smith PH, Foster EM, Boyd LA, Brown JKM (1996) The early development of *Erysiphe pisi* on *Pisum sativum* L. Plant Pathol 45: 302-309.
7. Bernier CC, Hanounik SB, Hussein MM, Mohamed HA (1993) Field manual of common Faba bean diseases in the Nile Valley. International Center for Agricultural Research in the Dry Areas (ICARDA), Beirut, Lebanon.
8. Wheeler JB (1969) An introduction to plant diseases. Wiley, London. 347 p.
9. Campbell CL, Madden VL (1990) Introduction to plant disease epidemiology. John Wiley and Sons, New York, USA.
10. SAS Institute (1998) SAS/STAT guide for personal computers, version 6. (12thedn). Cary, NC: SAS Institute.
11. Van der Plank JE (1963) Epidemiology of plant disease. New York and London Academic publishers.
12. Warkentin TD, Rashid KY, Xue AG (1996) Fungicidal control of powdery mildew in field pea. Can J Plant Sci 76: 933-935.
13. Ateet M, Bhupendra B, Raju PA, Sagar GC, Swati S (2015) Efficacy assessment of treatment methods against powdery mildew disease of pea (*Pisum sativum* L.) caused by *Erysiphe pisi* var. pisi. World J Agric Res 3: 185-191.
14. Sharma KD (2000) Management of pea powdery mildew in trans-Himalayan region. Indian J Agric Sci 70: 50-52.
15. Gritton ET, Ebert RD (1975) Interaction of planting date and powdery mildew on pea plant performance. Am Soc Horti Sci 100: 137-142.
16. Shah VK, Shakya SM, Gautam DM, Shristava A (2007) Effect of sowing time and row spacing on yield and quantity of seed of pea crop at Rampur, Chitwan, IAAS Research Advances Vol. 1. Institute of Agriculture and Animal Sciences, Rampur, Chitwan, Nepal, 59-64.
17. Fondevilla R (2012) Powdery mildew control in pea. A review. Agron Sustain Dev 32: 401-409.
18. Munjal RL, Chenulu VV, Hora TS (1963) Assessment of losses due to powdery mildew (*Erysiphe polygoni*) on pea. Indian Phytopathol 19: 260-267.

19. Dengjin B, Yantai G, Tom W (2011) Yields in mixtures of resistant and susceptible field pea cultivars infested with powdery mildew-defining thresholds for a possible strategy for preserving resistance. Can J Plant Sci 91: 873-880.

20. Dixon GR (1987) Powdery mildew of vegetables and allied crops. In: Speaure DM (Ed). Powdery Mildew. Academic Press, London, UK.

4

Effect of Integrated Climate Change Resilient Cultural Practices on Faba Bean Rust (*Uromyces viciae-fabae*) Epidemics in Hararghe Highlands, Ethiopia

Habtamu Terefe[1*], Chemeda Fininsa[1], Samuel Sahile[2] and Kindie Tesfaye[3]

[1]*School of Plant Sciences, Haramaya University, P.O.Box 138, Dire Dawa, Ethiopia*
[2]*Natural and Computational Science, University of Gondar, Ethiopia*
[3]*The International Maize and Wheat Improvement Center (CIMMYT), Addis Ababa, Ethiopia*

Abstract

Climate variability due to increasing temperature and erratic precipitation could affect faba bean rust disease epidemics and the crop productivity. Rust caused by *Uromyces viciae-fabae* is one of the serious foliar diseases of faba bean in Ethiopia. Field studies were conducted at Haramaya and Arbarakate during 2012 and 2013 to assess effects of integrated climate change resilient cultural practices on rust epidemics in the Hararghe highlands of Ethiopia. Three climate change resilient cultural practices: intercropping, compost application and furrow planting alone and in integration were evaluated using Degaga and Bulga-70 faba bean varieties and Melkassa-IV maize variety. Treatments were factorial arranged in a randomized complete block design with three replications. Faba bean-maize row intercropping and intercropping integrated treatments significantly reduced disease severity, AUDPC and disease progress rate. These treatments reduced rust mean severity by up to 36.5% (2012) and 27.4% (2013) at Haramaya, and up to 27% in 2013 at Arbarakate on both varieties as compared to sole planting. Compost fertilization also led to slow epidemic progression of rust and significantly reduced disease parameters when integrated with maize row intercropping. Compost fertilization in row intercropping recorded the lowest (23.1%) final mean disease severity and the highest (36.5%) percentage reduction in mean disease severity compared to sole cropping at Haramaya in 2012. The trend was similar in 2013 at both locations. Degaga had the lowest rust disease parameters studied compared to Bulga-70 at both locations over years. The overall results indicated that integrated climate change resilient cultural practices were effective to slow the epidemics of rust and to increase faba bean productivity. Hence, integrated climate change resilient cultural practices along with other crop management systems are recommended in the study areas.

Keywords: Cultural practices; Epidemics; Rust; *Uromyces viciae-fabae*; *Vicia faba*

Introduction

Faba bean (*Vicia faba* L.) is an important pulse crop produced in the world for both human diet and animal feed as source of protein and carbohydrate. It is also an excellent complement of crop rotations for fixing atmospheric N and as green manure [1]. China is the largest producer of faba beans in the world and in Africa, Egypt, Sudan, Ethiopia and Morocco are the dominant producers of faba bean [2]. In Ethiopia, faba bean production is estimated to account for 3.94% of the total grain production [3]. However, yields of faba beans have seen more fluctuations than area harvested and the world cultivated area has decreased in the last 50 years [4]. Climate variability, diseases, weeds and other pests are the major constraints of faba bean production. Diseases have always been the major limiting factors [5] and faba bean is susceptible to several pathogenic fungi, the major ones include rust (*Uromyces viciae-fabae* (Pers.) J. SchrÖt.), chocolate spot (*Botrytis fabae* Sard.) [6] and recently faba bean gall (*Olpidium viciae*) in Ethiopia [7].

Faba bean rust is a major disease of faba bean in almost every area in the world where faba bean is grown [8,9] that can cause up to 70% of yield loss in early infection [8]. The disease is severe and influences yield in areas like the Middle East, North Africa and parts of Australia [9]. It is also widely distributed in Ethiopia [10]. In Ethiopia, rust is devastating next to chocolate spot, and depending on severity of infection, it can cause a seed yield loss ranging from 2 to 15% in lower altitudes and 14-21% for intermediate altitudes [6], the loss may exceed the stated ranges in the present day conditions. Yield loss could be even higher when in mixed infection with chocolate spot disease [11].

Climate variability due to increased temperature and erratic precipitation over time increase susceptibility of faba bean and could also favour disease development. Faba bean production, which is seriously affected by diseases and parasitic weeds, are also worsened by climate change [12]. In faba bean, climate variability in the form of water stress, for example, decreases the final leaf area [13], net photosynthesis [14], light use efficiency [13], pod retention and filling and distorting hormonal balance [15]. Food legume growers are experiencing frequent droughts due to climate change and variability. Drought predisposes resistant varieties to be easily attacked by pathogens, which are not problems during normal growing seasons and new diseases may happen [16] and could decrease grain yield. Thus, climate change and associated changes in disease scenarios will demand changes in crop and disease management strategies. But such studies could be more difficult to undertake in conditions where historical weather and crop disease data are not available and where available facilities are not enabling to generate sound data.

Hence, climate change effects studies could be approached through climate change resilient crop management practices. These are practices that enhance the capacity of an ecological system to absorb

***Corresponding author:** Habtamu Terefe, School of Plant Sciences, Haramaya University, P.O.Box 138, Dire Dawa, Ethiopia
E-mail: habmam21@gmail.com

stresses and maintain its organizational structure and productivity, the capacity for self-organization, and the ability to adapt to stress and change following a perturbation [17]. They are generally designed to reduce climate change and its impacts in order to sustain agricultural crop production. Thus, a "resilient" agroecosystem would be capable of providing food production, when challenged by severe drought or by erratic rainfall [18] and better prepared for future climate change effects (used in the sense of mitigating/adaptive strategy in this paper).

The most important climate change resilient crop management practices include enhancing functional crop diversification to adjust to changing temperature and precipitation patterns [19] in the form of intercropping [20,21], effective soil nutrient management [22], and efficient soil moisture conservation via furrow planting and mulching [23,24] to reduce risk from crop failure due to climate change. Integrating these climate change resilient cultural practices for the management of crop diseases and sustainable crop production has a dual role for understanding the effects of climate change and the role of these cropping strategies for mitigation or adaptation. However, despite the significance of crop diseases in limiting crop productions and food supply, there has been limited field-based empirical research to assess the potential effects of climate change on plant diseases [25,26]. The integral role of climate change resilient cultural practices for the management of plant diseases and sustaining crop production in the face of climate change is not well addressed. Therefore, the objective of this study was to assess the effects of integrated climate change resilient cultural practices on the epidemics of faba bean rust in Hararghe highlands, Ethiopia.

Materials and Methods

Experimental sites

Field experiments were conducted at two locations in the 2012 and 2013 main cropping seasons. The 2012 field experiment was conducted at Haramaya University main campus experimental field on a sandy clay loam soil [27]. The station is located at 9°26'N and 42°3'E with an altitude of 2006 m.a.s.l. The highest mean annual rainfall for the location is 790 mm with mean minimum and maximum temperatures of 14.0°C and 23.4°C, respectively. The 2013 field experiment was conducted both at Haramaya University and Arbarakate Farmers' Training Center (FTC) on clay vertisol soil during the main cropping season. Arbarakate FTC is located at 9°2.86'N and 40°54.79'E with an altitude of 2274 m.a.s.l. in West Hararghe zone. Arbarakate was characterized by extended higher precipitation (estimated to exceed 1300 mm per annum) and many rainy days than Haramaya during the cropping periods with mean daily temperatures ranging between 13.1 and 17.5°C.

Experimental sites' weather data

Daily maximum and minimum temperatures (°C), relative humidity (%), and total rainfall (mm) were obtained for Haramaya University experimental site for both seasons from its weather station. The weather data obtained for Arbarakate from the nearby stations were found unrepresentative. However, the weather trend at Arbarakate was characterized by extended period of rainfall and many rainy days and relatively mild temperature levels. The daily mean minimum and maximum temperatures of Arbarakate were derived using the Adiabatic Lapse Rate Model [28] from nearby meteorological stations; and the daily minimum temperatures range from 5.31 to 12.43°C and the daily mean maximum temperatures range from 20.17 to 22.61°C (June to November) in 2013. The monthly total rainfall and the monthly average temperature in the cropping seasons are presented in Table 1.

Experimental materials

Planting material: The two faba bean varieties used in this study were Degaga (moderately resistant to major faba bean diseases) and Bulga-70 (moderately susceptible) and their characteristic features are presented in Table 2. Both faba bean varieties were obtained from Holleta Agricultural Research Center, Ethiopia. The maize variety used as a component crop was Melkassa-IV (*ECA-EE-36*), which was obtained from Melkassa Agricultural Research Center, Ethiopia. Melkassa-IV was released in 2006 with an agronomic attribute: area of adaptation (altitude of 1000-1600 meters above sea level, rainfall of 500-700 mm annual rainfall), early maturing (105 days) and a production potential of 2-4 t ha^{-1}.

Compost: The compost used in this study to substitute the application of mineral fertilizer was mainly made of a pile of khat (*Catha edulis* Forsk) residues collected from the nearby market of Awaday, eastern Ethiopia. Well-decomposed and matured compost was air-dried and sieved. Composite random samples were taken for chemical analysis before application. The compost constituted organic carbon (8.01%), organic matter (13.80%), total nitrogen (0.69%), available phosphorus (234.80 mg kg^{-1}) and C:N ratio of 11.61. In the experiment, the compost was row applied to a depth of 10-15 cm at the rate of 8 t ha^{-1} and mixed with the soil a week before maize planting and four weeks in 2012 and three weeks in 2013 before faba bean planting. Furrows were prepared by digging about 20 cm deep rows once the faba bean was planted and established as seedling, and rain water was made to stagnate.

Treatments, experimental design and procedure

Three on-farm based climate resilient cultural practices (crop diversification in the form of intercropping, moisture conservation as planting in furrows and soil nutrient management as compost application), two faba bean varieties and one open pollinated Melkassa-IV maize variety were used in this study. Thus, the treatments included faba bean-maize row intercropping, furrow planting, compost application and sole faba bean row planting. The treatments were applied solely and in integration with each other (Table 3). A total of 16 treatments (for both faba bean varieties) were laid out in a randomized complete block design in a factorial arrangement with three replications. In a gross plot size of 4 m×3.2 m, a 1 maize: 1 faba bean planting pattern of row intercropping was maintained by planting maize rows spaced 0.80 m apart and planting one row of faba bean between the two maize rows. In the row intercropping, 5 rows of maize were intercropped with 4 rows of faba bean variety each at the center of the two maize rows per plot. In addition, sole faba bean row planting was included as experimental treatment, which was planted at 0.40 m×0.10 m inter-row and intra-row spacing, respectively. In case of sole faba bean row planting there were 10 rows per plot. In the intercrops, maize was planted three weeks in 2012 and two weeks in 2013 prior to faba bean planting. The spacing between blocks was 1.5 m and that between plots was 1 m.

Sowing of maize was done manually by planting two seeds per hill, which were later thinned to one plant per hill. The faba bean varieties were also manually planted. Maize was planted at Haramaya on 21 June 2012 and on 27 June 2013; and at Arbarakate on 3 July 2013. Faba bean was planted at Haramaya on 11 July 2012 and on 12 July 2013; and at Arbarakate on 16 July 2013. The crops were grown without application of any chemical fertilizer and no artificial pathogen inoculation was performed. Weeding and other agronomic practices were done properly and uniformly as per the recommendations to grow a successful crop.

Cropping month	Mean of temperature (°C)			Monthly rainfall (mm)	
	Haramaya		Arbarakate	Haramaya	
	2012	2013	2013	2012	2013
June	19.97	19.30	17.52	0.00	15.80
July	18.56	17.63	15.81	214.00	215.40
August	18.90	18.25	16.48	149.50	185.10
September	18.73	18.43	16.62	105.00	142.10
October	15.50	16.82	15.47	4.60	71.60
November	14.68	15.04	13.14	0.50	81.70
Mean	17.72	17.58	15.84	78.93	118.62

Table 1: Monthly mean temperature (°C) and monthly total rainfall (mm) during faba bean growing periods at Haramaya and Arbarakate, Ethiopia in 2012 and 2013 main cropping seasons.

Faba bean variety	Year of release	Area of adaptation		Maturity (days)	Seed size (g)	Yield (t/ha)	
		Altitude (m)	Annual rain-fall (mm)			On station	On farm
Degaga	2002	1800-3000	800-1100	116-135	400-450	2.5-5.0	2.0-4.5
Bulga-70	1994	2300-3000	800-1100	143-150	400-450	2.0-4.5	1.5-3.5

Table 2: Characteristic features of faba bean varieties used for the field experiment at Haramaya and Arbarakate during the 2012 and 2013 main cropping seasons.

S.No.	Treatment	Treatment combination description
1	SP	Sole faba bean row planting (control)
2	FP	Furrow faba bean planting
3	CA	Faba bean planting using compost application (compost fertilization)
4	RI	Faba bean-maize row intercropping
5	FP+CA	Faba bean furrow planting with compost application
6	FP+RI	Faba bean furrow planting in faba bean-maize row intercropping
7	CA+RI	Faba bean planting using compost application in faba bean-maize row intercropping
8	FP+CA+RI	Faba bean furrow planting with compost application in faba bean-maize row intercropping

Table 3: Treatment combinations used for faba bean field experiments at Haramaya and Arbarakate during 2012 and 2013 main cropping seasons and their respective descriptions.

Disease severity assessment

Disease severity was assessed six times at Haramaya and four times at Arbarakate on weekly intervals starting from the first visible disease symptoms both in 2012 and 2013. For disease severity assessments, 12 plants were randomly selected from central rows of each plot and tagged prior to disease symptom appearance. Disease severity assessment began at 50 days after planting (DAP) in 2012 and 44 DAP in 2013 at Haramaya. At Arbarakate, disease severity recording began from 65 DAP onwards during 2013. Disease severity was scored using a 1-9 scale of ICARDA [29]; where, 1=no pustules or very small non-sporulating flecks; 3=few scattered pustules on leaves, few or no pustules on stem; 5=pustules common on leaves, some pustules on stem; 7=pustules very common on leaves, many pustules on stem; and 9=extensive pustules on the leaves, petioles and stem, many leaves dead and plant defoliated. Disease severity scores were converted into percentage severity index (PSI) for analysis [30]; where,

$$PSI = \frac{\text{Sum of Numerical Ratings} \times 100}{\text{Number of Plants Scored} \times \text{Maximun Score on Scale}}$$

From disease severity data, the area under disease progress curves (AUDPC) in %-days were calculated as used in Campbell and Madden [31]:

$$AUDPC = \sum_{i-1}^{n-1} (0.5(X_i + X_{i+1}))(t_{i+1} - t_i)$$

where, X_i=percentage of disease severity index (PSI) of disease at i[th] assessment; t_i=time of the ith assessment in days from the first assessment date; and n=total number of disease assessments.

AUDPC was calculated separately for disease assessments made on different DAP for each climate change resilience cultural practices used and the control treatment. Since the epidemic period of the two locations varied in 2013, AUDPC were standardized by dividing the values by the epidemic duration of the respective locations [31]. The epidemic periods were 35 days at Haramaya and 21 days at Arbarakate; and AUDPC values were standardized accordingly.

Data analysis

Analysis of variance (ANOVA) was run for disease severity data and AUDPC values of both faba bean varieties to determine treatment effects on each disease parameter in each year across locations using SAS GLM Procedure [32]. Mean separations were made using LSD at 0.05 probability level. To determine the disease progress rate from the linear regression, a Logistic model, ln[(y/1-y)] [33], was used to estimate the disease progression from each separate treatment. The transformed data were regressed over time, DAP to determine the disease progress rate. The slope of the regression line estimated the disease progress rate. Regression was computed using Minitab (Release 15.0 for windows®, 2007). The two locations and seasons were considered as different environments because of heterogeneity of variances tested using Bartlett's test [34]. As a result, data were not combined for analysis.

Results

Rust severity

Faba bean rust appeared during the flowering growth stage of both faba bean varieties both in 2012 and 2013 at Haramaya and during pod formation growth stage at Arbarakate in 2013 cropping season. The mean disease severity of faba bean rust in the two cropping seasons was significantly different (*P* ≤ *0.05*) between some of the climate change resilient cultural practices and the control, among some of the climate change resilient cultural practices used and between varieties both at Haramaya and Arbarakate (Table 4). In both cropping seasons, mean disease severity assessments at different DAP showed significant variation between treatments starting from 57 DAP in 2012 and 51 DAP in 2013 at Haramaya in the disease epidemic periods. Whereas at Arbarakate, it was started from 65 DAP during 2013. Higher rust severity was observed on both faba bean varieties in 2013 than in 2012 and it was severe after pod filling growth stage at Haramaya.

The lowest final mean disease severity in 2012 was recorded on both faba bean varieties at Haramaya from integrated climate change resilient cultural practices treated plots as compared to sole planting. A similar trend was obtained in 2013 on both varieties at both experimental locations. Intercropping and intercropping integrated climate change resilient cultural practices (furrow planting in row intercropping and/or compost fertilization in row intercropping and/or furrow planting with compost fertilization in row intercropping or referred as intercropping integrated treatments hereafter unless stated) had the lowest mean disease severity records on both varieties and locations over years in comparison to sole planting. On the final disease severity assessment days, intercropping integrated treatments treated plots recorded up to 23.14% compared to 36.42% of sole plot in 2012 and 32.72% compared to 45.06% of sole plot in 2013 on both faba bean varieties at Haramaya. At Arbarakate, the trend was 16.67% as compared to 22.84% of sole plot on both varieties in 2013.

Thus, intercropping integrated treatment treated plots were found to highly reduce disease severity of rust compared to control plots. The highest mean disease severity reductions reached 36.46% (2012) and 27.39% (2013) on both varieties at Haramaya. Similarly, the reduction was up to 27.01% on both varieties studied at Arbarakate. Although there was no consistent significant variation among compost fertilization, furrow planting, furrow planting along with compost fertilization and sole planting, compost fertilization and furrow planting with compost fertilization had lower faba bean rust severity on both varieties in 2012 and 2013 at Haramaya. In 2012, compost fertilization and furrow planting with compost fertilization lowered the final mean disease severity of faba bean rust in the range between 9.31 and 10.16% at Haramaya. A similar trend was also indicated in 2013 on both varieties at both locations (Table 3). Moreover, at both locations and seasons, the overall mean disease severity records showed that the two faba bean varieties were varied significantly. Degaga variety had lower mean disease severities than Bulga-70 variety studied. The interaction between faba bean varieties and climate change resilient cultural practices used was not significant during both cropping seasons at both locations.

Standardized area under disease progress curve (rAUDPC)

rAUDPC values calculated from disease severity assessed at different DAP on both faba bean varieties for both locations and cropping seasons significantly ($P \leq 0.05$) varied between some of the climate change resilient cultural practices and the control, among some of the climate change resilient cultural practices used and between faba bean varieties studied (Table 4). rAUDPC values were lower on intercropped and intercropping integrated treatments treated plots than on other treatments. In 2012, sole plots had the highest (29.50%-days) rAUDPC values, while the lowest (19.84%-days) rAUDPC values were calculated from compost fertilization in row intercropping treated plots. The overall values indicated that intercropping and intercropping integrated treatments treated plots showed consistent reduction in rAUDPC values. In 2013, a similar trend was also calculated for the sole cropped and integrated climate change resilient cultural practices treated plots for both varieties and locations. Compost fertilization with or without furrow planting also lowered rAUDPC values of faba bean rust on both faba bean varieties in 2012 and 2013 at Haramaya and in 2013 at Arbarakate areas.

Disease progress rate

Disease progress rates and parameter estimates of faba bean rust are tabulated in Tables 5-7. The disease progress rates computed from mean disease severity records showed variations among treatments used in both faba bean varieties, locations and seasons. Disease progress rates of Degaga variety ranged from 0.0182 to 0.0288 units/day in 2012 and from 0.0340 to 0.0456 units/day in 2013 (Table 5); whereas for Bulga-70, the rates ranged from 0.0234 and 0.0331 units/day in 2012 and 0.0461 and 0.0546 units/day in 2013 at Haramaya (Table 6). The rates were also from 0.0158 to 0.0309 units/day for Degaga and 0.0279 to 0.0412 unit/day for Bulga-70 in 2013 at Arbarakate (Table 7). The disease progress rate was relatively higher at Haramaya in 2013 and relatively fast on both varieties in 2012 and 2013 at Haramaya than at Arbarakate. The variety Bulga-70 had higher disease progress rate than the variety Degaga in both cropping seasons at both locations. It was also observed that disease progressed relatively at faster rates on sole and non-intercropped and non-intercropping integrated treatments treated plots across locations and over years for both faba bean varieties. The results indicated that the rate at which faba bean rust progressed was slower when climate change resilient cultural practices were applied in integration than the untreated plots.

Treatment [1]	Haramaya				Arbarakate	
	2012		2013		2013	
	PSI [2]	rAUDPC [3]	PSI [2]	rAUDPC [3]	PSI [2]	rAUDPC [3]
Variety						
Bulga-70	31.79a	26.07a	42.67a	24.59a	21.99a	16.20a
Degaga	27.01b	22.18b	34.34b	20.92b	17.60b	14.30b
Mean	29.40	24.13	38.51	22.76	19.80	15.25
LSD (0.05)	1.09	0.76	1.14		0.57	0.46
Cultural practice						
SP	36.42a	29.50a	45.06a	26.48a	22.84a	18.06a
FP	34.57ab	28.30ab	44.45b	25.68a	22.22ab	16.87b
CA	32.72b	27.01b	41.98b	24.54b	21.60b	16.36b
RI	25.00cd	20.37cd	33.95c	20.56c	17.90c	13.94c
FP + CA	33.03b	27.59b	41.98b	24.35b	21.60b	16.05b
FP + RI	26.55c	21.30c	34.57c	20.77c	18.21c	13.99c
CA + RI	23.14d	19.07d	32.72c	19.60d	16.67d	13.17c
FP + CA + RI	23.77d	19.84cd	33.34c	20.05cd	17.29cd	13.58c
Mean	29.40	24.12	38.51	22.75	19.79	15.25
LSD (0.05)	2.19	1.52	2.27	0.93	1.15	0.91
CV (%)	6.31	5.36	5.00	3.48	4.92	5.07

[1]SP: Sole Planting (Control); FP: Furrow Planting; CA: Compost Application; RI: Row Intercropping; FP+CA: Furrow Planting with Compost Application; FP+RI: Furrow Planting in Row Intercropping; CA+RI: Compost Application in Row Intercropping; and FP+CA+RI: Furrow Plating with Compost Application in Row Intercropping.

[2]Percent severity index on 85 days after planting (DAP) in 2012 and 79 DAP in 2013 at Haramaya and on 86 DAP at Arbarakate during 2013 main cropping season.

[3]rAUDPC, standard area under disease progress curve of faba bean rust.

Means in each column followed by the same letter are not significantly different according to the least significant difference test at 5% probability level.

Table 4: Effects of climate change resilient cultural practices on faba bean rust (*Uromyces viciae-fabae*) severity (%) and standard area under disease progress curve (%-days) at Haramaya during 2012 and 2013 and at Arbarakate during 2013 main cropping seasons.

Year	Treatment¹	Percent severity²		Intercept³	SE of intercept⁴	Disease progress rate (Logit/day)	SE of Rate⁴	R^2 (%)⁵
		PSI_i	PSI_f					
	SP	14.19	33.33	-3.14	0.2439	0.0288	0.0033	81.4
	FP	15.43	31.48	-2.85	0.2131	0.0244	0.0029	80.4
	CA	14.19	29.63	-3.05	0.2036	0.0256	0.0028	83.2
	RI	12.34	22.84	-3.03	0.1647	0.0212	0.0022	83.8
2012	FP+CA	13.58	30.25	-3.07	0.2586	0.0265	0.0035	76.6
	FP+RI	11.73	24.69	-3.23	0.2213	0.0246	0.0030	79.4
	CA+RI	11.73	21.60	-3.03	0.1762	0.0200	0.0024	80.2
	FP+CA+RI	12.96	22.22	-2.83	0.1817	0.0182	0.0025	75.8
	SP	12.96	39.51	-4.36	0.1477	0.0438	0.0020	96.5
	FP	11.73	38.89	-4.54	0.1257	0.0456	0.0017	97.7
	CA	11.73	37.04	-4.41	0.0951	0.0431	0.0013	98.5
	RI	11.11	30.86	-4.22	0.1193	0.0375	0.0016	96.9
2013	FP+CA	11.73	37.04	-4.42	0.1187	0.0432	0.0016	97.7
	FP+RI	11.73	31.48	-4.13	0.1636	0.0365	0.0022	94.0
	CA+RI	11.73	29.63	-4.01	0.1684	0.0340	0.0023	92.8
	FP+CA+RI	11.11	30.25	-4.15	0.1369	0.0362	0.0019	95.7

¹SP: Sole Planting (Control); FP: Furrow Planting; CA: Compost Application; RI: Row Intercropping; FP+CA: Furrow Planting with Compost Application; FP+RI: Furrow Planting in Row Intercropping; CA+RI: Compost Application in Row Intercropping; and FP+CA+RI: Furrow Plating with Compost Application in Row Intercropping. Parameter estimates are from a linear regression of ln(y/(1-y)) disease severity (PSI) proportions at different days after planting (DAP). ²Initial and final disease severity (PSI) of rust recorded at 50 DAP and 85 DAP in 2012 and at 44 DAP and 79 DAP in 2013, respectively. ³Intercept of the regression equation. ⁴Standard error of parameter estimates. ⁵Coefficient of determination of the logistic model.

Table 5: Mean initial (PSI_i) and final (PSI_f) severity index and parameter estimates of faba bean rust (*Uromyces viciae-fabae*) on Degaga variety at Haramaya, Ethiopia during 2012 and 2013 main cropping seasons.

Year	Treatment¹	Percent severity²		Intercept³	SE of intercept⁴	Disease progress rate (Logit/day)	SE of Rate⁴	R^2 (%)⁵
		PSI_i	PSI_f					
	SP	15.43	39.51	-3.23	0.3264	0.0331	0.0044	76.2
	FP	14.81	37.66	-3.18	0.3053	0.0314	0.0042	76.8
	CA	14.81	35.81	-3.07	0.2959	0.0296	0.0040	75.7
	RI	15.43	35.81	-3.01	0.2880	0.0289	0.0039	75.8
2012	FP+CA	12.96	27.16	-3.21	0.1912	0.0261	0.0026	85.4
	FP+RI	14.20	28.40	-2.98	0.2199	0.0237	0.0030	78.4
	CA+RI	12.34	24.68	-3.09	0.1990	0.0234	0.0027	81.3
	FP+CA+RI	11.73	25.31	-3.20	0.2264	0.0251	0.0031	79.4
	SP	12.96	50.62	-4.93	0.1512	0.0546	0.0021	97.6
	FP	12.96	50.00	-4.92	0.1645	0.0538	0.0022	97.1
	CA	12.34	46.92	-4.90	0.1376	0.0527	0.0019	97.9
	RI	11.11	37.04	-4.74	0.1700	0.0471	0.0023	96.1
2013	FP+CA	12.34	46.91	-4.97	0.1246	0.0534	0.0017	98.3
	FP+RI	11.11	37.66	-4.78	0.1443	0.0480	0.0020	97.2
	CA+RI	11.11	35.80	-4.72	0.1627	0.0461	0.0022	96.2
	FP+CA+RI	11.11	36.42	-4.74	0.1475	0.0467	0.0020	97.0

¹SP: Sole Planting (Control); FP: Furrow Planting; CA: Compost Application; RI: Row Intercropping; FP+CA: Furrow Planting with Compost Application; FP+RI: Furrow Planting in Row Intercropping; CA+RI: Compost Application in Row Intercropping; and FP+CA+RI: Furrow Plating with Compost Application in Row Intercropping. Parameter estimates are from a linear regression of ln(y/(1-y)) disease severity (PSI) proportions at different days after planting (DAP). ²Initial and final disease severity (PSI) of rust recorded at 50 DAP and 85 DAP in 2012 and at 44 DAP and 79 DAP in 2013, respectively. ³Intercept of the regression equation. ⁴Standard error of parameter estimates. ⁵Coefficient of determination of the logistic model.

Table 6: Mean initial (PSI_i) and final (PSI_f) severity index and parameter estimates of faba bean rust (*Uromyces viciae-fabae*) on Bulga-70 variety at Haramaya, Ethiopia during 2012 and 2013 main cropping seasons.

Discussion

The overall results of the study indicated that severity of rust was higher and rapidly increasing at the later stages of the epidemic period at Haramaya both in 2012 and 2013. However, in both cropping seasons at Haramaya and Arbarakate, rust severity, rAUDPC and disease progress rate were reduced and grain yield per unit area was increased by integrated climate change resilient cultural practices compared to sole planting. Among the resilient cultural practices, intercrops and intercropping integrated treatments had the lowest disease parameters of faba bean rust and chocolate spot as well [35] as compared to sole cropping. Such effects could be reduced faba bean density due to intercropping and maize acting as a physical barrier that might hamper inoculum spread and disease progress. In addition, intercrops might have also modified the microclimate by modifying the density of host plants thereby changing canopy microenvironment.

Previous studies indicated that deploying crop diversity in the form of intercropping is one way of introducing more biodiversity into agroecosystems; and results from intercropping studies showed that higher species richness may be associated with significant reduction in the negative impacts of diseases [36,37] and weeds [38]. Intercropping also limits the places where pests can find optimal foraging or reproductive conditions [39]. Similarly, mixtures play a major role in

Variety	Treatment[1]	Percent severity[2]		Intercept[3]	SE of intercept[4]	Disease progress rate (Logit/day)	SE of Rate[4]	R^2 (%)[5]
		PSI_i	PSI_f					
	SP	12.34	20.37	-3.76	0.2052	0.0276	0.0027	90.6
	FP	11.11	19.75	-4.08	0.2319	0.0309	0.0030	90.5
	CA	11.11	19.14	-3.98	0.3233	0.0293	0.0042	81.2
	RI	11.11	16.05	-3.44	0.2021	0.0205	0.0026	84.4
Degaga	FP+CA	11.11	19.14	-4.06	0.1770	0.0301	0.0023	93.9
	FP+RI	11.73	16.05	-3.19	0.2609	0.0172	0.0034	69.3
	CA+RI	11.11	14.82	-3.15	0.1490	0.0158	0.0019	85.7
	FP+CA+RI	11.11	15.44	-3.26	0.2021	0.0177	0.0026	80.2
	SP	12.96	25.31	-4.43	0.2423	0.0391	0.0032	93.3
	FP	12.34	24.69	-4.58	0.3533	0.0398	0.0046	87.0
	CA	12.34	24.07	-4.51	0.2838	0.0385	0.0037	90.8
	RI	11.73	19.75	-3.96	0.2796	0.0286	0.0036	84.7
	FP+CA	11.73	24.07	-4.75	0.2097	0.0412	0.0027	95.4
Bulga-70	FP+RI	11.11	20.37	-4.33	0.1676	0.0336	0.0022	95.6
	CA+RI	11.11	18.52	-3.97	0.2360	0.0279	0.0031	88.2
	FP+CA+RI	11.11	19.14	-4.13	0.2865	0.0303	0.0037	85.6

[1]SP: Sole Planting (Control); FP: Furrow Planting; CA: Compost Application; RI: Row Intercropping; FP+CA: Furrow Planting with Compost Application; FP+RI: Furrow Planting in Row Intercropping; CA+RI: Compost Application in Row Intercropping; and FP+CA+RI: Furrow Plating with Compost Application in Row Intercropping. Parameter estimates are from a linear regression of ln(y/(1-y)) disease severity (PSI) proportions at different days after planting (DAP). [2]Initial and final disease severity (PSI) of rust recorded at 65 DAP and 86 DAP in 2013, respectively. [3]Intercept of the regression equation. [4]Standard error of parameter estimates. [5]Coefficient of determination of the logistic model.

Table 7: Mean initial (PSI_i) and final (PSI_f) severity index and parameter estimates of faba bean rust (*Uromyces viciae-fabae*) on Degaga and Bulga-70 varieties at Arbarakate, Ethiopia during 2013 main cropping season.

reducing the efficiency of the pathogen through the dilution effect [40] and mixed crop species can also delay the onset of diseases by reducing the spread of disease carrying spores and by modifying environmental conditions to less favorable to the spread of certain pathogens [9,41]. In addition, under Ethiopian conditions, mixed cropping has also been reported to reduce disease severity of faba bean rust [42].

In faba bean-maize intercrops and intercrop integrated treatments of this study, the population of faba bean per plot was reduced by more than half. This could modify the microclimate of the faba bean canopy in that there was free air-circulation, low leaf wetness and reduced damp sites. Likewise, Biddle and Catline [43] stated that densely planted faba beans encourage humid microclimate within the canopy, thereby, encouraging infection and spore production in the presence of warm temperatures and light film of moisture on the leaf surface. Thus, Fernández-Aparicio et al. [44] noted that intercropping faba bean with cereals has been proposed as a means to lessen the incidence of faba bean rust. The cereal favors aeration and prevents the formation of a dense faba bean canopy that might enhance disease damage. Reddy [45] also indicated that varietal mixtures reduce disease epidemics by reducing the spatial density of susceptible plants where the deposition probability of released spores on susceptible tissue from a lesion is reduced.

It was also observed that compost fertilization alone and in integration with furrow planting and row intercropping in particular highly reduced faba bean rust and decreased chocolate spot severity [35]. Slight increase in epidemic development on the most integrated treatment was obtained compared to compost fertilization in row intercropping, which might be due to the presence of furrows that could slightly increase humidity under that canopy late in the cropping season. Compost fertilization might have enhanced the health and vigority of plants that could have increased plant chances to withstand pathogen attack and to activate the host defense system. Neher et al. [46] found that compost amended soils reduced disease severity of ear blight on brassicas compared to the bare soil. Haggag and Saber [47] reported that compost teas significantly reduced disease incidence and population counts of alternaria blight and significantly increased the activities of both peroxidase, β-1,3-glucanase and chitinase that could increase plant resistance both under greenhouse and field planted tomato and onion. Similar results were also observed by Sang et al. [48] against Phytophthora capsici in pepper plants by compost water extracts and the test again activates expression of pathogenesis-related genes and peroxide generation in the leaves and lignin accumulation in the stems.

The epidemics of faba bean rust was appeared early and higher at Haramaya than at Arbarakate areas. This could be associated with the weather conditions and the altitude differences of the two locations. Arbarakate was characterized by many rainy days with extended period of rainfall and mild temperature (15.8 to 16.6°C) during the cropping season which might have delayed the onset of rust and its epidemics. Haramaya was relatively warm (temperature ranging from 14.7 to 19.8 °C) with high relative humidity and fair rainfall distribution. Moreover, the results also demonstrated that rust severity was relatively higher late in the epidemic period during 2013 than 2012 at Haramaya. This might be partially explained by early termination of rainfall that would in turn reduce leaf wetness and infection in 2012.

Supporting the current study, Hawthorne et al. [49] stated that rust infection is favored by humid and warm temperatures. This infection can occur following six hours of leaf wetness. The development of both primary and secondary inoculum sources of faba bean rust are also influenced by environmental factors. Such that cloudy weather with high humidity and 17-22°C favors development of the disease [9]. That is, spore production is encouraged by high humidity and warm temperatures and once spores are released and deposited on a susceptible host crop, germination occurs quickly in the presence of a light film of moisture on the leaf surface to cause infection [43]. Of course, Dipak et al. [50] also found that rainy days are negatively correlated with disease development of *Uromyces viciae-fabae;* which could be the most probably reason for the delayed onset of faba bean rust at Arbarakate in 2013.

Faba bean rust epidemics might also be associated with altitude in which Arbarakate recorded lower rust severity than Haramaya since the former location is more highland than the later. In accordance with this study, a survey conducted by Shifa et al. [51] in Hararghe highlands of Ethiopia in the 2009 cropping season to determine the incidence and severity of faba bean rust, and its association with environmental factors and cultural practices found that the incidence and severity of faba bean rust showed higher association with altitudes. The results indicated that those surveyed locations with an altitude above 2450 m.a.s.l had relatively low incidence and severity than locations below 2450 m.a.s.l. The variation could be partly due to the difference in the relative warmness of locations, as faba bean rust epidemic is lower in lower and intermediate altitudes (<2300 m.a.s.l) and usually late in the season [52].

Conclusions

Climate change resilient cultural practices alone and in integration found effective to slow the epidemic progression of faba bean rust and improve crop productivity in the prevailing climate change effects. Intercropping integrated climate change resilient cultural practices highly reduced disease parameters of faba bean rust. Similarly, compost fertilization of the soil with or without row intercropping also plays an important role to manage faba bean rust. These practices could also be employed as an option in climate resilient agriculture to mitigate climate change and variability impacts in subsistence farming systems. It is, therefore, promising to grow faba bean with these climate change resilient cultural practices (maize row intercropping and compost fertilization in row intercropping in particular) in addition to using host resistance and other crop management strategies to manage faba bean rust in Hararghe highlands. Further studies on integrated control of rust should continue that include host resistance and cultivar mixtures in the system. Moreover, the mechanisms through which compost fertilization reduces severity of foliar diseases should also be thoroughly investigated.

Acknowledgements

The study was financed by the Swedish International Development Agency (SIDA) and Haramaya University, Ethiopia. We thank Mr. Getahun Tessema, Berhanu Asefaw and Tefera Birhanu for their assistance during field follow up and data collection; and staff members of West Hararghe Bureau of Agriculture and Rural Development who facilitated the allotment of experimental land at Arbarakate. We are also very grateful to farmers and daily laborers who involved in the field experiments.

References

1. Salmeron JIC, Avila C, Torres AM (2010) Faba bean and its importance in food security in developing countries. International conference on food security and climate change in dry areas, 1-4 February 2010. Amman, Jordan p: 13.
2. Akibode S, Maredia M (2011) Global and regional trends in production, trade and consumption of food legume crops. Report submitted to SPIA, 27 March 2011. Michigan State University, USA pp: 1-87.
3. CSA (Central Statistics Authority) (2014) Agricultural sample survey (2013/2014). Report on area and production of major crops. Statistical Bulletin 532, Central Statistical Authority, Addis Ababa, Ethiopia pp. 10-20.
4. Rosegrant MW (2010) Impacts of climate change on food security and livelihoods. In: Solh M, Saxena MC (eds) Food security and climate change in the dry areas. Proceedings of International Conference, 1-4 February, 2010. Amman, Jordan pp: 24-27.
5. Agegnehu G, Ghizaw A, Sinebo W (2006) Yield performance and land-use efficiency of barley and faba bean mixed cropping in Ethiopian highlands. Eur J Agron 25: 202-207.
6. Dereje G, Tesfaye B (1993) Faba bean diseases in Ethiopia. In: Asfaw T, Geletu B, Saxena MC, Solh MB (eds) Cool season food legumes of Ethiopia. Proceedings of the first national cool season food legumes review conference, 16-20 December 1993. Addis Ababa, Ethiopia pp: 328-345.
7. Endale H, Gezahegn G, Tadesse S, Nigussie T, Beyene B, et al. (2014) Faba bean gall: a new threat for faba bean (*Vicia faba*) production in Ethiopia. Adv Crop Sci Tech 2: 1-5.
8. Torres AM, Roman B, Avila CM, Satovic Z, Rubiales D, et al. (2006) Faba bean breeding for resistance against biotic stresses: Towards application of marker technology. Euphy 147: 67-80.
9. Stoddard FL, Nicholas AH, Rubiales D, Thomas J, Villegas-Fernández AM (2010) Integrated pest management in faba bean. Field Crops Res 115: 308-318.
10. Berhanu B, Getachew M, Teshome G, Temesgen B (2003) Faba Bean and Field Pea Diseases Research in Ethiopia. In: Ali K, Gemechu K, Ahmed S, Malhotra R, Beniwal S, et al. (eds) Food and forage legumes of Ethiopia: progress and prospects. Proceedings of the workshop on food and forage legumes, 22-26 September 2003. Addis Ababa, Ethiopia pp: 221-227.
11. MacLeod B (2006) Faba Bean: Rust disease. Farmnote 114/96, Department of Agriculture, Government of Western Australia.
12. Khan HR, Paull JG, Siddique KHM, Stoddard FL (2010) Faba bean breeding for drought-affected environments: A Physiological and agronomic perspective. Field Crops Res 115: 279-286.
13. Costa CL, Morison J, Dennett M (1997) Effects of water stress on photosynthesis, respiration and growth of faba bean (*Vicia faba* L.) growing under field conditions. Revista Brasileira de Agrometeorologia 5: 9-16.
14. Hura T, Hura K, Grzesiak M, Rzepka A (2007) Effect of long-term drought stress on leaf gas exchange and fluorescence parameters in C3 and C4 plants. Acta Physiol Plant 29: 103-113.
15. Karamanos AJ, Gimenez C (1991) Physiological factors limiting growth and yield of faba. Options Méditerranéennes 10: 79-90.
16. Ahmed S, Muhammad I, Kumar S, Malhotra R, Maalouf F (2011) Impact of Climate Change and variability on diseases of food legumes in the dry areas. International Center for Agricultural Research in the Dry Areas (ICARDA), Aleppo, Syria pp: 157-165.
17. Cabell JF, Oelofse M (2012) An indicator framework for assessing agroecosystem resilience. Ecology and Society 17: 1-13.
18. Heal G (2000) Nature and the marketplace: capturing the value of ecosystem services. Island Press, Washington, D.C.
19. NRC (National Research Council) (2010) Adapting to the impacts of climate change. National Research Council. Academic Press, Washington, DC.
20. Tilahun T, Minale L, Alemayehu A (2012) Role of maize (*Zea mays* L.)- faba bean (*Vicia faba* L.) intercropping planting pattern on productivity and nitrogen use efficiency of maize in northwestern Ethiopia highlands. Int Res J Agri Sci and Soil Sci 2: 102-112.
21. Workayehu T (2014) Legume-based cropping for sustainable production, economic benefit and reducing climate change impacts in southern Ethiopia. J Agri Crop Res 2: 11-21.
22. Katungi E, Farrow A, Chianu J, Sperling L, Beebe S (2009) Common bean in Eastern and Southern Africa: a situation and outlook analysis of targeting breeding and delivery efforts to improve the livelihoods of poor in drought prone areas through ICRISAT. Baseline research report. Kampala, Uganda pp: 1-126.
23. Wang Q, Zhang E, Li F, Li F (2008) Runoff efficiency and the technique of micro water harvesting with ridges and furrows for potato production in semi-arid areas. Water Res Manag 22: 1431-1443.
24. Zhao H, Xiong YC, Li FM, Wang RY, Qiang SC, et al. (2012) Plastic film mulch for half growing-season maximized WUE and yield of potato via moisture-temperature improvement in a semi-arid agroecosystem. Agri Water Manag 104: 68-78.
25. Coakley SM, Scherm H (1996) Plant Disease in changing global environment. App Biol 45: 227-238.
26. Garrett KA, Dendy SP, Frank EE, Rouse MN, Travers SE (2006) Climate change effects on plant disease: genomes to ecosystems. Annu Rev Phytopathol 44: 489-509.
27. Gelgelo B (2012) Response of improved potato (Solanum tuberosum L.) varieties to nitrogen application in Eastern Ethiopia. M.Sc. Thesis, Haramaya University, Haramaya, Ethiopia p: 69.

28. Brunt D (2007) The adiabatic lapse-rate for dry and saturated air. Quarterly J Royal Meteo Soci 59: 351-360.

29. ICARDA (International Center for Agricultural Research in the Dry Areas) (1986) Screening techniques for disease resistance in faba beans. International Center for Agricultural Research in the Dry Areas (ICARDA), Aleppo, Syria pp: 1-59.

30. Wheeler BEJ (1969) An Introduction to plant diseases. Wiley and Sons, London.

31. Campbell CL, Madden LV (1990) Introduction to plant disease epidemiology. Raleigh, North Carolina, Wooster Ohio.

32. SAS Institute (2001) SAS/STAT User's Guide, Version 8.2. Cary, NC, SAS Institute Inc, USA.

33. Van der Plank JE (1963) Plant diseases: epidemics and control. Academic Press, London.

34. Gomez KA, Gomez AA (1984) Statistical procedures for agricultural research. (2nd edn). John Wiley and Sons Inc, New York.

35. Terefe H, Fininsa C, Sahile S, Dejene M, Tesfaye K (2015) Effect of integrated cultural practices on the epidemics of chocolate spot (Botrytis fabae) of faba bean (*Vicia faba*) in Hararghe highlands, Ethiopia. Glob J Pests Dis Crop Prot 3: 113-123.

36. Fininsa C (1996) Effect of intercropping bean with maize on bean common bacterial blight and rust diseases. Int J Pest Manag 42: 51-54.

37. Bannon FJ, Cooke BM (1998) Studies on dispersal of *Septoria tritici* pycnidiospores in wheat-clover intercrops. Plant Pathol 47: 49-56.

38. Hauggaard-Nielsen H, Ambus P, Jensen ES (2001) Interspecific competition, N use and interference with weeds in pea-barley intercropping. Field Crops Res 70: 101-109.

39. Lithourgidis AS, Derdas CA, Damalas CA, Vlachostergios DN (2011) Annual intercrops: an alternative pathway for sustainable agriculture. Aust J Crop Sci 5: 396-410.

40. Mundt CC (2002) Use of multiline cultivars and cultivar mixtures for disease management. Annu Rev Phytopathol 40: 381-410.

41. Altieri MA (1999) The ecological role of biodiversity in agroecosystems. Agri Eco and Env 74: 19-31.

42. AARC (Adet Agricultural Research Center) (2000) Research progress report. Adet Agricultural Research Center, Adet, Ethiopia.

43. Biddle AJ, Cattlin ND (2007) Pests, diseases and disorders of peas and beans: a color handbook. Manson Publishing Ltd, London.

44. Fernández-Aparicio M, Rubiales D, Flores F, Hauggard-Nielsen H (2006) Effects of sowing density, nitrogen availability and cop mixtures on faba bean rust (Uromyces viciae-fabae) infection. In: Avila CM, Cubero JI, Moreno MT, Suso MJ, Torres AM (eds) International workshop on faba bean breeding and agronomy, Co´rdoba, Spain pp: 143-147.

45. Reddy PP (2013) Recent advances in crop protection. Springer, New York.

46. Neher DA, Weicht TR, Dunseith P (2014) Compost for management of weed seeds, pathogen, and early blight on Brassicas in organic farmer fields. Agroec Sust Food Sys 39: 3-18.

47. Haggag WM, Saber MSM (2007) Suppression of early blight on tomato and purple blight on onion by foliar sprays of aerated and non-aerated compost teas. J Food Agri and Env 5: 302-309.

48. Sang MK, Kim JG, Kim KD (2010) Biocontrol activity and induction of systemic resistance in pepper by compost water extracts against *Phytophthora capsici*. Phytopath 100: 774-783.

49. Hawthorne W, Bretag T, Raynes M, Davidson J, Kimber R, et al. (2004) Faba bean diseases management strategy for the southern region pp: 1-4.

50. Dipak S, Tripathi HS, Kumar SA (2012) Influence of environmental factors on development of field pea rust caused by Uromyces viciae-fabae. J Plant Dis Sci 7: 13-17.

51. Shifa H, Hussien T, Sakhuja PK (2011) Association of faba bean rust (Uromyces viciae-fabae) with environmental factors and cultural practices in the Hararghe highlands, Eastern Ethiopia. East African J Sci 5: 58-68.

52. Nigussie T (1991) Expansion of rust focus in faba beans (*Vicia faba* L.). Wageningen Agricultural University, Wageningen, The Netherlands.

5

Effect of Temperature on Growth and Sporulation of *Botrytis fabae*, and Resistance Reactions of Faba Bean against the Pathogen

Habtamu Terefe[1]*, Chemeda Fininsa[1], Samuel Sahile[2] and Kindie Tesfaye[3]

[1]*School of Plant Sciences, Haramaya University, P.O. Box 138, Dire Dawa, Ethiopia*
[2]*College of Natural and Computational Science, University of Gondar, P.O. Box 196, Gondar, Ethiopia*
[3]*The International Maize and Wheat Improvement Center (CIMMYT), Addis Ababa, Ethiopia*

Abstract

Chocolate spot (*Botrytis fabae)* is a devastating disease of faba bean and reduces its production and productivity. Three controlled condition experiments were conducted to assess the effect of temperature on growth and sporulation of *B. fabae*, and faba bean resistance reaction against the pathogen using a single *B. fabae* isolate and Degaga and Bulga-70 faba bean varieties. For cultural experiment, a circular block of actively growing *B. fabae* mycelia was placed on faba bean dextrose agar medium and arranged in a completely randomized design (CRD) with four replications. For resistance reaction evaluation, fresh culture of isolate suspension was prepared (2×10^5 spores ml^{-1}) and inoculated on to three weeks-old faba bean seedling detached leaves and the whole plant. Inoculated leaflets and seedlings were factorial arranged in a CRD with four replications. Both sets were incubated at 20, 22, 24 and 26°C. The maximum (84.00 mm) radial growth on 5 days after inoculation (DAI); average conidial size (24.86 × 16.32 µm), sporulation (2.48×10^3 conidia ml^{-1}) on 12DAI and growth rate (1.058 mm day^{-1}) were recorded at 22°C. The least values of these parameters and nil sporulation were obtained from 26°C. The highest average lesion size (17.67 mm in Degaga and 22.83 mm in Bulga-70), AUDPC for lesion sizes (30.92 mm in Degaga and 42.08 mm in Bulga-70) and severity (2.13 score) values were recorded at 22°C on 5DAI in detached leaf test. Infection and disease development was reduced at 26°C. The trend was similar in the whole plant test. Such parameters were linearly increased with temperature to maximum and declined progressively in both reaction evaluation tests. The two evaluation experiments indicated that the optimum temperature for *B. fabae* growth, sporulation, infection and disease development was at 22°C.

Keywords: *Botrytis fabae*; Chocolate spot; Detached leaf test; Mycelial growth; Sporulation; Temperature; *Vicia faba*; Whole plant test

Introduction

Faba bean (*Vicia faba* L.) is the third most important food legume in the world [1]. Faba bean has high nutritional value and thus, it is a rich available source of food for human beings and feed for animals [2]. Faba bean is used as an excellent component of crop rotation and green manure to improve soil fertility [3]. Over the last century, however, there has been a steady reduction in the cultivated area of faba bean in many countries due to several reasons. Under Ethiopian conditions, biological limitations include inherently low grain yielding potential of the indigenous cultivars and susceptibility to biotic and abiotic stresses [4]. Among biotic stresses, diseases have always been the major limiting factors for faba bean cultivation. The major ones include ascochyta blight (*Ascochyta fabae* Speg.), rust [*Uromyces viciae-fabae* (Pers.) J. SchrÖt.] and chocolate spot (*Botrytis fabae* Sard.) in Ethiopia [5].

Chocolate spot is an important disease of worldwide distribution, which causes high yield loss [6,7]. The prevalence of chocolate spot in main faba bean growing areas of Ethiopia is about 94.6% [8] and the disease caused up to 34% and 61% yield losses on tolerant and susceptible faba bean varieties, respectively [9]. Sahile et al. [10] reported a yield loss of up to 68% on the variety CS20DK and the local cultivars of faba bean in northern Ethiopia. A number of factors, such as physiological and environmental conditions, are known to influence plant pathosystems, either by affecting the host, the pathogen, or their interaction [11] and hence crop yield loss. For instance, climate influences the pathogen and host environments separately and in interaction throughout the period of crop growth from infection to host death [12].

Temperature affects plant resistance against a disease due to interactions of temperature with some corresponding gene pairs [13]. Sillero et al. [14] reported the interactions of temperature with some *Uromyces viciae-fabae*: *Vicia faba* gene combinations. They also showed that host-pathogen pairs responded differently to varying temperatures. It is well known that temperature governs the rate of reproduction of fungi and the physiological conditions of the host and has a marked effect on the incidence of diseases [12]. Temperature also affects the growth and aggressiveness of pathogens and expression of disease symptoms in the plants [15]. Moreover, it has been demonstrated that inoculum density and host physiology have been closely related with temperature and disease development [16]. Thus, plants and pathogens require optimum temperature ranges to grow and carry out their physiological activities. Temperature ranging from 18 to 27°C for faba bean growth [17] and between 15 and 23°C for *B. fabae* development [18] are reported. Severity of chocolate spot is favored between 92-100% relative humidity and 15-20°C temperature [19].

Although inoculum density, leaf wetness periods, relative humidity and temperature [5,20], host age and resistance reaction [11] influence

***Corresponding author:** Habtamu Terefe, School of Plant Sciences, Haramaya University, P.O. Box 138, Dire Dawa, Ethiopia, E-mail: habmam21@gmail.com

development of chocolate spot severity, quantitative relationships among these variables have not been reported. The effect of temperature could be addressed using detached leaf technique [21] and whole plant tests [7] under controlled conditions. A detached leaf test has been widely used for resistance screenings, analyzing components of resistance and assaying different factors affecting the response of *V. faba* to *B. fabae* [21,22]. Whole plant screenings are also widely used in the analysis other legume pathosystems [7].

Knowledge of the interaction of host and pathogen with environment factors has a practical significance because the environment could alter cultivars resistance and pathogen pathogenicity. Moreover, the effect of temperature and other factors on the development of a plant disease after infection depends on the specific host-pathogen combination. The role of leaf wetness, relative humidity, raining frequency and temperature on infectivity and subsequent development of chocolate spot on faba bean has been inferred from field conditions [5], which are difficult to address interaction effects. Climatic dynamics also indicated that temperature affects host resistance; and is known to be of great importance in the process of infection [20,23]. However, its relationship with resistance has hardly been the subject of research; and with regard to faba bean/*B. fabae*, there are very few studies concerning the relationship between resistance response and temperature [11,24]. In addition, pathogen responses to temperature *in vitro* may be used as an indirect measure of adaptation to a particular environment [25] and can provide useful information on epidemics development and associated management strategies under rising temperatures of climate change scenarios. Therefore, the objectives of this study were to assess the effects of temperature on the growth and sporulation of *B. fabae*, and resistance reactions in faba bean against *B. fabae* under controlled conditions.

Materials and Methods

Pathogen isolation and culturing

The *B. fabae* isolate used throughout this study was isolated from naturally infected faba bean plants cultivated in Haramaya University (Ethiopia) crop research site that showed typical chocolate spot symptoms. Infected leaflets with advanced margins of chocolate spot lesions were surface-sterilized by 5% sodium hypochlorite solution for 3 minutes and dried with a sterile blotting paper. The patch specimens were placed on the surface of potato dextrose agar (PDA) medium in Petri dishes [26]. The dishes with 2-3 mm piece of infected leaflets were kept in a glass case at room temperature (18-20°C) under 12 h day/night alternating cycles using fluorescent light and examining them 5-7 days after inoculation for emerging fungal colonies [26]. Following 5-7 days of incubation, fragments of the edges of freshly growing mycelia were transferred into new dishes of faba bean dextrose agar, FDA [27]. The dish with the fungus was again incubated at room temperature (18-20°C) under 12 h of day/night alternating cycles of light and examined 3-5 days after incubation. The mycelia were sub-cultured several times, until the pure cultures were obtained.

Spore production

The procedure used by Zakrzewska [27] was followed to produce an abundantly sporulating mycelium. The *B. fabae* pure cultures from FDA medium were transferred to MnPDA medium (PDA medium supplemented with 20 g of faba bean seed meal per 1 L of the medium). Petri dishes with the fungus were arranged in stacks and incubated at room temperature, for 3-5 days until the mycelia began to grow. The cultures were exposed to the cycles of 12 h of light/darkness again to induce sporulation. The dishes with sporulating *B. fabae* were transferred again to a room with natural light 6 days after such treatment, and they remained there for 4-5 days. In this manner, the ability of sporulation was confirmed before evaluation was commenced.

Evaluation of growth and sporulation of *B. fabae*

An agar disc, 6 mm in diameter, was taken from the actively growing margin of 10 days-old culture using a sterile cork borer, and placed in the center of a 9 cm Petri dish containing 20 ml of FDA. Inoculation was made on MnPDA medium for sporulation evaluation. The inoculated FDA and MnPDA Petri plates were arranged together in completely randomized design with four replications and incubated at four temperature (20, 22, 24 and 26°C) levels. Visual observations were made with regard to colony growth two days after inoculation (DAI). The colony diameters (mm) were measured in two directions at right angles to each other at every 24 h interval until the mycelium fully covered the Petri dish. Colony morphology, texture and shape were characterized at full plate colony growth (6-10 DAI).

Conidial size, sporulation and sclerotial production were estimated from 12 days-old culture for each temperature level per plate. For conidial size, the length and the width of 30 conidia per sample were measured. Each plate was flooded with 10 ml of sterile distilled water and its entire surface was gently rubbed with a glass rod several times to release all the spores. The spore suspension obtained was filtered through two layers of sterile gauze and was poured into a small beaker, the plate rinsed thoroughly, and the final volume was adjusted to 20 ml by adding sterile distilled water [28]. Sporulation was determined under the microscope by counting 4 samples (0.1 ml each) per replicate. Number and size of spores were counted and measured using the Malassez haemacytometer slide and micrometer under an optical microscope field of vision (10x eyepiece and 40x objective), respectively. The experiment was repeated twice.

Evaluation of faba bean resistance reaction against *B. fabae*

Faba bean varieties Degaga (moderately resistant) and Bulga-70 (susceptible) were used for both detached leaf and whole plant tests to investigate the reaction in both varieties against *B. fabae* at different temperature levels under controlled conditions. Both varieties were collected from Holleta Agricultural Research Center, Ethiopia.

Detached leaf test (Plant material, inoculum, and inoculation): The plant material was prepared by growing the two faba bean varieties in a growth chamber at Plant Protection Laboratory, Haramaya University. Seeds were surface-disinfected in 5% sodium hypochlorite solution for 3 minutes and rinsed three times with distilled sterile water. Six seeds of each faba bean variety were separately planted in 14 cm diameter pots filled with a sieved and autoclave-sterilized arable loam soil, peat and sand 3:1:1 (v:v:v) proportion. Germinated seedlings were thinned to four plants per pot. The seedlings were exposed to a temperature of 20/16°C (day/night) [21] and grown for 3 weeks with a photoperiod of 14 h of visible light (150 $\mu mol\ m^{-2}\ s^{-1}$ photon flux density) and 10 h of darkness [11].

Inoculum was prepared from *B. fabae* previously isolated and maintained as described in Section 2.1. The fungal culture was transferred to chrysanthemum (*Chrysanthemum sinense* Sabine) flower medium (4 g of dried chrysanthemum flower + 0.5 g of dextrose + 15 ml of distilled water were blended together in a 250 ml flask. Then the content was autoclaved for 30 minutes) to further induce abundant sporulation. The flasks containing inoculated medium were tightly sealed and left at room temperature (18-20°C) with alternating cycles of 12 h of light from a 40 W fluorescent tube and 12 h of darkness for 13-15 days. A spore suspension was prepared from 15 days-old cultures of the isolate. The spores were dislodged by scraping the surface of the medium with a bent glass rod and a sterile needle and washed out of the surface of the medium with distilled sterile water (10-15 ml). The suspension was filtered through two layers of sterile cheesecloth 30 minutes after stirring, to remove mycelium fragments. The resulting suspension of *B. fabae* spores was adjusted to the concentration of 2×10^5 spores ml^{-1} with the help of a Malassez haemacytometer under 10 fields of optical microscope. Finally, Tween-20 (0.03% v/v) was added to the suspension.

Leaves of sample plants were collected from each variety two hours prior to inoculation. Fully-expanded leaflets of similar physiological age were excised from the 5th node position and immediately laid flat on a moistened sterile double filter paper immersed in glucose solution (0.4% w/v) laid on sterile 10 cm Petri plates. Small, humid pieces of cotton were put at the end of the leaflet petioles to maintain cells at maximum turgescence [22]. A drop of (2×10^5 spores ml^{-1}) spore suspension was placed on each half leaflet of each variety. Leaflets inoculated with distilled sterile water served as control. The Petri plates were covered to maintain high moisture, and incubated at room temperature (18-20°C) over night. Plates were distributed the next day to each separate incubator adjusted at four temperature (20, 22, 24 and 26°C) levels and incubated for six days. The cotton was moistened with 1 ml of distilled sterile water every 24 h to ensure an environment of high humidity. The experiment was factorial arranged in a CRD with 4 replications (Petri dishes containing two leaflets per faba bean variety). Temperature and variety were considered as main effects and the experiment was repeated twice.

Whole plant test (Plant material, inoculum, and inoculation): Similar procedures were followed related to planting material and inoculum preparation as in Section 2.4.1. Plants of each variety were exposed to each respective temperature level prior to inoculation. Three weeks-old plants (4–6 expanded leaves) were sprayed with 1.5 ml (2×10^5 spores ml^{-1}) spore suspension per plant to run-off using an atomizer. The treated pots were factorial arranged in a completely randomized design with four replications and kept in darkness overnight at room temperature in an incubation chamber. Then they were transferred to the experimental run of the growth chamber adjusted at 20, 22, 24 and 26°C with a photoperiod of 14 h of visible light and 10 h of darkness, where the relative humidity was maintained on average over 90%, following the modified procedure of Villegas-Fernandez et al. [11]. The setup was sprayed with a mist of water three times a day to ensure adequate moisture till the end of the trial period. The experiment was not repeated.

Disease assessment: For detached leaf test, leaflets were assessed for their susceptibility to *B. fabae* infection by measuring the expansion of the lesions daily to 5 DAI. Lesion size (LS) was recorded on 3, 4 and 5 DAI. The average lesion size (ALS) was calculated for each leaflet at each evaluation time as the mean of the sizes of lesions of the two leaflets measured. The average measurements of the lesion sizes (considered as a single observation) on the two inoculated leaflets in each Petri dish were used for statistical data analysis. Area under disease progress curve (AUDPC) for lesion sizes was computed for each incubation temperature for both faba bean varieties. AUDPC was determined with the expansion of lesion sizes over time [29] as:

$$AUDPC=\sum_{i=1}^{n} 1/2\left[(y_{i+1}+y_i)(x_{i+1}-x_i)\right]$$

where x_i=time (days); y_i=lesion size at the day i; and n=total number of lesion symptom observations.

Disease severity (DS) assessment was evaluated on 5 DAI using a 1-4 scoring scale [2,7]. In the whole plant test, LS and DS were assessed from nine leaflets per plant starting from 5 DAI. Three plants per pot and four pots per variety were randomly taken for disease parameters assessment. Disease severity recordings were made on 5, 8, 11, 14, 17 and 20 DAI. LS assessments were made on 5, 9, 13, 17 and 21 DAI. The values of each leaflet DS and LS were expressed as means per plant to carry out the statistical analysis. Disease severity was assessed as the percentage of the total leaf surface covered with chocolate spot lesions on each expanded leaflet separately at regular intervals using a 0–9 scale Ding et al. [30], where, 0=no visible infection on leaves; 1=a few dot-like accounting for less than 5% of total leaf area; 3=discrete spots less than 2 mm in diameter (6–25% of leaf area); 5=numerous scattered spots with a few linkages, diameter 3–5 mm (26–50% of leaf area) with a little defoliation; 7=confluent spot lesions (51–75% of leaf area), mild sporulation, half the leaves dead or defoliated; 9=complete destruction of the larger leaves (covering more than 76% of leaf area), abundant sporulation, heavy defoliation and plants darkened and dead. The DS data were converted to percentage severity index (PSI) according to Wheeler [31]:

$$PSI=\frac{\text{Sum of Numerical Ratings} \times 100}{\text{Number of Plants Scored} \times \text{Maximum Score on Scale}}$$

Based on LS and DS data, area under disease progress curve (AUDPC) was computed for size of spots and disease scores according to the formula used by Madden and Hughes [29].

Data analysis

Data from two runs of experiments were pooled after confirming homogeneity of variances for growth and sporulation evaluation. Analysis of Variance (ANOVA) was performed to determine effects of incubation temperature on colony radial growth rate, sporulation and conidial size. The numbers of conidia ml^{-1} were analyzed after logarithmic transformation of the values obtained [32]. Regression analysis of diameters of colony radial growth against time after inoculation were performed and the slopes were used as measures of growth rates (mm day^{-1}) for each temperature treatment [33]. For resistance reaction evaluation, data on incubation period, LS, DS and AUDPC for both ALS and DS were analysed using ANOVA, to know the effects of incubation temperature on the growth of the pathogen and development of chocolate spot and faba bean resistance reaction against *B. fabae*. Regression analysis of lesion expansion against time after inoculation was performed and the slope was considered as the measure of rate (mm day^{-1}) of chocolate spot using both detached leaf and whole plant tests [34]. In all cases, ANOVA was run using SAS GLM Procedure [35]. Treatment mean separations were done using the least significant difference (LSD) test at 0.05 probability level. Bartlett's variance homogeneity tests were performed for each variable before combining data over the two runs of experiments both in cultural study and detached leaf test [36].

Results

Effect of temperature on growth and sporulation of *B. fabae*

The radial growth, conidial formation and growth rate of the isolate at different incubation temperatures is presented in Table 1. The results showed a significant (*$P \leq 0.05$*) reduction in the mycelial growth at 26°C compared to the temperatures at 20, 22 and 24°C. Significant difference was measured among incubation temperatures starting from 72 h after inoculation periods. Radial mycelium growth increased from 27.57 mm at 26°C to 84 mm at 22°C on 120 h after inoculation. Radial growth rate was affected by temperature. The isolate grew faster (1.058 mm day^{-1}) at 22°C and relatively slower (0.317 mm day^{-1}) at 26°C than at other temperatures tested. The radial growth followed a linear increasing trend at each incubation temperature over time (Figure 1).

Mycelial growth patterns included both light and dense extending mycelium. Highly dense mycelium with grey semi-concentric rings associated with black sclerotia was observed at both 20 and 22°C at later incubation period. Such characteristics were intermediate at 24°C, but a very thin mycelium with a very slow extending rate without grey bands (associated with colour of conidia) was obtained at 26°C. The colony colour was more or less similar at 20, 22 and 24°C (Figure 2). However, very clear grey colouration at the later growth period was observed only at 20 and 22°C, due to rate of sporulation and distribution of conidia. On the other hand, a very odd whitish colony (composed of white mycelium) was observed nearly in the whole cycle of the incubation period at 26°C. Growth resumed when inoculum plugs were placed on both FDA and PDA media and incubated at the optimum (22°C) temperature identified.

The isolate highly (2.476 × 10^{3} ml^{-1}) sporulated at 22°C followed by 20 and 24°C temperature levels. No sporulation was recorded at 26°C

Temperature (°C)	Radial colony growth (mm) after different time periods (h) of incubation[1]					Radial growth rate (mm day^{-1}) [2]	R^2 (%)
	48	72	96	108	120		
20	8.20^{a}	24.90^{b}	44.33^{b}	64.60^{b}	80.67^{b}	0.995	96.5
22	8.67^{a}	30.97^{a}	52.43^{a}	73.37^{a}	84.00^{a}	1.058	98.6
24	6.77^{a}	22.97^{b}	38.77^{c}	54.70^{c}	69.10^{c}	0.845	96.6
26	6.13^{a}	12.07^{c}	21.90^{d}	25.70^{d}	27.57^{d}	0.317	94.7
Mean	7.44	22.73	39.36	54.59	65.34	0.804	
LSD (0.05)	2.80	4.30	4.56	4.57	2.89		
CV (%)	20.0	10.05	6.15	4.45	2.35		

Table 1: *In vitro* effect of incubation temperature (°C) on radial colony growth (mm) and radial growth rate (mm day^{-1}) of *Botrytis fabae* cultured on faba bean dextrose agar (FDA) medium, and conidial formation of *B. fabae* isolate cultured on MnPDA.
[1]Radial growth was determined as average of two runs of experiments on radial colony growth of *B. fabae*.
[2]Linear radial growth rates were estimated as the slope of the following function: Colony diameter = Radial growth rate x time + b.
Means of colony radial growth in the same column followed by the same letters are not significantly different ($P \leq 0.05$). The values in the table are based on untransformed data of radial mycelial growth from two runs of experiments.

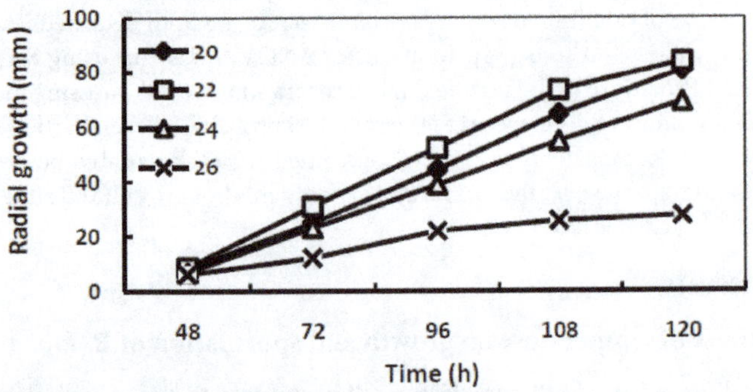

Figure 1: *In vitro* effect of incubation temperature (°C) on radial colony growth of *Botrytis fabae* on faba bean dextrose agar (FDA) medium (values are means of pooled data from two runs of the experiment).

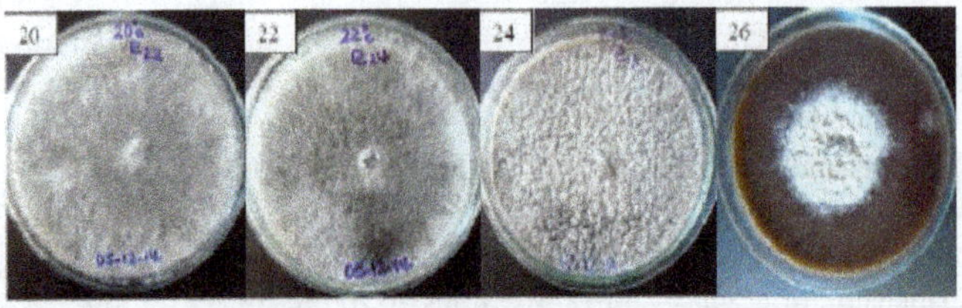

Figure 2: *In vitro* effect of incubation temperature (°C) on morphology (conidation and colour) of *Botrytis fabae* on MnPDA medium 10 days after inoculation and incubation.

Temperature (°C)	Sporulation (x10^{3}/ml) [2]					
	Length		Width		Rough data	Transformed data
	Mean	Range[3]	Mean	Range[3]		
20	20.24^{b}	16.47-23.53	13.18^{b}	11.77-14.12	266	2.426^{b}
22	24.86^{a}	21.18-28.24	16.32^{a}	11.77-18.82	298	2.476^{a}
24	19.54^{c}	13.79-22.99	11.58^{c}	6.90-18.39	202	2.307^{c}
26	0.00^{d}	-	0.00^{d}	-	0	0.000^{d}
LSD (0.05)	0.26		0.35			0.022
CV (%)	1.06		2.21			0.658

Table 2: *In vitro* effect of incubation temperature (°C) on conidial size (μm) and sporulation of *Botrytis fabae* isolate cultured on MnPDA medium.
[1]Average of 30 readings.
[2]Conidial production/sporulation was observed 12 days after inoculation. Variables "Sporulation ml^{-1} (x 1000)" were analysed after logarithmic transformation [log (x + 1)].
[3]Indicated absence of sporulation and hence, no conidial dimension record at 26 °C.
Means of conidial dimensions in the same column followed by the same letters are not significantly different ($P \leq 0.05$).

on 12 DAI (Table 2). The mean size of conidia varied both in length and width. The conidial length of the isolate ranged from 21.18 to 28.24 μm while conidial width ranged from 11.77 to 18.82 μm at 22°C. The longest (24.86 μm) mean conidial length and the widest (16.32 μm) mean conidial thickness were measured from the isolate incubated at 22°C (Table 2).

Effect of temperature on faba bean resistance reaction against *Botrytis fabae*

Detached leaf test: First characteristic symptoms of chocolate spot lesions appeared at the site of inoculation 24 h after inoculation on both faba bean varieties, especially on Bulga-70 at 20, 22 and 24°C. Forty eight hours after inoculation, LS aggressively increased on leaflets at the first three temperature levels on both faba bean varieties. At these temperatures, lesions enlarged rapidly and centrally deep black with brown margin spots that fused with time to form larger lesions. However, at 26°C, only water-soaked-like symptoms, which were followed by clear and visible symptoms, were seen with growing lesions 96 h after inoculation. The incubation period (IP) was relatively longer at 26°C and appeared shorter at 20 to 24°C but did not show any significant variation among incubation temperatures and between varieties (data not shown).

Lesion sizes on leaflets of both inoculated faba bean varieties showed significant ($P \leq 0.05$) differences among incubation temperatures, between varieties and temperature × variety interaction for ALS and AUDPC for lesion sizes starting from 72 h after inoculation (Table 3). The variety Degaga showed significantly smaller lesions and AUDPC values than those expressed by Bulga-70 variety at all incubation temperature levels. The highest ALS and AUDPC values on both varieties were recorded from 22°C on all recording DAI. On 5 DAI, the

Temperature (°C)	Average lesion size (mm)					Average lesion size (mm)				
	Degaga					Bulga-70				
	72 h	96 h	120 h	AUDPC[1]	Rate[2]	72 h	96 h	120 h	AUDPC[1]	Rate[2]
20	11.83^{bcd}	13.83^{bc}	15.50^{bc}	27.50^{bc}	0.099	16.17^{a}	20.50^{a}	22.33^{a}	39.75^{a}	0.173
22	12.83^{bc}	15.67^{b}	17.67^{b}	30.92^{b}	0.132	17.33^{a}	22.00^{a}	22.83^{a}	42.08^{a}	0.159
24	10.50^{d}	11.83^{cd}	14.67^{c}	24.42^{c}	0.107	13.17^{b}	15.17^{b}	17.83^{b}	30.67^{b}	0.123
26	8.17^{e}	10.00^{d}	10.83^{d}	19.50^{d}	0.074	10.83^{cd}	13.50^{bc}	14.83^{c}	26.33^{bc}	0.111
Mean	10.83	12.83	14.67	25.59	0.103	14.38	17.79	19.46	34.71	0.142
CV (%)	21.90	21.92	18.26	19.51		21.90	21.92	18.26	19.51	
LSD (0.05)	**	**	**	**		**	**	**	**	

Table 3: *In vitro* effect of incubation temperature (°C) on reactions of two faba bean varieties against *Botrytis fabae* using detached leaf test.
[1]Area under disease progress curves for lesion sizes of chocolate spot in the detached leaf test.
[2]Linear lesion size expansion rates were estimated as the slope of the following function: Lesion diameter = a + Lesion growth rate x time (based on untransformed data).
** The presence of highly significant difference at $P \leq 0.05$ probability level.
Means of lesion size and AUDPC in the same column with the same letters are not statistically different ($P \leq 0.05$). The values in the table are based on untransformed data of lesion expansion from two runs of experiments.

highest (22.83 mm) ALS and AUDPC (42.08 mm-days) were obtained from 22°C on Bulga-70. Comparably, the highest (17.67 mm) ALS and AUDPC (30.92 mm-days) were recorded from the same temperature level on Degaga variety. The lowest values of both parameters were recorded from 26°C and un-inoculated control of both varieties. The ALS and AUDPC values increased from 20 to 22°C with significant differences between the varieties and decreased as temperature increased beyond 22°C.

The rate of lesion growth increased rapidly on Bulga-70 as compared to the variety Degaga. The average rate of lesion expansions varied from 0.0744 mm day^{-1} at 26°C to 0.132 mm day^{-1} at 22°C on variety Degaga and from 0.111 mm day^{-1} (26°C) to 0.173 mm day^{-1} (20°C) on variety Bulga-70 (Table 3). Reaction differences between varieties and among incubation temperatures in disease severity scores became significantly ($P \leq 0.05$) different on 120 h (5 DAI) (Table 4). Chocolate spot severity on both faba bean varieties was also significantly different from their respective un-inoculated controls (which were scored 1 on the scale) but without any significant interaction effect. The highest (2.13) mean severity was scored at 22°C and the lowest (1.67) was at 26°C. In all the incubation temperatures studied, the variety Bulga-70 had higher mean chocolate spot severity than the variety Degaga.

Whole plant test: Faba bean plants inoculated with *B. fabae* conidia during seedling stages and maintained at different incubation temperature levels developed chocolate spot symptoms. All infected plants showed typical symptoms of small brown necrotic flecks clearly visible on leaves by the next DAI, which evolved into typical chocolate spot lesions 24 h later on both faba bean varieties at 20, 22, and 24°C. Small necrotic flecks appeared 32 h after inoculation on the variety Bulga-70 and on 48 h after inoculation on the variety Degaga at 26°C. Aggressive lesions (which were developed progressively and the nearby smaller flecks of spots begin to coalesce) were visible 48 h after inoculation on both varieties at 20, 22 and 24°C. Symptoms appeared earlier at both 20 and 22°C 16 h after inoculation. In contrast, it took slightly longer (24 h) for symptoms to develop at 24°C and 48 h and more at 26°C after inoculation (data not shown). No control plants developed chocolate spot symptoms. There was no infection observed on new leaves that emerged after the plants were inoculated.

The lesion development and percent leaf damage by the *B. fabae* isolate on the two faba bean varieties showed a significant ($P \leq 0.05$) variation among incubation temperatures, varieties and variety × temperature interaction for ALS, mean DS and AUDPC (Tables 5 and 6). The lesion growth was least on both faba bean varieties at 26°C, 1.45 mm (Degaga) and 1.47 mm (Bulga-70) on 21 DAI. The highest (2.89 mm on Degaga and 6.33 mm on Bulga-70) mean values of lesion sizes were recorded on 21 DAI at 22°C. AUDPC for lesion sizes ranged from 25.43 to 46.85 mm-days (Degaga) and 27.49 to 113.89 mm-days (Bulga-70). The variety Bulga-70 had the largest (6.33 mm) ALS at 22°C on 21 DAI while the lowest (2.89 mm) ALS was measured from the variety Degaga at the same temperature and DAI.

The final mean DS values ranged from 14.79 to 21.61% for Degaga and 20.37 to 32.10% for Bulga-70 on 20 DAI (Table 6). The overall final mean DS at all incubation temperatures revealed that Bulga-70 scored the highest (26.70%) DS than Degaga, which scored 17.74% on 20 DAI. Both faba bean varieties showed similar trends in DS progress, in which DS increased from 20 to 22°C and decreased beyond 22 to 26°C. Similarly, the AUDPC values for severity exhibited a similar trend to DS in that the highest (361.75%-days on Degaga and 556.81%-days on Bulga-70) AUDPC values for severity were recorded at 22°C. Progress of chocolate spot on whole plants was much slower than in experiment on excised leaves. The disease progress rates were significantly lower on the variety Degaga than on Bulga-70 at all incubation temperatures. Moreover, significant variety × temperature interactions for the studied disease parameters indicated that the inherent ability of varieties to express resistance reaction against *B. fabae* was not the same at all incubation temperatures.

Discussion

This study has indicated differences in the effects of temperatures on mycelial growth, sporulation, conidial size, sclerotial formation and morphology of *B. fabae*. Radial growth, sporulation, conidial size and sclerotial formation of the fungus increased with increase in temperature and reached maximum at 22°C but progressively declined thereafter. However, no sporulation was detected at 26°C. Effects of temperature on pathogenic fungi growth, sporulation, sclerotial formation and morphology is well documented [37,38]. Fernández et al. [39] found that temperature highly affected the mycelial growth of *B. cinerea* isolates and discriminate isolates based on their temperature optima. Pefoura et al. [32] showed that radial growth of *Trachysphaera fructigena* decreased to minimum at higher temperatures, which can be considered as lethal for radial growth of the pathogen. The investigators found that sporulation increased to optimum temperature and then declined till nil at higher temperature levels.

Similarly, Sehajpal and Singh [40] noted that temperature of 20 ± 1°C was the best for mycelial growth of *Botrytis gladiolorum*; the least was observed at 30 ± 1°C. No conidial and sclerotial production was recorded at lower and extreme temperatures. The rate of mycelial

Treatment	Severity(score)
Variety	
Bulga-70	3.17^{a}
Degaga	2.42^{b}
Control	1.00^{c}
Mean	2.20
LSD (0.05)	0.25
Temperature (°C)	
20	1.96^{ab}
22	2.13^{a}
24	1.83^{bc}
26	1.67^{c}
Mean	1.90
LSD (0.05)	0.25
CV (%)	15.70

Table 4: *In vitro* effect of incubation temperature (°C) on resistance reactions of faba bean (*Vicia faba*) against *Botrytis fabae* isolate using detached leaf test on two faba bean varieties.
Mean disease severity based on 1-4 rating scale for detached leaf test [7,27] where 1 = highly resistant, no infection or very small flecks (1-25% necrosis); 2 = resistant, necrotic flecks with few small lesions (26-50% necrosis), and very poor sporulation; 3 = moderately resistant, medium coalesced lesions (51-75% necrosis) with intermediate sporulation; and 4 = susceptible, large coalesced lesions (76-100% necrosis) with abundant sporulation.
Means of leaf disease severity of the same letters are not statistically different ($P \leq 0.05$).

Faba bean variety	Temperature (°C)	Average lesion size (mm)						
		Days after inoculation, DAI						
		5	9	13	17	21	AUDPC[1]	Rate[2]
Degaga	20	1.61^{d}	1.78^{d}	1.92^{d}	2.50^{c}	2.73^{d}	41.81^{d}	0.295
	22	2.01^{c}	2.12^{c}	2.22^{c}	2.58^{c}	2.89^{c}	46.85^{c}	0.221
	24	1.08^{e}	1.17^{fg}	1.25^{f}	1.36^{e}	2.14^{f}	26.94^{fg}	0.231
	26	0.83^{f}	1.06^{g}	1.45^{e}	1.45^{e}	1.45^{g}	25.43^{g}	0.161
	Mean	1.38	1.53	1.71	1.97	2.30	35.26	0.227
Bulga-70	20	4.08^{b}	4.97^{b}	5.39^{b}	5.72^{b}	5.86^{b}	105.28^{b}	0.431
	22	4.58^{a}	5.30^{a}	5.92^{a}	6.11^{a}	6.33^{a}	113.89^{a}	0.429
	24	1.19^{e}	1.61^{e}	1.97^{d}	2.19^{d}	2.55^{e}	38.26^{e}	0.331
	26	1.14^{e}	1.25^{f}	1.47^{e}	1.47^{e}	1.47^{g}	27.49^{f}	0.089
	Mean	2.75	3.28	3.69	3.87	4.05	71.23	0.320
	CV (%)	10.69	9.58	9.11	7.38	5.85	4.70	
	LSD (0.05)	**	**	**	**	**	**	

Table 5: *In vitro* effect of incubation temperature (°C) on lesion size of two faba bean varieties against *Botrytis fabae* using whole plant test.
[1]Area under disease progress curves of lesion size of chocolate spot in the whole plant test.
[2]Linear lesion size expansion rates were estimated as the slope of linear regression of the following function: Lesion diameter = a + Lesion growth rate x time (based on untransformed data).
** The presence of highly significant difference at $P \leq 0.05$ probability level.
Means of lesion sizes and AUDPC in the same column followed by the same letters are not statistically different ($P \leq 0.05$). Values are based on untransformed data.

growth of *Sphaeropsis pyriputrescens* increased as temperature increased up to 20°C and then decreased rapidly as temperature increased. Slight changes in colony morphology were observed at lower and higher temperatures than the optimum temperature [41]. Fernando et al. [42] also reported that *Corynespora cassiicola* sporulated freely on PDA at 10 to 35 °C with a peak at 30 °C. However, no sporulation or growth of the colonies of the isolates was observed at temperatures below 5 and above 35°C.

This study demonstrated that temperature also strongly influenced infection due to *B. fabae* and development of chocolate spot in faba bean. Temperature affected incubation period, lesion expansion, percent leaf damage, AUDPC and rate of chocolate spot progress on both detached leaf and whole plant tests in faba bean. When temperature was raised from 20 to 22°C, lesion expansion, percent leaf damage and AUDPC linearly increased, while these parameters decreased with increase in temperature beyond 22°C. Incubation period was delayed at the highest (26°C) temperature tested. Tu [43] found that disease severity in beans inoculated with *Colletotrichum lindemuthianum* was greater at temperatures ranging from 20 to 24°C than at lower or higher temperatures. Disease severity in round-leaved mallow inoculated with *Colletotrichum gloeosporioides* [44] and in soybean infected with *Colletotrichum truncatum* [45] increased with increase in temperatures between 10 and 25°C and sharply decreased at 30°C. With regard to incubation period, Xu [46] and Xu and Robinson [47] noted that the median incubation period was longer at lower and higher temperatures than at intermediate temperature ranges in rose and hawthorn powdery mildew.

Concerning such optimum-type relationships between lesion sizes and temperature levels, similar results were found by Kuruppu and Schneider [48] who reported that lesion development of aerial blight in soybean increased with increasing temperature (from 21 to 29°C) but decreased at 33°C, where lesions did not develop at all at 37°C or above during day temperatures in growth chambers with high humidity. Pedersen and Morrall [49] reported that lesion counts due to ascochya blight in lentil indicated that the optimal temperature for infection ranged from 10 to 15°C and that very little infection occurred at 25°C and it appeared to be less favorable for disease development than the lower temperatures. Similarly, lesions of *Cercosporidium personata* leaf spot of peanut on detached leaves of all genotype were largest, developed most rapidly, and sporulated most profusely at 24°C. Few infections occurred at 28 and 32°C regardless of duration of the high relative humidity [50].

In the whole plant test of this study, plants grown at 26°C also showed heat scorching, stunting, marginal diebacks of leaflets, and rapid senesces even without disease symptoms. These findings suggest that this temperature regime might have adversely affected the physiology and growth of the plants, and that the reduction in lesion size and disease severity might not be due to increased resistance. Chongo

Faba bean variety	Temperature (°C)	Leaf area damage (%)						
		Days after inoculation, DAI						
		5	8	11	14	17	20	AUDPC[1]
Degaga	20	14.20^{bc}	14.81^{cd}	15.43^{de}	17.28^{d}	17.28^{e}	18.52^{e}	324.69^{e}
	22	16.05^{b}	16.67^{bc}	17.29^{d}	17.90^{d}	19.76^{d}	21.61^{d}	361.75^{d}
	24	12.35^{cd}	12.96^{de}	12.96^{f}	14.82^{ef}	15.43^{f}	16.05^{f}	281.48^{f}
	26	11.11^{d}	11.73^{d}	12.35^{f}	12.96^{f}	14.81^{f}	14.79^{f}	259.26^{f}
	Mean	13.43	14.04	14.51	15.74	16.82	17.74	306.80
Bulga-70	20	19.14^{a}	23.46^{a}	24.08^{b}	25.93^{b}	27.78^{b}	29.63^{b}	502.50^{b}
	22	20.37^{a}	25.31^{a}	26.55^{a}	29.63^{a}	31.48^{a}	32.10^{a}	556.81^{a}
	24	16.05^{b}	18.52^{b}	20.37^{c}	22.84^{c}	24.69^{c}	24.69^{c}	427.19^{c}
	26	11.73^{d}	13.58^{de}	14.17^{ef}	16.05^{de}	19.14^{d}	20.37^{d}	315.95^{e}
	Mean	16.82	20.22	21.29	23.61	25.77	26.70	450.61
	CV (%)	20.00	20.39	18.05	16.70	8.86	10.90	12.75
	LSD (0.05)	**	**	**	**	**		**

Table 6: *In vitro* effect of incubation temperature (°C) on leaf area damage (severity) of two faba bean varieties against *Botrytis fabae* using whole plant test
[1]Area under disease progress curves of percent disease severity of chocolate spot in the whole plant test.
** The presence of highly significant difference at $P \leq 0.05$ probability level.
Means of disease severity and AUDPC in the same column followed by the same letters are not statistically different ($P \leq 0.05$).

and Bernier [51] observed a similar pattern in beans inoculated with *Colletotrichum lindemuthianum* at higher temperatures. The nearly complete inhibition of infection and development of chocolate spot on faba bean at higher (26°C) temperature was unexpected. However, increasing temperature could affect the pathogen to produce more propagules for subsequent infection and thus conidia must survive exposures to high temperature and low relative humidity that occur between deposition and favorable infection periods.

The current study pointed out that the most optimum temperature for *B. fabae* growth and sporulation as well as chocolate spot infection and development was between 20 and 22°C. In fact, temperatures around 22°C are near those existing in the high prevalence zones of the disease in Ethiopia where the study was conducted under natural conditions. The absence of sporulation and its reduced radial growth, infection and disease development at 26°C could explain the reduced pathogenicity at this temperature; but interactions are more complex in the natural environment where multiple climatological and biological factors vary simultaneously [52]. Harrison [20] indicated that *B. fabae* normally infects faba bean when temperatures are mild (15-22°C) with high relative humidity; and the most risk factors associated with chocolate spot infection under field conditions include mild temperature (20°C) and humid conditions [53]. Dereje [18] also found that humid and warm (10-23°C) with frequent rainy weather conditions are favorable for the development of chocolate spot epidemics, and progression rate reduces late in the season.

Reduced infection and disease development in the present study could be due to several factors including inhibition of mycelial growth and spore mortality at higher temperatures, genetic background of the faba bean varieties used, isolate virulence, leaf-wetness duration, and the interaction among such factors. Higher temperature during epidemic development could also affect the pathogen's ability to produce more propagules. Previous studies showed that an increase in temperature beyond the optimum decreases pustule production on leek leaves by *Puccinia allii* infection, suggesting that high spore densities were required for successful infection at higher temperatures [54]. In strawberry leaves inoculated with *Colletotrichum acutatum* conidia and incubated at different temperatures with continuous wetness, the number of germinated conidia tended to decrease with increasing temperature due to cell lysis [55]. Xu [46] also found that low rate of disease development in rose powdery mildew at supra-optimal temperatures is likely due to higher mortality of spores.

Moreover, Christiansen and Lewis [56] indicated that when high-temperature stress is exacerbated, plants show symptoms including wilting, leaf burn, leaf folding, and abscission, and changes in physiological responses. Such changes are expected to occur in the current and future climate dynamics. These changes will certainly affect susceptibility to pathogens, though the wide range of changes may make interactions difficult to predict; and challenging to discriminate between temperature effects on host resistance genes versus effects on pathogen virulence [23]. Hence, we did not directly attempt to partition the effect of temperature on the pathogen virulence gene(s) versus the faba bean resistance gene(s). Rather to the host-pathogen interaction and these have to be further elucidated with known resistance gene(s) in the host and virulence gene(s) in the pathogen using a wide range of temperature levels and other environmental factors. In connection with the impact of temperature on resistance expression, Fetch [57] noted the loss of resistance genes in oat lines at higher temperature and that observation on rust reaffirms the complex interaction between host, pathogen, and environment that determine the host response to pathogen infection. Similarly, in leaf rust-wheat combination under variable temperatures, a change in reaction may neither be a direct response to temperature on the part of the host's resistance alone nor on the part of the parasite's pathogenicity genes alone. Thus, temperature most likely affects the host-parasite interaction [58].

Conclusions

The current results indicated that temperature strongly influenced growth and sporulation of *B. fabae* and infection and development of chocolate spot in faba bean varieties. Resistance reactions in both faba bean varieties against *B. fabae* appeared temperature dependent. Temperature at 22°C was the optimum temperature for the growth, sporulation, infection and disease development; whereas, low disease infection and nil sporulation were recorded at 26°C. Infection and disease development were more severe on detached leaf than on the whole plants inoculated. Both detached leaf and whole plant tests were promising tools for assessing the response of faba bean plants to *B. fabae* under different incubation temperatures. The methods are also useful for studying various aspects of chocolate spot and behavior of *B. fabae* under controlled conditions to elucidate epidemiological attributes under natural conditions. Such studies can improve our understanding of conditions required for epidemic onset, disease progress rate over time and eventual decline of epidemics; and could allow us to predict the initiation and potential severity of chocolate spot epidemics in the growing season. Absence of sporulation and highly reduced disease levels at 26°C seem to imply that the projected temperature rise by 2.1°C in Ethiopia my disfavor *B. fabae* pathogenicity; and may not be a threat to faba bean production in the highland agro-ecologies of Ethiopia in the future climate change scenarios. However, it is difficult to exactly predict the effect of increasing temperature on the host/pathogen and their interactions under controlled conditions. Rising temperature is a gradual process that gives time-window for adjustment and interaction with other dynamic climate variables may influence the effect of temperature. Therefore, investigations on the effects of temperature on infection and disease development in faba bean due to *B. fabae* has to be further tested in greenhouse conditions to enable valid comparisons with field conditions. Moreover, thorough investigations on influence of temperature on the pathogen virulence gene, host resistance gene and on host-pathogen interactions have to be elucidated with known host resistance gene(s) and pathogen virulence gene(s) using a wide range of host cultivars, pathogen isolates, temperature levels and other environmental factors as well as their interactions.

Acknowledgments

The study was financed by the Swedish International Development Agency (SIDA), Sweden and Haramaya University, Ethiopia. We are thankful to Haimanot Bizuneh, Marta Wondimu, Addisalem Yosef and Yegile G/Mariam, School of Plant Sciences, Plant Protection Program, Haramaya University, for their assistance in laboratory works. We thank Holleta Agricultural Research Center for accessing us chrysanthemum flower collections.

References

1. Torres AM, Román B, Avila CM, Satovic Z, Rubiales D, et al. (2006) Faba bean breeding for resistance against biotic stresses: Towards application of marker technology. Euphytica 147: 67-80.

2. Sahile S, Sakhuja PK, Fininsa C, Ahmed S (2011) Potential antagonistic fungal species from Ethiopia for biological control of chocolate spot disease of faba bean. African Crop Science Journal 19: 213-225.

3. Bendahmane BS, Mahiout D, Benzohra IE, Benkada MY (2012) Antagonism of three Trichoderma Species against Botrytis fabae and B. cinerea, the causal agents of chocolate spot of faba bean (Vicia faba L.) in Algeria. World Applied Sciences Journal 17: 278-283.

4. Mussa J, Gorfu D, Keneni G (2008) Procedures of faba bean improvement through hybridization. Technical Manual No. 21, Ethiopian Institute of Agricultural Research. Addis Ababa, Ethiopia.

5. Dereje G, Tesfaye B (1993) Faba bean diseases in Ethiopia. In: Asfaw T, Geletu B, Saxena MC, Solh MB (eds) Cool-season food legumes of Ethiopia. Proceedings of the 1st 5 national cool-season food legumes review conference, 16-20 December, 1993. Addis Ababa, Ethiopia.

6. Stoddard FL, Nicholas AH, Rubiales D, Thomas J, Villegas-Fernández AM (2010) Integrated pest management in faba bean. Field Crops Research 115: 308-318.

7. Tivoli B, Baranger A, Avila CM, Banniza S, Barbetti M, et al. (2006) Screening techniques and sources of resistance to foliar diseases caused by major necrotrophic fungi in grain legumes. Euphytica 147: 223-253.

8. Endale H, Gezahegn G, Tadesse S, Nigussie T, Beyene B, et al. (2014) Faba bean gall: a new threat for faba bean (Vicia faba) production in Ethiopia. Advances in Crop Science and Technology 2: 1-5.

9. Dereje G, Yaynu H (2001) Yield loss of crops due to plant diseases in Ethiopia. Pest Management Journal of Ethiopia 5: 55-67.

10. Sahile S, Chemeda F, Sakhuja PK, Seid A (2010) Yield loss of faba bean (Vicia faba) due to chocolate spot (Botrytis fabae) in sole and mixed cropping systems in Ethiopia. Archives of Phytopathology and Plant Protection 43: 1144-1159.

11. Villegas-Fernández AM, Sillero JC, Rubiales D (2011) Screening faba bean for chocolate spot resistance: evaluation methods and effects of age of host tissue and temperature. European Journal of Plant Pathology.

12. Dixon GR (2012) Climate change - impact on crop growth and food production, and plant pathogens. Canadian Journal of Plant Pathology 34: 362-379.

13. Webb KM, Oña I, Bai J, Garrett KA, Mew T, et al. (2010) A benefit of high temperature: increased effectiveness of a rice bacterial blight disease resistance gene. New Phytol 185: 568-576.

14. Sillero JC, Morenoa MT, Rubiales D (2000) Characterization of new sources of resistance to Uromyces viciae-fabae in a germplasm collection of Vicia faba. Plant Pathology 49: 389-395.

15. Singh D, Yadav DK, Sinha S, Choudhary G (2014) Effect of temperature, cultivars, injury of root and inoculums load of Ralstonia solanacearum to cause bacterial wilt of tomato. Archives of Phytopathology and Plant Protection 47: 1574-1583.

16. Griffiths E, Amin SM (1977) Effects of Botrytis fabae infection and mechanical defoliation on seed yield of field beans (Vicia faba). Annals of Applied Biology 86: 359-367.

17. Saber HA, Revri R, Abdallah HM (2000) Faba bean (Vicia faba L.): cultural practices and integrated pest management.

18. Dereje G (1993) Studies on the epidemiology of chocolate spot (Botrytis fabae Sard.) of faba bean (Vicia faba L.). M.Sc. Thesis, Alemaya University of Agriculture. Alemaya, Ethiopia. pp. 30-70.

19. Harrison JG (1984) Effect of humidity on infection of field bean leaves by Botrytis fabae and germination of conidia. Transactions of the British Mycological Society 82: 245-248.

20. Harrison JG (1988) The biology of Botrytis spp. on Vicia beans and chocolate spot disease-a review. Plant Pathology 37: 168-201.

21. Bouhassan A, Sadiki M, Tivoli B, Portapuglia A (2004) Influence of growth stage and leaf age on expression of the components or partial resistance of faba bean to Botrytis fabae Sard. Phytopathologia Mediterranea 43: 318-324.

22. Bouhassan A, Sadiki M, Tivoli B, El Khiati N (2003) Analysis by detached leaf assay of components of partial resistance of faba bean (Vicia faba L.) to chocolate spot caused by Botrytis fabae Sard. Phytopathologia Mediterranea 42: 183-190.

23. Garrett KA, Dendy SP, Frank EE, Rouse MN, Travers SE (2006) Climate change effects on plant disease: genomes to ecosystems. Annu Rev Phytopathol 44: 489-509.

24. Bouhassan A, Sadiki M, Tivoli B (2007) Effets de la temperature et de la dose de la inoculum sur les composantes de la resistance partielle de la fève au Botrytis fabae Sard. Acta Botanica Gallica 154: 53-62.

25. Brasier CM, Webber JF (1987) Positive correlation between in vitro growth rate and pathogenesis in Ophiostoma ulmi. Plant Pathology 36: 462-466.

26. ICARDA (International Center for Agricultural Research in the Dry Areas), 1986. Screening techniques for disease resistance in faba beans. International Center for Agricultural Research in the Dry Areas (ICARDA), Aleppo, Syria.

27. Zakrzewska E (2004) Reaction of morphological types of faba bean to infection with Ascochyta fabae Speg. and Botrytis fabae Sard. Plant Breeding and Seed Science 49: 3-7.

28. Hmouni A, Hajlaoui MR, Mlaiki A (1996) Resistance de Botrytis cinerea aux benzimidazoles et aux dicarboximides dans les cultures de tomate en Tunisie. OEPP/EPPO Bulletin 26: 697-705.

29. Madden LV, Hughes G (1995) Plant disease incidence: distributions, heterogeneity, and temporal analysis. Annu Rev Phytopathol 33: 529-564.

30. Ding G, Xung L, Oifang G, Pingxi L, Dazaho Y, et al. (1993) Evaluation and screening of faba bean germoplasm in China. Fabis Newsletter 32: 8-10.

31. Wheeler BEJ (1969) An introduction to plant diseases. Wiley and Sons, London.

32. Pefoura AM, Ouamba AJK, Nkenfou C, Nguidjo O, Dongmo R (2007) Influence of the temperature on radial growth and sporulation of Trachysphaera fructigena, causal agent of the Musa cigar end rot disease. African Crop Science Conference Proceedings 8: 849-852.

33. Ramirez ML, Chulze SN, Magan N (2004) Impact of osmotic and matric water stress on germination, growth, mycelial water potentials and endogenous accumulation of sugars and sugar alcohols in Fusarium graminearum. Mycologia 96: 470-478.

34. Berger RD, Filho AB, Amorim L (1997) Lesion expansion as an epidemic component. Phytopathology 87: 1005-1013.

35. SAS Institute (2001) SAS/STAT Users Guide, Version 8.2. SAS Institute Inc., Cary, NC, USA.

36. Gomez KA, Gomez AA (1984) Statistical procedures for agricultural research. (2ndedn, John Wiley and Sons Inc., New York.

37. Gaston TNR, Appolinaire LJ, Jean MCP, Ajong FD (2014) Effect of different pH and temperature levels on in vitro growth and sporulation of Phytophthora colocasiae, taro leaf blight pathogen. International Journal of Agronomy and Agricultural Research 4: 202-206.

38. Gupta V, Sharma AK (2013) Assessment of optimum temperature of Trichoderma harzianum by monitoring radial growth and population dynamics in different compost manures under different temperature. Octa Journal of Biosciences 1: 151-157.

39. Fernandez JG, Fernandez-Baldo MA, Sansone G, Calvente V, Benuzzi D, et al. (2014) Effect of temperature on the morphological characteristics of Botrytis cinerea and its correlated with the genetic variability. Journal of Coastal Life Medicine 2: 543-548.

40. Sehajpal PK, Singh PJ (2014) Effect of temperature on growth, sporulation and sclerotial formation of the fungus Botrytis gladiolorum timm. in different culture media and standardization of inoculum load of the fungus for generation of disease. International Journal of Research 1: 772-779.

41. Kim YK, Xiao CL, Rogers JD (2005) Influence of culture media and environmental factors on mycelial growth and pycnidial production of Sphaeropsis pyriputrescens. Mycologia 97: 25-32.

42. Fernando THPS, Jayasinghe CK, Wijesundera RLC, Siriwardane D (2012) Some factors affecting in vitro production, germination and viability of conidia of Corynespora cassiicola from Hevea brasiliensis. Journal of the National Science Foundation of Sri Lanka 40: 241-249.

43. Tu JC (1992) Colletotrichum lindemuthianum on bean: population dynamics of the pathogen and breeding for resistance. In: Bailey JA, Jeger MJ (eds) Colletotrichum: Biology, Pathology and Control. Redwood Press, Ltd., Melksham, England, pp. 203-224.

44. Makowski RMD (1993) Effect of inoculum concentration, temperature, dew period and plant growth stage on disease of round leaved mallow and velvet leaf by Colletotrichum gloeosporioides f.sp. Malvae. Phytopathology 83: 1229-1234.

45. Khan M, Sinclair JB (1991) Effect of soil temperature on infection of soybean roots by sclerotia-forming isolates of Colletotrichum truncatum. Plant Disease 75: 1282-1285.

46. Xu X-M (1999) Effects of temperature on the length of the incubation period of rose powdery mildew (Sphaerotheca pannosa var. rosae). European Journal of Plant Pathology 105: 13-21.

47. Xu X-M, Robinson JD (2000) Effects of temperature on the incubation and latent periods of hawthorn powdery mildew (Podosphaera clandestina). Plant Pathology 49: 791-797.

48. Kuruppu PU, Schneider RW (2001) Temperature effects on development of aerial blight in soybean. Phytopathology 91: S51.

49. Pedersen EA, Morrall RAA (1994) Effects of cultivar, leaf wetness duration, temperature, and growth stage on infection and development of Ascochyta blight in lentil. Phytopathology 84: 1024-1030.

50. Shew BB, Beute MK, Wynne JC (1988) Effects of temperature and relative humidity on expression of resistance to Cercosporidium personatum in peanut. Phytopathology 78: 493-498.

51. Chongo G, Bernier CC (2000) Effects of host, inoculum concentration, wetness duration, growth stage, and temperature on anthracnose of lentil. Plant Disease 84: 544-548.

52. Coakley SM, Scherm H, Chakraborty S (1999) Climate change and plant disease management. Annu Rev Phytopathol 37: 399-426.

53. Stoddard FL, Nicholas AH, Rubiales D, Thomas J, Villegas-Fernandez AM (2010) Integrated pest management in faba bean. Field Crops Research 115: 308-318.

54. Gilles T, Kennedy R (2003) Effects of an Interaction between Inoculum Density and Temperature on Germination of Puccinia allii Urediniospores and Leek Rust Progress. Phytopathology 93: 413-420.

55. Leandro LF, Gleason ML, Nutter FW, Wegulo SN, Dixon PM (2003) Influence of Temperature and Wetness Duration on Conidia and Appressoria of Colletotrichum acutatum on Symptomless Strawberry Leaves. Phytopathology 93: 513-520.

56. Christiansen MN, Lewis CF (1982) Breeding plants for less favorable environments. Wiley, New York.

57. Fetch TGJr (2006) Effect of temperature on the expression of seedling resistance to Puccinia graminis f. sp. avenae in oat. Canadian Journal of Plant Pathology 28: 558-565.

58. Kaul K, Shaner G (1989) Effect of temperature on adult-plant resistance to leaf rust in wheat. Phytopathology 79: 391-394.

6

Biocontrol of Rhizoctonia Root Rot in Tomato and Enhancement of Plant Growth using *Rhizobacteria* Naturally associated to Tomato

Nada Ouhaibi-Ben Abdeljalil[1,2*], Jessica Vallance[3,4], Jonathan Gerbore[5], Emilie Bruez[3,4], Guilherme Martins[6], Patrice Rey[3,4] and Mejda Daami-Remadi[2]

[1]*Higher Agronomic Institute of Chott-Mariem, Sousse University, 4042-Chott Mariem, Tunisia*
[2]*UR13AGR09-Integrated Horticultural Production in the Tunisian Centre-East, Regional Centre of Research on Horticulture and Organic Agriculture, University of Sousse, 4042, Chott-Mariem, Tunisia*
[3]*INRA, UMR1065 Santé et Agroécologie du Vignoble (SAVE), ISVV, F-33140 Villenave d'Ornon, France*
[4]*Université de Bordeaux, Bordeaux Sciences Agro, ISVV, UMR1065 SAVE, F-33140 Villenave d'Ornon, France*
[5]*BIOVITIS, 15400 Saint Etienne de Chomeil, France*
[6]*USC Oenologie-INRA, Université Bordeaux Segalen, Bordeaux Sciences Agro, ISVV, Villenave d'Ornon, France*

Abstract

In the present study, 25 rhizobacterial isolates, obtained from rhizosphere of healthy tomato plants collected from various tomato-growing sites in Tunisia, were tested *in vitro* and *in vivo* against *Rhizoctonia solani*. This bacterial collection, composed of isolates belonging to *Bacillus* spp., *Enterobacter cloacae, Chryseobacterium jejuense*, and *Klebsiella pneumoniae,* was assessed for its antifungal potential against *R. solani* the causative agent of Rhizoctonia Root Rot disease in various crops including tomato. Antifungal activity of diffusible and volatile metabolites derived from these isolates was tested against target pathogen using dual and distance culture bioassays, respectively. Growth inhibition rates, recorded after 5 days of incubation at 25°C, depended significantly upon tested bacterial isolates and screening methods and reached 34-59% and 18-45% for diffusible and volatile metabolites, respectively. The screening of disease-suppressive and plant growth-promoting abilities of these tomato-associated rhizobacteria showed 47-100% decrease in disease severity and significant increments in plant height by 62-76%, roots fresh weight by 53-86%, and aerial parts fresh weight by 34-67% compared to pathogen-inoculated and untreated control. *B. thuringiensis* B2 (KU158884), *B. subtilis* B10 (KT921327) and *E. cloacae* B16 (KT921429) were found to be the most efficient isolates in decreasing *R. solani* radial growth, suppressing disease severity, and enhancing plant growth.

Keywords: Antifungal metabolites; Biocontrol; Disease severity; Growth promotion; Mycelial growth; *Rhizoctonia solani*; Tomato

Introduction

Tomato (*Solanum lycopersicum* L.) is the second most important vegetable crop worldwide after potato based on grown areas [1]. In Tunisia, it is a strategic and an economically relevant crop. However, this crop is still threatened by serious wilting and root-rotting pathogens both in greenhouse and open-field growing systems [2,3]. The most widely grown tomato cultivars were susceptible to soilborne infections and especially to Rhizoctonia Root Rot disease caused by *Rhizoctonia solani* (Kühn). This pathogen is mostly known as a damping-off agent but is also responsible for collar and root rots and eventual death of severely diseased plants leading to significant crop yield loss [4-6].

Efficient disease control is difficult due to the various host range of the causative agent, the persistence of its resting structure (sclerotia) in soil, the lack of genetic resistance and to the limited efficacy of chemical fungicides [7]. Such issues have focalized research efforts on development of environmentally safe, long lasting and effective alternatives such as biological control [8].

Several biocontrol agents (BCAs) were reported to be effective in the bio-suppression of *R. solani* on various crops. The most efficient bacterial agents used for the biomanagement of Rhizoctonia Root Rot disease belonged mainly to the genera *Bacillus* [9-12], *Pseudomonas* [6], *Enterobacter* [13], *Serratia* [14], *Burkholderia* [12,15] and *Streptomyces* [16].

Among the group of BCAs, plant growth promoting rhizobacteria (or PGPR) have been widely used for the bio-suppression of various soilborne diseases [17]. In fact, PGPR strains can display disease-suppressive effects against various crown, root and foliar diseases through direct inhibition of target pathogens or indirectly via the induction of systemic resistance (ISR) which is active throughout the entire plant [18-20]. PGPR-treated plants showed enhanced emergence potential and increased vegetative and root growth [17,21,22].

In our previous studies, a collection of 25 rhizobacterial isolates, obtained from rhizospheric soils collected around healthy tomato plants and belonging to *Bacillus, Chryseobacterium, Enterobacter*, and *Klebsiella* genera, was morphologically, biochemically, molecularly, and metabolically characterized [23] and screened for its capacity to suppress *Sclerotinia sclerotiorum in vitro* and *in vivo.* Interesting results were obtained where these isolates had significantly protected tomato plants from Sclerotinia Stem Rot disease and enhanced growth of pathogen-inoculated plants [24]. In the current investigation, the same collection of isolates will be assessed for its antifungal potential against *R. solani* mycelial growth and its capacity to suppress Rhizoctonia Root Rot disease and to enhance growth of infected plants.

***Corresponding author:** Ouhaibi-Ben Abdeljalil N, UR13AGR09-Integrated Horticultural Production in the Tunisian Centre-East, Regional Centre of Research on Horticulture and Organic Agriculture, University of Sousse, 4042, Chott-Mariem, Tunisia, E-mail: nadouhaibi@hotmail.fr

Materials and Methods

Tomato cultivar and growth conditions

Tomato plants cv. Rio Grande seedlings were used for all *in vivo* bioassays. Seeds were disinfected with 5% sodium hypochlorite during 2 min, rinsed thrice with sterile distilled water (SDW) and air-dried. They were sown in disinfected alveolus plates and maintained under greenhouse conditions (30 ± 4°C; 13/11 h light/dark photoperiod). Seedlings were regularly watered to avoid water stress.

Pathogen origin and growth conditions

R. solani isolate used in the present study was originally isolated from tomato plants exhibiting severe Rhizoctonia Root Rot infection. Pathogen cultures were gratefully provided by the Laboratory of Plant Pathology at the Regional Centre of Research on Horticulture and Organic Agriculture of Chott-Mariem, Tunisia. Pathogen was grown onto Potato Dextrose Agar (PDA) medium amended with streptomycin sulfate (300 mg/L w/v) and incubated at 25°C for 5 days before use.

Rhizobacterial collection tested and growth conditions

The 25 bacterial isolates used in the current study were originally recovered from the rhizospheric soils of apparently healthy and vigorous tomato plants grown in various infested tomato fields. They were identified using morphological, biochemical and molecular tools. They were also characterized for antibiotic producing ability (Bacillomycin D and fengycin A) and PGPR traits such as IAA detection, siderophore production, phosphate solubilization. Their main traits were previously detailed [23].

Rhizobacterial stock cultures were stored at -20°C in Luria Bertani (LB) broth amended with 15% glycerol. Bacterial cultures used for the different tests were previously grown for 48 h onto Nutrient Agar (NA) and incubated at 28°C.

Suspensions of bacterial cells used for plant challenge were prepared as previously described and adjusted to approximately 10^8 cells/mL using an haemocytometer [24].

Screening of the antifungal potential of tomato-associated rhizobacteria against *Rhizoctonia solani*

The antifungal activity of the 25 rhizobacterial isolates against *R. solani* was screened *in vitro* using dual culture and distance culture bioassays for elucidating the suppressive effects of their diffusible and volatile compounds, respectively.

Dual culture assay

R. solani 5 day-old cultures were used for this bioassay. Agar plugs (5 mm in diameter) were cut using a sterile cork borer and placed at one side of a Petri plate (9 cm in diameter) containing PDA medium. At the opposite side, 10 µL of a bacterial cell suspension (10^8 cells/mL) were dropped into a well (5 mm in diameter) performed using sterile cork borer in Petri plates containing PDA. Control plates were challenged with pathogen plugs and bacterial suspension was replaced by a same volume of SDW. Plates were maintained at 25°C for 5 days. Three plates were used per each individual treatment. The diameter of pathogen colony and the inhibition zone were measured and the percentage of inhibition of pathogen growth was calculated as previously described [23].

Distance culture assay

Antifungal activity of volatile metabolites of the tomato-associated rhizobacteria against *R. solani* was assessed using the distance culture assay also known as the sealed plate method. For this test, 10 µL of 48 h-old bacterial culture adjusted to 10^8 cells/mL were dropped into wells (5 mm in diameter) performed using sterile cork borer in Petri plates containing NA medium. A second PDA Petri plate was challenged with pathogen plug only (5 mm in diameter). Both half plates were wrapped together with parafilm to seal in the bacterial volatile compounds. For control plates, pathogen-challenged half plate was inverted over a half one containing NA only. The paired plates were incubated at 25°C for 5 days. Three plates were used per each individual treatment. After the incubation period, the diameter of pathogen colony was measured and the percentage of growth inhibition was calculated as previously described [24].

Assessment of Rhizoctonia Root Rot-suppressive and plant growth-promoting abilities

The ability of the 25 rhizobacterial isolates to limit the *in vivo* expression of *R. solani* and to enhance plant growth was screened based on pot experiments maintained under greenhouse conditions. Rhizobacteria and pathogen cultures were prepared as described above. Tomato cv. Rio Grande seedlings (at the two-true-leaf growth stage), grown in alveolus plates, were watered at the collar level with 30 mL of a suspension of bacterial cells (adjusted to 10^8 cells/mL). Seven days post bacterial treatment, 30 mL of *R. solani* inoculum (mycelial fragments) were poured at the same level to each seedling. Control seedlings were watered with SDW only. One day post pathogen challenge, seedlings were transplanted into pots (16 cm in diameter) filled with peat previously infected with 40 ml of fungal inoculum. A reminder bacterial treatment was performed 24 h post-transplanting [24]. Overall, the bioassay included a positive control (pathogen-free and rhizobacteria-free seedlings), a negative control (*R. solani*-inoculated and untreated seedlings) and 25 treatments consisting of tomato seedlings pathogen-challenged and individually treated with the tested 25 rhizobacterial isolates.

Two months after transplanting, tomato plants were uprooted and washed for eliminating the adhering peat. Three growth parameters (plant height and aerial parts and roots fresh weights) were recorded. Disease severity on collars and roots was estimated based on a 0-5 scale depending on root browning extent on the whole root system where: 0 = no symptom, 1 = 0-25% of root browning, 2 = 26-50% of root browning, 3 = 51-75% of root browning, 4 = 76-100% of root browning, and 5= plant death [24]. Disease incidence was also calculated for each individual treatment by dividing the number of symptomatic plants over the total number of plants.

Statistical analysis

The results were subjected to one-way analysis of variance and means separations were carried out using the Duncan's Multiple Range test at $P \leq 0.05$. ANOVA was performed using SPSS version 16.0. Experiments were conducted according to a completely randomized design both for the *in vitro* (26 individual treatments, 3 replications) and the *in vivo* trials (27 individual treatments, 5 replications). Correlation analyses between Rhizoctonia Root Rot severity and plant growth parameters were carried out using Pearson's correlation analysis at $P \leq 0.05$.

Results

Antifungal activity of diffusible metabolites from tomato-associated rhizobacteria toward *R. solani*

ANOVA analysis indicated that the diameter of *R. solani* colony, noted after 5 days of incubation at 25°C, depended significantly (at $P \leq 0.05$) upon bacterial treatments tested. In fact, data given in Table 1 showed that all the 25 rhizobacterial isolates had significantly (at $P \leq 0.05$) lowered pathogen mycelial growth over the control. The percentage of growth reduction, versus the untreated control, varied between 34.44 and 59.26% depending on isolates and exceeded 40% using 18 isolates out of the 25 tested.

B. thuringiensis B2 (KU158884), *B. subtilis* B10 (KT921327), *E. cloacae* B16 (KT921429) (Figure 1) and *B. subtilis* B6 (KT921427) isolates were found to be the most active in inhibiting *R. solani* radial growth by 48.89-59.26%.

This assay also showed that some tested rhizobacterial isolates led to the formation of inhibition zones when dual cultured with *R. solani.* Dimension of this zone ranged between 3 and 10.3 mm depending on isolates and was more than 5 mm when *R. solani* was dual-cultured with 13 out of the 25 isolates tested (Table 1). The largest inhibition zones, of about 8.3-10.3 mm, were induced by *E. cloacae* B16 (KT921429), *B. subtilis* (B14), and *B. megaterium* B24 (KT923048).

Antifungal activity of volatile metabolites from tomato-associated rhizobacteria toward *R. solani*

Data analysis revealed that the diameter of pathogen colony, recorded after 5 days of incubation at 25°C, varied significantly ($P \leq 0.05$) upon bacterial treatments tested. In fact, as shown in Table 1, *R. solani* growth was lowered by 18.52 to 45.37% over control due to the inhibitory effects of volatile metabolites released by the rhizobacterial

Bacterial treatment	Isolate	Diffusible metabolites			Volatile metabolites	
		Colony diameter (mm)	Growth Inhibition (%)	Inhibition zone (mm)	Colony diameter (mm)	Growth Inhibition (%)
Bacillus megaterium	B1	51.0^{bc}	43.34	3.3^{cd}	54.67^{fg}	39.26
B. thuringiensis	B2	36.67^{d}	59.26	5.3^{abc}	50.0^{g}	44.44
Enterobacter cloacae	B3	49.5^{bc}	45	4.7^{bcd}	57.5^{defg}	36.11
E. cloacae	B4	53.67^{bc}	40.37	4.0^{bcd}	62.83^{bcdef}	30.18
B. megaterium	B5	52.83^{bc}	41.3	4.7^{cd}	66.83^{bcde}	25.74
B. subtilis	B6	46.0^{cd}	48.89	3.3^{bcd}	56.67^{defg}	37.03
B. amyloliquefaciens	B7	50.0^{bc}	44.44	4.0^{bcd}	67.5^{bcd}	25
B. subtilis	B8	48.33^{bc}	46.3	4.3^{bcd}	64.17^{bcdef}	28.7
B. amyloliquefaciens	B9	59.0^{b}	34.44	6.0^{abc}	60.33^{cdef}	32.96
B. subtilis	B10	36.67^{d}	59.26	5.7^{abc}	49.17^{g}	45.37
Chryseobacterium jejuense	B11	52.0^{bc}	42.22	5.3^{abc}	64.5^{bcdef}	28.34
Klebsiella pneumoniae	B12	50.5^{bc}	43.89	3.3^{cd}	63.5^{bcdef}	29.44
B. amyloliquefaciens	B13	48.83^{bc}	45.74	5.0^{bcd}	63.33^{bcdef}	29.63
B. subtilis	B14	56.17^{bc}	37.6	9.0^{bc}	60.83^{cdef}	32.4
B. amyloliquefaciens	B15	56.83^{bc}	36.85	6.0^{abc}	62.0^{cdef}	31.11
E. cloacae	B16	37.0^{d}	58.89	10.3^{a}	49.17^{g}	45.37
B. subtilis	B17	58.83^{b}	34.63	4.0^{bcd}	60.5^{cdef}	32.78
B. amyloliquefaciens	B18	56.0^{bc}	37.78	7.0^{abc}	69.17^{bc}	23.15
B. subtilis	B19	53.83^{bc}	40.19	6.7^{abc}	55.83^{efg}	37.96
B. subtilis	B20	50.83^{bc}	43.52	4.0^{bcd}	62.67^{bcfef}	30.37
B. amyloliquefaciens	B21	50.33^{bc}	44.07	7.3^{abc}	73.33^{b}	18.52
B. amyloliquefaciens	B22	51.0^{bc}	43.34	6.0^{abc}	63.83^{bcdef}	29.07
B. thuringiensis	B23	58.33^{b}	35.19	4.7^{bcd}	65.0^{bcdef}	27.78
B. megaterium	B24	54.17^{bc}	39.81	8.3^{abc}	65.0^{bcdef}	27.78
B. subtilis	B25	53.67^{bc}	40.37	3.0^{cd}	63.83^{bcdef}	29.07
Untreated control		90.0^{a}	0	0.0^{d}	90.0^{a}	0

Table 1: Effects of diffusible and volatile metabolites released by tomato-associated rhizobacteria against *Rhizoctonia solani* growth noted after 5 days of incubation at 28°C. For each parameter, values followed by the same letter are not significantly different according to Duncan's Multiple Range test (at $P \leq 0.05$).

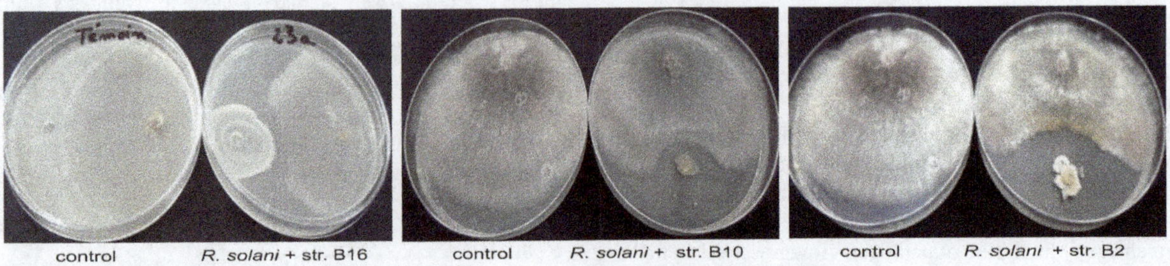

str. B2: *Bacillus thuringiensis* (KU158884); str. B10: *B. subtilis* (KT921327); str. B16: *Enterobacter cloacae* (KT921429).

Figure 1: Inhibition of *Rhizoctonia solani* mycelial growth induced by three tomato-associated rhizobacterial isolates noted after 5 days of incubation at 25°C as compared to the untreated control.

isolates tested. Pathogen growth decrease exceeded 30% with 13 out of the 25 tested. Volatiles from *E. cloacae* B16 (KT921429), *B. subtilis* B10 (KT921327), *B. thuringiensis* B2 (KU158884) and, at a lesser extent, those from *B. megaterium* B1 (KU168423), *B. subtilis* B19 (KT921430), and *E. cloaceae* B3 (KT923049) were the most effective against *R. solani* leading to 36-45% lower radial growth relative to control.

Suppression of Rhizoctonia Root Rot disease using tomato-associated rhizobacteria

Disease incidence, calculated 60 days post-transplanting and estimated based on the presence of typical root browning symptoms whatever their levels of extent, varied from 0 to 100% depending on bacterial isolates used for seedling treatments (Table 2). Higher disease incidence (100%) records seemed to be more associated to root browning indexes ranging between 1 and 2.4.

Rhizoctonia Root Rot severity depended significantly ($P \leq 0.05$) upon tested bacterial treatments. As shown in Table 2, this parameter ranged from 0 to 2.4 (using 0-5 scale) for all rhizobacteria-based treatments and these disease severity scores were significantly ($P \leq 0.05$) lower than that recorded on pathogen-challenged and untreated control plants (disease index 4.6).

It should be highlighted that disease index values did not exceed 1 on tomato plants treated with 19 rhizobacterial isolates and were less than 0.5 with 13 out of the 25 tested indicating approximately total suppression of disease development. Moreover, compared to *R. solani*-inoculated and untreated control, 21 out of the 25 tested isolates led to more than 70% decrease in disease severity and this percent ranged between 47 and 53% for the four remaining ones (Table 2). These results demonstrated the capacity of the rhizobacterial collection tested to decrease Rhizoctonia Root Rot development and severity.

It should be mentioned that, interestingly, isolates *B. thuringiensis* B2 (KU158884), *B. subtilis* B8 (KU158885), *B. subtilis* B10 (KT921327), *E. cloacae* B16 (KT921429), and *B. amyloliquefaciens* B21 (KT923047) had totally suppressed Rhizoctonia Root Rot disease. Furthermore, treatments of *R. solani*-inoculated plants using *B. megaterium* B1 (KU168423) and B5 (KT923054), *E. cloacae* B3 (KT923049) and B4 (KT923050), *B. subtilis* B6 (KT921427), B14 (KU161090), B17 (KT923055), and B19 (KT921430), *B. thuringiensis* B23 (KT923056), *B. amyloliquefaciens* B7 (KT921428) and B15 (KT923051) had significantly limited Rhizoctonia Root Rot severity by 86-96% relative to pathogen-inoculated and untreated control. *K. pneumoniae* B12 (KT921328) and *C. jejuense* B11 (KU158886) lowered disease severity by 73 and 83%, respectively, if compared to *R. solani*-inoculated and untreated control (Table 2).

These results indicate that this collection of native tomato-associated rhizobacteria exhibited variable and strong disease suppressive abilities.

Bacterial treatment	Isolate	Disease Incidence (%)	Disease severity	Root fresh weight (g)	Aerial part fresh weight (g)	Plant height (cm)
Bacillus megaterium	B1	40	0.4^{fgh} (91.30)[1]	9^{ab} (84.44)[2]	20.4^{cd} (53.93)[2]	56^{bcdef} (72.5)[2]
B. thuringiensis	B2	0	0.0^{h} (100.0)	10.2^{a} (86.27)	25.4^{ab} (62.99)	63^{ab} (75.56)
Enterobacter cloacae	B3	60	0.6^{efgh} (86.95)	5.3^{defghi} (73.58)	15.8^{efgh} (40.51)	46^{ghij} (66.53)
E. cloacae	B4	60	0.2^{gh} (95.65)	5.5^{defghi} (72.55)	18.1^{cdefgh} (48.06)	50^{defghi} (69.2)
B. megaterium	B5	40	0.4^{fgh} (91.30)	4.8^{ghijk} (70.83)	17.1^{defgh} (45.02)	52^{defg} (70.39)
B. subtilis	B6	20	0.6^{efgh} (86.95)	5.1^{fghij} (74.55)	15.7^{fgh} (40.12)	53^{cdefg} (70.94)
B. amyloliquefaciens	B7	60	0.6^{efgh} (86.95)	5.4^{defghi} (74.07)	16.8^{defgh} (44.04)	49^{efghij} (68.57)
B. subtilis	B8	0	0.0^{h} (100.0)	6.7^{cde} (79.10)	22.3^{bc} (57.85)	58^{abcd} (73.45)
B. amyloliquefaciens	B9	100	1^{ef} (78.26)	5^{fghij} (72.0)	18.5^{cdefg} (49.18)	52^{defg} (70.38)
B. subtilis	B10	0	0.0^{h} (100.0)	9.6^{a} (85.42)	26^{ab} (63.84)	61^{abc} (74.75)
Chryseobacterium jejuense	B11	60	0.8^{efg} (82.61)	5.5^{defghi} (74.55)	18.5^{cdefg} (49.19)	52^{defg} (70.38)
Klebsiella pneumoniae	B12	100	1.2^{de} (73.91)	6.5^{cdef} (78.46)	18.2^{cdefgh} (48.35)	56^{bcdef} (72.5)
B. amyloliquefaciens	B13	100	2.4^{bc} (47.82)	3.4^{klm} (58.82)	14.4^{ghi} (34.72)	43^{hij} (64.18)
B. subtilis	B14	40	0.4^{fgh} (91.30)	4.4^{hijkl} (68.18)	17.8^{cdefgh} (47.19)	48^{fghij} (67.92)
B. amyloliquefaciens	B15	20	0.2^{gh} (95.65)	7.8^{bc} (82.05)	21^{cd} (55.23)	55^{bcdef} (72.0)
E. cloacae	B16	0	0.0^{h} (100.0)	10^{a} (86.0)	28.7^{a} (67.25)	65^{a} (76.31)
B. subtilis	B17	20	0.2^{gh} (95.65)	5.1^{efghij} (72.55)	18.9^{cdef} (50.26)	51^{defgh} (69.80)
B. amyloliquefaciens	B18	100	2.2^{bc} (52.17)	3.6^{jklm} (61.11)	9.3^{k} (-1.07)	42^{ij} (63.33)
B. subtilis	B19	60	0.4^{fgh} (91.30)	6.4^{cdefg} (78.12)	18.9^{cdef} (50.26)	55^{bcdef} (72.0)
B. subtilis	B20	100	1.8^{cd} (60.87)	4.2^{hijkl} (66.67)	10^{jk} (6.0)	46^{ghij} (66.52)
B. amyloliquefaciens	B21	0	0.0^{h} (100.0)	6.8^{cd} (79.41)	20.3^{cde} (53.67)	58^{abcd} (73.45)
B. amyloliquefaciens	B22	100	2.4^{bc} (47.83)	3^{lm} (53.33)	11^{ijk} (14.55)	42^{ij} (63.33)
B. thuringiensis	B23	60	0.4^{fgh} (91.30)	5.8^{defgh} (75.86)	19.7^{cdef} (52.29)	57^{bcde} (72.98)
B. megaterium	B24	80	0.8^{efg} (82.61)	4.1^{ijkl} (65.85)	17.9^{cdefgh} (47.48)	54^{cdefg} (71.48)
B. subtilis	B25	100	2.4^{bc} (47.83)	2.3^{mn} (39.13)	19.9^{cdef} (52.76)	41^{j} (62.44)
Untreated control	-	40	0.4^{fgh} (91.30)	2.9^{lm} (51.73)	13.8^{hij} (31.88)	32^{k} (51.87)
R. solani- inoculated control	-	100	4.6^{a}	1.4^{n} (0)	9.4^{k} (0.0)	15.4^{l} (0)

Table 2: Effects of tomato-associated rhizobacteria on incidence and severity of Rhizoctonia Root Rot disease and growth of tomato plants noted 60 days post-planting. Rhizoctonia Root Rot severity was assessed using a 0-5 scale where: 0=no symptom; 1= 0-25% of root browning; 2= 26 % - 50% of root browning; 3= 51% - 75% of root browning; 4= 76% - 100% of root browning and 5= 100% of root browning [24].
1: Values in parenthesis indicate the percentage (in %) of decrease in disease severity as compared to *R. solani*- inoculated and untreated control plants.
2: Values in parenthesis indicate the percentage (in %) of increase in plant growth parameters as compared to *R. solani*- inoculated and untreated control plants.
For each parameter, values followed by the same letter are not significantly different according to Duncan's Multiple Range test (at $P \leq 0.05$).

Improvement of growth of *R. solani*-inoculated tomato plants using tomato-associated rhizobacteria

The 25 bacterial isolates, naturally associated to tomato, were screened for their plant growth-promoting effects based on various growth parameters and their data were compared to those of the untreated control plants (*R. solani*-inoculated or disease-free controls). ANOVA analysis indicated that the plant height and the aerial parts and roots fresh weights depended significantly ($P \leq 0.05$) upon tested bacterial treatments. Their comparative abilities to promote growth of above- and below-ground plant parts were detailed below.

Plant height increase

All the rhizobacterial isolates had significantly ($P \leq 0.05$) augmented the plant height of *R. solani*-inoculated and treated plants over the inoculated and untreated ones (Table 2). This increase ranged between 62.44 and 76.31% and exceeded 70% using 15 out of the 25 isolates tested.

The highest plant height increments, of about 73-76% compared to the inoculated and untreated control, were recorded on pathogen-inoculated plants treated with *E. cloacae* B16 (KT921429), *B. thuringiensis* B2 (KU158884), *B. subtilis* B10 (KT921327) and B8 (KU158885), and *B. amyloliquefaciens* B21 (KT923047). Moreover, treatments of tomato plants using these four isolates led to significant enhancement of their plant height by 51, 49, 47, and 45%, respectively, compared to disease-free and untreated control (Table 2).

Aerial parts fresh weight increase

Data given in Table 2 showed that plant treatment with the majority of tested isolates, excluding two isolates of *B. amyloliquefaciens* namely B18 (KT923052) and B22 (KT923053) and one isolate of *B. subtilis* B20 (KT921431), led to significant ($P \leq 0.05$) increase in the aerial parts fresh weight relative to *R. solani*-inoculated and untreated control. Enhancements recorded in the aerial parts fresh weight ranged between 34.72 and 67.25% depending on tested bacterial treatments and exceeded 50 and 60% using 11 and 3 isolates out of the 25 tested, respectively.

Based on their capacity to increase the aerial part growth of tomato plants already infected with *R. solani*, *E. cloacae* B16 (KT921429), *B. subtilis* B10 (KT921327), and *B. thuringiensis* B2 (KU158884) were found to be the most promising PGPR candidates generating 67.25, 63.84, and 62.99% increase in this parameter over the inoculated control, respectively. Moreover, increments in the fresh weight of the aerial parts allowed by these isolates were significantly higher than that recorded on pathogen-free and untreated control plants.

Plants treated with *E. cloacae* B16 (KT921429), *B. subtilis* B10 (KT921327), and *B. thuringiensis* B2 (KU158884) showed significant increments in the aerial parts growth by 52, 47, and 46%, respectively, compared to disease-free control and untreated control (Table 2).

Roots fresh weight increase

Data presented in Table 2 revealed that the majority of the tested isolates, excepting *B. subtilis* B25 (KU161091), had significantly ($P \leq 0.05$) improved root development. In fact, root fresh weight increment, compared to pathogen-inoculated and untreated control plants, ranged from 53.33 to 86.27% depending on tested bacterial treatments and reached up to 70% using 18 isolates.

The greatest root growth-promoting effects, expressed by more than 84% increase in the root fresh weight, were obtained using *B. thuringiensis* B2 (KU158884), *E. cloacae* B16 (KT921429), *B. megaterium* B1 (KU168423), and *B. subtilis* B10 (KT921327) (Figure 2). Furthermore, plants infected with *R. solani* and treated with these four isolates exhibited 68-71% significantly higher root growth relative to pathogen-free and untreated control ones. This indicates that these tomato-associated bacterial isolates have additionally bio-fertilizing benefits.

Correlation between Rhizoctonia Root Rot severity and plant growth parameters

Pearson's correlation analysis demonstrated that plant height was significantly and negatively related to root browning index (r = -0.773; P = 4.9843 E-28) indicating that increased disease severity led to plant stunting if compared to pathogen-free control ones. Similar correlations were noted between the fresh weights of the aerial parts (r = -0.608; P = 5.5636 E-15) and roots (r = -0.675; P = 2.8624 E-19) and disease severity.

This analysis indicated that the lowered Rhizoctonia Root Rot severity on tomato plants, allowed using rhizobacteria-based treatments, was linked to the registered growth promotion.

R. solani + str.B16   *R. solani* + str.B10   *R. solani* + str. B2   Untreated-control   *R. solani* -inoculated control

str. B2: *Bacillus thuringiensis* (KU158884); str. B10: *B. subtilis* (KT921327); str. B16: *Enterobacter cloacae* (KT921429).

Figure 2: Suppression of Rhizoctonia Root Rot severity and increased root growth in tomato cv. Rio Grande achieved using three tomato-associated rhizobacterial isolates compared to *R. solani*-inoculated control and to disease-free untreated control noted 60 days post-transplanting.

Discussion

Recently, rhizobacteria have gained more attention because of their successful ability to colonize roots and to the broad spectrum of their metabolites involved in disease biocontrol and/or growth enhancement such as antibiotics, lytic enzymes, siderophore, and phytohormones [17,25-27]. In the present study, 25 rhizobacterial isolates, naturally associated to tomato plants and recovered from rhizospheric soils removed from tomato-producing sites Tunisia, were assessed against *R. solani*. This same rhizobacteria collection was previously tested and was shown able to inhibit *S. sclerotiorum* mycelial growth and myceliogenic germination of its sclerotia, to suppress Sclerotinia Stem Rot disease and to improve tomato growth [24].

Based on *in vitro* findings, diffusible and volatile metabolites from tested rhizobacterial isolates displayed antifungal activity against *R. solani* where *B. thuringiensis* B2, *B. subtilis* B6, *B. subtilis* B10, and *E. cloacae* B16 (KT921429) were found to be the most bioactive agents based on both dual and distance culture assays. According to Adesina et al. [28], the root-associated bacteria have an antagonistic potential towards *Rhizoctonia* spp. In fact, used as whole cell suspensions or cell-free culture filtrates, they displayed suppressive effects against root rot disease of tomato caused by *R. solani*. *Bacillus* spp. have been reported as effective biocontrol agents against *R. solani* in several other studies [10,29-31]. These findings are also in agreement with previous studies reporting the capacity of *B. subtilis* and *B. amyloliquefaciens* to control various fungal plant pathogens including *R. solani* using diffusible and/or volatile metabolites [32-36].

Antibiotic production by bacterial antagonists is an essential component in the biological control of fungal phytopathogens [37] and cyclic lipopeptide antibiotics, in particular, are able to suppress various phytopathogenic fungi including *R. solani* [38,39]. This antibiotic production ability within *Bacillus* spp. and their extensive uses as biocontrol agents have been reported in many reviews [40,41]. In fact, *B. subtilis* and *B. thuringiensis* synthesized at least five lipopeptide antibiotics including bacillomycin [10,42]. However, Mandal et al. [43] demonstrated that *E. cloacae* can produce kurstakin, iturin, surfactin and fengycin probably involved in its antifungal activity displayed toward *R. solani*. In fact, *E. cloacae* isolates tested in the current investigation were previously demonstrated able to produce fengycin A and/or bacillomycin D [23]. *Chryseobacterium* species were frequently encountered in soils and were effective against various soilborne pathogens [44,45]. However, their growth inhibitory effects vary depending on species and isolates. *C. jejuense* B11 used in the present work was previously found to be a bacillomycin D-producing agent [23]. However, in other studies [45,46], *C. wanjuense* KJ9C8 and *Chryseobacterium* species were found able to produce other antifungal compounds like hydrogen cyanide (or HCN) but not antibiotics.

The *in vivo* screening of the ability of the 25 rhizobacterial isolates to suppress Rhizoctonia Root Rot severity revealed that this bacterial collection contained interesting biocontrol agents. The most effective isolates allowing total disease suppression (i.e. having 0 as disease index) were *B. thuringiensis* B2, *B. subtilis* B10, *E. cloacae* B16, *B. subtilis* B8, and *B. amyloliquefaciens* B21. It should be highlighted that the isolates B2, B10 and B16 were previously found to be the most effective in suppressing Sclerotinia Stem Rot in tomato and in promoting plant growth [24]. Moreover, in the present investigation, *B. subtilis* B8 and *B. amyloliquefaciens* B21 isolates had totally suppressed Rhizoctonia Root Rot and, interestingly, had also enhanced by more than 70% the root fresh weight and plant height. However, in their *in vitro* screening, they had significantly decreased *R. solani* radial growth by 44-46 and 19-29%, relative to the untreated control, using dual and distance culture bioassays compared to 58-59 and 45% achieved using the three most effective isolates. These both isolates were previously shown to be fengycin A- and bacillomycin D-producing agents and also able to produce IAA and to solubilize phosphate [23]. Thus, these properties may explain their *in vivo* efficacy but further investigations are still required to elucidate their probable unknown features. Overall, all tested isolates were efficient in controlling Rhizoctonia Root Rot but with a varying degree. In fact, in our previous study, among the rhizobacterial isolates screened for detection of fengycin A and bacillomycin D biosynthesis genes, 20 were able to produce at least one of these antibiotics and 15 isolates were positive for both antibiotic biosynthesis genes [23]. This finding may explain the recorded disease suppression achieved using these isolates.

During the ten last years, several workers underlined the ability of bacterial isolates belonging mainly to *Pseudomonas* [6,47,48] and *Bacillus* genera [9-12,49,50] to control *R. solani* but to our knowledge, few reports are available on use of *C. jejuense*, *E. cloacae* and *K. pneumoniae* against this pathogen. Disease suppression achieved using *Chryseobacterium* B12 is in accordance with Krause et al. [44] findings where *C gleum* was shown to be a putative biocontrol agent able to suppress Rhizoctonia damping-off on several plants.

The current study clearly demonstrated that all treatments performed using the rhizobacterial isolates had significantly increased plant growth parameters i.e. plant height by 62-73%, aerial part fresh weight by 34-67%, and root fresh weight by 53-86%. Thus, these findings showed the additional growth-promoting effects exhibited by the rhizobacterial collection when challenged to tomato plants already infected with the pathogen. According to Ahmad et al. [51], an efficient biocontrol agent is generally equipped with several tools allowing both plant growth promotion and target pathogen inhibition due to their efficient root colonization, phytohormone production ability and nutrient competition. Regarding plant growth-promoting (PGP) properties characterizing this bacterial collection, our previous study demonstrated that 20 isolates among the 25 tested were able to produce siderophore, 18 had solubilized phosphate, 19 were capable to synthesize IAA, and that interestingly 13 isolates showed positive response to at least two PGP traits [23].

Other effects lead to growth promotion during PGPR-plant interactions such as increased root permeability, enhanced ability to survive in strict competitive niche and inhibition of harmful microorganisms [52]. In the present investigation, among the tested rhizobacteria collection, the most promising strains combining both disease suppressive and growth-promoting abilities were *B. subtilis* B10, *B. thuringiensis* B2 and *E. cloacae* B16. Also, interestingly, *C. jujuense* B12 and *K. pneumoniae* B11 isolates had also reduced Rhizoctonia Root Rot severity and augmented root growth and plant height by more 70% compared to pathogen-inoculated control. These tomato-associated agents are able to produce antibiotic lipopeptides, IAA, and siderophore and to solubilize phosphate [23]. *Bacillus* and *Pseudomonas* species are extensively reported as phosphate solubilizing and as IAA- and siderophore-producing agents compared to the other species of rhizobacteria [53]. Additionally, Almaghrabi et al. [54] noted significant increment in shoot dry weight, plant height, and yield in tomato plants treated with *B. amyloliquefaciens*, *B. subtilis* and *B. cereus* and other rhizobacterial species such as *Serratia marcescens*, *Pseudomonas putida*, *P. fluorescens*.

In addition, in the current study, *Enterobacter*, *Chryseobacterium* and *Klebsiella* isolates cumulated the three PGP traits. Also, previous studies demonstrated that *Chryseobacterium* species represent an important bacterial group associated with plants and displaying interesting plant-growth promoting activities [55-57]. *C. balustinum* pepper-associated rhizobacteria also showed PGP properties and had improved aerial surface, aerial length and the dry weight of the above- and underground plant parts [58]. Contrarily, reports on *Klebsiella* species as PGP agents are relatively rare. In fact, Ahemad and Khan [59] found that *Klebsiella* sp. strain PS19 solubilized the inorganic phosphate considerably, produced IAA and siderophores. *K. oxytoca* Rs-5, isolated from Chinese saline cotton fields are able to attenuate salt stress, to enhance plant growth and to release IAA [60]. Sachdev et al. [61] found that IAA-producing *Klebsiella* strains had significantly improved root length and shoot height of infected wheat seedlings relative to control.

Conclusion

The present investigation clearly demonstrated that tomato rhizospheric soils removed from tomato-growing sites of Tunisia harbor interesting biocontrol candidates belonging to *Bacillus*, *Enterobacter*, *Chryseobacterium*, and *Klebsiella* genera. These tomato-associated rhizobacteria displayed Rhizoctonia Root Rot suppression ability and plant growth promoting capacity. Thereby, they could be developed as biofertilizing and biocontrol agents once their effectiveness demonstrated under field conditions and in different tomato-producing sites. Further studies will be focused on the assessment of their disease-suppressive effects against Rhizoctonia Root Rot disease and their plant growth-promoting potential when used as consortia and the eventual shifts in rhizosphere microbial community occurring after their release.

Acknowledgments

This work was funded by the Ministry of Higher Education Scientific Research in Tunisia through the budget assigned to UR13AGR09-Integrated Horticultural Production in the Tunisian Centre-East, Regional Centre of Research on Horticulture and Organic Agriculture (CRRHAB) of Chott-Mariem, Tunisia, and INRA Bordeaux-France through the budget allocated to SAVE UMR. Authors thank the research teams of UMR SAVE / INRA Bordeaux-France and CFBP / INRA Anger-Nante team for their hospitality, their guidance and support.

References

1. Olaniyi JO, Akanbi WB, Adejumo TA, Akande OG (2010) Growth, fruit yield and nutritional quality of tomato varieties. Afr J Food Sci 4: 398-402.
2. Jabnoun-Khiareddine H, El-Mohamedy RSR, Abdel-Kareem F, Aydi Ben Abdallah R, Gueddes-Chahed M, et al. (2015) Variation in chitosan and salicylic acid efficacy towards soil-borne and air-borne fungi and their suppressive effect of tomato wilt severity. J Plant Pathol Microbiol 6: 325.
3. Jabnoun-Khiareddine H, Aydi Ben Abdallah R, El-Mohamedy RSR, Abdel-Kareem F, Gueddes-Chahed M, Hajlaoui A, et al. (2016) Comparative efficacy of potassium salts against soil-borne and air-borne fungi and their ability to suppress tomato wilt and fruit rots. J Microb Biochem Technol 8: 45-55.
4. Anderson NA (1982) The genetics and pathology of *Rhizoctonia solani*. Ann Rev Phytopathol 20: 329-347.
5. Jiskani MM, Pathan MA, Wagan KH, Imran M, Abro H (2007) Studies on the control of tomato damping-off disease caused by *Rhizoctonia solani* Kühn. Pak J Bot 39: 2749-2754.
6. Arora NK, Khare E, Oh H, Kang SC, Maheshwari DK (2008) Diverse mechanisms adopted by fluorescent Pseudomonas PGC2 during the inhibition of *Rhizoctonia solani* and *Phytophthora capsici*. World J Microbiol Biotechnol 24: 581-585.
7. Zachow C, Grosch R, Berg G (2011) Impact of biotic and a-biotic parameters on structure and function of microbial communities living on sclerotia of the soil-borne pathogenic fungus *Rhizoctonia solani*. Appl Soil Ecol 48: 193-200.
8. Curtis DF, Lima G, Vitullo D, Cicco DV (2010) Biocontrol of *Rhizoctonia solani* and *Sclerotium rolfsii* on tomato by delivering antagonistic bacteria through a drip irrigation system. Crop Prot 29: 663-670.
9. Marín VM, Olvera HAL, Coronado CFS, Alférez BP, Ramos LHM, et al. (2008) Antagonistic activity of selected strains of *Bacillus thuringiensis* against *Rhizoctonia solani* of chili pepper. Afr J Biotechnol 7: 1271-1276.
10. Killani AS, Abaidoo RC, Akintokun AK, Abiala MA (2011) Antagonistic effect of indigenous *Bacillus subtilis* on root-/soil-borne fungal pathogens of cowpea. Researcher 3: 11-18.
11. Huang X, Zhang N, Yong X, Yang X, Shen Q (2012) Biocontrol of *Rhizoctonia solani* damping-off disease in cucumber with *Bacillus pumilus* SQR-N43. Microbiol Res 167: 135-143.
12. Szczech M, Shoda M (2004) Biocontrol of Rhizoctonia damping-off of tomato by *Bacillus subtilis* combined with *Burkholderia cepacia*. J Phytopathol 152: 549-555.
13. Velusamy P, Kim KY (2011) Chitinolytic activity of Enterobacter sp. KB3 antagonistic to *Rhizoctonia solani* and its role in the degradation of living fungal hyphae. Int Res J Microbiol 2: 206-214.
14. El Khaldi R, Daami-Remadi M, Chérif M (2016) Biological control of stem canker and black scurf on potato by date palm compost and its associated fungi. J Phytopathol 164: 40-51.
15. Elshafie HS, Camel, I, Racioppi R, Scrano L, Iacobellis NS, et al. (2012) In vitro antifungal activity of *Burkholderia gladioli* pv. *agaricicola* against some phytopathogenic fungi. Int J Mol Sci 13: 16291-16302.
16. Goudjal Y, Toumatia O, Yekkour A, Sabaou N, Mathieu F, et al. (2014) Biocontrol of *Rhizoctonia solani* damping-off and promotion of tomato plant growth by endophytic actinomycetes isolated from native plants of Algerian Sahara. Microbiol Res 169: 59-65.
17. Lugtenberg B, Kamilova F (2009) Plant-growth-promoting rhizobacteria. Annu Rev Microbiol 63: 541-556.
18. Van der Ent S, Verhagen BW, Van Doorn R, Bakker D, Verlaan MG, et al. (2008) MYB72 is required in early signaling steps of rhizobacteria-induced systemic resistance in Arabidopsis. Plant Physiol 146: 1293-1304.
19. Labuschagne N, Pretorius T, Idris AH (2010) Plant growth promoting rhizobacteria as biocontrol agents against soilborne plant diseases. Microbiol Monog 18: 211-230.
20. Saharan BS, Nehra V (2011) Plant growth promoting rhizobacteria: A critical review. Life Sci Med Res 21: 1-30.
21. Shahab S, Ahmed N, Khan NS (2009) Indole acetic acid production and enhance growth promotion by indigenous PSBs. Afr J Agric Res 11: 1312-1316.
22. Kumar A, Kumar A, Devi S, Patil S, Payal CH, et al. (2012) Isolation, screening and characterization of bacteria from rhizospheric soils for different plant growth promotion (PGP) activities: an in vitro study. Recent Res Sci Technol 4: 1-5.
23. Abdeljalil NOB, Vallance J, Gerbore J, Bruez E, Martins G, et al. (2016) Characterization of tomato-associated rhizobacteria recovered from various tomato-growing sites in Tunisia. J Plant Pathol Microbiol 7: 351.
24. Abdeljalil NOB, Vallance J, Gerbore J, Rey P, Daami-Remadi M (2016) Bio-suppression of Sclerotinia Stem Rot of tomato and biostimulation of plant growth using tomato-associated rhizobacteria. J Plant Pathol Microbiol 7: 331.
25. Arias AA, Ongena M, Halimi B, Lara Y, Brans A, et al. (2009) *Bacillus amyloliquefaciens* GA1 as a source of potent antibiotics and other secondary metabolites for biocontrol of plant pathogens. Microb Cell Fact 8: 63-70.
26. Ahemad M, Kibret M (2014) Mechanisms and applications of plant growth promoting rhizobacteria: Current perspective. Journal of King Saud University - Science 26: 1-20
27. Saraf M, Pandya U, Thakkar A (2014) Role of allelochemicals in plant growth promoting rhizobacteria for biocontrol of phytopathogens. Microbiol Res 169: 18-29.
28. Adesina MF, Grosch R, Lembke A, Vatchev TD, Smalla K (2009) In vitro antagonists of Rhizoctonia solani tested on lettuce: rhizosphere competence, biocontrol efficiency and rhizosphere microbial community response. FEMS Microbiol Ecol 69: 62-74.
29. Montealegre JR, Reys R, Perez LM, Herrera R, Silva P, et al. (2003) Selection of bioantagonistic bacteria to be used in biological control of *Rhizoctonia solani* in tomato. Elect J Biotechnol 6: 115-127.
30. Lahlali R, Bajii M, Jijakli MH (2007) Isolation and evaluation of bacteria and fungi as biological control agents against *Rhizoctonia solani*. Commun Agric

Appl Biol Sci 72: 937-982.

31. Príncipe A, Alvarez F, Castro MG, Zacchi LF, Fischer SE, et al. (2007) Biocontrol and PGPR features in native strains isolated from saline soils of Argentina. Curr Microbiol 55: 314-322.

32. Fiddaman PJ, Rossall S (1994) Effect of substrate on the production of antifungal volatiles from *Bacillus subtilis*. J Appl Bacteriol 76: 395-405.

33. Leifert C, Li H, Chidburee S, Hampson S, Workman S, et al. (1995) Antibiotic production and biocontrol activity by *Bacillus subtilis* CL27 and *Bacillus pumilus* CL45. J Appl Bacteriol 78: 97-108.

34. Yoshida S, Hiradate S, Tsukamoto T, Hatakeda K, Shirata A (2001) Antimicrobial Activity of Culture Filtrate of *Bacillus amyloliquefaciens* RC-2 Isolated from Mulberry Leaves. Phytopathology 91: 181-187.

35. Kai M, Effmert U, Berg G, Piechulla B (2007) Volatiles of bacterial antagonists inhibit mycelial growth of the plant pathogen *Rhizoctonia solani*. Arch Microbiol 187: 351-360.

36. Liu WW, Mu W, Zhu BW, Du YW, Liu F (2008) Antagonistic Activities of Volatiles from four strains of *Bacillus* spp. and *Paenibacillus* spp. against soil-borne plant pathogens. Agric Sci China 7: 1104-1114.

37. Stein T (2005) Bacillus subtilis antibiotics: structures, syntheses and specific functions. Mol Microbiol 56: 845-857.

38. Nagorska K, Bikowski M, Obuchowski, M (2007) Multicelluar behaviour and production of a wide variety of toxic substances support usage of *Bacillus subtilis* as a powerful biocontrol agent. Acta Biochim Pol 54: 495-508.

39. Ongena M, Jacques P (2008) Bacillus lipopeptides: versatile weapons for plant disease biocontrol. Trends Microbiol 16: 115-125.

40. Singh JS, Pandey VC, Singh DP (2011) Efficient soil microorganisms: A new dimension for sustainable agriculture and environmental development. Agric Ecosys Environ 140: 339-353.

41. Yuan J, Raza W, Shen Q, Huang Q (2012) Antifungal activity of *Bacillus amyloliquefaciens* NJN-6 volatile compounds against *Fusarium oxysporum* f. sp. *cubense*. Appl Environ Microbiol 78: 5942-5944.

42. From C, Pukall R, Schumann P, Hormazábal V, Granum PE (2005) Toxin-producing ability among Bacillus spp. outside the *Bacillus cereus* group. Appl Environ Microbiol 71: 1178-1183.

43. Mandal SM, Sharma S, Pinnaka AK, Kumari A, Korpole S (2013) Isolation and characterization of diverse antimicrobial lipopeptides produced by Citrobacter and Enterobacter. BMC Microbiol 13: 152.

44. Krause MS, Madden LV, Hoitink HA (2001) Effect of potting mix microbial carrying capacity on biological control of rhizoctonia damping-off of radish and rhizoctonia crown and root rot of poinsettia. Phytopathology 91: 1116-1123.

45. Kim HS, Sang MK, Jung HW, Jeun YC, Myung IS, et al. (2012) Identification and characterization of *Chryseobacterium wanjuense* strain KJ9C8 as a biocontrol agent of Phytophthora blight of pepper. Crop Prot 32: 129-137.

46. Park MS, Jung SR, Lee KH, Lee MS, Do JO, et al. (2006) *Chryseobacterium soldanellicola* sp. nov. and *Chryseobacterium taeanense* sp. nov., isolated from roots of sand-dune plants. Int J Syst Evol Microbiol 56: 433-438.

47. Andersen JB, Koch B, Nielsen TH, Sørensen D, Hansen M, et al. (2003) Surface motility in *Pseudomonas* sp. DSS73 is required for efficient biological containment of the root-pathogenic microfungi *Rhizoctonia solani* and *Pythium ultimum*. Microbiology 149: 37-46.

48. Berta G, Sampo S, Gamalero E, Massa N, Lemanceau P (2005) Suppression of Rhizoctonia root-rot of tomato by Glomus mossae BEG12 and *Pseudomonas fluorescens* A6RI is associated with their effect on the pathogen growth and on the root morphogenesis. Eur J Plant Pathol 111: 279-288.

49. Asaka O, Shoda M (1996) Biocontrol of *Rhizoctonia solani* Damping-Off of Tomato with *Bacillus subtilis* RB14. Appl Environ Microbiol 62: 4081-4085.

50. Khedher SB, Kilani-Feki O, Dammak M, Jabnoun-Khiareddine H, Daami-Remadi M, et al. (2015) Efficacy of *Bacillus subtilis* V26 as a biological control agent against *Rhizoctonia solani* on potato. C R Biol 338: 784-792.

51. Ahmad F, Ahmad I, Khan MS (2008) Screening of free-living rhizospheric bacteria for their multiple plant growth promoting activities. Microbiol Res 63: 173-181.

52. Enebak SA, Carey WA (2000) Evidence for induced systemic protection to fusiform rust in loblolly pine by plant growth-promoting rhizobacteria. Plant Dis 84: 306-308.

53. Gechemba OR, Budambula NLM, Makonde HM, Julius M, Matiru VN (2015) Potentially beneficial rhizobacteria associated with banana plants in Juja, Kenya. J Bio Env Sci 7: 181-188.

54. Almaghrabi OA, Massoud SI, Abdelmoneim TS (2013) Influence of inoculation with plant growth promoting rhizobacteria (PGPR) on tomato plant growth and nematode reproduction under greenhouse conditions. Saudi J Biol Sci 20: 57-61.

55. Anderson M, Habiger J (2012) Characterization and identification of productivity-associated rhizobacteria in wheat. Appl Environ Microbiol 78: 4434-4446.

56. Brown SD, Utturkar SM, Klingeman DM, Johnson CM, Martin SL, et al. (2012) Twenty-one genome sequences from Pseudomonas species and 19 genome sequences from diverse bacteria isolated from the rhizosphere and endosphere of Populus deltoides. J Bacteriol 194: 5991-5993.

57. Dardanelli MS, Manyani H, González-Barroso S, Rodríguez-Carvajal MA, Gil-Serrano AM, et al. (2010) Effect of the presence of the plant growth promoting rhizobacterium (PGPR) *Chryseobacterium balustinun* Aur9 and salt stress in the pattern of flavonoids exuded by soybean roots. Plant Soil 328: 483-493.

58. Cezón R, Manero GFJ, Probanza A, Ramos B, García LJA (2003) Effect of two plant growth-promoting rhizobacteria on the germination and growth of pepper seedlings (*Capsicum annum*) cv. Roxy. Arch Agron Soil Sci 49: 593-603.

59. Ahemad M, Khan MS (2011) Effects of insecticides on plant-growth-promoting activities of phosphate solubilizing *rhizobacterium Klebsiella* sp. strain PS19. Pestic Biochem Phys 100: 51-56.

60. Yue HT, Mo WP, Zheng YY, Li CH, Li H (2007) The salt stress relief and growth promotion effect of Rs-5 on cotton. Plant Soil 297: 139-145.

61. Sachdev DP, Chaudhari HG, Kasture VM, Dhavale DP, Chopade BA (2009) Isolation and characterization of indole acetic acid (IAA) producing *Klebsiella pneumoniae* strains from rhizosphere of wheat (*Triticum aestivum*) and their effect on plant growth. Indian J Exp Biol 47: 993-1000.

Evaluation of Local Isolates of *Trichoderma* Spp. against Black Root Rot (*Fusarium solani*) on Faba Bean

Eshetu Belete[1]*, Amare Ayalew[2] and Seid Ahmed[3]

[1]*Department of Plant Sciences, P.O. Box 1145, Wollo University, Dessie, Ethiopia*
[2]*School of Plant Sciences, P.O. Box 241, Haramaya University, Ethiopia*
[3]*International Center for Agricultural Research in the Dry Areas, P O Box 5689, Addis Ababa, Ethiopia*

Abstract

Faba bean (*Vicia fabae L.*) is one of the most important pulse crops in Ethiopia and is now cultivated on large areas in many countries. In most growing areas, however, the production of the crop is constrained by several disease infections, including fungal diseases. Black root rot caused by *Fusarium solani* is the most destructive disease of faba bean. The antagonistic potentials of locally isolated *Trichoderma* spp. from rhizosphere soils of faba bean plants in the highlands of northeastern Ethiopia were evaluated against *F. solani*, responsible for black root rot. All isolates of *Trichoderma* spp. had strong biological control activity against *F. solani in vitro* as well as *in vivo* pot experiment. Under dual culture, the percentage of mycelial growth inhibition of *F. solani* by the *Trichoderma* ranged from 33.9 to 67.0%. The highest (67.0%) inhibition was obtained from isolate TS036, while the lowest (33.9%) by isolate TS015. Pathogen-inoculated faba bean plants grown in pots that were treated with antagonists had taller plant heights and biomass than the *Trichoderma* untreated control inoculated with *F. solani*. The *Trichoderma* isolates significantly reduced black root rot severity on faba bean seedlings with disease reduction ranging from 64.4 to 74.6% over control. Use of *Trichoderma* species can be a potential source of biological control agents for the management of black root rot in faba beans grown in the region. Hence, the potential *Trichoderma* isolates under field condition might used as a components in the integrated management of *F. solani* that caused faba bean black root rot in the highlands of northeastern Ethiopia.

Keywords: Antagonist; Biocontrol; *Fusarium solani*; Inhibition; Rhizosphere; *Trichoderma*

Introduction

Faba bean (*Vicia fabae L.*) is one of the most important pulse crops in the world and is used as a source of protein in human diet for substitute of animal protein, animal feed, generate incomes and improve soil fertility [1]. In Ethiopia, black root rot of faba bean caused by *Fusarium solani* is the most widespread and destructive disease in black clay soils, where water-logging is severe [2,3] with yield reduction of up to 45% [4]. A number of management options have been used in minimizing the effects. These include use of resistant varieties; application of cultural practices, such as crop rotation and furrow planting to drain out excess water [2,5,6].

Biological control of plant diseases is considered as one of the viable alternative methods to manage plant diseases [7,8]. Application of fungicides is not economical in the long term because they pollute the environment, leave harmful residues and can lead to the development of resistant strains of the pathogen with repeated use [9]. However, use of biological control is safe, nonhazardous for human, farm animals and avoids environmental pollution [7,10]. The application of biological controls using antagonistic microorganisms has proved to be successful for controlling various plant diseases in many countries [11]. Most of these studies were on the control of root and soil-borne plant pathogens and, to a lesser extent of foliar pathogens [12].

One of the most important biological control agents is the use of *Trichoderma* spp. that are most frequently isolated soil fungi and present in plant root ecosystems [12,13]. They colonize the root and rhizosphere of plant and suppress plant pathogens by different mechanisms, such as competition, mycoparasitism, antibiosis production and induced systemic resistance, improvement of the plant health by promote plant growth, and stimulation of root growth [13-17].

The antagonistic activity of the genus *Trichoderma* to *F. solani* has been widely demonstrated [18]. Application of *T. harzianum* as seed treatment significantly reduced the incidence of damping-off diseases of faba bean, lentil, and chickpea, when planted in a soil naturally contaminated with *Fusarium* spp. and *R. solani* [19]. *Trichoderma harzianum*, *T. koningii* and *T. viride* are reported to improve seedling emergence and health of runner bean (*Phseous coccineus*) when applied as seed treatments [20,21]. *Trichoderma harzianum* introduced to the rhizosphere of faba bean plants significantly reduced root rot incidence more than the fungicide Rizolex-T [22].

Little work has been done on *Trichoderma* spp. as antagonists in the control of faba bean black root rot in Ethiopia. So far, not much previous research has been done on the use of biological control agents for the control of black root rot of faba beans in Ethiopia. The strain of *Trichoderma viride*, tested in the laboratory, proved to be an effective antagonistic activity against *Fusarium solani*, black root rot of faba bean [23,24]. Therefore, the objective of the present study was to identify native *Trichoderma* species and to evaluate their antagonistic effects against *Fusarium solani* on faba bean.

***Corresponding author:** Eshetu Belete, Department of Plant Sciences, P.O. Box 1145, Wollo University, Dessie, Ethiopia
E-mail: ebelete70@gmail.com

Materials and Methods

Isolation of *Trichoderma* spp.

Soil samples were collected from different healthy faba bean growing fields in the highlands of northeastern Ethiopia (Table 1). One hundred gram rhizosphere soil was collected into each sterile plastic bag and kept in the refrigerator at the Plant Pathology Laboratory, Haramaya University, for further analysis. Isolation of antagonistic *Trichoderma* spp. from rhizosphere soil was made using serial dilution technique [25]. Each composite soil sample was thoroughly mixed and pulverized by means of mortar and pestle, and passed through a 0.5 mm soil screen sieve before 1 g was suspended in 9 ml sterile distilled water. The suspensions were made homogeneous by agitation using a vortex mixer and further serial dilutions of 10^{-2}, 10^{-3} and 10^{-4}.

One milliliter of serially diluted suspension from each dilution was pipetted into potato dextrose agar (PDA) medium and the petri plates were thoroughly mixed by gently swirling in clockwise and anti-clockwise direction to uniformly spread the suspension. Isolates of *Trichoderma* colonies were picked for antagonism studies after incubating the plates at 25 ± 1°C for 48 h. and re-streaked on a new plate but of the same media to obtain pure colonies. Nineteen *Trichoderma* isolates were identified according to Kubicek and Harman, [26] based on their conidial morphology, color and texture, and growth characteristics and the isolates were confirmed to be *Trichoderma koningii* (17 isolates) and *Trichoderma viride* (2 isolates) (Table 1).

In vitro screening test

The antagonistic effects of each *Trichoderma* sp. against *F. solani* were tested using dual culture technique [10,27]. The tested isolates of *Trichoderma* spp. were grown on potato dextrose agar (PDA) medium at 20°C, for 6-days and used as inocula. Disks from each isolate of *Trichoderma* spp. (5 mm in diameter) were inoculated on PDA medium in one side in Petri plate and the opposite side was inoculated by *F. solani* inocula. The inoculated plate with *F. solani* only but without *Trichoderma* treatment was used as control. The treatments were replicated three times and arranged in a completely randomized design (CRD) and the experiment was repeated. Four and five days after incubation periods at 25 ± 2°C, data on growth inhibition zone and colony diameter were recorded for each plate. The radii of the fungal colony towards and away from the antagonistic colony were measured and the percentage growth inhibition was calculated [28] as follows:

$$\% \text{ Inhibition} = \frac{(R\text{-}r)}{R} \times 100$$

where, *R* is the maximum radius of the fungal colony away from the antagonist colony, and *r* is the radius of the fungal colony opposite the antagonist colony

Greenhouse test

On the bases of *in-vitro* results, nine *Trichoderma* isolates were tested to assess their effects on black root rot development on faba bean seedlings. *Trichoderma* isolates were grown on the plates and the spores were washed from the plates with sterile distilled water and the concentration was adjusted to 10^6 conidia ml^{-1}. Sterilized faba bean seeds of the susceptible faba bean variety CS-20DK were dipped into the suspension of each *Trichoderma* spp. for five hrs. Seeds inoculated with sterile water only (i.e. without antagonists) were used as control. The inoculated seeds (5 seeds per 15 cm diameter pot) were planted in sterile black soil (clay 70%: silt 16%: sand 14%) and kept in the greenhouse at 24-28°C during day and 15-20°C at night. After a week, each pot was artificially drenched with aggressive isolate of *F. solani* at a rate of 10^6 conidia ml^{-1}. The experiment was conducted in a randomized complete block design (RCBD) with three replications and the experiment was repeated.

One month after planting, plants were removed from the soil and the roots were washed with sterile distilled water and the root rot severity was scored by assessing necrotic lesions on the roots and hypocotyls using a rating scale of 0-4, where 0=hypocotyls and roots white and firm, no root pruning; 1=slightly brown or discolored hypocotyls and roots, slight root pruning; 2=moderately discolored hypocotyls and roots, extensive root pruning; 3=darkly discolored hypocotyls and roots, hypocotyls completely collapse or, severe root pruning; 4=dead or dying plant [29] (McFadden et al.), Based on the disease severity data, percentage of root rot suppression was calculated [30] as follows:

$$\% \text{ Suppression} = \frac{A\text{-}B}{A} \times 100$$

where, *A* is the disease severity exhibited in the root region due to *F. solani* alone and *B* is the disease severity exhibited in the root region after inoculation with both the pathogen and bacterial antagonists. Data of plant height and biomass was also recorded.

Data analyses

Statistical analyses were performed using General Linear Modeling (GLM) procedure of SAS® System for Windows Version 9.1 software [31]. Severity ratings were normalized before analysis using square root transformation with the formula $(X + 0.5)^{1/2}$, where X is the severity rating of root rot and 0.5 is constant number added [32]. The least significant difference (LSD) test at 5% level of significance was used to separate treatment means for each measured parameter.

Results

In vitro screening test

Isolates of *Trichoderma* species was found to occur on faba bean rhizospher soil in the three districts of northeastern highlands of Ethiopia (Table 2 and Apendix C3); however, the distribution of isolates was found to vary among the districts. Highest number of isolates was obtained from Jamma, followed by Woreillu and Delanta. The elevation of the sampled areas varied between 2551 and 3017 meters above sea level.

The efficacy of local *Trichoderma* isolates in inhibiting the mycelial growth of *F. solani* in dual culture was determined on PDA medium. Results of dual culture tests clearly showed that all the isolated

Isolates	*Trichoderma* species
TS004, TS007, TS015, TS018, TS019, TS022, TS025A, TS027, TS030, TS032, TS037, TS041A, TS047, TS050, TS058, TS064, TS090	*Trichoderma koningii*
TS010, TS036	*Trichoderma viride*

Table 1: Identification of the local *Trichoderma* isolates.

Districts	Altitude	Grid reference	Potential antagonistic *Tricoderma* isolates
Delanta	2885	50°E, 12°N	TS090
Jamma	2551-2621	52-53°E, 11°N	TS004, TS007, TS010, TS015, TS018, TS019, TS022, TS025A, TS027, TS030, TS032, TS036, TS037, TS041A,
Woreillu	2627-2725	54°E, 11°N	TS047, TS050, TS058, TS064,

Table 2: Occurrence of *Trichoderma* isolates on the soil rhizospher of faba bean plants in the highlands of northeastern Ethiopia.

Trichoderma spp. significantly ($P \leq 0.05$) inhibited the radial growth of *F. solani* at varying degrees (Table 3 and Figure 1). These *Trichoderma* isolates were able to inhibit the mycelial growth of *F. solani* by the range of 33.9 to 67%. Nine isolates significantly inhibited the colony mycelial growth of *F. solani* the most promising isolates resulting in more than 50% inhibition. Maximum inhibition zone (67.0%) was exhibited by the isolate TS036, followed by TS025A (65.9%) and TS050 (63%), while the lowest (33.9%) inhibition was due to the isolate TS015. Generally, the antagonists inhibited the mycelial growth of *F. solani* but could not overgrow the pathogen until three to four days. However, five days later, the *Trichoderma* overgrew the pathogen and wholly occupied the medium.

Greenhouse test

The application of antagonists reduced significantly ($P \leq 0.05$) the extent of black root rot infection in comparison to *F. solani* alone (without *Trichoderma* isolates) inoculated plants (Table 4). The effectiveness of the antagonists ranged from 69 to 74% disease suppression over the control. The maximum (74%) control or suppression of black root rot was observed in bean plants treated with isolate TS025 and TS050, while the lowest (64.4%) was with isolate TS058.

In the present study, all antagonists significantly enhanced the height and biomass of bean seedlings as compared to bean seedlings inoculated with *F. solani* alone (Table 4 and Figure 2). The maximum plant height was observed in plants treated with isolates TS036 (38.5 cm), TS019 (38.4 cm) and TS007 (37.9 cm) followed by isolates TS010 (36.8 cm), TS050 (36.7 cm), TS058 (36.6 cm) and TS022 (34.5 cm), while the shortest (33.4 cm) plant height was observed in plants treated with isolate TS025. *Trichoderma* treated seedlings increased in fresh biomass ranging from 24.1 to 40.5% over the control. Among the potential biological control agents in this study, isolates TS019, TS036, and TS010 resulted in 40.5, 37.8 and 37.4%, respectively, increased fresh biomass over control. The least reduction in plant fresh weight was observed in plants treated with TS047 (42.4 g). Better overall growth of seedlings indicated the efficiency of *Trichoderma* antagonists in controlling faba bean black root rot.

Isolates	*Trichoderma* radius (mm)	Inhibition (mm)	Inhibition (%)
TS004	49.7 ± 0.1b-d	19.3 ± 0.1a	40.2 ± 0.1f
TS007	52.3 ± 0.2ab	13.3 ± 0.1c	55.1 ± 0.2d
TS010	54.0 ± 0.1a	12.7 ± 0.2cd	60.5 ± 5.1b-d
TS015	50.3 ± 0.2bc	20.7 ± 0.1a	33.9 ± 2.7g
TS018	50.0 ± 0.2b-d	16.7 ± 0.2b	47.4 ± 5.0e
TS019	52.3 ± 0.2ab	11.3 ± 0.2d-f	61.4 ± 4.4a-c
TS022	50.0 ± 0.1b-d	12.3 ± 0.2c-e	60.7 ± 3.8b-d
TS025A	51.7 ± 0.2ab	10.7 ± 0.1ef	65.9 ± 1.2ab
TS027	47.0 ± 0.2de	19.3 ± 0.1a	34.8 ± 1.7fg
TS030	45.7 ± 0.5e	19.3 ± 0.2a	37.0 ± 2.8fg
TS032	47.0 ± 0.3de	20.0 ± 0.0a	34.8 ± 1.3fg
TS036	49.6 ± 0.1b-d	10.3 ± 0.1f	67.0 ± 1.6a
TS037	49.7 ± 0.1b-c	16.0 ± 0.1b	47.2 ± 3.5e
TS041A	48.0 ± 0.3c-e	17.0 ± 0.2b	40.1 ± 5.9f
TS047	51.3 ± 0.2ab	12.0 ± 0.1c-f	60.4 ± 3.2b-d
TS050	49.3 ± 0.2c-d	11.3 ± 0.1d-f	63.0 ± 0.5a-c
TS058	49.3 ± 0.1b-d	12.3 ± 0.1c-e	58.6 ± 5.8cd
TS064	49.7 ± 0.1b-d	19.6 ± 0.1a	34.3 ± 2.0fg
TS090	49.3 ± 0.1b-d	20.0 ± 0.0a	35.9 ± 5.1fg
Control	-	31.6	-
CV%	3.89	7.05	7.11

Values in parentheses are square root transformed $(x + 0.5)^{1/2}$ values.
Means within column followed by the same letter(s) are not significantly different from each other at 5% level of significance.

Table 3: Effects of *Trichoderma* isolates on the radial growth (inhibition) of *Fusarium solani.*

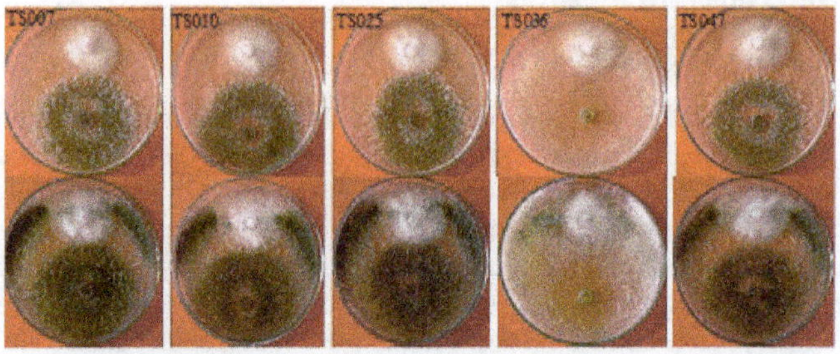

Figure 1: Differences *In vitro* antagonistic effect of selected *Trichoderma* isolates against *Fusarium solani* using dual culture method four (top) and five (bottom) days after inoculation. At the bottom of the plates overwhelming growth of *Trichoderma* isolate on *F. solani*, which indicates parasitic interaction. Heavy sporulation by *Trichoderma* is seen on and in vicinity of *F. solani* colony. Each perti dish has *F. solani* at the upper and *Trichoderma* at the lower side of the petri plate.

Isolate code	Disease severity (0-4 scale)	Disease suppression (%)	Plant height (cm)	Plant biomass	
				Fresh weight (g)	Dry weight (g)
TS007	1.20 (1.30)b	69.5	37.9 ± 3.2a	46.8 ± 7.4a-c	6.8 ± 1.6b
TS010	1.07 (1.23)b	72.8	36.8 ± 4.5ab	51.4 ± 3.7ab	7.7 ± 0.5ab
TS019	1.07 (1.23)b	72.8	38.4 ± 0.9a	54.1 ± 2.7a	8.5 ± 0.2a
TS022	1.06 (1.23)b	73.0	34.5 ± 1.6ab	47.1 ± 3.3a-c	7.0 ± 0.5b
TS025	1.00 (1.20)b	74.6	33.4 ± 2.4b	48.3 ± 3.7a-c	7.2 ± 0.5b
TS036	1.07 (1.23)b	72.8	38.5 ± 2.1a	51.8 ± 7.3ab	7.2 ± 0.8b
TS047	1.13 (1.27)b	71.3	33.5 ± 2.4b	42.4 ± 2.1c	6.9 ± 0.3b
TS050	1.00 (1.20)b	74.6	36.7 ± 1.6ab	49.5 ± 3.1a-c	7.3 ± 0.5b
TS058	1.40 (1.37)b	64.4	36.6 ± 1.5ab	45.5 ± 3.6bc	6.7 ± 0.4b
Control	3.93 (2.10)a	0.0	28.8 ± 1.9c	32.2 ± 2.9d	4.8 ± 0.6c
Mean	1.39		35.5	46.9	7.0
CV%	8.84		6.76	9.48	8.92

Values in parentheses are square root transformed $(x + 0.5)^{1/2}$ values.
Means within column followed by the same letter(s) are not significantly different from each other at 5% level of significance.

Table 4: Efficacies of potential *Trichoderma* isolates on faba bean black root rot under greenhouse condition.

Figure 2: Greenhouse pot experiment illustrating the efficacy of *Trichoderma* isolates in the suppression of black root rot infection in four weeks old faba bean seedlings. Faba bean plants inoculated with bioagents (A) showed better shoot and root biomass than plants treated with *F. solani* alone (B).

Discussion

Trichoderma is known antagonist of plant pathogens, and has been shown to be very efficient biological control agent of several soil-borne plant pathogenic fungi [12,13,33]. *Trichoderma* is a good candidate for biological control due to the different modes of action that inhibit the growth of other fungi. The study was undertaken to determine the potentials of locally isolated antagonistic *Trichoderma* spp. that perform as biological control agents for the management of *F. solani*, which is responsible for black root rot on faba bean.

The results revealed that the isolates of *Trichoderma* spp., which were obtained from the rhizosphere soil of healthy bean plants, had effective biological control activity against *Fusarium solani* under *in vitro* and *in vivo* pot experiments. The potentiality of *Trichoderma* spp. as biological control agents of phytopathogenic fungi in several crops is well known, especially to *Fusarium solani* infection [34]. *T. harzianum*, *T. koningii* and *T. viride* protected the germinating bean seedlings against *Fusarium* spp. and *R. solani* infection [19,35].

Antagonistic capability of all *Trichoderma* isolates showed inhibitory effect against mycelial growth of *F. solani* in dual *in vitro* testing. The percentage of mycelial growth inhibition by the *Trichoderma* isolates against *F. solani* varied between 33.9 and 67.0%. Some isolates were highly inhibitory to *F. solani* growth, whereas others showed only lower activity. In the medium, the *Trichoderma* isolates grew much faster and suppressed the growth of *F. solani in vitro*. The competition mechanism of *Trichoderma* depends on their rapid growth rate limiting nutrients and space for *F. solani* and this may produce inhibition of *F. solani* growth up to 67%. Effective biological control agents inhibit the growth of the target organisms through their ability to grow much faster than the pathogenic fungi thus competing efficiently for space and nutrients [13]. Starvation is the most common cause of death for microorganisms, so that competition for limiting nutrients results in biological control of fungal phytopathogens. Competition is effective when the pathogen conidia need exogenous nutrients for germination and germ-tube elongation [36].

A second mechanism of pathogen control that *Trichoderma* displayed was mycoparasitism. The mycoparasitic activity of the tested *Trichoderma* isolates was detected morphologically by subsequent profuse sporulation of *Trichoderma* and its ability to overgrow upon the mycelial growth of *F. solani* in culture which may indicate its ability to directly parasitize the pathogen. *Trichoderma* species exert biological control against fungal phytopathogens either indirectly by competing for nutrients and space, modifying the environmental conditions, or promoting plant growth and plant defensive mechanisms and antibiosis, or directly by mechanisms such as mycoparasitism [15,16,37].

The *Trichoderma* isolates, which showed high efficacy *in vitro* study, also significantly reduced the root rot severity in the greenhouse. The effective colonization of faba bean roots by *Trichoderma* isolates might have contributed to their capability to inhibit *F. solani* infection on faba bean roots. The effect of antagonist on the faba bean plant growth under pot condition revealed that faba bean seedlings grown in *Trichoderma*-treated soils had taller plant height and fresh weight than *F. solani* alone inoculated faba bean plants. The isolates significantly reduced black root rot severity on faba bean seedlings with disease reduction ranging from 64.4 to 74.6% over control. Application of the antagonist *Trichoderma* spp. as seed treatment significantly reduced the incidence of *Fusarium* spp. and *R. solani* in some leguminous crops [19,35], and grew readily along with the developing root system of the treated plants [35,37,38]. This might due to modification of the rooting system.

Conclusion

All antagonist species of *Trichoderma* isolates proved to be effective in controlling *Fusarium solani*, both under laboratory and pot conditions. The results indicate that the selected *Trichoderma* species could be potential sources of antagonistic agents for the management of black root rot on faba bean grown in the highlands of northeastern Ethiopia, where biological control agents can be used as one of the components in the integrated management of the disease. However, applying biological control agents in the field is influenced by many environmental, biological and physical factors. So, it desirable to evaluate further the biological control potential of the *Trichoderma* spp. under field conditions.

Acknowledgments

This research was financed by Ministry of Education (MOE), Ethiopia. The authors thank all Plant Pathology Laboratory technicians, Haramaya University, for their assistance in data collection and laboratory works.

References

1. Tilaye A, Getachew T, Demtsu B (1994) Genetics and breeding of faba bean. In: Tilaye, Asfaw (eds.), Cool Season Food Legumes in Ethiopia. Addis Ababa, Ethiopia, pp. 97-121.
2. Gorfu D, Beshir T (1994) Faba bean disease in Ethiopia. In: Asfaw Tilaye (ed) Cool Season Food Legumes in Ethiopia. Addis Ababa, Ethiopia, pp 328-345.
3. Bekele B, Muhammed G, Gelano T, Belayneh T (2003) Faba Bean and Fieldpea Diseases Research. In: Ali K (ed) Food and Forage Legumes: Progress and Prospects. Progress of the Workshop on Food and Forage Legumes. Addis Ababa, Ethiopia, pp. 221-227.
4. PPRC (Plant Protection Research Center) (1996) Progress Report for the period of 1995/96. PPRC. Ambo, Ethiopia. p. 53.
5. Beshir T (1996) Evaluation of faba bean cultivars for resistance to black root rot (Fusarium solani). In: Nile Vally and Red Sea Regional Program on Cool-Season Food Legumes and Cereals. Cairo, Egypt, pp. 72-76.
6. SARC (Sirinka Agricultural Research Center) (2005) Crop Protection Research Progress Report. SARC. Sirinka, Woldia.
7. Barakat R, Al-Masri MI (2005) Biological control of gray mold disease (Botrytis cinerea) on tomato and bean plants by using local isolates of Trichoderma harzianum. Dirasat, Agricultural Sciences 32: 145-156.
8. Pal KK, Gardener BM (2006) Biological Control of Plant Pathogens. The Plant Health Instructor pp. 1-25.
9. Vinale F, Sivasithamparam K, Ghisalberti EL, Marra R, Lorito SL (2008) Trichoderma - plant pathogens interactions. Soil Biology & Biochemistry 40: 1-10.
10. Abdel-Kadir MM, El-Mougy NS, Ashour AMA (2002) Suppression of root rot incidence in faba bean fields by using certain isolates of Trichoderma. Egypt Journal of Phytopathology 30: 15-25.
11. Janisiewicz WJ, Tworkoski TJ, Sharer C (2000) Characterizing the mechanism of biological control of postharvest diseases on fruits with a simple method to study competition for nutrients. Phytopathology 90: 1196-1200.
12. Whipps JM, Lumsden RD (2001) Commercial use of fungi as plant disease biological control agents: status and prospects. In: Butt TM, Jackson C, Magan N (eds) Fungi as Biocontrol Agents: Progress, Problems and Potential, CABI Publishing: Wallingford, UK, p. 390.
13. Harman GE, Howell CR, Viterbo A, Chet I, Lorito M (2004) Trichoderma species--opportunistic, avirulent plant symbionts. Nat Rev Microbiol 2: 43-56.
14. Harman GE (2000) Myths and dogmas of biocontrol: changes in perceptions derived from research on Trichoderma harzianum T- 22. Plant Disease 84: 377-393.
15. Howell CR (2003) Mechanisms employed by Trichoderma species in the biological control of plant diseases: The History and Evolution of Current Concepts. Plant Disease 87: 1-10.
16. Benítez T, Rincón AM, Limón MC, Codón AC (2004) Biocontrol mechanisms of Trichoderma strains. Int Microbiol 7: 249-260.
17. Mohidden FA, Khan MR, Khan SM, Bhat BH (2012) Why Trichoderma is considered superhero (super-fungus) against the evil parasite? Plant Pathology Journal 9: 92-102.

18. Lewis JA, Larkin RP, Rogers DL (1998) A formulation of Trichoderma and Gliocladium to reduce damping-off caused by Rhizoctonia solani and saprophytic growth of the pathogen in soilless mix. Plant Diseases 82: 501-506.

19. Abou-Zeid NM, Arafa MK, Attia S (2003) Biological control of pre- and post-emergence diseases on faba bean, lentil and chickpea in Egypt. Egypt Journal of Agriculture Research 81: 1491-1503.

20. Pieta D, Pastucha A, Patkowska E (2003) The use of antagonistic microorganisms in biological control of bean diseases. Sodininlysteir Darzininkyste 22: 401-406.

21. Pieta D, Pastucha A (2004) Biological methods of protecting common bean (Phaseolus ulgaris, L.) Folia Universitaris Agriculturae Stetinensis Agricultura 95: 301-305.

22. Abdel-Kader MM (1997) Field application of Trichoderma harzianum as biocide for control of bean root rot. Egypt Journal of Phytopathology 25: 19-25.

23. Belayneh T (2003) Evaluation of native Trichoderma isolates of Ethiopia for biological control of Fusarium solani of faba bean. MSc. Thesis Presented to the School of Graduate Studies, Haramaya University. p. 70.

24. Beshir T (1999) Evaluation of the potential of Trichoderma viride as biological control agent of root rot disease, Fusarium solani, of faba bean. Pest Management Journal of Ethiopia 3: 91-94.

25. Waksman SA (1922) A Method for Counting the Number of Fungi in the Soil. J Bacteriol 7: 339-341.

26. Kubicek CP, Harman GE (2002) Trichoderma and Gliocladium: Basic Biology, Taxonomy and Genetics. I: 278.

27. Coskuntuna A, Āzer N (2008) Biological control of onion basal rot using Trichoderma harzianum and induction of antifungal compounds in onion set following seed treatment. Crop Protection 27: 330-336.

28. Abeysinghe S (2007) Biological control of Fusarium solani f.sp. phaseoli the causal agent of root rot of bean, using Bacillus subtilis CA32 and Trichoderma harzianum RU01. Ruhuna Journal of Science 2: 82-88.

29. McFadden W, Hall R, Phillips LG (1989) Relations of initial inoculum density to severity of fusarium root rot of white bean in commercial fields. Canadian Journal of Plant Pathology 11: 122-126.

30. Villajuan-Abgona R, Kagayama K, Hyakumachi M (1996) Biocontrol of Rhizoctonia damping-off of cucumber by non-pathogenic binucleate Rhizoctonia. Europian Journal of Plant Pathology 102: 227-235.

31. SAS Institute (2004) SAS/STATA User Guide for Personal Computers Version 9.1 edition. SAS Institute. Carry, NC, USA.

32. Gomez KA, Gomez AA (1984) Statistical procedures for agricultural research, 2ndEd. John Wiley and Sons, New York.

33. Rosa DR, Herrera CJL (2009) Evaluation of Trichoderma spp. as biocontrol agents against avocado white root rot. Biological Control 51: 66-71.

34. Rojo FG, Reynoso MM, Sofia MF, Chulze N, Torres AM (2007) Biological control by Trichoderma species of Fusarium solani causing peanut brown root rot under field conditions. Crop Protection 26: 549-555.

35. Raats PAC (2012) Effect of Trichoderma species on damping-off disease incidence, some plant enzymes activity and nutritional status of bean plants. International Journal of Agriculture and Environment 2: 13-25.

36. Elad Y (2000) Biological Control of Foliar Pathogens by Means of Trichoderma harzianum and Potential Modes of Action. Crop Protection 19: 709-714.

37. Harman GE (2006) Overview of Mechanisms and Uses of Trichoderma spp. Phytopathology 96: 190-194.

38. Howell CR, Hanson LE, Stipanovic RD, Puckhaber LS (2000) Induction of Terpenoid Synthesis in Cotton Roots and Control of Rhizoctonia solani by Seed Treatment with Trichoderma virens. Phytopathology 90: 248-252.

8

Effect of Seed Treatment on Incidence and Severity of Garlic White Rot (*Sclerotium cepivorum* Berk) in the Highland Area of South Tigray, North Ethiopia

Harnet Abrha[1]*, Alem Gebretsadik[2], Girmay Tesfay[3] and Girmay Gebresamuel[4]

[1]*Tigray Agricultural Research Institute, Alamata Agricultural Research Center, P.O.Box-56 Alamata, Ethiopia*

[2]*Capacity Building for Scaling up of Evidence-Based Best Practices in Agricultural Production in Ethiopia (CASCAPE) project, Mekelle, Ethiopia*

[3]*Department of Land Resources Management and Environmental Protection, Mekelle University, Ethiopia*

[4]*Department of Natural Resources Economics and Management, Mekelle University, Ethiopia*

Abstract

A participatory evaluation of integrated white rot management was conducted for two years during the 2013 and 2014 main cropping seasons (Julys-October) in Emba-Alaje and Enda-Mokoni Woredas of South Tigray. The sites selected for the trial were potential for garlic production and many farmers used this crop as main source of income. However, the productivity of the crop is declining from time to time due to white rot infestation. Hence, this study was conducted to evaluate hot water treatment and chemical application on controlling white rot infestation. During the first year, three treatments were used for the experiment: chemical treatment (Apron star 42 WS), hot water treatment at 46°C and farmers practice (control). During the second demonstration year additional treatment hot treatment plus chemical treatment included. The first year results indicated that lowest white rot incidence (25%) and number of cloves per hectare (2670) infected by white rot was recorded from hot water treated plots followed by chemical treatment (29% incidence) and (4240 infected cloves) per hectare. White rot incidence and number of plants per hectare infected by white rot was very high in the control plots as exhibited by 45% and 7910 plants, respectively. Significantly higher marketable yield (126.09 qt/ha) was obtained from hot water treated plots, while the lowest marketable yield (96.576 qt/ha) was obtained from the control or farmers practice plots. About 30.5% yield advantage was obtained from hot water treated plots compared to the control plot. The same result was also recorded during the second year demonstration. In Ayba and Atsela Kebeles the highest marketable yield and low white rot incidence and severity were recorded from hot water treated plots. Hot water treatment has also received higher acceptances and was ranked first by the farmer's participated in the research for its low cost and higher response Unlike the two Kebeles, incidence of white rot in Simret Kebele was 100% in all treatments, however hot water treatment preferred by participant farmers by its accessibility. Therefore, from the result we can conclude that based on the accessibility, environmental benefits and farmer's preferences hot water treatment is promising practice for white rot treatment in the intervention Kebeles.

Keywords: Apron star; Farmers practice; Garlic; Hot water; White rot

Introduction

Garlic (*Allium sativum*) belongs to the Alliaceae family, in the same family with onions, shallots and leeks. Garlic is one of the most important crops widely cultivated throughout the world including Ethiopia. It is the second most widely cultivated Allium species next to onion [1]. In Ethiopia, a total of 10,690.41 ha of land was under garlic production during the 2011 main cropping season, taking up about 6.64% of land area covered by all vegetable crops at country level and yielding about 128,440.94 tons of those cultivated by small scale farmers, contributing about 7.42% to the total country level vegetable crop production [2].

Onion and garlic white rot caused by *Sclerotium cepivorum* is a major production threat of garlic and onion in the Tigray Regional state of Ethiopia. During favorable weather condition and when susceptible varieties are in the production system, the disease can cause 100% yield loss [3]. The highland areas of south Tigray are potential areas for cultivation of garlic, many farmers in the particular zone widely cultivate this crop and used as the main source of income [4]. However, the productivity of the crop is declining from time to time and farmers are forced to lose the profit that they used to earn from sale of the produce of this commodity. Because of this, farmers are discouraged to grow Garlic. There is no resistant variety available in Ethiopia to tackle the disease. However, Alamata agriculture research center was conducting adaption trail of different garlic varieties including a local check namely Bora local. The result indicated that, the highest bulb yield (62.79 and 27.90 qt/ha) was recorded from Bora local at Fala and Ayba *Kebeles*, respectively and it was moderately white rot resistance as compared to the released varieties.

Fungicides are among the most effective options for white rot management. Systemic as well as non-systemic fungicides significantly reduced incidence of white rot, its progress rate, severity, and thereby improved the yield of the crop. It has been reported that the use of procymidone as seed and soil treatment reduced disease incidence up to 75–95% [5]. The fungicide apron star 42 WS is the most effective

***Corresponding author:** Harnet Abrha, Tigray Agricultural Research Institute, Alamata Agricultural Research Center, P.O.Box-56 Alamata, Ethiopia
E-mail: hany7mn@gmail.com

means of controlling many soil and seed borne fungal diseases. It is also effective in controlling white rot of garlic under Ethiopian condition and is ecologically safe since it is used as clove or seed treatment. Alamata Agriculture Research Centre annual report also indicated that onion seed treated by apron star showed 12% yield increment under farmers' field than the untreated seed at Ofla *Woreda* in Zata *Kebele* [6].

Cultural management practices like avoidance and sanitation are recommended. Cull bulbs, litter, and soil should not be moved from infested to non-infested fields. Equipment should be cleaned before moving from one field to another. The most effective way to avoid introducing the disease is to plant only clean stock from known origins that have no history of white rot. However, the fungus is vulnerable at temperatures above 46°C, thus dipping seed of garlic in hot water before planting can greatly reduce the amount of pathogen and is a good preventative measure, although it may not completely eradicate the fungus [7]. However, temperatures above 49°C may kill the garlic, so careful temperature control is essential. This study was conducted to evaluate the performance of hot water and apron star treatment on controlling white rot infestation of local garlic variety managed by typical smallholder farmers. This research was intended to provide farmers with alternative way of dealing white root rot disease using local resource.

Objective

To enhance production and productivity of garlic through integrated management of white rot disease.

Materials and Methods

Area description

Demonstration was conducted in 3 major garlic growing highland *Kebeles* of Emba-Alaje and Enda Mekoni *Woredas* of southern zone of Tigray region. The *Kebeles* included were Atsela and Ayba from Emba-Alaje, and Simret from Enda Mekoni *Woredas*. Atsela, Ayba and Simret, *Kebeles* are located at 1255.733'N and 039°31.606'E; 12°53.495'N and 039°32.120'; and 12°43.576'N and 039°30.355'E, respectively. The *Kebeles* have altitudes of 2463, 2724, and 2509 m.a.s.l, respectively. The intervention *Kebeles* experience a bi-modal type of rainfall and the mean annual rainfall ranges from 300 to 1600 mm. Eutric cambisols with eutric regosols/lithosols are found in Emba-Alaje and chromic vertisols with eutric cambisols is found in Enda Mekoni *Woreda*. Site selection was made based on the potential of the area for garlic production and availability of white rot disease.

Experimental design and procedure

The trial was conducted during 2013 and 2014 cropping seasons. A total of 30 farmers (10 in first year and 20 in second years) participated on the demonstration considering availability of the white rot disease on specific farm and farmer's willingness to participate in demonstration. Participant farmers in each *Kebele* were considered as replication and all treatments were applied per farm. Each participating farmer allocated an area of 10m × 31m and 4.8m × 21.5m in first and second years, respectively. Well sprouted Bora local garlic variety was used as planting material for the demonstration. Management of the farms was based on agreed recommended agronomic practices.

Treatments used in the experiments

The treatments used in the experiments were:

Treatment one: Chemical/apron star treatment: The required amount of cloves to be treated was weighed as rate of 600 kg/ha and poured in wet table powders of Apron Star 42 WS with a rate of 3 gm apron star and 10 ml water for one kg garlic seed. Chemical treated garlic clove was planted with spacing of 10 cm between plants and 30 cm between rows. Inorganic fertilizers, diammonium phosphate (DAP) and urea were applied at the rate of 200 kg/ha DAP and 150 kg/ha Urea. The whole amount of phosphorus fertilizer was applied at planting, whereas half rate of urea was applied during planting and remaining was applied 30-45 days after planting.

Treatment two: Hot water treatment: Garlic cloves weighted as rate of 600 kg/ha and dipped in hot water at 46°C for 20 minutes. The treated garlic clove was planted with spacing of 10 cm between plants and 30 cm between rows. Inorganic fertilizers, diammonium phosphate (DAP) and urea were applied at the rate of 200 kg/ha DAP and 150 kg/ha Urea. The whole amount of phosphorus fertilizer was applied at planting, whereas half rate of urea was applied during planting and remaining was applied 30-45 days after planting.

Treatment three: Hot water followed by chemical treatment: First garlic cloves was weighted in the rate of 600 kg/ha and dipped in hot water at 46°C for 20 minutes followed by poured in wet table powders of Apron Star 42 WS with a rate of 3 gm apron star and 10 ml water for 1 kg garlic seed. Hot water and chemical treated garlic clove was planted with spacing of 10 cm between plants and 30 cm between rows. Inorganic fertilizers, diammonium phosphate (DAP) and urea were applied at the rate of 200 kg/ha DAP and 150 kg/ha Urea. The whole amount of phosphorus fertilizer was applied at planting, whereas half rate of urea was applied during planting and remaining was applied 30-45 days after planting.

Treatment four: Control plot managed based on farmer's practices and the other management practices (fertilizer, sanitation, irrigation interval, weeding and cultivation) were applied equally according to the recommendations for all treatments.

Data collection and statistical analysis

Disease, yield and yield components data like, white rot and rust severity and incidence, plant height, leaf number, number of bulbs infected and non-infected by white rot, infected/unmarketable yield and health/marketable yield were collected according to their standard procedure. The different traits were recorded at different growth stages of the crop and analyzed using GenStat version16 computer software.

Results and Discussion

White root rot and rust intensity

During the first year, as compared to the control plots, incidence and severity of white root rot was very low on hot water treated plots followed by chemical treated plots. The incidence of white rot was 45% in farmers/control plots and the number of plants per hectare infected by white root rot was 7910. In contrast, incidence and number of plants infected by white root rot was 25% and 2670 for hot water treated plots and 29% and 4240 for apron star treated plots, respectively. These values are very low compared what was found under control plots demonstrating both management practices are promising for the area. This result were in line with the findings of Davis [8] who revealed that dipping garlic clove in hot water will greatly reduce the amount of pathogen and is a good preventative measure although it may not completely eradicate the fungus (Figure 1).

In the case of rust occurrence, it was observed on all plots but the intensity was higher on farmer's practises plots. Incidence of rust

Figure 1: Healthy clove harvested from hot water treated plots.

was low on hot water treated plots followed by chemical treated plots. However, less severity was observed on chemical treated plot (Table 1). The incidence of rust on the chemical treated and hot water treated plots was 92 and 84 percent, respectively with a severity of 25 and 31% on chemical treated and hot water treated plots, respectively. On the other hand, incidence and severity of rust on the control plots were 97 and 51%, respectively (Figure 2).

During the second production season, intensities of white rot and rust were varied from location to location depending on the climatic condition of each area (Table 2). In line with this, Pinto et al. [9] also reported incidence and severity of white rot will depend both on the state of development of the host and on the suitability of soil conditions (mainly temperature) for the development of the pathogen and the host root system and, therefore, the production of exudates. A maximum (100%) possible rust incidence was noted in Ayba and Simret *Kebeles* for all the treatment combinations at both locations. This indicated that rust was important for garlic crop as the treatments used have no effect on the incidence of the disease. Significant difference of disease severity was noted at Ayba but not at Simret Kebele. Hot water treatment has scored the highest severity percentage (58.21%), whereas the lowest was for chemical treatment at Ayba Kebele. In Simret *Kebele*, the highest rust severity percentage was recorded on the farmers plot. Same result was also observed from last year demonstration, in Ayba Kebele where less severity of rust was recorded on chemical treated plot. Even though severity of rust disease was higher in hot-water treated plots in Ayba *Kebele*, the availability of disease has not significantly affected yield of garlic, because occurrence was started at the crop maturity stage. The incidence of white rot was significantly different among treatments at Ayba *Kebele* but not-significant at Atsela and Simret *Kebeles* and it has sustained 100% incidence in Simret *Kebele*. Significantly higher incidence was noted from farmers practice plots than other treatments at Ayba and higher numeric increase at Atsela. The severity of white rot was non-significant at all locations for the treatments. Reduced severity was noted from hot water plus chemical treated plots but not significantly different with hot water treated plots. However, at Simret *Kebele* the lowest severity was recorded from plots planted to chemical treated cloves. Delgadillo et al. [10] also reported that reduction of Allium white rot severity has been achieved by use of fungicides. The two year demonstration result revealed that the two diseases (white rot and rust) were major concern, as it is expected to have a major economic significance in the Tigray highlands, especially with poor crop management practices.

Plant height (cm) and leaf number per plant

The highest and significantly different plant height and leaf number per plant (80.31cm and 10.76) was recorded from hot water treated plots followed by chemical treated plots for the values of 78.38 cm and 10.47 plant height and leaf number per plant, respectively. In contrast the lowest and significantly different plant height and average leaf number per plant (76.68 cm and 9.88) also recorded from the control plots (farmers practise plots), respectively (Table 3).

Garlic bulb yield (qt/ha)

In the first year demonstration, statistically higher health (marketable) yield (126.09 qt/ha) was obtained from hot water treated plots followed by chemical treated plots for the value of 111.761 qt/ha. In contrary, significantly lower health/ marketable yield (96.576 qt/ha) was recorded from the farmers practice/control plots. On the other hand, control plots had significantly higher rate of infested bulb yield by white rot (34.707 qt/ha), while the lowest was recorded from hot water and chemical treated plots 16.587 and 21.505 qt/ha respectively. A total of 30.5% yield advantage was obtained from hot water treated plot than farmers practice plot in health/marketable yield situations.

The 2014 production season result obtained on yield data is presented in Table 4. The present result indicated that the highest health bulb yield (87.08 qt/ha) in Ayba *Kebele* was obtained from chemical plus hot water treated plots followed by hot water treated plots which recorded a mean marketable yield of (77.46 qt/ha). However, in the case of Atsela Kebele numerically the highest health yield (57.93 qt/ha) was recorded from hot water treated plots followed

Treatments	Number of Plants/ha infected by white rot	White root rot Incidence (%)	Rust Severity (%)	Rust Incidence (%)
Apron Star Treatment	4240^b	29^b	25^c	92^a
Hot Water Treatment	2670^b	25^b	31^b	84^a
Farmers Practice	7910^a	45^a	51^a	97^a
LSD (5%) CV (%)	2310.5 49	6.12 45	5.12 15	13.54 15

*Values connected by the same letter across a column are not statistically different at 5% significant level

Table 1: Effect of seed treatment on white rot and rust intensity of garlic in 2013 production season.

Figure 2: Infected clove by white rot from control plots.

Treatments	*Kebeles* (demonstration sites)									
	Ayba				Atsela		Simret			
	Rust		White root rot		White root rot		Rust		White root rot	
	INC (%)	SEV (%)	INC (%)	SEV (%)	INC (%)	SEV (%)	INC (%)	SEV (%)	INC (%)	SEV (%)
Chemical	100	41.96^{b}	3.91^{b}	31.1^{a}	16.28^{a}	10.2^{a}	100	10 ^{a}	100	29.17^{a}
Hot water	100	58.21^{a}	3.72^{b}	28.4^{a}	15.09 ^{a}	9.1^{a}	100	8.3^{a}	100	50.83^{a}
Hot water+chemical	100	46.07^{ab}	3.32^{b}	22.6^{a}	14.61^{a}	9.0^{a}	100	7^{a}	100	42.50^{a}
Control	100	51.71^{ab}	20.59^{a}	33.7^{a}	18.30^{a}	10.3^{a}	100	15^{a}	100	33.33^{a}
LSD	100	13.51	10.29	15.43	7.21	5.139	100	9.67	100	31.38
CV	24.3	24.3	18	33.4	32.6	28.6	28	28	23.3	23.3

Table 2: Effect of seed treatment on white rot and rust intensities of garlic in Atsela, Ayba and Simret *Kebeles* in 2014 production season. *Values connected by the same letter across a column are not statistically different at 5% significant level.

Treatments	Average plant Height (cm)	Average leaf Number per Plant	Marketable (health) yield qt/ha	Unmarketable (infected by white rot) yield qt /ha
Chemical treatment	78.38^{b}	10.47^{b}	111.761^{b}	21.505^{b}
Hot water treatment	80.31^{a}	10.76^{a}	126.090^{a}	16.587^{b}
Farmers practice	76.68^{c}	9.88^{c}	96.576^{c}	34.707^{a}
LSD (5%)	0.86	0.20	8.34	6.63
CV (%)	1.17	2.014	7.96	29.08

Table 3: Effect of seed treatment on yield and yield component of garlic in 2013 production season. *Values connected by the same letter across a column are not statistically different at 5% significant level.

Treatments	Atsela		Ayba		Simret
	Healthy yield qt/ha	Infected yield qt/ha	Healthy yield qt/ha	Infected yield qt/ha	Total yield
Chemical treatment	44.49^{a}	13.208^{ab}	59.71^{c}	23.54^{a}	39.21^{a}
Hot water treatment	57.93^{a}	9.505a^{b}	77.46^{b}	14.50^{b}	35.47^{ab}
Chemical+hot water treatment	57.44^{a}	7.547^{b}	87.08^{a}	5.500^{c}	32.00^{b}
Farmers practice	38.40^{a}	17.933^{a}	38.75^{d}	30.08^{a}	29.99^{b}
LSD (5%)	22.83	9.1372	9.19	8.7328	6.715
CV (%)	34.35	34.35	9.07	9.07	9.8

Table 4: Effect of seed treatment on yield and yield component of garlic in 2014 production season in Atsela, Ayba and Simret Kebeles. *Values connect by the same letter across a column are not statistically different at 5% significant level.

by chemical plus hot water treated plots for the value of 57.44 qt/ha but there was no statistically significant difference among all treatments. Unlike the two Kebeles, incidence of white rot in Simret was 100% in all farmer's plots and treatments. However, in terms of total yield, the highest (39.21 qt/ha) total yield was recorded from chemical treated plots followed by hot water treated plots for the value of (35.47 qt/ha), although no statistical difference was observed. In contrary, bulb yield was significant lower (38.75 and 38.40 Sqt/ha) in the control plots in Ayba and Atsela *Kebeles*, respectively.

On the other hand, significantly higher white rot infected cloves/unmarketable yield of 30.08 and 17.93 qt/ha was found from the control plots in Ayba and Atsela *Kebeles*, respectively. However, in both *Kebeles* the lowest white rot infected clove was reported from chemical+hot water treated plots which showed 5.500 and 7.547 qt/ha unmarketable yield followed by hot water treated plots for the mean yield of 14.50 and 9.505 qt/ha in Ayba and Atsela *Kebeles*, respectively. The lowest white rot infected yield recorded from the treated plots indicated that seed treatment and improved management practices like recommended seed rate, recommended fertilizer rate and method of planting have decreased incidence and severity of white rot significantly than the control plots. Zeray and Mohammed [11] also reported that, garlic white rot is highly associated with crop density and planting time of the crop season.

Conclusion

The production of cash crop like garlic is proved to be income generating activity for farmers, especially for those who have limited cultivated land. Recognizing its importance and total area under garlic production has increased from time to time. In spite of its importance and increased in production, productivity is generally low due to different factors including garlic white rot disease. For minimizing the severity and incidence of this disease on-farm demonstration was conducted for two years under different garlic growing areas of southern Tigray.

The first season demonstration result showed that the highest marketable/healthy yield and low incidence and severity of white rot were recorded from hot water treated plots. In contrary, the lowest marketable yield and the highest infected yield were recorded from control plots. The highest and significantly different plant height and leaf number was also reported from hot water treated plots followed by chemical treated plots. Hot water treatment was also selected and ranked first by participant farmers from the other treatments. Beside its importance on controlling the incidence and severity of the disease white rot and recorded higher marketable yield farmers have preferred the technology by its accessibility, easiness and environmental friendly.

Same result was also observed in the second season, as compare to the other plots, incidence and severity of white rot was very low on hot water plus chemical treated plots but not significantly different with hot water treated plots, respectively. However, in the case of Simret *Kebele*, even though the incidence of white rot was 100% in all demonstration plots, comparatively the highest yield was recorded from chemical treated plots. The present result indicated that, white rot intensity and the efficacy of the treatments were varied from location to location depending on the climatic condition of each *Kebele* and occurrence of the disease. Therefore, site specific recommendation based on the efficiency of the treatments and environmental condition of each *Kebele* is crucial.

Therefore, based on the availability, environmental friendliness and farmers preference of the technology, hot water treatment is feasible for all intervention *Kebeles* and practising of proper agronomic practises (site selection and land preparation, planting at the right time, weeding and hoeing, crop rotation, use of disease free planting material, keeping optimum irrigation interval, harvesting and curing) promote the effectiveness and productivity of the recommended technology.

Acknowledgement

The study was part of CASCAPE project (Capacity Building for Scaling up of Evidence-Based Best Practices in Agricultural Production in Ethiopia). We thank CASCAPE project for covering the entire cost of the research. We also thank all the research assistants of CASCAPE project working in the experimental sites for their participation in data collection and organization the participation of local farmers. We are grateful to all farmers who have participated in this research for contributing their lands and sharing necessary information. Last but not least we thank Alamata Agriculture Research center pathology researchers Mr. Teklay Abebe and Mr. Muruts Legesse who have participated in data collection and technical support.

References

1. Rubatzky VE, Yamaguchi M (1997) World vegetable, In Chapman and hall, (eds.) Principles, production and nutritive values, (2edsn). International Thomson publishing, New York. USA. pp: 843.
2. Central Statistical Authority (2012) Agricultural sample survey 2010/2011. Report on area and production for major crops (Private peasant holding, main season). Statistical Bulletin A.A, Ethiopia.
3. Zeray S (2011) Distribution and Management of Garlic White Rot in Northern Ethiopia: distribution, significance and management of garlic white rot in northern Ethiopia, Ambo University, Ambo Ethiopia.
4. HarnetA, Yibrah G (2015) Evaluating Local GarlicAccessions using Multivariate Analysis Based on agro-morphological Characters in Southern Tigray, Ethiopia. Journal of Natural Sciences Research 5: 211-216.
5. Stewart A, Fullerton RA (1991) Additional studies on the chemical control of onion white rot (*Sclerotium cepivorum* Berk.) in New Zealand, New Zealand Journal of Crop and Horticultural Science, 19: 129-134.
6. Alamata Agriculture Research Center (2012) Horticulture case team annual problem appraisal report .
7. Tamire Z, Chemeda F, Parshotum K, Sakhuja PK, Ahmed S (2007a) Association of white rot (*Sclerotium cepivorum*) of garlic with environmental factors and cultural practices in the North Shewa highlands of Ethiopia. Journal of Crop Protection 26: 1566-1573.
8. Davis RM (2010) UC IPM Pest Management Guidelines: onion and garlic.
9. Pinto CMF, Maffía LA, Berger RD, Mizubutiw ES, Casali WD (1998) Progress of white rot on garlic cultivars planted at different times. Plant Disease 82: 1142-1146.
10. Delgadillo SF, Zavaleta-Mejía E, Osada Kawasoe S, Arévalo Valenzuela A, González-Hernández V, et al. (2002) Densidad de inóculo de Sclerotium cepivorum ysu control mediante Tebuconazole enajo (*Allium sativum L.*). RevistaFitotecnia Mexicana 25: 349-354.
11. Zeray S, Mohammed Y (2012) Survey on Distribution and Significance of Garlic White Rot (*Sclerotium cepivorum Berk*) in East and southeast Tigray Highlands, Northern Ethiopia. Journal of Applied Science 3: 43-56.

Fungal Diseases of Trees in Forest Nurseries of Indore, India

Hemant Pathak*, Saurabh Maru, Satya HN and Silawat SC

Forest Research and Extension Circle, Indore, Madhya Pradesh, India

Abstract

The forest nurseries, maintained by Forest Research and Extension Circle, Indore Department of Madhya Pradesh in Indore and Dewas Dist. have many tree species. During a routine survey of nurseries, 8 tree species were found infected by fungal pathogens. The infected species showed Leaf Spot disease during the winter and rainy season. The survey was conducted at 8 nurseries in the region and the incidence of fungal disease commonly found, were recorded. The fungal pathogens were identified and studied in relation to respective environment.

Keywords : Nursery disease; Fungal pathogens; Tree species

Introduction

The forest serves as a source for timber, fuel, fodder and minor forest produce to human along with conserving soil & water, moderating climate, offering food & shelter for wildlife and adding to the aesthetic value & recreational needs of man. There is a close relationship of plants and the environment. In the natural forest, productivity is generally low due to inherent slow growth of the species and mixed composition where all species may not be valuable. One of the interesting areas of the forest department is to know the causes and mechanism of a disease outbreak. Particularly in plantation, where, due to drastic changes in ecosystems, catastrophic losses may occur in the event of an outbreak of a disease. The forest nurseries have been observed with such conditions and serves as sample to understand the ecological metamorphosis. In the forest nurseries, different type of diseases such as Damping off, Root rot, Stem rot, leaf curl, Wilt, Canker, Rust, Decay etc. can be found. In the present study, we are observed specific and wide spread pathological problems of forest nurseries of Indore Dist. due to fungal pathogens in trees.

Materials and Methods

Sampling sites

The forest nurseries of Indore district were selected to observe fungus infected plant species. The head quarter of Research and Extension Circle of forest is situated at Malwa Demo Nursery in Khandwa Road, Indore. The Malwa demo (near Soyabean Research Center), Devi Ahilya (near Reseidency club), Varahmihir (near Govt. Malhar Asharam H.S. School), Kishanpura (near National Highway 59), Omkar (near Indore- Icchapur Road), Dr. Bhimrao Ambedkar (near Balaji Ka Mandir Badgonda), Paras (near Punjapura Village) and Chandrakesar (near Chandrakesar Dam) nurseries were selected for sample collection and about 20-30 different types of plant species were observed for fungal pathogens. Location of nurseries shown on the map (by Google Maps Search) of Indore and Dewas Dist. as Red marked. The Selected nurseries were observed in month of September 2013 to February 2014 (Figure 1).

Collection and Identification

The infected plant parts of these species were collected in transparent airtight polythene bags from nurseries and identified for causal organism or disease type by observing them in Compound microscope and Dissecting microscope. The fungal species were identified based on their habit, diseases symptoms and spores identification. Plants species as *Madhuca latifolia* (Roxb.) Chev., *Pongamia pinnata* (L.) Pierre, *Tectona grandis* L.f., *Terminalia arjuana* W. & A., *Terminalia elliptica* Willd., *Ficus racemosa* L., *Ficus benghalensis, Azadirachta indica* A. Juss. and *Delbergia latifolia* Roxb. found infected with fungus.

Result and Discussion

In the present study, the nurseries were observed after the rainy season. During this season, the conditions were favorable for fungal growth. After observing different nurseries, following plant species were commonly found infected.

Madhuca latifolia

The *Madhuca latifolia* (Roxb.) Chev. commonly known as Mahua is an economically and medicinally important plant growing throughout the subtropical region of the Indo-Pakistan subcontinent. Large numbers of Mahua trees were found in the Madhya Pradesh [1]. In the current study, parasitic fungi *Scopella echinulata* rust was observed on host *Madhuca latifolia* (Roxb.) Chev. lower leaf surface. Rust has dark brown telutospores on lower surface of leaf. Telutospores were observed are spiny and irregular round in shape [2]. The rust formed dark chocolate color pustules on lower surface of the leaf (Figure 2).

Pongamia pinnata

It is medicinally important plant, particularly in Ayurvedic medicine in bronchitis, whooping cough, rheumatic joints and diabetes. Leaves used as a medicated bath for relieving rheumatic pain and for cleaning ulcer in gonorrhea. The pongam seed oil has a bitter taste and considered as non edible oil. It is used as fuel for cooking and lamps, lubricant, tanning leather making soap and as illuminating oil. The seed oil contains karanjin, an active ingredient with important biological attributes [3]. A parasitic fungus *Fusicladium pongamiae* observed on host *Pongamia pinnata* (L.) Pierre, lower leaf surface. It causes black patches on lower surface of leaves and sometime leaf

***Corresponding author:** Hemant Pathak, Forest Research and Extension Circle, Indore, Madhya Pradesh, India
E-mail: shemantpathak777@gmail.com

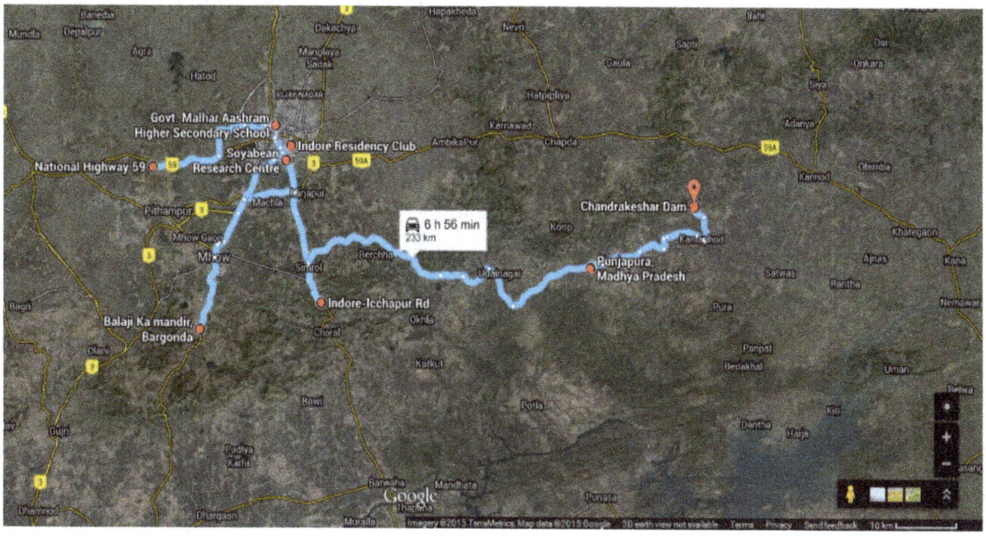

Figure 1: Location of Nurseries.

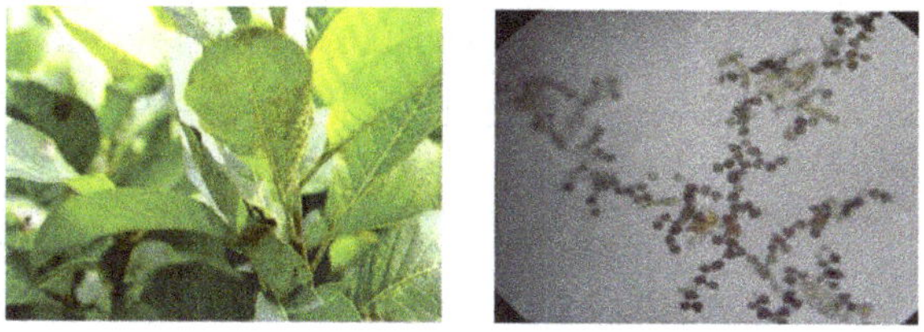

Figure 2: Rust disease caused by *Scopella echinulata* (10 × 10) on *Madhuca latifolia.*

curled (Figure 3). Fungal spores were found on lower leaf surface are long, one and two cell capsule shaped.

Tectona grandis

Tectona grandis L.f. is a large deciduous tree with a rounded crown and, under favorable conditions, a tall, clean cylindrical bole of more than 25 m. Leaves are broad, elliptical or obovate and usually 30 to 60 cm long. Over most of its range, teak occurs in moist and dry deciduous forests. Teak (*Tectona grandis* L.f.) is one of the world's premier hardwood timbers, rightly famous for its mellow color, fine grain and durability [4].

Olivea tectonae, the teak leaf rust, found throughout all the observed nurseries. It is also occurs, the range of distribution of the host in India. The disease commonly found in forest nurseries. Leaves were commonly attacked, usually from October to February. Uredial and telial sori orange, yellow and observed on the lower surface of the leaf [5]. The upper surface of the leaves showed gray appearance due to the formation of flecks (Figure 4) Infection due to teak leaf rust causes premature defoliation in forest nurseries. Early defoliation resulted retardation in plant growth.

Terminalia arjuna

Terminalia arjuna occurs commonly throughout India along rivers, streams, ravines and dry water courses. *Terminalia arjuna* is a deciduous large sized fluted tree to 30 m tall and 2-2.5 m diameter in width. It is one of the most versatile medicinal plants. In forest nurseries height of *Terminalia arjuna* is about 6-7 ft. the wood is used for building, agricultural implements, carts and boats. The bark is used for tanning.

The fungal *Sphaceloma terminaliae* were found parasitic on host *Terminalia arjuna*. This showed rust like symptoms on host plant. It causes small spots on leaves. Fruiting body of rust presented on lower surface of leaves (Figure 5). Rust produce white creamy color pustules on lower surface of leaves. Telutospores of rust were found round shaped, whitish and one celled.

Delbergia latifolia

D. latifolia is native to the Indian Subcontinent and Southern Iran and best known internationally as deciduous premier timber species. Its wood is used in the construction industry and fuel.

Powdery mildews, one of the common plant pathogen forming colonies on the leaves and tender portions of many of the economically important plants. They are distributed in the tropical, subtropical and temperate regions of the world. *Ovulariopsis sissoo* causes powdery mildew on lower leaf surface of *D. latifolia* (Figure 6).The fungus produces persistent, dense mycelium on the lower surface of *D. latifolia* leaves. Spores were found are club shaped one celled and transparent. Infected leaves may become distorted, turn yellow with small to large patches of green, and fall prematurely. The disease on *D. latifolia* also observed [6].

Azadirachta indica

Azadirachta indica (neem) is an herbal plant widely distributed in our subcontinent during all seasons. For thousands of years the beneficial properties of *A.indica* A. Juss have been recognized in the Indian traditional medicine. Each part of the neem tree has some medicinal property. Neem leaves, bark extracts and neem oil are commonly used for therapeutic purpose [7]. During a disease observation in nurseries powdery mildew found on upper leaf surface of *A. indica* plants. *Acrosporium* (*Oidium*) sp. found parasitic on this plant and causes powdery mildew on upper leaf surface (Figure 7). Spores were found are whitish capsule shaped and one celled. These powdery mildews mostly observed in the month of October- February. Fungal species on *A. indica* in khandesh of Maharashtra also observed [8].

Ficus benghalensis

F. benghalensis is widely cultivated in the tropics. It is cultivated

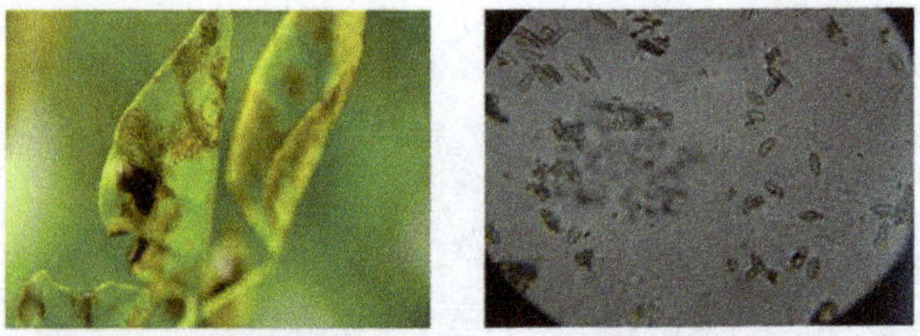

Figure 3: Rust disease caused by *Fusicladium pongamiae* (10 × 10) on *Pongamia pinnata.*

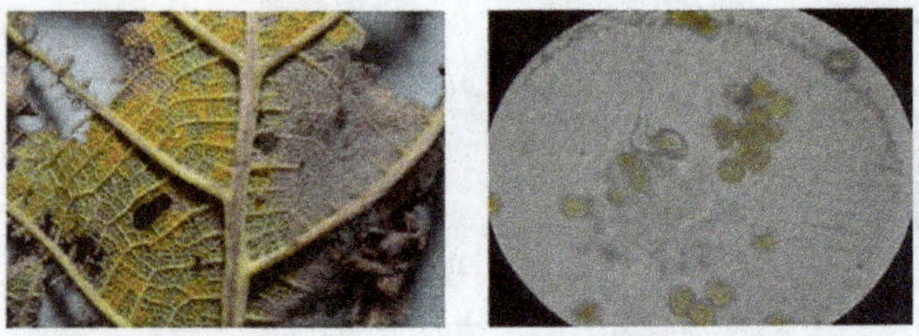

Figure 4: Rust disease caused by *Olivea tectonae* (10 × 10) on *Tectona grandis.*

Figure 5: Rust disease caused by *Sphaceloma terminaliae* (10 × 10) on *Terminalia arjuna.*

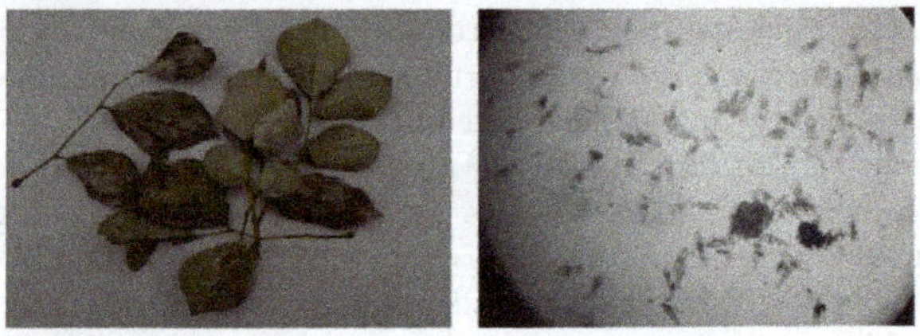

Figure 6: Powdery mildew caused by *Ovulariopsis sissoo* (10 × 10) on *Delbergia latifolia.*

Figure 7: Powdery mildew caused by *Acrosporium (Oidium)* (10 × 10) on *Azadirachta indica.*

in India and has not had its associated wasp introduced and therefore has not yet spread from initial plantings. *F. benghalensis* is the world's largest tree in terms of its spread with some old trees covering over an acre of ground. The roots of *ficus benghalensis* are given in obstinate vomiting and an infusion of its bark is considered as a tonic and astringent and is also used in diarrhea, dysentery and diabetes [9].

The *ficus benghalensis* plant species found infected in nurseries. Fungal species *Septoria arcuata* found parasitic on ficus plants [10]. The leaves of infected plants have dark black spot on lower surface of leaves (Figure 8). Spores were found are yellowish long rod shaped one celled. In growing stage the cells divide and forming 2-3 celled club shaped structure. After some time of infection the infected area of leaves become dead and discarded from leaves.

Ficus racemosa

Ficus racemosa Linn. (Moraceae) is an evergreen, moderate to large sized spreading, lactiferous, deciduous tree, without much prominent aerial roots found throughout the greater part of India in most localities and is often cultivated in villages for its edible fruit. Different parts of *F. racemosa* are traditionally used as fodder, edible and ceremonial. All parts of this plant (leaves, fruits, bark, latex, and sap of the root) are medicinally important in the traditional system of medicine in India [11].

The fungal *Cerotellium fici* were found parasitic on host *Ficus racemosa*. It causes small dark brown spots on the lower surface of leaves (Figure 9). Fruiting body presented on lower surface of leaves. The spores have single celled, rounded and spiny outer layer. The disease also observed in Patna, India [12].

Suggestion to prevent and control of the pathogenic outbreak

Fungicide based approach: Rust or fungal is very hard to treat. Fungicides such as Mancozeb or Triforine may help but may never eradicate the disease. Some organic preventative solutions are available and Sulfur powder is known to stop the growth of rust and any other fungal species. In the studies going on at our center showed significant pesticide activity against rust by a Neem oil based bio-pesticide.

Non fungicide based approach: High standards of hygiene and good soil drainage and careful watering may minimize problems. Any appearance of rust or fungal must be immediately dealt with by removing and burning all affected leaves.

Conclusion

The plant pathogen causes significant losses in forests. Fungal pathogens are one of the most common resistant types of pathogens. Thus, a study on fungal pathogens and finding solutions to eradicate

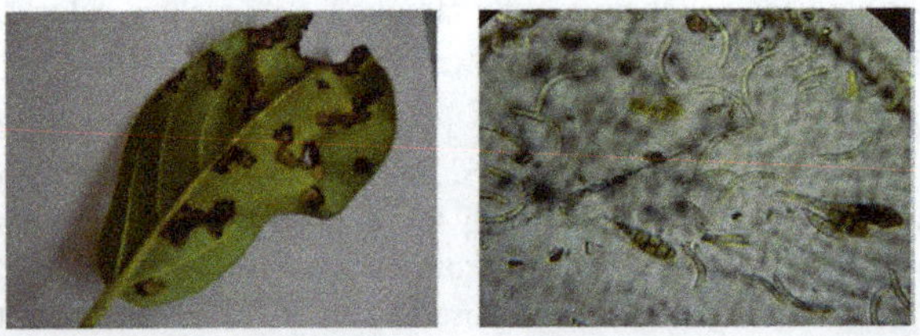

Figure 8: Shot holes caused by *Septoria arcuata* (10 × 10) on *F. benghalensis.*

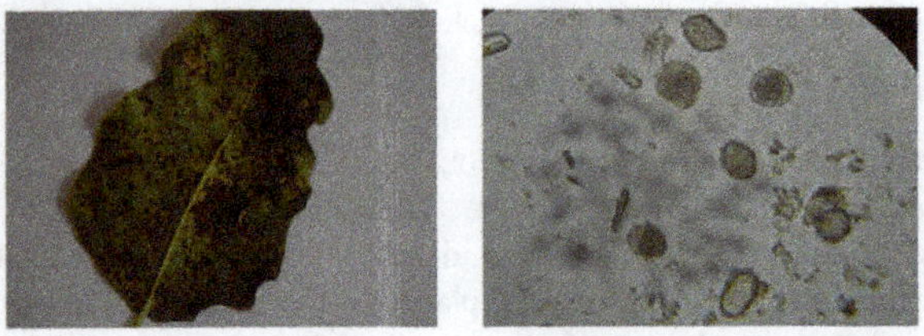

Figure 9: Rust disease caused by *Cerotellium fici* (10 × 10) on *Ficus religiosa.*

as well as prevent their infection is of utmost importance. In the current study, we concluded various types of fungal disease present in forest nurseries of Indore dist. Rust and Powdery Mildew diseases commonly found in all nurseries and some fungal pathogens *Septoria arcuata, Fusicladium pongamiae, Cerotellium fici,* showed leaf spots on infected plant species. The current survey may help forester or nursery coordinators to understand about occurrence timing, specific infecting organism, and host plant of fungal pathogens. The knowledge of the given pathogen can be useful in reducing the fungal diseases and infection in forest nursery.

Acknowledgement

Authors are thankful to Forest Research and Extension Circle, Indore for conducting experiments.

References

1. Patel Pushpendra K, Prajapati Narendra K, Dubey BK (2012) *Madhuca indica*: a review of its medicinal property. International journal of pharmaceutical science and research 3: 1285-1293.
2. Hemant P, Maru S, Satya HN, Silawat SC (2014) Field study of antifungal activity of some plant extracts on host *Madhuca latifolia* in forest nursery of Indore. Journal of Trop Forestry 30: 73-81.
3. Gordhanbhai KH (2010) Studies on effect of medicated oil formulations of Karanj oil (*Derris Indica*), Ph.D. Thesis, C.U. Shah College of pharmacy & research, Saurashtra University, India.
4. Pandey D, Brown C (2000) Teak: a global overview. Unasylva 201 Vol. 51.
5. Bakshi BK (1976) Forest pathology- principle and practice in forestry, controller of publication, Delhi, 252.
6. Thite SV, Aparadh VT, Kore BA (2013) Effect of Powdery Mildew Infection on Mineral Status of Dalbergia sissoo Roxb. ex DC. International Journal of Research in Pharmaceutical and Biomedical Sciences
7. Mohammad Asif (2013) A Review on Spermicidal Activities of *Azadirachta indica*. Journal of Pharmacognosy and Phytochemistry 1: 5.
8. Pawar VP, Vidya A, Patil (2011) Occurrence of powdery mildew on some wild plants from khandesh region of maharashtra state. Recent Research in Science and Technology 3: 94-95.
9. Mandal SG, Shete RV, Kore KJ, Otari KV, Kale BN, et al. (2010) Review: Indian national tree (*Ficus bengalensis*). International Journal Of Pharmacy & Life Sciences 1: 268-273.
10. Butler EJ (1997) Fungi of India, Biotech book Publication Delhi, page-220.
11. Baby J, Justin RS (2010) Phytopharmacological properties of *Ficus racemosa* linn-overview. International journal of pharmaceutical science and research 3: 134-138.
12. Sinha JN, Singh AP (1992) Two new host records from India. Journal of Applied Biology 2: 105.

10

Effect of Age on Susceptibility of Groundnut Plants to *Sclerotium rolfsii* Sacc. Caused Stem Rot Disease

Bekriwala TH[1], Kedar Nath[2]* and Chaudhary DA[1]

[1]*Department of Plant Pathology, N. M. College of Agriculture, Navsari Agricultural University, Navsari, India*
[2]*Regional Rice Research Station, Navsari, Agricultural University, Vyara, India*

Abstract

Stem rot of groundnut caused by *Sclerotium rolfsii* (Sacc.) is a soil borne disease favoured in humid and warmer soil condition at all growth stages. Our objective was to determine how plants ages affect susceptibility of plants exposed to *Sclerotium rolfsii.* Groundnut seeds were grown in pots containing sterilized soil. Groundnut plants were inoculated 0, 15, 30, 45 and 60 days after sowing (DAS) by actively mycelium and sclerotia developed on sorghum grains places near the seeds/plants. Stem rot developed in all inoculated plants but severity decreased with increasing plant age at inoculation. Highest disease severity (79.04%) was recorded in 45 DAS inoculated plants. Whereas plants inoculated 0 DAS may cause pre-emergence rotting and few plants emerged. Plants were inoculated at 15, 30 and 60 DAS developed stem rot symptoms. Our findings suggest that plants are more susceptible to infection at early development stages (0-45 DAS). However, susceptibility to stem infection was reduced after 45 DAS of inoculation. Moreover, young stage of maturity was more susceptible to *S. rolfsii.*

Keywords: Groundnut; *Sclerotium rolfsii;* Stem rot; Disease severity; Plant ages

Introduction

Groundnut (*Arachis hypogea* L.) is commonly called peanut, goober pea goober, pindad jack nut, manila nut, pygmy nut, pignut and monky nut [1]. It is also known as 'king of oil seeds [2]. It has wide range of cultivation tropical and subtropical countries in the world. Groundnut is an important oilseed crop of India, grown extensively in various parts of the country in both Kharif and Rabi/Summer seasons. Groundnut plants suffer from several diseases caused by fungi, viruses, bacteria and nematodes resulting yield losses. *Sclerotium rolfsii* Sacc. is a soil born pathogen have wide host range (>500) including agricultural and horticultural crops [3,4]. Groundnut plants infected by *S. rolfsii* caused stem rot, root rot, sclerotial wilt, [5] and stem and pod rot [6]. Stem rot also known as white mold or southern blight, is a devastating soil borne disease in the India. Stem rot has been observed, where moisture and temperature conditions are sufficiently high to allow the growth and survival of *S. rolfsii*. Groundnut plants were infected by *S. rolfsii* at all growth stage including the germinating stage of the seed causing pre-emergence rot and young plant shown stem rot. The time taken for wilting varied from 8 to 15 days. The younger plants were found more susceptible as the infection was more and rapid [7].

The stem and pod rot caused by *S. rolfsii* Sacc. is major constraints and potential to reduces summer groundnut production in South Gujarat region. The objective of this study was to determine how plants ages affect susceptibility of plants exposed to *Sclerotium rolfsii.*

Materials and Methods

The stem rot pathogen *Sclerotium rolfsii* was isolated from tissue-segmented method from groundnut plants with typical showing stem rot symptoms collected from collected the Regional Rice Research Station N.A.U., Vyara farm and farmers field of Tapi district during 2015-2016.

Infected stem tissues were surface sterilized with 0.1% $HgCl_2$ (1 g/lit) for 1miunte followed by three subsequent washing with sterilized distilled water in aseptic condition. The sterilized pieces were then transferred aseptically under laminar airflow on sterilized Petri plates containing 20 ml potato dextrose Agar (PDA) medium. The Petri plates were incubated in biological oxygen demand (BOD) at 27°C to 2°C temperature for optimum growth. The fungal hyphae developing from the infected tissues were sub-cultured aseptically on PDA media containing in Petri plates. Thus, pure culture was obtained by hyphal tip method and microscopically examined for identification and it was further purified by using single sclerotial body. The culture was maintained on PDA slants for further investigations.

Identification of the pathogen causing stem rot of groundnut was carried out by studying the cultural and morphological characters were recorded right from initiation of mycelial growth till the period of 15 days. The morphological characters *viz.*, mycelia growth and sclerotial formation, its size, shape and colour were studied under low power magnification (10X) from 10 days old culture of *S. rolfsii* and were compared with identification key described in "Illustrated Genera of Imperfect Fungi" [8]. The pathogenicity test of the pathogen was also carried out in pots by stem inoculation technique as described [9].

Preparation of inoculums

The pathogen *Sclerotium rolfsii* was multiplied on sorghum grains (200 g) soaked overnight in water for pot experiment. About 100 g of soaked sorghum grains were taken in 500 ml capacity saline bottles tightly plugged. The bottles were then sterilized for 20 min at 121°C. After sterilization the sorghum seeds in saline bottles were inoculated with 5 mm mycelial disc from 7-day-old pure culture of *S. rolfsii* at each bottle and bottles were incubated for a 15 days at 27°C ± 2°C for proper mycelial growth.

Experiment was conducted at Regional Rice Research Station, N.A.U., Vyara during the year 2015-2016 under pot condition. Five

***Corresponding author:** Kedar Nath, Regional Rice Research Station, Navsari Agricultural University, Vyara-394650, India
E-mail: drkdkushwaha@nau.in

stages i.e. 0, 15, 30, 45and 60 DAS of the groundnut plants were taken for their susceptible reaction against stem rot causal pathogen *S. rolfsii*. These stages of plants were maintained in the eighteen plastic pots of 15 × 30 cm diameter replicated in three times and filled with sterilized soil. In each pot 10 seeds of groundnut (cv. GJG-9) was shown and fertilizer dose applied as per recommended. After raising all the respective stages, the sorghum grain inoculums were added at near the stem up to 4-5 grain on each plants of groundnut. Inoculated pots were kept in open place for observation and the pots were irrigated as when required. Stem rot disease severity was made at 15, 30, 45, 60 and 75 days after inoculation at respective stages, number of plants showed typical symptoms i.e. stem rot, lesion of stem, weathering of leaf and dead plants due to *S. rolfsii* was observed and per cent disease incidence was calculated using formula [10] (Table 1).

$$\text{Disease incidence (\%)} = \frac{\text{No of infected plant}}{\text{Total no. of observed plants}} \times 100$$

Disease rating	Description
1	Healthy
2	Lesions on stem only
3	Up to 25% of the plant symptomic (wilt, dead or dying)
4	26% to 50% of the plant symptomic
5	>50% of the plant symptomic
Disease severity (Ds) was calculated as [12]	

Table 1: Symptoms on groundnut plants were observed as per 1-5 rating scale [11].

Sr. No	Treatments (Days)	Disease incidence	PDI*
1	0	100	25.71 (5.10)[a]
2	15	100	69.36 (8.35)
3	30	100	74.45 (8.64)
4	45	100	79.04 (8.89)
5	60	100	49.68 (7.06)
6	Control	0.00	0.00 (0.70)
S. Em. ±			0.29
C.D. at 5%			0.96
C.V. %			8.35

*=Average of three replications. a=Figures in parentheses are the corresponding square root transformed values + 0.5 added

Table 2: Effect of age of groundnut plants on stem rot disease development in pot conditions.

$$\text{Disease severity} = \frac{\sum ab}{AK} \times 100$$

Where, a=No. of disease plants having the same degree of infection, b=Degree of infection, A=Total no. of examine plant, K=Highest degree of infection

Result and Discussion

Stem rot fungal pathogen showed white fluffy mycelium growth appearance on PDA medium. Microscopic view of mycelium was hyaline, branched, compact with septet and clamp connection. Initially sclerotia formation was observed 4 days after incubation and continued till 7 day old, numerous round to oval, globose or irregular mustard seed like sclerotia were produced. Initially, white colored sclerotia were formed then their color changed from white to off-white, light brown and dark brown as they attained maturity within 10-12 days. However, dark brown and black coloured sclerotia survived for longer times. The change color of sclerotia might also be due to utilization/exhaustion of nutrients. Also, found that sclerotia of some pathogen showed shiny appearance due to presence of gummy material on surface. All the above morphological characteristics of fungus was identified as *Sclerotium rolfsii* Sacc. and further confirmed with identification key described in "Illustrated Genera of Imperfect Fungi" [8,11-17]. Proved pathogenicity on 15 days olds groundnut plants (cv. GJG-9) under pot conditions. 4 days after of inoculation, the first symptoms were observed as water soaked brown to dark brown spots at basal portion of plants. The leaves of infected plants gradually yellowing and dry up. The professed white cottony growth of the fungus was also observed near collar region of the plant. Mycelium on stem/soil produced naked mustard seed like white sclerotia, later become dark brown. The collar region was weakened by the pathogen which resulted in to withering and death of the plant. Re-isolation of pathogen was done from inoculate infected plants and proved pathogenic nature of fungus. Un-inoculated seedlings did not develop any symptoms

To find out the susceptible stage of the groundnut to stem rot disease development, an experiment was conducted in pot conditions (Figure 1). The results are presented in the Table 2 depicted in Plate-I. The results revealed that there was no difference in disease severity percentage among the different stage of plant. Fourty five day old plant had maximum 79.04% disease severity was recorded followed by

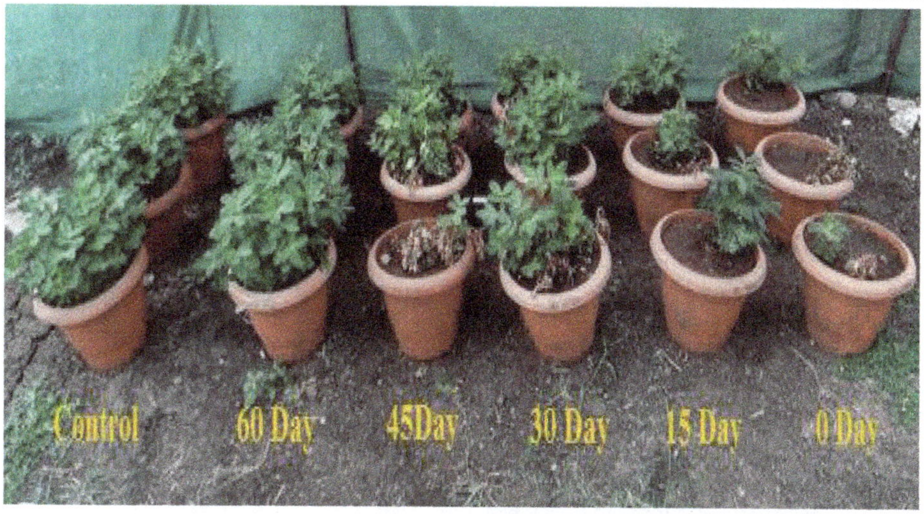

Figure 1: Effect of age of groundnut plants on stem rot disease development in pot conditions.

30 and 15 days old plants with 74.45% and 69.36% disease severity, respectively. Least disease severity was recorded in 0 days old plants with 25.71% whereas; 60 days old plant had 49.68% disease severity. However, few plants were emerged from inoculums incorporated with seeds at the time of sowing. It may due to low germination or plant emergence may due to production of organic acid by *S. rolfsii*, which are toxic to living cell. Therefore, this result was used to identified most susceptible stages to evaluation of genotypes under artificial conditions.

The present findings revealed that groundnut plants were infected by *S. rolfsii* at all stage of plant growth from seed germinating to harvesting. Germinating stage of the seed causing pre-emergence rot and the susceptibility of groundnut plants against *S. rolfsii* was decreased with the increase in the age of groundnut plants. Our finding described the young stage of maturity was more susceptible against *S. rolfsii*, groundnut plants were infected by *S. rolfsii* at all growth stage of the plant including the germinating stage of the seed causing pre-emergence rot [7]. The younger plants were found more susceptible as the infection was more and rapid. The time taken for wilting varied from 8 to 15 days. Disease severity was decreased as the age of plant increased. Maximum plant mortality due to *S. rolfsii* was recorded in 15 days old groundnut seedling followed by 30 days old plants [18]. They also found that least mortality was recorded in 105 days old plants and susceptibility of groundnut seedling to *S. rolfsii* decrease with the increase in the age of groundnut plants. 10 days old plants were more susceptible to collar rot infection (80.00%) followed by 15 days old groundnut plants (75.00%). Plant mortality was increased with the increase in age of plant from 5 days (30.00%) to 10 days (80.00%), but it decreased thereafter i.e. at 15 days (75.00%), 20 days (65.00%) and 25 days (37.50%) plants were mortile [19]. Moreover, susceptibility or resistance of plants to stem rot disease is often influenced by their age. The per cent plant killing increased with increased in age up to (5 to 10 days) but it was decreased beyond 15 days, also in chick pea plant [20] and peppermint [21]. Hence, further studies are in progress to manage this disease at early stage of the crop growth.

Conclusion

Initially *S. rolfsii* appeared as white fluffy mycelium growth on PDA as well as around the basal portion of stem than it produced light brown and dark brown round to oval, globose or irregular mustard seed like sclerotia were produced. Groundnut plants were infected by *S. rolfsii* at all growth stages of plant from seed germinating to maturity. But the younger plants were found more susceptible to infection by *S. rolfsii* caused highest plant mortality results to reduced pod yield. Maximum (79.04%) disease severity was recorded in 45 days old groundnut plants whereas, 30 and 15 days old plants had 74.45% and 69.36% disease severity, respectively. Few plants emerged from inoculums was incorporated with seeds at the time of sowing. Low germination or plant emergence may due to production of organic acid by *S. rolfsii* which are toxic to living cell. Germinating stage of the seed causing pre-emergence rot and the susceptibility of groundnut plants against *S. rolfsii* was decreased with the increase in the age of groundnut plants.

Acknowledgements

We are grateful to Dr. Vipul P. Patel, Associate Research Scientist, Regional Rice Research Station, N.A.U., Vyara, for providing necessary facilities required during this research work. We also thank to Dr. V.A. Solanki, Professor and Head, Dept. of Plant Pathology, N.M. College of Agriculture, Navsari, for providing Technical guidance during my M.Sc. research program.

References

1. Rathnakumar AL, Singh R, Parmar DL, Misra JB (2013) Groundnut: A crop profile and compendium of varieties notified in India. Art India Press, Gujarat, India.
2. Aycock R (1966) Stem and other diseases caused by *Sclerotium rolfsii*. NC Agric Exp sta Tech Bull pp: 174-202.
3. Priya RS, Chinnusamy C, Manicaksundaram P, Babu C (2013) A review on weed management in groundnut (*Arachis hypogaea* L.). Int J Agri Sci Res 3: 163-172.
4. Domsch KH, Gams W, Anderson TH (1980) Compendium of soil fungi. Academic Press, London, New York.
5. Chohan JS (1974) Recent advances in diseases of groundnut in India. In Current trends in plant pathology, 171-184. In: Raychaudhuri SP, Verma JP (Eds.) Lucknow University Press, Lucknow, Uttar Pradesh.
6. Mehan VK, Mayee CD, Brenneman TB, McDonald D (1995) Stem and pod rots of groundnut. Information Bulletin no 44: 28.
7. Patil MB, Rane MS (1983) Studies on host range effect of plant age on susceptibility and varietal reaction of groundnut to *Sclerotium rolfsii*. Indian J Mycol Plant Pathol 13: 183-186.
8. Barnett HL, Hunter BB (1998) Illustrated genera of imperfect fungi. (4thedn.) Published by Am Phytolo Soci p: 196.
9. Patil MB, Patil GD, Wani PV (1977) Varietal reaction of groundnut against *Sclerotium rolfsii*. Indian phytopath 30: 562.
10. Kokalis-Burelle N, Porter DM, Rodriguez-Kabana R, Smith, DH, Subrahmanyamm P (1997) Compendium of peanut diseases (2ndedn). APS Press, St. Paul.
11. Shokes FM, Rhogalski K, Gorbet DW, Brenneman TB, Berger DA (1996) Techniques for inoculation of peanut with *Sclerotium rolfsii* in the greenhouse and field. Peanut Sci 23: 124-128.
12. Filion M, St-Arnau M, Jabaji-Hare SH (2003) Quantification of *Fusarium solani* f. sp. *Phaseoli* in mycorrhizal bean plants and surrounding mycorhizosphere soil using real time polymerase chain reaction and direct isolations on selective media. Phytopathology 93: 229-235.
13. Anahosur KH (2001) Integrated management of potato *Sclerotium* wilt caused by *Sclerotium rolfsii*. Indian Phytopath 54: 158-166.
14. Kokub D, Azam F, Hassan A, Ansar M, Asad MJ, et al. (2007) Comparative growth, morphology and molecular characterization of indigenous *Sclerotium rolfsii* strains isolated from different locations of Pakistan. Pak J Bot 39: 1848-1866.
15. Chaurasia S, Chaurasia A, Chaurasia S, Chaurasia S (2014) Pathological studies of *Sclerotium rolfsii* causing food-rot disease of Brinjal (*Solanum melongena* Linn.). Int J Phar Life Sci 5: 3257-3264.
16. Kumar MR, Santhoshi MVM, Giridhra TK, Reddy KR (2014) Cultural and morphological variability *Sclerotium rolfsii* isolates infecting groundnut and its reaction to some fungicidal. Int J Curr Microbiol App Sci 3: 553-561
17. Rakholiya KB, Jadeja KB (2011) Morphological diversity of *Sclerotium rolfsii* caused stem and pod rot of groundnut. J Mycol Pl Pathol 41: 500-504.
18. Kulkarni SA, Kulkarni S, Anahosur KH (1994) Effect of age of groundnut plant to infection of *Sclerotium rolfsii* Sacc. a causal agent of stem rot disease. Karnataka J Agrl Sci **7**: 367-268.
19. Nathawat BDS, Patel DS, Singh RP, Partap M (2014) Effect of different plant age on incidence and varietal screening against of collar rot in groundnut. Trends Biosci 7: 580-581.
20. Hussain A, Iqbal SM, Ayub N, Zahid MA (2006) Factors affecting development of collar rot disease in chickpea. Pak J Bot 38: 211-216.
21. Muthukumar A, Venkatesh A (2013) Occurrence, virulence, inoculum density and plant age of *Sclerotium rolfsii* Sacc. causing collar rot of peppermint. J Path Microb 4: 2-4.

11

Assessment of the Inhibitory Activity of Resin from *Juniperus procera* against the Mycilium of *Pyrofomes demidoffi*

Dagnew Bitew*

Department of Microbial, Cellular and Molecular Biology, College of Natural Science, Addis Ababa University, Ethiopia

Abstract

Juniperus procera is an evergreen dioecious more seldom monoecius tree, which belongs to the family Cupressaceae and the only Juniper species, which is found in the mountains of East Africa and it is an important indigenous forest tree species in Ethiopia. However *J. procera* is subjected to a serious attack by the slow growing white heart-rot fungus, *Pyrofomes demidoffii*. Resin has been reported to be active against wood colonizing fungi, however, the role of resin of *J. procera* in protecting the tree from the attack of *P. demidoffii* is not known. Therefore this study has initiated to assess the inhibitory effect of resin from J.*procera* against a white rot fungi *P. demidoffii*. Resin and basidiocarps of *P. demidoffii* was collected from infected *J. procera* trees at Menagesha Suba Forest situated 30 km South West of Addis Ababa, Ethiopia. The antifungal activity of *J. procera* resin against *P. demidoffii* was tested using agar dilution assay technique and an impressive result was observed. The MIC value of resin extract was within the range of 5 to 6 mg/100 ml MEAP. Phytochemical screening test for resin extract was done and it revealed the presence of alkaloids, saponins, terpenoids, phenolic compounds, flavenoides and fixed oils and the absence of carbohydrate, glycoside, steroids and fats. The crude extract of *J. procera* resin was fractionated into six fractions with column chromatography using different organic solvents and all fractions run on Thin Layer Chromatography (TLC) plate using benzene: methanol (18:6) and benzene: ethanol: ammonia (18:2:1). All fractions gave different retention factor (Rf) in each developing solvents.

Keywords: Antifungal; Conifers; Fungicide; Heart rot; Phytochemical

Introduction

Juniperus procera is an evergreen dioecious, more seldom monoecius tree, which belongs to the family Cupressaceae and the only Juniper species, which is found in the mountains of East Africa and it is one of the two indigenous conifer species which found in Ethiopia [1].

It is naturally found in the central high lands of Ethiopia. Within the dry afromontane forests of the Ethiopian highlands, it dominates mainly between altitudes from 1800-3200 m above sea level, with the mean annual rainfall ranges from 500 to 1100 mm millimeters [2].

Natural forest is both an ecosystem and a resource: as an ecosystem it integrates diverse fauna, flora and the physical environment; as a resource it has various economic, ecological and social values [3]. *J. procera* forests of the afromontane areas of Ethiopia have considerable economic value at a local and national level [4]. It is a very important source of wood for timber and fuel. Its wood is hard and resistant to termites and fungal diseases. Because of these distinctive qualities, it is highly valued for the construction of houses, internal structures of churches, furniture and for poles. The tree is planted as ornamental and for its shade in homesteads [5]. Its fruits have also some medicinal values for curing headaches, skin diseases etc. Generally *J. procera* is a multipurpose tree, which results from its drought tolerant and easily adaptability [5].

Forest decline affects the ecosystem as well as the usefulness of forest as a resource [6]. Reports of forest decline and tree mortality have increased in recent years and are presently considered to be a major threat in temperate ecosystems [7].

In Ethiopia forest decline is identified as an important problem [8]. *J. procera* forest once covered a large part of the country. However, as a consequence of long-lasting and persistent human influence, it has been considerably depleted and reduced to some isolated patches [9].

Heart rot caused by the wood rot fungus known as *Pyrofomes demidoffii* also contribute a lot for the decline of *J. procera* forest [10]. Now it is included on the IUCN red list of endangered species [11].

Forest decline due to tree diseases is characterized by the interactions of predisposing abiotic factors and biotic agents that come together in an orderly fashion resulting in tree death [12]. Abiotic factors include natural branch thinning due to shading, pruning wounds, vandalism, frost or drought cracks, fire, lightning, insects that bore into the trunk or branches; knives and bicycles, automobiles and damage from machinery or construction which creates suitable condition for biotic factors [13]. Forest decline diseases play a direct role in the destruction of natural resources. Of which, fungal pathogens cause the most important losses [14]. As an example *J. procera* forest is subjected to a serious attack by the white heart-rot fungus, *P. demidoffii* [10].

P. demidoffii is the only serious pathogen of *Junipers sp*. It attacks the heart wood of the tree and creates cavities of various sizes and in case of serious infestation a large tree may be reduced to a mere shell [15]. It produce large, perennial, ungulate, adnate and broadly attached, brightly colored a woody upper surface, hoof-shaped fruiting body (conk) that is dark brown to nearly black in color. The lower (pore) surface is bright yellowish to orange-brown or pale brick-red with pores angular to round, 2-3 per mm, occasionally larger in old specimens [10].

P. demidoffii is spread by airborne spores that infect host trees

***Corresponding author:** Dagnew Bitew, Department of Microbial, Cellular and Molecular Biology, College of Natural Science, Addis Ababa University, Ethiopia E-mail: btdagne@gmail.com

through injuries and wounds. Generally heart rot fungi do not penetrate sound trees but require an opening into the heartwood or exposed dead sapwood adjacent to heartwood is a potential site for fungi to become established [15,16].

Plants use several strategies to defend against damage caused by herbivores and pathogens. The defense system of conifers to biotic agents such as pathogens and herbivores consists of: (i) a constitutive; (ii) a preformed resin response; and (iii) an induced resin response that develops simultaneously and complements each other. The constitutive and preformed mechanisms exist in the absence of a pathogen and include tough outer bark, several classes of secondary metabolites and an elaborate network of resin ducts. However, the induced response and mechanisms are activated only when a pathogen or herbivore attacks the tree and consists of several classes of secondary metabolites and uses the network of pre-formed resin ducts [12].

Higher plants produce hundreds and thousands of different secondary metabolites, which are active against plant pathogens [17]. Many secondary metabolites of wood are toxic or inhibitory to pathogenic fungi [18]. Often, plant secondary metabolites may be referred to as plant natural products, in which case they illicit effects on other organisms [19].

Resin is a secondary metabolite secreted by many plants, particularly coniferous trees and it is one of the main lines of chemical and physical defense system in plant against herbivores and pathogens by physically cleanse wounds, expelling invaders and sealing off damaged tissue [20-22].

The term "resin" also encompasses synthetic substances of similar mechanical properties thick liquids that harden into transparent solids. Other liquid compounds found in plants or exuded by plants, such as sap, latex, or mucilage, are sometimes confused with resin, but are not chemically the same. Saps, in particular, serve as nutritive function that resins do not. However, resins consist primarily of secondary metabolites or compounds that apparently play no role in the primary physiology of a plant [23].

In nature, many secondary metabolites play an important role in the protection of plants as antibacterial, antiviral, antifungal, insecticides and also against herbivores by reducing their appetite for such plants [24]. These active ingredients can be synthesized, making new chemical pesticides, or used in the form of extracts. It is expected that these measures will provide new tools for controlling this disease in the context of integrated pest management strategies [25].

Integrated Pest Management (IPM) means the careful consideration of all available pest control techniques and subsequent integration of appropriate measures that discourage the development of pest populations and keep pesticides and other interventions to levels that are economically justified and reduce or minimize risks to human health and the environment. IPM emphasizes the growth of a healthy crop with the least possible disruption to agro-ecosystems and encourages natural pest control mechanisms [26].

Despite the enormous socioeconomic importance of *J. procera*, it is declining at an alarming rate [4]. Therefore, the purpose of this study is to assess the inhibitory activity *J. procera* resin against its wood colonizing fungus, *P. demidoffii*, which will have a crucial role to prevent the decline *J. procera* by the infestation of *P. demidoffii*.

The hypothesis of this research is, based on critical observation of the presence of fruiting body of *P. demidoffii* in old *J. procera* trees, which have no or little amount of resin unlike that of young trees, which is not infested and have more resin. However, the role of the resin in protecting the tree from the pathogen is not known.

Materials and Methods

Collection of resin from *J. procera* tree

This study was conducted in Mycology Laboratory, Department of Microbial, Cellular and Molecular Biology, College of Natural Science, Addis Ababa University. Resin from *J. procera* was collected from Menagesha Suba Forest situated 30 km South West of Addis Ababa, Ethiopia. The resin was collected from living trees with and without decline symptom using autoclaved Stikine, ependroff and bottle, and dried at room temperature.

Collection and cultivation of *P. demidoffii*

Basidiocarps of *P. demidoffii* were collected from *J. procera* from the same forest using axe, basket and paper bag, so as to characterize and cultivate the fungus in the laboratory. Pieces of basidiocarp were transferred aseptically onto 20 ml of malt extract agar (MEA) medium (OXOID). Inoculated plates were incubated at 25°C until young hyphae emerge, from which pieces of agar culture blocks were transferred to the same medium to obtain pure culture [27]. Although, its host was a diagnostic feature, the fungus was characterized morphologically by observing its hyphae and spore under light microscope, oil immersion magnification.

Evaluation of culture media

P. demidoffii is a slow growing *J. procera* inhabiting fungus; and there is no formal data which shows in which media it grows faster. In order to identify good growing medium for *P. demidoffii* 4 mm inoculum plug of it were taken using sterile cork borer and placed on ten different media Corn Meal Agar (CMA) medium (OXOID), Corn Meal Dextrose Agar (CMDA) (OXOID), Malt Extract Agar (MEA), Czapek Dox Agar (CDA) medium (OXOID), Potato Dextrose Agar (PDA) medium (OXOID), Oat Meal Agar (OMA) (Rolled oats 30.0 g and agar 20.0 g 1,000.0 ml distilled water), Lenionian agar medium (LAM) (Peptone 0.625 g, Maltose 6.25 g, Malt extract 6.25 g, KH_2PO_4 1.25 g, $MgSO_4$+$7H_2O$ 0.625 g and Agar 20.00 g per 1,000.0 ml distilled water), Malt Extract Agar with Peptone (MEAP) (Malt extract 20.0 g, Peptone 1.0 g, Glucose 20.0 g, Agar 20.0 g per 1,000.0 ml distilled water), Malt Extract Agar with Juniper Extract (MEAJE) (Malt extract 20.0 g, juniper extract 10 g, Glucose 20.0 g, Agar 20.0 g per 1,000.0 ml distilled water) and Juniper Extract (JE) (Juniper extract 20.0 g, Agar 20.0 g per 1,000.0 ml distilled water) and the colony diameters were traced in four days interval in each medium.

Solvent selection for the extraction of resin

The selection of the solvent was based on its ability to dissolve the resin completely. In addition the cost and availability of the solvent were taken into account. Different six solvents (cold water, hot water, methanol, ethanol, ethyl acetate and chloroform) were used to dissolve the resin, that gives clue, as which solvent is better to extract the resin material. The resin was suspended in solvents listed above separately and shaken for about 6 hours on the shaker.

Preparation of resin extracts

The resin material collected from trees was air dried under shade, in Mycology Laboratory, Department of Microbial, Cellular and Molecular Biology, College of Natural Science, Addis Ababa University. Then after drying the resin powder was dissolved by four different organic solvents, which dissolve the resin in different degrees during solvent selection for extraction of resin (chloroform, Ethanol, Ethyl acetate and methanol) after 6 hours of shaking on the shaker and dried using Rota Vapor to test its antifungal activity.

Antifungal assays

Effect of crude resin extract on fungal mycelium growth: The inhibitory effect of resin against the mycelial growth of test fungal species was undertaken using agar dilution method [28]. Different concentrations of resin sample were weighed (1.5 mg, 2.5 mg, 3 mg, 4 mg, 5 mg and 6 mg) and dissolved in different solvents to test against *P. demidoffii*. Solvents were removed using Rota Vapor and each concentration of resin extract was added into a separate 250 ml flasks which contains 100 ml of MEAP, on which *P. demidoffii* was grown better during attempting media formulation, and poured in 90 mm Petri plate. Pure media was used as a control. Then inoculum plugs (4 mm diameter) of test fungi were cut using a cork borer from the actively growing margin of the source of fungus and transferred to the centre of each study plate within 1 h after the resin had been poured and incubated at 25°C. At the end of the incubation period antifungal activity of the resin was evaluated by measuring the diameter growth of the fungi (including diameter of the disk) using ruler. Growths diameters were traced at four days interval about five times after inoculation [12]. Data were expressed as the means of diameter growth, growth rate and percentage of inhibition, of the growth of the fungi in the presence of any of the concentrations of resin. Percentage of mycelial inhibition was calculated in relation to the control using the following formula: % of inhibition=[1 - (fungal growth in experimental test/Control growth)] x 100 [29].

And the growth rate was calculated using the formula: GR=G2 - G1/T2 - T1. Where: G2=growth attained at the final time, G1=Growth attained at the initial time, T2=Final time and T1=initial time [28].

Determination of minimum inhibitory concentration (MIC) of crude resin and fungicides

Based on the preliminary antifungal screening, ethyl acetate and methanol extract of resin revealed potent antifungal activity. Thus the MIC value of ethyl acetate and methanol extracts of resin and those commercial fungicides were undertaken using Agar dilution method [12]. Different concentrations of resin were weighted (4 mg, 5 mg, 6 mg, 7 mg and 8 mg) dissolved in ethyl acetate and methanol and dried using Rota Vapor. And different concentrations of fungicides (0.1 mg, 0.25 mg and 0.5 mg, 0.75 mg and 1 mg similar for each fungicide) were also weighted. Then each concentration of resin and fungicides were separately added into separate 250 ml flasks containing 100 ml of MEAP after autoclaved and cooled and poured in 90 mm Petri plate. *P. demidoffii* was grown on MEAP medium for 10-14 days at 25°C.

Then inoculum plug (4 mm diameter) of *P. demidoffii* was cut using a cork borer from the actively growing margin of the source of fungus and transferred to the centre of each study plate within 1 h after the resin had been poured and incubated for 20 days at 25°C. At the end of the incubation period, the plates were evaluated for the presence or absence of fungal growth. The MIC was the minimum concentration of resin extract and fungicides that inhibited the growth of the test organism completely.

The MFC (Minimum Fungicidal Concentration) was determined by subculturing the inoculum plug from the plate which did not show any growth during the MIC test by placing on new MEAP plates to determine whether the resin and fungicides were fungistatic or fungicidal against the pathogen. The MFC was demonstrated when no growth of the pathogen occurs on the subcultured medium [30,31].

Identification of different ingredients in resin extract

The identification of different ingredients from resin of *J. procera* involved both simple, preliminary phytochemical tests, and complex methods, thin layer and column chromatography. Silica gel (60-200 mesh) and silica gel (Kieselgel GF254 plates) (Merck) were used for column and thin layer chromatography respectively [31].

Preliminary phytochemical screen: Qualitative phytochemical tests of *J. procera* resin extract were carried out to identify the class of compounds by colour tests. A 10 g resin material was dissolved in methanol and dried using rotary evaporator. The resin extract was subjected to various standard phytochemical tests as follows.

1 *Test for alkaloids:* Hager's test was used to test the presence of alkaloids. To 1 ml of extract, 3 ml of Hager"s reagent (saturated aqueous solution of picric acid) was added, and yellow colored precipitate indicated the presence of alkaloids [32].

2. *Test for carbohydrates:* The presence of carbohydrate and reduced sugar was tested using Molisch"s test [33]. Two (2) ml of the crude extract resin was transferred into 15 ml test tube to which 1 ml of α-napthol solution was added and mixed. Then 1 ml of concentrated sulphuric acid was added through the sides of the test tube. Purple or reddish violet color at the junction of the two liquids reveals the presence of carbohydrates.

3. *Test for glycosides:* Baljet"s test was used for glycosides; to 1 ml of the test extract, 1 ml of sodium picrate solution was added and the yellow to orange color reveals the presence of glycoside [34].

4. *Test for steroids:* Salkowski test was followed to detect the presence of steroids. One (1) mg crude extract resin was dissolved in chloroform (3 ml) and equal volume of concentrated sulphuric acid was added. Formation of bluish red to cherry color in chloroform layer and green fluorescence in the acid layer represents the steroidal components [35].

5. *Test for flavonoides:* Alkaline reagent test was followed to test the presence of flavonoides. A few drops of sodium hydroxide solution was added into 2 ml test extract; formation of an intense yellow color, which turned to colorless on addition of few drops of dilute hydrochloric acid indicated the presence of flavonoides [34].

6. *Test for saponins:* To 1 ml of extract taken in a measuring jar, 9 ml of distilled water was added and shaken vigorously for 15 secound and extracts were allowed to stand for 10 min. Formation of stable foam indicates the presence of saponins [35].

7. *Test for fixed oils and fat:* Spot and saponification tests were used for fixed oils and fats [32].

Spot test: a small quantity of resin extracts was pressed between the filter paper. Oil stains on paper indicates the presence of fixed oils.

Saponification test: To 1 ml of the extract few drops of 0.5 N alcoholic potassium hydroxide were added along with a drop of phenolphthalein. The mixture was heated on a water bath for one and half hours. The formation of soap or partial neutralization of alkali indicated the presence of fixed oils and fats.

8. *Test for tannins and phenolic compounds:* Three phytochemical screening methods were used for tannins and phenolic compounds. To 2 ml of each extract, a few drops of 10% lead acetate were added.

The appearance of white precipitate indicates the presence of tannins [34]. To 1 ml of the extract, 2 ml ferric chloride solution was added; formation of a dark blue or greenish black color product shows the presence of tannins [32].

Alkaline reagent test: To 1 ml of extract 1 ml of sodium hydroxide was added gives yellow to red precipitate within short time showed the presence of tannins and phenolic compounds [32].

9. *Test for terpenoids (Salkowski test):* Five ml of extract, 2 ml of chloroform was mixed and 3 ml of concentrated H_2SO_4 was carefully added to form a layer. A reddish brown coloration of the inter face show positive results for the presence of terpenoids [5].

Column chromatography: A 36 cm long glass column was used and it was filled up to 25 cm with silica gel. Twenty (20) gram of resin was dissolved in methanol, which showed higher inhibitory activity during attempting inhibitory activity of crude extract of the resin at a concentration of 5 mg/100 ml MEAP and dried using rotary evaporator. 60-200 mesh size silica gel suspended in methanol was used as a stationary phase. Then after the dried crude resin extract was fractionated into six fractions using different organic solvents were used as eluting solvent gradually with increasing order of polarity of solvents from non- polar to polar solvents (chloroform, ethyl acetate, ethanol and methanol) which is tabulated (Table 1) [22]. And all fractions were dried using Rota Vapor. Then 3 mg /100 ml MEAP of each fraction was tested for its antifungal activity against *P. demidoffii* using the same procedure and methods that were used to test the antifungal activity of crude extract as mentioned in section 4.6.1.

Thin layer chromatography (TLC): All fractionated samples by column chromatography were dried using Rota Vapor. Resin extracts dissolved in methanol at a concentration of 2 mg/ml and were applied on TLC plates. The chromatograms were run using benzene and methanol (18:6) and benzene: ethanol: ammonia (18:2:1) as running solvent. At the end of the run, spots and bands on TLC plates were visualized under UV lights. Retention factor (Rf), which is a relative scale of the distance of the compound moved, calculated as Distance traveled by solute/distance traveled by solvent [31].

Data analysis

All the measurements were done in triplicate and the results were presented as mean ± SD. The statistical analysis was performed by one-way analysis of variance (ANOVA) followed by Post Hoc Multiple Comparison Tests using statistical software (SPSS) package version 17.0 for windows and P values < 0.05 were considered as significant.

Result and Discussion

Media formulation

P. demidoffii was cultivated on different medium and the growth diameter was evaluated in four days interval in each media (Table 2). As indicated in Table 2, based on the growth diameter of *P. demidoffii* MEAP was a good medium for the cultivation of this fungus with growth diameter of 58.00 ± 0.2 mm followed by LAM, with growth diameter of 57.00 ± 0.17 mm after 24 days of incubation at 25°C. CDA was the least with growth diameter 17.50 ± 0.02 after the same period of incubation. The growth of the fungus was thick in MEAP while its growth in LAM was too thin. Concerning the growth medium, there is no formal data on optimization of media for the cultivation of *P. demidoffii*.

Antifungal assay

Antifungal activity of resin (crude extract) on *P. demidoffii*: Organic solvent extracts of resin were tested against *P. demidoffii* using agar dilution method and the result is documented in Table 3. Resin from *J. procera* significantly inhibited the growth of the fungus (*P. demidoffii*). The study in ref. [12] on the effects of oleoresins in

Fraction	Percentage of solvent	Volume of solvent used
Fraction 1	100% $CHCl_3$	50 ml
Fraction 2	25% EtOAC: 75% $CHCl_3$	50 ml
Fraction 3	50% EtOAC: 50% $CHCl_3$	50 ml
Fraction 3	50% EtOAC: 50% $CHCl_3$	50 ml
Fraction 4	100% EtOAC	50 ml
Fraction 5	50% EtOAC: 50% MeOH	50 ml
Fraction 6	100% MeOH	50 ml

Table 1: Organic solvents used as eluting solvent in column chromatography.
Note: $CHCl_3$: chloroform; EtOAC: Ethyl acetate; MeOH: Methanol.

Culture media	Measurement days					
	4th day	8th day	12th day	16th day	20th day	24th day
CMA	5.50 ± 0.05[d]	9.00 ± 0.05[e]	12.00 ± 0.05[c]	17.50 ± 0.05[d]	19.00 ± 0.1[d]	22.00 ± 0.25[d]
CMDA	8.50 ± 0.05[c]	13.20 ± 0.02[d]	15.00 ± 0.05[b]	18.00 ± 0.1[d]	18.80 ± 0.05[d]	20.00 ± 0.13[e]
CDA	3.50 ± 0.05[e]	5.60 ± 0.03[f]	8.50 ± 0.05[d]	12.50 ± 0.05[f]	14.00 ± 0.2[e]	17.50 ± 0.02[f]
JE	2.40 ± 0.02[f]	2.90 ± 0.02[g]	5.60 ± 0.02[e]	15.00 ± 0.1[e]	19.00 ± 0.17[d]	21.50 ± 0.05[e]
LAM	11.00 ± 0.1[a]	17.00 ± 0.05[b]	30.50 ± 0.05[a]	39.00 ± 0.1[a]	51.00 ± 0.17[a]	57.00 ± 0.17[a]
MEA	11.00 ± 0.1[b]	16.50 ± 0.07[b]	30.00 ± 0.5[a]	36.50 ± 0.05[b]	45.00 ± 0.36[b]	52.00 ± 0.2[b]
MEAJE	3.20 ± 0.32[e]	5.30 ± 0.03[f]	9.30 ± 0.03[c]	16.00 ± 0.1[d]	22.50 ± 0.08[c]	25.30 ± 0.17[d]
MEAP	11.00 ± 1.15[a]	18.00 ± 0.1[a]	31.00 ± 0.1[a]	39.00 ± 0.26[a]	52.00 ± 0.17[a]	58.00 ± 0.2[a]
OMA	10.30 ± 1.05[b]	13.70 ± 0.02[c]	16.50 ± 0.05[b]	20.50 ± 0.05[c]	23.30 ± 0.01[c]	26.30 ± 0.17[d]
PDA	10.00 ± 1[b]	14.20 ± 0.02[c]	17.50 ± 0.04[b]	22.50 ± 0.25[c]	24.30 ± 0.02[c]	28.50 ± 0.05[c]

Table 2: The diameter growth of *P. demidoffii* (including the diameter of inoculum disc) in mm in each media and measurement date.
Note: Results are the mean values ± standard deviations.
Means within the column under a parameter having a common letter do not differ significantly at (p<0.05).

an in vitro growth of fungi associated with pine decline showed that resin inhibited all tested fungi. The report in ref. [28] indicated that phytochemicals present in the *Larrea tridentata* resin have a powerful antifungal activity against *Phytophthora capsici* Leo. Which cause chili pepper (*Capsicum annuum* L.) wilt disease?

As shown in Table 3 the resin dissolved by methanol showed higher inhibitory activity against *P. demidoffii* with complete mycelial growth inhibition (100%) at the concentration of 5 mg/100 ml MEAP followed by ethyl acetate extract which completely inhibited the mycelia growth of the pathogen at 6 mg/100 ml MEAP concentration after 20 days of incubation at 25°C. Ethanol and chloroform extract were less effective with 66.19 ± 0.02 and 60.49 ± 0.02 percent of average mycelial inhibition of the pathogen respectively at 6 mg/100 ml MEAP after the same period of incubation.

In terms of growth diameter and growth rate the growth diameter and growth rate of *P. demidoffii* were also highly reduced as compare to the control (Figure 1). The average growth diameter and growth rate of the *P. demidoffii* on control medium was 30.60 ± 0.01 mm and 2.25 ± 0.03 mm respectively, but in resin containing medium was ranged from complete inhibition (no growth) to 26.30 ± 0.04 mm respectively at highest and lowest concentration of resin extract, which depending on the solvent systems used.

Inhibitory activity of resin fractions again the mycelium of *Pyrofomes demidoffii*: In this study both crude and fractions poweders of the resin showed great inhibitory effect on the mycelial growth and growth rate of *P. demidoffii* (Table 4). Three (3) mg per 100 MEAP of each fraction was tested against *P. demidoffii* using agar dilution method. As indicated from below table fraction 3, 4 and 5 inhibited the mycelia a growth of the pathogen completely (100% mycelial inhibition). Next to those potent fractions fraction 2 inhibited the

Solvent system		Concentration of resin mg/100 ml MEAP						
		6	5	4	3	2.5	1.5	Control
Chloroform ($CHCl_3$)	GD (mm)	12.30 ± 0.01^{g}	13.80 ± 0.01^{f}	16.30 ± 0.02^{e}	18.60 ± 0.03^{d}	20.80 ± 0.06^{c}	26.30 ± 0.04^{b}	30.60 ± 0.01^{a}
	GR (mm)	0.9 ± 0.1^{c}	1 ± 0.5^{bc}	1.2 ± 0.1^{bc}	1.2 ± 0.2^{bc}	1.3 ± 0.1^{b}	1.9 ± 0.1^{a}	2.25 ± 0.03^{a}
	%INH	60.49 ± 0.02^{a}	50.17 ± 0.03^{b}	46.16 ± 0.4c	37.32 ± 0.03^{d}	29.35 ± 0.1^{e}	13.18 ± 0.03^{f}	0.00
Ethanol (EtOH)	GD (mm)	9.40 ± 0.02^{e}	12.40 ± 0.01^{de}	15.40 ± 0.01^{cd}	17.20 ± 0.02^{c}	18.60 ± 0.02^{c}	22.00 ± 0.2^{b}	30.60 ± 0.01^{a}
	GR (mm)	0.6 ± 0.1^{d}	0.9 ± 0.1^{cd}	1 ± 0.6^{c}	1.08 ± 0.02^{c}	1.14 ± 0.03^{c}	1.56 ± 0.03^{b}	2.25 ± 0.03^{a}
	%INH	66.19 ± 0.02^{a}	57.08 ± 0.02^{b}	49.04 ± 0.26^{c}	40.28 ± 0.48^{d}	33.68 ± 0.03^{e}	23.82 ± 0.02^{f}	0.00
Ethyl acetate (EtOAC)	GD (mm)	-	11.30 ± 0.01^{f}	13.10 ± 0.02^{e}	16.90 ± 0.01^{d}	19.20 ± 0.03^{c}	22.90 ± 0.01^{b}	30.60 ± 0.01^{a}
	GR (mm)	-	0.76 ± 0.02^{e}	0.77 ± 0.03^{e}	1.14 ± 0.02^{d}	1.3 ± 0.02^{c}	1.55 ± 0.03^{b}	2.25 ± 0.03^{a}
Methanol (MeOH)	GD (mm)	-	-	5.00 ± 0.1^{e}	9.00 ± 0.1^{d}	11.60 ± 0.02^{c}	13.30 ± 0.03^{b}	30.60 ± 0.01^{a}
	GR (mm)	-	-	0.10 ± 0.05^{d}	0.57 ± 0.01^{c}	0.65 ± 0.03^{b}	0.68 ± 0.02^{b}	2.25 ± 0.03^{a}
	%INH	100^{a}	100^{a}	79.85 ± 0.02^{b}	67.76 ± 0.03^{c}	58.36 ± 0.03^{d}	51.06 ± 0.00^{e}	0.00

Table 3: *In vitro* inhibitory *effect* of resin from *J. procera* in the growth diameter, growth rate and percentage mycelial growth inhibition of *P. demidoffii*, dissolved by different solvent after 20 days of incubation at 25°C.

Note: Results are the mean values ± standard deviations. Means within the column under a parameter having a common letter do not differ significantly at ($p<0.05$). %INH: percentage of mycelial inhibition. GD: Growth Diameter GR: Growth Rate -: No growth (100% inhibition).

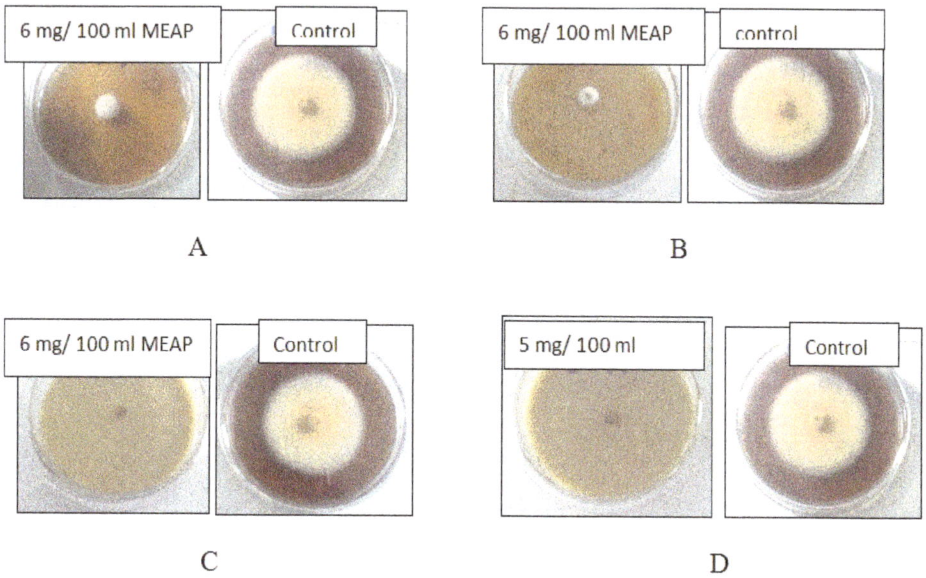

Figure 1: The inhibitory effect of resin extracted by different organic solvent against the mycelium of *P. demidoffii* Chloroform (A), Ethanol (B), Ethyl acetate(C) and Methanol (D) extract of resin.

Fractions	Growth diameter (mm)	Growth rate (mm)	% inhibition
Fraction 1	10.5 ± 0.05[c]	0.5 ± 9[c]	51.3 ± 0.1[c]
Fraction 2	5.5 ± 0.01[d]	0.49 ± 0.1[d]	75.85 ± 0.05[b]
Fraction 3	-	-	100[a]
Fraction 4	-	-	100[a]
Fraction 5	-	-	100 [a]
Fraction 6	13.7 ± 0.03[b]	0.92 ± 0.2[b]	35.5 ± 0.1[d]
Control	22.7 ± 0.05[a]	1.92 ± 0.03[a]	0.00

Table 4: Inhibitory effect of fractionated resin in growth diameter, growth rate and percentage mycelial growth inhibition of *P. demidoffii* after 18 days of incubation at 25°C.

Note: Results are the mean values ± standard deviations.

Means within the column under a parameter having a common letter do not differ significantly at ($p<0.05$).

mycelial growth of the pathogen by 75.85 ± 0.05 %. Fraction 6 was the most ineffective that inhibited the pathogen by 35%.

In case of growth diameter and growth rate, the growth diameter and growth rate of the pathogen was highly reduced as compared to the control. The average growth diameter of the pathogen was 10.5 ± 0.05 mm, 5.5 ± 0.01 mm, no growth, no growth, no growth, 13.7 ± 0.03 mm and 22.7 ± 0.05 mm respectively on fraction 1, 2, 3, 4, 5, 6, and control medium (Table 4 and Figure 2). This gives an impressive clue for the increment of antifungal activity of bioactive elements within the resin when they are semi purified. The study in refs. [36,37] has confirmed that the strong antifungal property of semi purified extracts of several plants extracts.

Minimum inhibitory concentration (MIC)

The Minimum Inhibitory Concentration assay was also employed to evaluate the effectiveness of the resin to inhibit the growth of *P. demidoffii*. MIC value or resin was less as compared with commercial fungicides, this is may be due to its impurity mean the resin extract contains many compounds which may or may not involve in the inhibitory activity of the pathogen.

MIC of resin was within the range of 5 to 6 mg/100 ml MEAP. The resin was fungistatic against the pathogen and the result is recorded on Table 5. When the fungus inoculum plug, which do not show growth, removed from plate containing resin and placed on a plate containing pure medium, the fungus respond the fungistatic activity of resin, by growing more rapidly. The study in ref. [23] showed the fungisatic activity of oleoresin against pine colonizing fungi.

Identification of active ingredient

Preliminary phytochemical screen: Qualitative phytochemical tests of *J. procera* resin extract revealed the presence alkaloids, saponins, terpenoides, phenolic compounds, flavonoides and fixed oils, and the absence of carbohydrates steroids, fats, glycosides and tannins (Table 6 and Figure 3). According to ref. [38] conifer resin contains several phytochemical groups, composed primarily terpenoids and phenolic compounds. Based on the study in ref. [38] coniferous trees have diverse group of secondary metabolic compounds including isoflavonoids, pterocarpans, stilbenes and saponins as inducible defense systems. Pathogen induced accumulation of related constitutive antimicrobial compounds in coniferous has been reported [39], including stilbenes and diterpenic resin acids [40], lignans and flavonoids [41]. The study in ref. [42] reported that the wood and leaf oils of *J. procera* from Ethiopia and showed that the leaf oil contains pinene and limonene with a small quantity of borneol.

The study conducted in ref. [31] on the phytochemical analyses of resin from *Spondias pinnata,* showed the presence of flavonoids and fixed oils and the absence of saponins and tannins. In contrast to this, the report of ref. [43] indicated that conifers resins contain terpenoids, flavonids, phenolics, lignans and saponins. The study in refs. [30,43] showed that alkaloids, saponins, flavonoids stilbenes and phenolic compounds are among the biologically active secondary metabolites of plants that found in the resin of many conifers trees including leaves, bark and wood of trees. In agreement to the report of refs. [42,43] but in contrast to ref. [31] the phytochemical analysis of resin from *J. procera* revealed the presence of saponins and the absence of tannins.

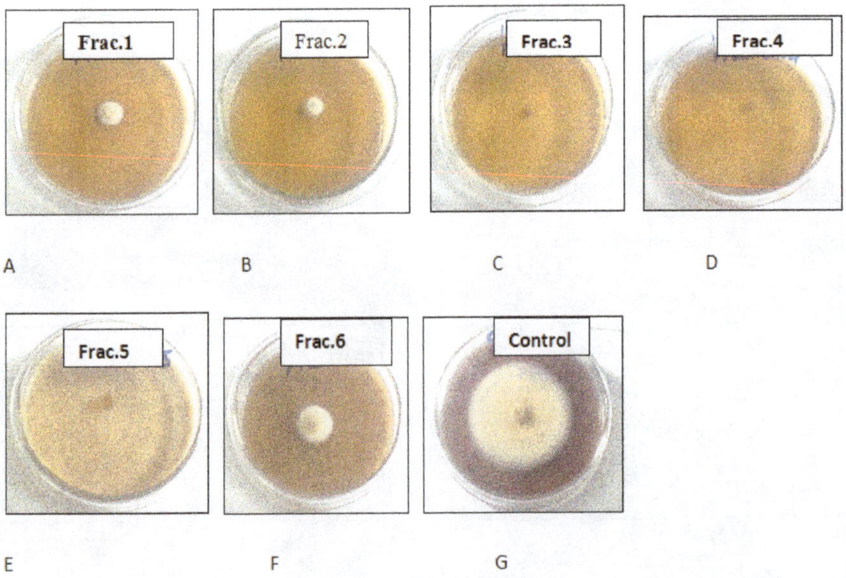

Figure 2: The antifungal activity of fractions against *P. demidoffii* A (frac.1), B (frac.2), C (frac.3), D (frac.4), E (frac.5), F (frac.6), G control) after 20 days of incubation at 25°C.

Antifungal substance	Concentration range for fungicides (mg/100 ml of MEAP)				Concentration range for resin in mg/100 ml of MEAP				
	0.1	0.25	0.5	1	4	5	6	7	8
Resin dissolved by MeOH					+	*	-	-	-
Resin dissolved by EtOAC)					+	+	*	-	-
Redomil Gold MZ	+	*	-	-					
Sancozeb 80 WP	+	+	*	-					
Cruzate R WP)	+	+	+	*					

Note: - (*) MIC. (+) Growth of fungus observed. (-) no growth of fungus.

Table 5: Minimum Inhibitory Concentration (MIC) for crude extracts resin and fungicides against *P. demidoffii.*

Phytochemicals	Presence (+)/absence (-)
Alkaloids	+
Saponins	+
Terpenoides	+
Phenolic compounds	+
Flavonoides	+
fixed oils	+
Carbohydrates	-
Steroids	-
Fats	-
Glycosides	-
Tannins	-

Table 6: Phytochemical constituents of resin from *Juniperus procera.*

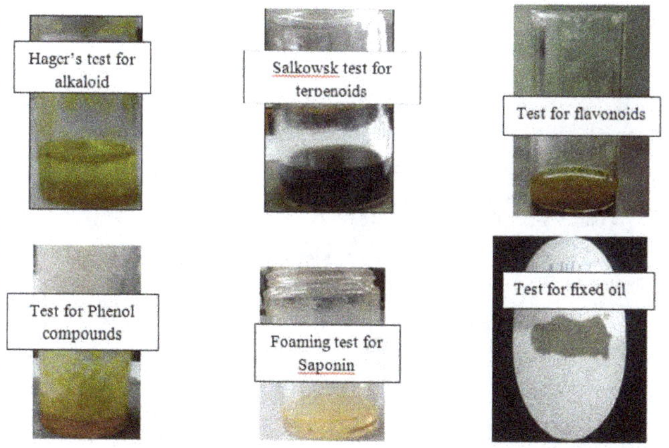

Figure 3: Preliminary phyotochemical test of resin for a class of compounds.

The difference on the composition of resin may be due to the difference in the plant species, fungal species and environment in which the tree grows. Although plants have the capacity to produce and store different amounts of secondary metabolites the properties and composition of these compounds may vary according to the producer organism's environment [44]. The observed antifungal activity is might be attributed to the presence of these bioactive compounds in the resin crude extract and the effect may be due to a single compound or may be a combined effect of different compounds present in the resin extract. The presence of alkaloids, saponins, terpenoids, phenolic compounds and flavonoids from resin extract is known to confer antifungal activity against plant pathogenic fungi [45].

Thin layer chromatography (TLC): Thin layer chromatography (TLC) was carried out with a moderately non-polar solvent system using benzene: methanol (18:6) and benzene: ethanol: ammonia (18:2:1). The Retention factor (RF) value of each fraction were calculated and tabulated below (Table 7). Bands and different spots were observed under UV lights. As shown in Table 7 each fraction gave different Rf values in two different solvent systems. This variation in Rf values of the fractions provides a very important clue in understanding of their polarity. The Rf values of phytochemicals in faction 1 were 0.68, 0.78 and 0.60, 0.72 respectively in two solvent system Benzene: Methanol (18:6) and Benzene: Ethanol: Ammonia (18:2:1), which are moderately non-polar. Whereas the Rf values phytochemicals in fraction 6 were 0.18, 0.25 and 0.42.0.38 respectively in two solvent system, Benzene: Methanol (18:6) and Benzene: Ethanol: Ammonia (18:2:1). The Rf values of phytochemicals in fraction 1 were greater than the Rf values of phytochemicals in fraction 6 in similar solvent system, which showed the non-polarity of phytochemicals within fraction 1 as compared to fraction 6. Compound showing high Rf value in less polar solvent system have low polarity and with less Rf value have high polarity [46].

Conclusion and Recommendations

Conclusion

The evaluation of growth diameter of *P. demidoffii* on diffserent media demonstrated, as MEAP is a good medium for the cultivation of *P. demidoffii.*

The results of this study showed that, resin from *J. procera* has significant inhibitory effect against *P. demidoffii.* Both fractions and crude extract of resin have a greatest inhibitory effect on *P. demidoffii.* It can be used as potential candidates for the formulation of bio fungicides, which may be useful in the treatment of *Juniper procera* heart rot, chocolate spot disease, tomato and coffee wilt disease.

TLC profiling reveals the presence of various phytochemicals within the resin of *J. procera.*

The results of phytochemical tests, of resin from *J. procera* showed the presence of alkaloids, saponins, terpenoids, phenolic compounds, flavonoides and fixed oils.

Recommendations

Further research is needed to purify bioactive elements, especially fraction 3, 4, and 5 in the resin

Only a few genera of fungal groups were used in the study. Antifungal tests are recommended to be done on more species which are important to plant health.

The results described above did not tell the possible inhibitory mechanisms of the resin, mechanisms of *P. demidoffii* for infestation of *J. procera* (enzymatic or phytotoxic) and method of extraction of

Fraction	Extraction solvent	Developing solvents	
		Benzene: Methanol (18:6)	Benzene: Ethanol: Ammonia (18:2:1)
Fraction 1	100% $CHCl_3$	0.68, 0.78	0.60, 0.72
Fraction 2	25% EtOAC:75% $CHCl_3$	0.50, 0.56	0.64
Fraction 3	50% EtOAC:50% $CHCl_3$	0.50, 0.55, 0.62	0.48, 0.60
Fraction 4	100% EtOAC	0.46, 0.62 , 0.67	0.45, 0.56
Fraction 5	50% EtOAC:50% MeOH	0.41, 0.52	0.50
Fraction 6	100% MeOH	0.18, 0.25	0.42.0.38

Note: $CHCl_3$: Chloroform EtOAC: Ethyl acetate MeOH: Methanol.

Table 7: Retention factor (RF) of different solvent fractions of resin extract through column chromatography.

enzymes or toxins that *P. demidoffii* use to attack *J. procera* tree, so it require further investigation.

References

1. Pohjonen V, Pukkala T (1992) Juniperus procera Hocht. ex Endl. in Ethiopian forestry. For Ecol and Mgt 49: 75-85.
2. Von Breitenbach F (1963) The indigenous Trees of Ethiopia. 2nd ed. Ethiopian Forestry Association, Addis Ababa, Ethiopia, pp. 305.
3. Gessesse D (2007) Forest Decline in South Central Ethiopia: Extent, history and process. Department of Physical Geography and Quaternary Geology Stockholm University, Stockholm, Sweden.
4. Nyssen J, Poessen J, Moeyersons J, Deckers J, Haile M, et al. (2004) Human impact on the environment in the Ethiopian and Eritrean highlands - a state of the art. Earth-Science Reviews 64: 273-320.
5. Chaffey DR (1982) South-west Ethiopia forest inventory project. A reconnaissance inventory of forests in south-west Ethiopia. Addis Ababa, Forestry and Wildlife Conservation and Development Authority.
6. Contreras-Hermosilla A (2000) The underlying causes of forest decline. Occasional Paper No. 30. Centre for International Forestry Research (CIFOR). Bogor.
7. Eckhardet LG, Weber AM, Menard RD, Jones JP, Hess NJ (2007) Insect-fungal complex associated with loblolly pine decline in central Alabama. For Sci 53: 84-92
8. EFAP (Ethiopian Forest Action Program) (1994) Ethiopian Forest Action Program Vol.2 Ministry of Natural Resources Development and Environmental Protection, Addis Ababa, Ethiopia.
9. Achalu Negussie (1995) Monographic review on Juniperus excelsa. Alemaya University of Agriculture, Faculty of Forestry, pp: 4.
10. Ryvarden L, Johansen I (1980) A Preliminary Polypore Flora of East Africa. Oslo: Fungi flora.
11. Tigabu M, Fjellström J, Odén P, Teketay D (2006) Germination of Juniperus procera seeds in response to stratification and smoke treatments, anddetection of insect-damaged seeds with VIS + NIR spectroscopy. New Forests 33: 155-169.
12. Eckhardt LG, Menared RD, Gray DE (2008) Effects of oleoresins and monoterpenes on in vitro growth of fungi associated with pine decline in the Southern United States. For Path 39: 157-167.
13. Scharpf RF, Goheen D (1993) Heart rots, In: Diseases of Pacific Coast Conifers Forest Service Agriculture Handbook, (Scharpf, R.F. eds). United States Department of Agriculture, Washington, DC. pp: 150-180
14. Abou-Zeid AM, Altalhi AD, Abd El-Fattah RI (2007) Fungal control of pathogenic fungi isolated from wild plants in Taif Governorate, Saudia Arabia. Roum Arch Microbiol Immunol 66: 90-96.
15. Gilbertson RL, Ryvarden L (1987) North American Polypores, Vols. I and II. Fungiora. Oslo, pp: 885
16. Pataky NR (1999) Wood rots and decays report on plant disease. RPD No. 642. University of Illinois at Urbana-Champaign
17. Cos P, Vlietinck AJ, Berghe DV, Maes L (2006) Anti-infective potential of natural products: how to develop a stronger in vitro 'proof-of-concept'. J Ethnopharmacol 106: 290-302.
18. Wagener WW, Davidson RW (1954) Heart rots in living trees. Bot Rev 20: 61-134.
19. Zwenger S, Basu C (2008) Plant terpenoids: applications and future potentials. Biotechnology and Molecular Biology Reviews 3: 001-007.
20. Berryman AA (1972) Resistance of conifers to invasion by bark beetle-fungus associations. Bio Science 22: 598-602.
21. Martin DM, Tholl D, Gershenzon J, Bohlmann J (2002) Methyl jasmonate induces traumatic resin ducts, terpenoid resin biosynthesis, and terpenoid accumulation in developing xylem of Norway spruce stems. Plant Physiol 129: 1003-1018.
22. Miller B, Madilao LL, Ralph S, Bohlmann J (2005) Insect-induced conifer defense. White pine weevil and methyl jasmonate induce traumatic resinosis, dnovo formed volatile emissions, and accumulation of terpenoid synthase and putative octadecanoid pathway transcripts in Sitka spruce. Plant Physiol 137: 369-382.
23. Chang SS, Ostic- Matijaesievic B, Hsieh OA, Huang CL (1977) Natural antioxidant from rosemany and sage. J Food Sci 42: 1102 - 1106.
24. Hajlaoui H, Trabelsi N, Noumi E, Snoussi M, Fallah H, et al. (2009) Biological activities of the essential oils and methanol extract of tow cultivated mint species (Mentha longifolia and Mentha pulegium) used in the Tunisian folkloric medicine. World J Microbiol Biotechnol 25: 2227-2238.
25. Lee CH, Lee HS (2005) Antifungal property of dihydroxyanthraquinones against phytopathogenic fungi. J Microbiol Biotechnol 15: 442-446.
26. Food and Agricultural Organization of United State (FAO) (2011) AGP - Integrated Pest Management fact sheet.
27. Bitew A (2010) Antibacterial and antifungal of culture filtrate extract of Pyrofomes demidoffii. Independent research, Addis Ababa University, Addis Ababa, Ethiopia.
28. Mojica-Marin V, Luna-Olvera HA, Sandoval-Coronado CF, Morales-Ramose LH, Gonzalez-Aguilar NA, et al. (2011) In vitro antifungal activity of "Gobernadora" (Larrea tridentata (D.C.) Coville) against Phytophthora capsici Leo. Afr J of Agric Res 6: 1058-1066.
29. Zadoks JC, Schein RD (1979) Epidemiology and plant disease management Oxford University Press. USA, pp: 427.
30. Celimene CC, Smith DR, Young RA, Stanosz GR (2001) In vitro inhibition of Sphaeropsis sapinea by natural stilbenes. Phytochemistry 56: 161-165.
31. Gupta VK, Roy A, Nigam V, Mukherjee K (2010) Antimicrobial activity of Spondias pinnata resin. Journal of Medicinal Plants Research 4: 1656-1661.
32. De S, Dey YN, Ghosh AK (2010) Phytochemical investigation and chromatographic evaluation of the different extracts of tuber of amorphaphallus paeoniifolius (araceae). Int J on Pharm and Biomed Res. 1: 150-157.
33. Njoku OV, Obi C (2009) Phytochemical constituents of some selected medicinal plants. African Journal of Pure and Applied Chemistry 3: 228-233.
34. Neelima N, Devidas NG, Sudhakar M, Kiran J (2011) A preliminary phytochemical investigation on the leavse of Solanum xanthocarpum. Int J of Res in Ayurveda and Pharm. 2: 845-850.
35. Harborne JB (1996) Phytochemical Methods. Chapman and Hall Ltd., London, Pp: 52-105.
36. Edeoga HO, Okwu DE, Mbaebie BO (2005) Phytochemical constituents of some Nigerian medicinal plants. Afr J of Biotechnol 4: 685-688.
37. Pretorius JC, Zietssman PC, Eksteen D (2002) Fungitoxic properties of selected South African plant species against plant pathogens of economic importance in agriculture. Ann. Appl Biol 141: 117-124.
38. Kopper BJ, Illman BL, Kersten PJ, Klepzig KD, Raffa KF (2005) Effects of diterpene acids on components of a conifer bark beetle-fungal interaction: tolerance by Ips pini and sensitivity by its associate Ophiostoma ips. Environ Entomol 34: 486-493.
39. Pearce RB (1996) Antimicrobial defences in the wood of living trees. New Phytol 132: 203- 233.
40. Bonello P, Heller W, Sandermann JH (1993) Ozone effects on root disease susceptibility and defence responses in mycorrhizal and nonmycorrhizal seedlings of Scots pine (Pinus sylvestris L.). New Phytol 124: 653- 663.
41. Adams RP (1990) Junipers procera East Africa: volatile leaf oil composition and putative relationship to J. excelsa. Biochem Syst and Ecol 18: 207- 210.
42. Raffa KF, Berryman AA (1982) Accumulation of monoterpenes and associated volatiles following fungal inoculation of grand fir with a fungus transmitted by the fir engraver Scolytus ventralis (Coleoptera: Scolytidae). Can Entomo 114: 797-810.
43. Lieutier F (2002) Mechanism of resistance in conifers and bark beetle attack strategies, In: Mechanisms and Development of Resistance in Trees to Insects, (Wagner, M. R.; Clancy, K. M.; Lieutier, F.; Paine, T. D. Ed). Dordrecht: Kluwer Academic Publishers pp: 31-78.
44. Firn RD, Jones CG (2003) Natural products--a simple model to explain chemical diversity. Nat Prod Rep 20: 382-391.
45. Farnsworth AC (1982) The role of ethnopharmacology drug development from plants. England Ciba: John Wiley and Sons. Pp: 2-10.
46. Siddiqui AA, Ali M (1997) Practical pharmaceutical disease 1st edn. CBS Publisher and distributions, New Delhi. Pp: 126-131.

12

Growth Inhibition Potentials of Leaf Extracts from Four Selected Euphorbiaceae against Fruit Rot Fungi of African Star Apple (*Chrysophyllum albidum* G. Don)

Ilondu EM[1] and Bosah BO[2]

[1]*Department of Botany, Faculty of Science, Delta State University, Abraka, Nigeria*

[2]*Department of Agronomy, Faculty of Agriculture, Delta State University, Asaba Campus, Nigeria*

Abstract

The efficacy of ethanolic leaf extracts from *Phyllanthus amarus, Euphorbia hirta, Euphorbia heterophylla* and *Acalypha fimbriata* in inhibiting the growth of post-harvest fruit rot fungi of *Chrysophyllum albidum* was investigated at the concentrations of 100, 80, 60 40 and 20 mg/ml *in-vitro*. The fungi isolated from rotted fruits and their frequency of occurrence includes *Aspergillus niger* (69.6%) and *Fusarium solani* (30.4%). These fungal isolates were cultured on different leaf extracts agar and their radial mycelia growth was observed. The antifungal activities increased with increase in concentrations of the plant extracts with *E. heterophylla* extract most effective in inhibiting the growth of *A. niger* while *A. fimbriata* extract was more effective in the inhibition of *F. solani* than other extracts. Phytochemical screening of the plant extracts revealed the presence of saponins, alkaloids, glycosides, terpenes, steroids, flavonoids, tannins and phenols. Gas Chromatography Mass Spectrometry (GC-MS) analysis revealed the presence a complex mixture of constituents ranging from 7 compounds in *E. hirta,* 10 compounds in *A. fimbriata,* 11 compounds in *E. heterophylla* and 14 compounds in *P. amarus*. The result of this study is an indication that these Euphorbiaceae could be a potential source of antifungal agents.

Keywords: Growth inhibition; Leaf extracts; Euphorbiaceae; Rot fungi; *Chrysophyllum albidum*

Introduction

Chrysophyllum albidum G. Don commonly called African star apple and locally called udara (Igbo), agbalumo (Yoruba) belongs to the family Sapotaceae [1]. It features prominently in the compound agro forestry system for fruit, food, cash income and other auxiliary uses including environmental purposes. It is also a tree that is common throughout the Tropical Central, East and West Africa regions for its sweet edible fruit and various ethnomedical uses [2].

C. albidum fruits are widely eaten in Southern Nigeria. The fruit is seasonal (December-March), when ripe. It is flattened seeds or sometimes fewer by abortion. The fruit is ovoid to sub-globose pointed at the apex and up to 6 cm long and 5 cm in diameter. The skin or peel is grey when immature turning orange red, pinkish or light yellow within the pulp having three to five seeds arranged as a star [3].

The fruit has been found to have the highest content of ascorbic acid with 1000 to 3330 µg of ascorbic acid per 100 gm of edible fruit or about 100 times that of oranges and 10 times of that of guava or cashew. It is also an excellent source of vitamins B and D as well as iron [4]. Umoh [5] and Ureigho [6] reported on the proximate composition, minerals and vitamins content of *Chrysophyllum albidum.*

The fruit has immense economic potential, especially following the report that jams that could compete with rasp berry jams and jellies could be made from it and it is eaten especially as snack by both young and old [2]. The fruits contain 90% anacadic acid, which is used industrially in protecting wood and as a source of resin. The fruits can be used in the preparation of wine, soft drink, jams and jellies [3,6].

The seed are used for local games; it is also a source of oil, which used for diverse purposes [7]. The seeds along with those of other Sapotaceae are used as anklets in dancing. It was also discovered in the removal of Ni^{2+} ions from synthetic wastewater [8]. The cotyledons are useful in the preparation of medicine for the treatment of infertility problems in both male and female; infertility due to the presence of abnormalities within the uterus and female tubes, abdominal pains in dysmenorrheal, secondary ammenorrhae in women (loss or absence of menstrual cycle). The seed cotyledon has been reported to possess anti-hyperglycemic and hypolipidemic effects [9].

Fungi have been reported to be associated with post harvest deterioration of agricultural products in Nigeria. However, *F. solani*, *L. theobromae*, *Rhizopus spp* and *A. flavus* have been reported to be associated with *C. albidum* [10]. Since most microbial spores are small in size and light, they could settle on the surface of African Star Apple fruits resulting in the range of microbial group isolated from them.

Preserving the freshness of these fruits for many days or months is therefore the problem, which most farmers and the traders seek to solve. Control of fruit rot by employing the use of local preservatives (plant extracts) like *Afromomum danielli, Afromomum melegueta* and chemical disinfectants like (parazone), sodium chloride and sodium benzoate at mild form has been suggested to reduce the losses due to storage moulds [10].

The objective of this study is therefore to isolate and identify fungi associated with *C. albidum* fruits rot in storage as well as to

***Corresponding author:** Ilondu EM, Department of Botany, Faculty of Science, Delta State University, Abraka, Nigeria, E-mail: martinailondu@yahoo.co.uk

determine the effects of various concentrations of ethanolic extracts of *Phyllanthus amarus, Euphorbia hirta, Euphorbia heterophylla* and *Acalypha fimbriata* on the identified fungi.

Materials and Methods

Collection of plant materials for the study

Mature healthy and rotted *C. albidum* fruits were purchased at Abraka Main Market, Delta State. Fresh and healthy leaves of *Euphorbia hirta, Euphorbia heterophylla Phyllantus amarus* and *Acalypha fimbriata* free from insect and pathogen attack were collected from different areas within Abraka community. Abraka (Ethiope East Local Government Area of Delta State lies within latitude 05° 47″N and longitude 06° 06″E of the Equator with an annual rainfall of 3,097.8 mm, annual relative humidity of 83% and annual mean temperature of 30.6°C [11]. The plants were identified using Akobundu and Agyakwa [12].

Isolation and identification of fungi

Isolation and identification of fungi from diseased *C. albidum* fruits was carried out using the method adopted from Ilondu [13]. Sections, 4 mm long, excised from the margins of diseased spot with sterile razor blade were surface-sterilized for 2 min in 2% aqueous solution of commercial bleach (sodium hypochlorite solution), rinsed in two changes of sterile distilled water. The disinfected tissue pieces were blotted between sterile Whatman No. 1 filter paper and aseptically plated on potato dextrose agar (PDA) plates (3 pieces per plate). The plates were then incubated at room temperature (32 ± 2°C) for five days. Any observed mycelial growth was repeatedly transferred to fresh PDA plates until pure cultures of isolates were obtained.

The frequency of isolations of the different types of fungi associated with *C. albidum* fruit rot diseases was determined. The number of times each fungus was encountered was recorded. The percentage frequency of occurrence was calculated with the formula below:

$$\frac{\text{Number of times a fungus was encountered}}{\text{Total fungal isolations}} \times \frac{100}{1}$$

Plant sample preparation and extraction procedures

The plants were collected into polyethylene bags and taken to the laboratory for processing. The leaves were separately plucked and rinsed in flowing tap water, shade dried on the bench in a ventilated section of the Department of Botany herbarium at ambient temperature (30°C ± 2) for two weeks [14]. Dried leaves were separately ground into powder using an electric blender before extraction. For extraction procedures, one hundred gram of each pulverized sample was put into Soxhlet extractor and three hundred milliliter of absolute ethanol (HPLC grade) was added and extracted for 8hrs for each batch of sample. The extracts were evaporated on a rotary evaporator at 40°C to remove excess alcohol. The solvent free extracts were stored at 4°C till needed.

Phytochemical tests

One gram of powdered sample was subjected to phytochemical test for alkaloid (Myers reagent), Flavonoids were determined by magnesium rebbon test, Sapoins by chloroform and H_2SO_4 tests, Tannins, by Ferric salt test, Sterol by Chloroform-acetic anhydride, Terpenes and phenols by following the procedures of Oyewale and Audu [15].

Extract analysis

GC-MS analysis was done at National Research Institute for Chemical Technology (NARICT) Zaria, Kaduna state, Nigeria. A SHIMADZU GCMS-QP 2010 Plus system was used. The GC-MS was operated under the following conditions: Column oven temperature: 70°C; Injection temperature: 250°C; Injection mode: split; Pressure: 104.1 kPa; Total flow: 6.2 ml/min; Column flow: 1.59 ml/min; Linear velocity: 46.3 cm/sec; Purge flow: 3.0 mL/min; and Split ratio: 1.0. The generated chromatogram was recorded. The identification of the components was carried out using the peak enrichment technique of reference compounds and computer matching with those of NIST.05 library mass spectrum [14,16].

Effect of extracts on fungal growth

Different concentrations (100, 80, 60 40 and 20 mg/ml) were prepared from each of the extracts. One millilitre of each level of concentration was aseptically incorporated into 20 ml of cool molten PDA in sterile test tube. Each medium was homogenized by gentle agitation before dispensing into sterile 9 cm Petri dishes. The control was set up using extract free PDA plates. The plates were allowed to set for 3 hr. The effect of the extracts on fungal growth was determined using the method of Chohan et al. [17]. This was done by inoculating at the Centre of 90 cm Petri plates with a mycelia disc (4 mm) obtained from the colony edge of 7-day old culture of the test fungi. Three replicates of both the control and PDA-extract plates per isolate were incubated at room temperature (28 ± 2°C) and radial growth was measured with a metric ruler daily for seven days. Colony diameter was taken as the means along two directions on two perpendicular lines drawn on the reverse of the plates. The percentage inhibition was calculated by the method of Ayodele et al. [18].

Data analysis

Data obtained were subjected to Analysis of Variance (ANOVA) using Statistical Package for Social Science (SPSS) version 17.0 and means were separated according to Duncan's Multiple Range Test (DMRT) at 5% probability level.

Results

The fungi isolated from the diseased *Chrysophyllum albidum* fruits were *Aspergillus niger* and *Fusarium solani. A. niger* occurred more frequently with 69.6% followed by *F. solani* with 30.4% (Table 1). The classes of natural products present in the plant investigated are shown in Table 2. Tannins, saponins, steroids and phenols were present in all

Fungal isolate	No of times isolated	Percentage frequency (%)	Pathogenicity of isolates
Aspergillus niger	80	69.6	+
Fusarium solani	35	30.4	+

Table 1: Percentage occurrence of fungi associated with *Chrysophyllum albidum*.

Phytochemicals	*P. amarus*	*E. hirta*	*E. heterophylla*	*A. fimbriata*
Saponins	+	+	+	+
Alkaloids	-	+	+	+
Tannins	+	+	+	+
Flavonoids	-	+	-	-
Steroids	+	+	+	+
Glycosides	-	+	-	+
Terpenes	-	+	-	+
Phenols	+	+	+	+

\+ = Presence
\- = Absence

Table 2: Phytochemical Screening of Plants used in the study.

the plants. Alkaloids were present in *E. hirta*, *E. heterophylla* and *A. fimbriata* except *P. amarus*. Flavoinoids was only present in *E. hirta*. Glycosides and terpenes were present only in *E. hirta* and *A. fimbriata*.

The gas chromatography profiles of the plants extracts used in the study were shown in Figures 1-4. The analysis of the extract revealed

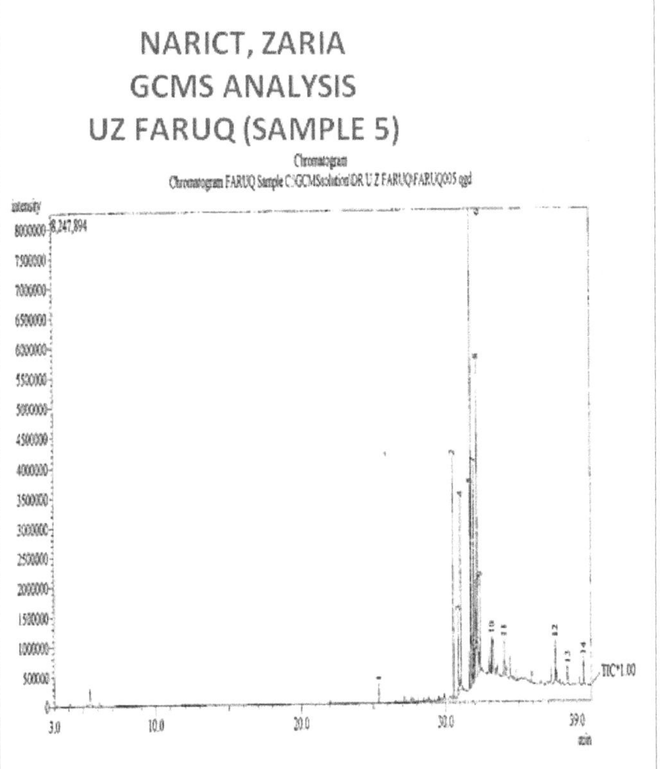

Figure 1: GC-MS Chromatogram for *P. amarus*.

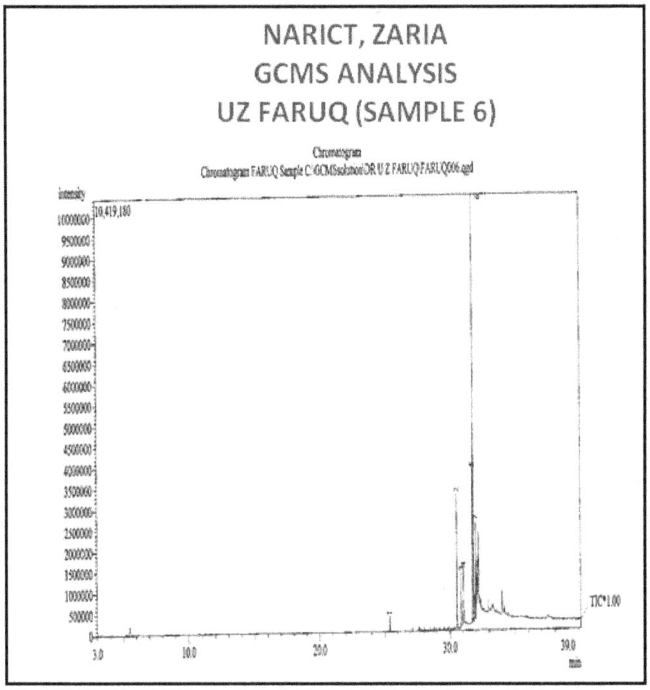

Figure 2: GC-MS Chromatogram for *E. hirta*.

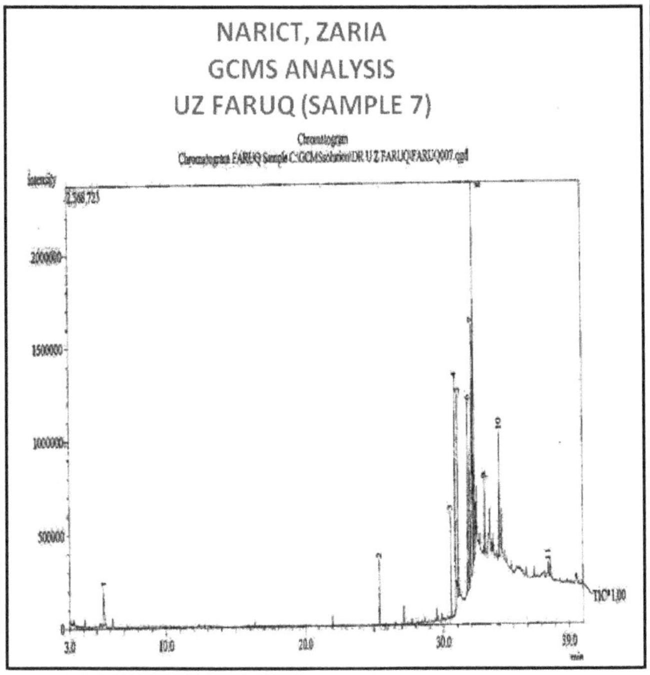

Figure 3: GC-MS Chromatogram for *E. heterophylla*.

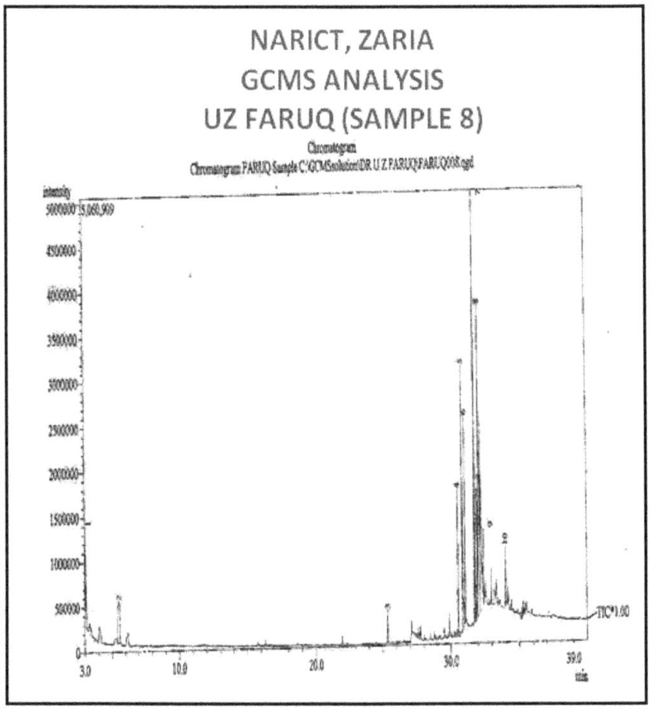

Figure 4: GC-MS Chromatogram for A. fimbriata.

complex mixture of constituents ranging from 7-14 compounds in the samples (Table 3).

Phenol 3,5-bis (1,1-dimethylethyl) were recorded in all plant. Hexadecanoid acid, methyl ester were recorded in all the plant except *A. fimbriata*. 10-Otadecenoic acid, methyl ester was recorded to be the most abundant of all the (14) compounds identified in *P. amarus*,

Plant Extracts	Peak No	Retention time (min)	% peak	Compound formula	Name of compound
P. amarus	1	25.381	1.09	$C_{14}H_{22}O$	Phenol3,5-bis(1,1-dimethylethyl)
	2	30.543	9.45	$C_{17}H_{34}O_2$	Hexadecanoic acid, methyl ester
	3	30.868	5.51	$C_{16}H_{36}O_2$	n-Hexadecanoic acid
	4	31.066	9.17	$C_{18}H_{36}O_2$	Hexadecanoic acid, ethyl ester
	5	31.745	9.50	$C_{19}H_{34}O_2$	11,14-octadecadienoc acid, methyl ester
	6	31.802	22.92	$C_{19}H_{34}O_2$	10-octadecenoic acid, methyl ester
	7	31.978	8.69	$C_{19}H_{38}O_2$	Octadecanoic acid, methyl ester
	8	32.223	19.33	$C_{20}H_{34}O_2$	9,12,15-octadealrienoic acid, ethyl ester (z,z,z)
	9	32.405	3.85	$C_{20}H_{40}O_2$	Octadecanoic acid, ethyl ester
	10	33.205	1.99	$C_{21}H_{42}O_2$	Eicosanoic acid, methyl ester
	11	34.017	2.31	$C_{18}H_{31}C_{10}$	9,12-octadecadienoyl chloride (z,z)
	12	37.501	2.71	$C_{22}H_{28}O_7$	Carissanol dimethyl ether
	13	38.330	1.20	$C_{34}H_{22}$	Dibenz(a,h) anthracene, 12-diphenyl
	14	39.463	2.29	$C_{10}H_{14}O_2$	Benzene, 4-ethyl-2-dimethoxy-
E. hirta	1	25.383	1.94	$C_{14}H_{22}O$	Phenol, 3,5-bis(1,1-dimethylethyl)
	2	30.542	12.68	$C_{17}H_{34}O_2$	Hexadecanoic acid, methyl ester
	3	30.869	8.85	$C_{16}H_{32}O_2$	n-hexadecanoic acid
	4	31.067	6.33	$C_{18}H_{36}O_2$	Hexadecanoic acid, ethyl ester
	5	31.747	16.27	$C_{19}H_{34}O_2$	9,12-octadecadienoic acid methyl ester
	6	31.806	44.9	$C_{19}H_{36}O_2$	9-octadecenoic acid (z)-, methyl ester
	7	31.98	9.33	$C_{19}H_{38}O_2$	Octadecanoic acid, methyl ester
E. heterophylla	1	5.496	4.78	C_8H_{10}	O-xylene
	2	25.386	3.47	$C_{14}H_{22}O$	Phenol3,5-bis(1,1-dimethylethyl)
	3	30.542	3.94	$C_{17}H_{34}O_2$	Hexadecanoic acid, methyl ester
	4	30.868	14.71	$C_{16}H_{32}O_2$	n-Hexadecanoic acid, methyl ester
	5	31.061	7.84	$C_{18}H_{36}O_2$	Hexadecanoic acid, ethyl ester
	6	31.798	9.22	$C_{19}H_{36}O_2$	10-Octadecenoic acid, methyl ester
	7	32.093	22.22	$C_{22}H_{42}O_2$	Erucic acid
	8	32.224	21.39	$C_{18}H_{32}O_2$	9,12-Octadecadienoica cid (z,z)-
	9	32.944	2.97	$C_{37}H_{74}NO_8P$	Hexadecanoic acid
	10	34.020	8.33	$C_{18}H_{34}O$	13-Octadecenal
	11	37.644	1.13	$C_{10}H_{14}O_2$	1,4-Benzenedimethanol, alpha, alpha, dimethyl
A. fimbriata	1	3.070	2.42	$C_5H_{12}O$	1-Butanol,3-methyl-/isopentyl alcohol
	2	5.43	10.52	C_8H_{10}	Xylene/benzene1,2-dimethyl
	3	25.372	1.70	$C_{14}H_{32}O$	Phenol3,5-bis(1,1-dimethyl ethyl)
	4	30.537	5.83	$C_{17}H_{34}O_2$	Pentadecanoic acid
	5	30.867	17.42	$C_{16}H_{32}O_2$	n-Hexadecanoic acid
	6	31.057	7.95	$C_{18}H_{36}O_2$	Hexadecanoic acid ethyl ester
	7	31.797	18.47	$C_{19}H_{36}O_2$	10-Octadecenoic acid, methyl ester
	8	32.093	28.88	$C_{22}H_{42}O_2$	Erucic acid
	9	32.401	3.25	$C_{20}H_{40}O_2$	Octadecanoic acid, ethyl ester
	10	34.015	3.56	$C_{16}H_{30}O$	Cis-9-Hexadecenal

Table 3: Major identified constituents of the plant extracts.

9-Octadecenoic acid (Z)-methyl ester was most abundant among the (7) compounds in *E. hirta*, Erucic acid was most abundant in *E. heterophylla* and *A. fimbriata*.

The two fungi were very sensitive to various concentrations of the plant extracts tested since the extracts significantly reduced the mycelia growth of the fungi at all concentrations (Table 4). However, the effectiveness of the plant extracts increased with increased in concentration and this was significantly different ($p<0.05$) when compared to the control. Similarly, percentage growth inhibition generally increased with increase in concentration of the leaf extracts when compared to the control. Although, the plant extracts could not give complete inhibition at the highest concentration tested, their effectiveness increased with increase concentrations.

There was no significant difference in the inhibitory effect of *P. amarus* on *A. niger* at the concentrations of 20 and 40 mg/ml, 60 and 80 mg/ml concentrations with *E. heterophylla* as well as 80 and 100 mg/ml concentrations with *A. fimbriata.* Similarly, there was no significant difference in the inhibitory effect of *A. fimbriata* extract on *F. solani* from 60-100 mg/ml concentrations (Table 5). *A. niger* was most sensitive to *E. heterophylla* followed by *A. fimbriata, P. amarus* and *E. hirta* respectively. Similarly, *F. solani* was most sensitive to *A. fimbriata* followed by *E. heterophylla, E. hirta* and *P. amarus.*

Discussion

The present study showed that two fungi were associated with post harvest fruit rot disease of *Chrysophyllum albidum, which* include *Aspergillus niger and Fusarium solani.* These fungi have previously been reported as fruit rot pathogens [13,19,20].

Aspergillus niger has the highest percentage occurrence of 69.6% followed by *F. solani* which is 30.4%. This was enhanced by the light

Extract conc. (mg/ml)	*P. amarus*		*E. hirta*		*E. heterophyta*		*A. Fimbriata*	
	A. niger	*F. solani*	*A. niger*	*F. solani*	*A. niger*	*F. solani*	*A. niger*	*F. solani*
0	4.30^{a}	4.30^{a}	4.30^{a}	4.30^{a}	4.30^{a}	4.30^{a}	4.30^{a}	4.30^{a}
20	2.07^{b}	3.1^{b}	2.83^{b}	2.83^{b}	1.53^{b}	1.90^{b}	1.87^{b}	1.53^{b}
40	1.95^{c}	2.23^{c}	2.40^{c}	1.93^{c}	0.95^{c}	1.73^{b}	1.40^{c}	0.58^{c}
60	1.13^{c}	1.85^{d}	1.97^{d}	1.63^{c}	0.72^{c}	0.85^{c}	0.92^{d}	0.42^{c}
80	0.92^{d}	1.27^{e}	1.50^{e}	1.03^{d}	0.62^{c}	0.62^{c}	0.75^{d}	0.30^{c}
100	0.84^{d}	0.92^{f}	1.03^{f}	0.82^{d}	0.47^{c}	0.42^{c}	0.60^{d}	0.22^{c}

Values with the same superscript(s) in the same column are not significantly different at P>0.05 by DMRT.

Table 4: Radial mycelia growth (cm) of fungi isolated from *Chrysophyllum albidum* fruits when exposed to various concentrations of plant leaf extracts.

Conc. (%)	*P. amarus*		*E. hirta*		*E. heterophylla*		*A. fimbriata*	
	F. Solani	*A. niger*	*F. Solani*	*A. niger*	*F. Solani*	*A. niger*	*F. Solani*	*A. niger*
0	0^{f}	0^{d}	0^{f}	0^{f}	0^{f}	0^{e}	0^{d}	0^{d}
20	27.91^{e}	51.86^{d}	45.81^{e}	34.19^{e}	55.51^{e}	64.42^{d}	64.42^{c}	56.51^{c}
40	48.14^{d}	54.65^{c}	55.43^{d}	44.18^{d}	59.77^{d}	77.91^{c}	86.49^{b}	67.44^{b}
60	56.98^{c}	73.72^{b}	62.09^{c}	54.19^{c}	80.37^{c}	83.43^{b}	90.23^{a}	78.75^{ab}
80	70.47^{b}	78.61^{a}	76.21^{b}	65.12^{b}	85.58^{b}	85.58^{b}	93.02^{a}	82.68^{a}
100	78.61^{a}	80.47^{a}	80.93^{a}	76.05^{a}	90.31^{a}	89.07^{a}	94.89^{a}	86.05^{a}

Values with the same superscript(s) in the same column are not significantly different at P>0.05 by DMRT.

Table 5: Percentage growth inhibition of fungal isolates from *Chrysophyllum albidum* fruits after exposure to varying concentrations of leaf extract of various plants.

spores, which are easily dispersed by wind. Similarly *Aspergillus species* are capable of utilizing an enormous variety of substrates as the result of large number of enzymes they produce [21].

Phytochemical screening of the plants revealed the presence of saponin, alkaloid, tanin, steroids, Phenols, terpenes, glycosides, and flavonoids. The presence of these secondary metabolites could be responsible for their antifungal activity. Egwin et al. have earlier demonstrated the presence of tannins in *Euphorbia hirta* and opined that it may account for its antimicrobial activity. Tannins have been reported to be toxic to bacteria, filamentous fungal and yeast [22]. Ogbo and Oyibo [19] reported that the presence of alkaloids, saponins and terpenoids in the extract of *Ocimum gratissimum* may have accounted for the broad spectrum of activities on the fungal isolate tested.

The analysis of the plant extract of the leaves in this study showed a complex mixture of constituents. The total number of compounds identified varied from 7-14 in all the plant samples. It is possible that these compounds identified in the plant extracts were responsible for the observed fungi-toxic effects in the study. Sunderham [23] reported that the toxic action of the plant extract of *E. heterophylla* is due to the combined action of its constituents this is similar to the observations of Ilondu [14,16].

Erucic acid was the highest constituent found in *E. heterophylla* (22.22%) and *A. fambiata* (28.88%) extracts. Antimicrobial activity of *Eruca sativa* seed oil has been reported to be due to higher concentration of erucic acid present in the oil [24,25]. Varied concentrations of fatty acid including their ethyl and methyl esters were found abundant in all the plant extracts. Several researchers have reported the antifungal activity of fatty acid and their ethyl and methyl esters against pathogenic fungi [14,16,26].

The percentage inhibition of the mycelia growth of the tested fungi was found to increase as concentration of the plant extracts increased. This may be as a result of the presence of the biologically active antimicrobial compounds of the extracts in higher quantity at lower dilutions, this findings is in consonance with the work of Fernadex et al. [27] who suggested that with increasing concentrations the antagonistic property of the extract increased.

The above result clearly confirms that the test fungi varied widely in the degree of their susceptibility to the extracts. The extract of *Euphorbia heterophylla* was the most effective of all the extracts in inhibiting the growth of *Aspergillus niger* followed by *Acalypha fimbriata, Phyllanthus amaraus* and *Euphorbia hirta.* While ethanolic extracts of *Acalypha fimbriata* was the most effective in inhibiting the growth of *Fusarium solani* followed by *Ephorbia heterophylla, Euphorbia hirta* and *Phyllanthus amarus.* Previous studies have shown that ethanolic leaf extracts of *E. hirta, E. heterophylla, A. fimbriata, P. amarus* and other species of these genera were capable of inhibiting the growth of bacteria, and fungi [13,23,28-32].

Conclusion

The result of this study is an indication that these Euphorbiaceae could be a potential source of antifungal agents. Knowledge of chemical constituents of non-economic plants is desirable because such information could be valuable in discovering new source of economic materials, which may be precursors for the synthesis of complex chemical substances. Such screening of various natural organic compounds and identification of active agents is the need of the century for the formulation of plant biofungicide and improvement of food security for the timing world population.

References

1. Ehiagbonare JE, Oniyibe HI, Okoegwale EE (2008) Studies on the isolation of normal and abnormal seedling of *Chrysophyllum albidum*: A step towards sustainable management of the taxon in the 21st century. Science Research Essay 3: 567-570.

2. Amusa NA, Ashaye OA, Oladapo MO (2003) Biodeterioration of African star apple (*Chrysophyllum albidum*) in storage and the effect on its food value. African Journal of Biotechnology 2: 56-59.

3. Agbogidi OM, Ilondu EM (2012) Heavy metal contents of *Gambaya albida* (Linn.) seedlings growth in soil contaminated with crude oil. Journal of Biological and Chemical Research 29: 320-325.

4. Bada SO (1997) Preliminary information on the ecology of *Chrysophyllum albidum* G.Don in West Central Africa. In: Denton OA, Ladipo DO, Adetoro MA and Sarumi MB (Eds.). Proceedings of a National Workshops on the potentials of star apple in Nigeria, pp:16-25.

5. Umoh IB (1998) Commonly used fruits in Nigeria In: Osagie, A.U. and Eka,

O.U. (Eds.). Nutritional quality of plant foods. Post-Harvest Research Unit, Department of Biochemistry, University of Benin, Benin City, Nigeria, pp: 84-120.

6. Ureigho UN (2010) Nutrient values of *Chrysophyllum albidum* Linn African star apple as a domestic income plantation species. African research review 4: 50-56.

7. Okoli BJ, Okere OS (2010) Antimicrobial activity of the phytochemical constituents of *Chrysophyllum albidum* G. Don-Holl (African star apple). Journal of Research In National Development 8: 1-7.

8. Oboh IO, Aluyor EO, Audu TOK (2009) Use of *Chrysophyllum albidum* for the removal of metal ions from aqueous solution. Scientific Research and Essay 4: 632-635.

9. Olorunnisola DS, Amao IS, Ehigie DO, Ajayi ZAF (2008) Antihyperglycernie and hyperlipidemic effect of ethanolic extracts of *Chrysophyllum albidum* seed cotyledon in Alloxan induced-diabetic rats. Research Journal of Applied Science 3: 123-127.

10. Kazeem-Ibrahim F, Asinwa IO, Iroko OA, Aiyeyika AK, Fapojuwomi OA (2013) Investigation of fungi associated with the spoilage of *Chrysophyllum albidum* (G.Don) fruits. WebPub Journal of Agricultural Research 1: 56-60.

11. Efe SI (2007) The climate of Delta State, Nigeria. In: Odemerho FO, Awaritefe OD, Atubi AO, Ugbomeh BA and Efe SI (Eds.) Delta State in Maps; Department of Geography and Regional Planning, Delta State University Abraka, Nigeria. pp: 24-30.

12. Akobundu IO, Agyakwa CW (1998) A Handbook of West African weeds 2nd edition. International Institute of Tropical Agriculture, Ibadan, Nigeria pp: 564.

13. Ilondu EM (2011) Evaluation of some aqueous plant extracts used in the control of pawpaw fruit (*Carica papaya* L.) rot fungi. Journal of Applied Biosciences 37: 2419-2424.

14. Ilondu EM (2013b) Phytochemical composition and efficacy of ethanolic leaf extracts of some Vernonia species against two phytopathogenic fungi. Journal of Biopesticides 6: 165-172.

15. Oyewale AO, Audu OT (2007) The medicinal potentials of aqueous and methanol extracts of six flora of tropical Africa. Journal of Chemical Society of Nigeria 32: 150-155.

16. Ilondu EM (2013a) Chemical constituents and comparative toxicity of Aspilia africana (pers) C.D Adams leaf extractives against two leafspot fungal isolates of Paw-paw (*Carica papaya* L.). Indian Journal of Science and Technology 6: 5242-5248.

17. Chohan S, Atiq R, mehmoo MA, Naz S, siddique B, et al. (2011) Efficacy of few plant extracts against *Fusarium oxysporium* F.sp gladioli, the cause of corm rot of gladiolus. Journal of Medicinal Plant Research 5: 3887-3890.

18. Ayodele SM, Ilondu EM, Onwubolu NC (2009) Antifungal properties of some locally used spices in Nigeria against some rot fungi. African Journal of Plant Science 3: 139-141.

19. Ogbo EM, Oyibo A (2008) Effects of three plant extracts (*Ocimum grattissimum, Acalypha wilkesiana* and *Acalypha macrostachya*) on post-harvest pathogen of Persea americana. Journal of Medicinal Plants Research 2: 311-315.

20. Ilondu EM, Echigeme MV, Isitohan FO (2015) Bioactive principles and antifungal activity of acetone extract fractions of three tropical spices against *Aspergillus niger* causing fruit rot of African pear (*Dacryodes edulis*). In: Book of Abstract and Programme, 40th Annual Conference of Nigerian Society for Plant Protection (NSPP) Abuja, pp: 36.

21. Alexopoulus CJ, Minms CW, Blackwell M (2002) Introductory mycology. 4th Edition. John Wiley and Sons Inc., Singapore, pp: 869.

22. Harborne BJ (1992) Phytochemical methods: A guide to modern techniques of Plant Analysis, 3rd Edition, Chapman and Hall, London, pp: 58.

23. Sunderham M (2010) Antimicrobial and Anticancer studies on *Euphorbia heterophylla*. Journal of Pharmacy Research 3: 2332-2333.

24. Khoobchandani M, Ojeswi BK, Ganesh N, Srivastava MM, Gabbanini S, et al. (2010) Antimicrobial properties and analytical profile of traditional *Eruca sativa* seed oil: Comparison with various aerial and root plant extracts. Food Chemistry 120: 217-224.

25. Gulfaraz M, Sadiq A, Tariq H, Imran M, Qureshi R, et al. (2011) Phytochemical analysis and antibacterial activity of *Eruca sativa* seed. Pakistan Journal of Botany 43: 1351-1359.

26. Chandrasekaran M, Senthilkumar A, Venkatesalu V (2011) Antibacterial and antifungal efficacy of fatty acid methyl esters from the leaves of Sesuvium portulacastrum L. European Review for Medicinal and Pharmacological Science 15: 775-780.

27. Fernández MA, García MD, Sáenz MT (1996) Antibacterial activity of the phenolic acids fractions of Scrophularia frutescens and Scrophularia sambucifolia. J Ethnopharmacol 53: 11-14.

28. Emele FE, Agbonlahor DE, Emakpare CI (1997) Antimcrobial activity of Euphorbia hirta leaves collected from two geographically dissimilar regions of Nigeria. Nigerian Journal of Microbiology 11: 5-10.

29. Brown PD, Izundu A (2004) Antibiotic resistance in clinical isolates of Pseudomonas aeruginosa in Jamaica. Rev Panam Salud Publica 16: 125-130.

30. Onyeke CC, Maduewesi JNC (2006) Evaluation of some plant extracts for the control of post-harvest fungal diseases of Banana (*Musa sapientus* Linn). Fruit in South-Eastern Nigeria. Nigerian Journal of Botany 19: 129-137.

31. Bhaskara KV (2010) Antibacterial and antifungal activities of Euphorbia hirta leaves. Journal of Pharmacy Research 3: 548-549.

32. Evans CE, Banso A, Samuel OA (2002) Efficacy of some nupe medicinal plants against *Salmonella typhi*: an in vitro study. J Ethnopharmacol 80: 21-24.

Effect of Four Mycorrhizal Products on *Fusarium* Root Rot on Different Vegetable Crops

Al-Hmoud G* and Al-Momany A

University of Jordan, Faculty of agriculture, Department of Plant Protection, Amman, Jordan

Abstract

Vesicular arbuscular mycorrhizal fungi (VAM) are symbiotic fungi which interact with the root system of higher plants by producing external and internal hyphae, vesicules and arbuscules. This study aimed to determine the efficiency of different vesicular arbuscular mycorrhizal fungi in improving plant resistance against *Fusarium oxysporum*. Four mycorrhizal products; Bacto_Prof, Endomyk_Basic, Endomyk_Conc and Endomyk_Prof were used. *Glomus intraradices* was the mycorrhizal fungus. All products reduced *Fusarium* infection by increasing the plant height and reducing the root infection. Bacto_Prof was the best product in the presence of *Fusarium* infection of tomato plants which showed an increase in plant height up to 44% and 154% in plant fresh root weight (FRW). *Fusarium* was reduced by 50% in Bacto_Prof treated tomato plants and mycorrhizal colonization was enhanced from 31% to 42% in *Fusarium* infected tomato seedlings. In pepper experiment; Endomyk_Basic was the best product in all treatments which enhanced mycorrhizal root colonization from 56% to 68%. *Fusarium* infection was suppressed in pepper treated plants with Endomyk. Basic from 2.45% to 1.5%. Mycorrhizal colonization with all products was enhanced by the presence of *Fusarium* more than in non-infected plants. In squash experiment Endomyk_Basic was the best product, but in root colonization Endomyk_Conc performed the best form 52% to 64%. From the results of this study, it was concluded that all mycorrhizal products were significantly inhibited *Fusarium* infection by enhanced and increased mycorrhizal root colonization, so enhanced plant growth and increased root volume.

Keywords: Symbiotic fungi; Mycorrhizal fungi; *Glomus intraradices*; *Fusarium*

Introduction

Mycorrhiza is a symbiotic association between a fungus and the root system of vascular plants belonging to phylum Glomeromycota [1,2] (www. Mycobank.org). Symbiosis, the mutual beneficial association. Mycorrhizal fungi are the most widespread fungal symbionts that colonizes root system of over 90% of plant species to the mutual benefit of both the plant host and fungus [3,4] either exteracellularly as in ectomycorrhizal fungi or intercellularly as in endomycorrhizal fungi [arbuscular mycorrhizal (AM) fungi]. Vesicular arbuscular mycorrhiza (VAM) colonizes plant roots and extends into the surrounding bulk soil to the root depletion zone around the root system [5]. Different crops exhibited different VAM species and different stages of fungal invasion ranging from hyphae, arbuscules and vesicles or combinations of all structures [6]. The major role of VA-fungi is to supply plant roots with phosphorus, because phosphorus is an extremely immobile element in soils [7]. A Vesicular-arbuscular mycorrhiza (VAM) fungus was first studied in Jordan by Dr. Al-Momany in 1993 [6].

However, the mutualistic symbiosis between arbuscular mycorrhizal fungi (AMF) and plant roots plays an important role in nutrient cycling in the ecosystem and can protect plants against soil-borne pathogens [4]. Two main groups of soil-borne pathogens have been studied: nematodes and fungi such as *Phytophthora* spp, *Verticillium* wilt and *Fusarium* wilt. In the presence of arbuscular mycorrhizal fungi (AMF), a reduction in pathogen population or in the disease severity on the host plant has been demonstrated [8]. Also, AMF protect unsupervised root from parasitic fungi by increase wall thickness in the cortical cells of the roots [9]. However, the efficacy of AMF as disease control agents depends on several abiotic and biotic factors, such as temperature, soil moisture, soil phosphorus content, mycorrhizal fungus, time of mycorrhizal inoculation, levels of mycorrhizal inoculum, and inoculum potential of pathogen [4].

The soil born pathogen *Fusarium oxysporum* infect the vascular system of vegetable crops severely especially in hot weather such as in Jordan Valley. *F. oxysporum* can persist in soil for long period of time as chlamydospores which enters the root system through wounds. The first symptoms appear as slight vein clearing on the outer younger leaflets, on older leaves epinasty caused by drooping of the petioles, followed by stunting of the plants, after that yellowing of the lower leaves and wilting of the whole plant then death occur.

Therefore, this work was conducted to determine the effect of mycorrhizal products in its recommended dose on the performance of infected tomato, pepper, squash and eggplant plants with *Fusarium oxysporum*.

Materials and Methods

Mycorrhizal products and *Fusarium* culture

Four different mycorrhizal products were exported from Terrabioscience UG. Bernbug-Germany; and each product was used in different doses according to the application rates and instructions indicated by the manufacturer; Bacto_Prof (1 g/L soil), Endomyk_Conc (1 g/L soil), Endomyk_Prof (2 g/L soil) and Endomyk_Basic (8 g/L soil). All the mycorrhizal products were in powder form except Endomyk_Basic, the mycorrhizae was produced in granule form.

***Corresponding author:** Al-Hmoud G, University of Jordan, Faculty of agriculture, Department of Plant Protection, Amman, Jordan
E-mail: Aeng_ghina2009@yahoo.com, momanyah@ju.edu.jo

Fusarium oxysporum f.sp. *lycopersici* was isolated from infested field soil planted with tomato, and *Fusarium oxysporum* f.sp. *cucumerinum* was isolated from infected field soil planted with cucumber in 2011 in Jordan Valley. *Fusarium oxysporum* f.sp. *lycopersici* was used to inoculate tomato, pepper and eggplant seedlings. While *Fusarium oxysporum* f.sp. *cucumerinum* was used for inoculate squash seedlings.

Glasshouse work

The effect of the four mycorrhizal products on *Fusarium* wilt (*Fusarium oxysporum*) was studied on different vegetable crops; tomato (*Solanum lycopersicum*) var. Majd, eggplant (*Solanum melongena*) var. Ajami, pepper (*Capsicum frutescens*) var. Hendi, and squash (*Cucurbita maxima*) var. Yasmina F1. The seedlings must be under drought conditions before using them. The *Fusarium* suspension was prepared in the same day of inoculation for the viability of micro and macro conidia. Seedlings were cleaned from peat moss around the root system to facilitate inoculation. *Fusarium* inoculum; PDA plates containing *Fusarium* growth were transferred to blender, mixed with distilled water for two minutes. The inoculum contained micro and macro conidia and mycelium. The concentration of *Fusarium* was determined by using a haemacytometer slide. 10^9 spores/ml was made to inoculate tomato, pepper and eggplant seedlings and 10^6 spores/ml used for squash seedlings. For each seedling, 50 ml of inoculum were added to the root zone of the four-week old mycorrhizal seedlings of tomato, pepper, eggplant and two-week old mycorrhizal seedlings of squash by pouring the suspension into a hole made around the root zone.

At the same time soil: sand mixture was treated with the normal recommended dose for each product to plant one seedling in each sterilized pot. For each crop, ten treatments were used; four products with *Fusarium*, four products without *Fusarium*, *Fusarium* alone and non-inoculated plants (control). For treatments using mycorrhizal products with *Fusarium*; half amount of the soil: sand: product mixture was transferred to the pot, then each seedling was dipped in 50 ml *Fusarium* suspension for two minutes and then add the rest of the suspension to the root zone, then completed the rest of the soil mixture. Each treatment was comprised of five replicate pots, grown under glasshouse conditions until harvest time (7-8 weeks). At the end of the experiment plants were taken to test them for Fusarial infectivity. Shoot length was measured; fresh shoot weight (FSW) and fresh root weights (FRW) were recorded. One gram of fine feeder roots from each treatment were taken to examine its mycorrhizal colonization, then put the fresh shoot and root of each treatment in separate closed paper bags, and dried for 24 hrs at 70°C in the oven to measure the dry shoot weight (DSW). All data were statistically analyzed by using the SAS program, comparison between means was done according to LSD at 5% level.

Evaluation of the experiment

Inoculation of mycorrhizal seedlings with *Fusarium oxysporum*: Fresh infected seedlings with *Fusarium* were examined from the crown area (creamy to brown color), cleaned by tap water from soil, cut to pieces, sterilized by 0.5% hypex, dried on sterilized filter paper and then four pieces were placed (2 × 2 cm) on potato dextrose agar (PDA) plates (39 gm/l), then kept for two weeks in the incubator at 25°C ± 2°C. *F. oxysporum* f.sp. *cucumerinum* needs higher incubation temperature than *F. oxysporum* f.sp. *lycopersici*; about 27-28°C. *Fusarium* growth appeared in different colors (white, pink, purple and sometimes yellow in *F. oxysporum* f.sp. *cucumerinum*). To make sure if it is *Fusarium*; microscopic slides from those different colors were prepared under the microscope. Micro and macro conidia ranging from one cell to four or five conidial cells were seen in huge numbers. Then pure inoculum of *Fusarium* was prepared using sterilized cork borer. In each PDA plate one or two *Fusarium* discs were placed, and plates were incubated for another two weeks at 24-26°C ± 2°C.

Evaluation of *Fusarium* infection: At the end of glasshouse experiments, plants were taken to the laboratory to re-isolate *Fusarium* on PDA plates to ensure the *Fusarium* infection. Above the crown area by 2 cm, small pieces (1 × 1 cm) were cut, washed by tap water, sterilized by 0.5% hypex, dried on sterilized filter paper and then placed four pieces in each plate. The plates were incubated at 25°C for 5-7 days. After the incubation period, infected pieces were counted and examined microscopically in each petridish to ensure the *Fusarium* infection.

Evaluation of VAM roots colonization: Roots were washed carefully to remove soil particles, then heated in 10% KOH at 90°C for half an hour to destroy cell cytoplasm, so the plant cell will be cleared, then washed for 2-3 minutes in running tap water gently, then stained with 0.05% trypan blue in lactophenol according to Philips and Hayman method (1970). Ten root segments with 1 cm length were mounted on each slide and examined microscopically. The incidence and intensity of root colonization with mycorrhiza were calculated using a scale of (0-10) where zero means no colonization, 5 means 50% mycorrhizal root colonization and 10 means 100% mycorrhizal root colonization [10], the readings were taken from the average percentage of thirty roots for each treatment.

Results

Tomato experiment

Effect of *Fusarium oxysporum* on tomato growth was presented in Table 1. There were no significant differences between plants inoculated with *Fusarium* and not inoculated with *Fusarium*; except in *Fusarium* severity. However; considering plant height parameters, Bacto_Prof was highly significant than all other products by 44%, and all mycorrhizal products were significantly higher than control plants by 29, 31 and 26% in Endomyk_Basic, Endomyk_Conc and Endomyk_Prof, respectively. The increase in plant FSW was 27, 23, 20 and 19% for the four products; Bacto_Prof, Endomyk_Basic, Endomyk_Conc and Endomyk_Prof in *Fusarium* treatments, respectively. However, the increase in plant FRW was enhanced more in mycorrhizal plants than in non-mycorrhizal plants by 154% for Bacto_Prof and Endomyk_Basic, while in Endomyk_Conc and Endomyk_Prof the increase was 134%. Infection percentage in non-mycorrhizal plants was the highest by 8% compared with the mycorrhizal plants by 4% for Bacto_Prof, Endomyk_Conc and Endomyk_Prof and 5% for Endomyk_Basic product.

Pepper experiment

Effect of *Fusarium oxysporum* on pepper growth was summarized in Table 2. There were significant differences in plant height in *Fusarium* treatments between mycorrhizal plants and non-mycorrhizal plants. Plant height of mycorrhizal plants infected with *Fusarium* in were improved by 20, 27, 21 and 25% by Bacto_Prof, Endomyk_Basic, Endomyk_Conc and Endomyk_Prof products, respectively. Plant FSW in *Fusarium* treatment with Endomyk_Basic was more significant than others by 115%. Plant DSW was enhanced significantly by all mycorrhizal products; Bacto_Prof, Endomyk_Conc and Endomyk_Prof by 56, 102, 85 and 98%, respectively compared to control plants

Treatment	*F. oxysporum*	Height (cm)	FSW (g)	DSW (g)	FRW (g)	Severity (cm)	Infection%
Bacto_Prof	+	51.6 a	10.622 a	1.868 ab	4.904 a	2.1a	4%
	-	50.2 ab	10.162 ab	1.734 ab	4.678 a	0b	0
Endomyk_Basic	+	46.4 abc	10.286 ab	1.844 ab	4.898 a	2.3a	5%
	-	45.4 bc	9.484 bc	2.006 a	4.798 a	0b	0
Endomyk_Conc	+	47.1 abc	10.00 ab	1.842 ab	4.5 a	1.8a	4%
	-	42.5 cd	9.734 bc	1.648 ab	4.614 a	0b	0
Endomyk_Prof	+	45.2 bc	9.96 ab	1.656 ab	4.568 a	2a	4%
	-	42 cd	9.722 bc	1.636 ab	4.438 a	0b	0
Control	+	35.9 e	8.36 d	1.146 c	1.932 b	2.8a	8%
	-	39.8 ed	9.076 cd	1.35 bc	2.808 b	0b	0

Values are average of five plants, values within each column followed by the same letter are not significantly different (P<0.05) according to LSD. (+) means inoculated with *F. oxysporum* while (-) means non-inoculated with *F. oxysporum*.

Table 1: Effect of *Fusarium oxysporum* with the recommended dose of mycorrhizal products on growth of tomato plants.

Treatment	*F. oxysporum*	Height (cm)	Fruit wt (g)	FSW (g)	DSW (g)	FRW (g)	Severity (cm)	Infection%
Bacto_Prof	+	36.1 b	0.368a	9.982c	1.004ab	5.586b	0.56a	1.55%
	-	35.5 b	0.058abc	9.828c	1.032ab	5.5b	0b	0
Endomyk_Basic	+	38.4 a	0 c	11.258 a	1.296 a	6.578 a	0.56a	1.50%
	-	38.5 a	0 c	11.182ab	1.27 a	6.294ab	0b	0
Endomyk_Conc	+	36.4 b	0.036ab	10.112c	1.186ab	5.958ab	0.48a	1.32%
	-	35.8 b	0.234ab	9.994 c	1.118ab	5.85ab	0b	0
Endomyk_Prof	+	37.7 a	0 c	10.59 abc	1.268 a	6.266ab	0.66a	1.75%
	-	36.6 a	0.064 b	10.35 bc	1.262 a	5.846ab	0b	0
Control	+	30.2 d	0.028 bc	5.234 e	0.642 c	3.936c	0.74a	2.45%
	-	33.6 c	0 c	7.04 f	0.858 bc	4.008c	0b	0

Values are average of five plants, values within each column followed by the same letter are not significantly different (P<0.05) according to LSD. (+) means inoculated with *F. oxysporum* while (-) means non-inoculated with *F. oxysporum*.

Table 2: Effect of *Fusarium oxysporum* with recommended dose of mycorrhizal products on pepper growth.

infected with *Fusarium*. Plant FRW, was increased significantly by 42, 67, 51 and 59% with Bacto Prof, Endomyk Conc and Endomyk Prof products, respectively. *Fusarium* infection percentage in pepper control plants was the highest 2.45%, compared to 1.32 and 1.5% in Endomyk_Conc and Endomyk_Basic treatments, respectively.

Eggplant experiment

Effect of *Fusarium oxysporum* on pepper growth was presented in Table 3. There were no significant differences between plants treated with four mycorrhizal products and inoculated with *Fusarium* and not inoculated with *Fusarium* referring to height, FSW, DSW and FRW. The four products were not significantly different from each other and all were significantly different from control plants infected with *Fusarium*. Plant height in *Fusarium* treatment with mycorrhizal products were enhanced by 14% for Bacto_Prof, 16% for Endomyk_Basic, Endomyk_Conc and Endomyk_Prof. Plant FSW in *Fusarium* treatment was raised 49% by Bacto_Prof and 51% by Endomyk_Basic, Endomyk_Conc and Endomyk_Prof over non-mycorrhizal plants. Fresh root weight in *Fusarium* treatment was increased 49, 50, 45 and 51% by Bacto_Prof, Endomyk_Basic, Endomyk_Conc and Endomyk_Prof, respectively. Infection percentage in control plants was the highest (5.80%) compared with 3.50, 2.22, 3.42 and 2.72% for Bacto_Prof, Endomyk_Basic, Endomyk_Conc and Endomyk_Prof, respectively.

Squash experiment

Effect of *Fusarium oxysporum* on squash growth was presented in Table 4. There were no significant differences between FSW in *Fusarium* inoculated mycorrhizal plants versus non-inoculated mycorrhizal plants. Plant height in *Fusarium* treatment with Endomyk_Basic was 24% higher than control plants inoculated with *Fusarium*, followed by 8, 7 and 6%, respectively for Endomyk_Prof, Endomyk_Conc and Bacto_Prof. In *Fusarium* treatment, FSW was increased 13% by Bacto_Prof and Endomyk_Prof, 23% by Endomyk_Basic and 12% by Endomyk_Conc. Squash FRW in *Fusarium* treatment was enhanced by 69, 88, 73 and 67% for Bacto_Prof, Endomyk_Basic, Endomyk_Conc and Endomyk_Prof, respectively compared to control plants infected with *Fusarium*. Infection percentage in non-mycorrhizal squash plants was the highest by 14.45% compared with 9.41, 10.13, 10.92 and 11.69% for Bacto_Prof, Endomyk_Basic, Endomyk_Conc and Endomyk_Prof products, respectively.

Evaluations of VAM root colonization

Mycorrhizal root colonization was more efficient with *Fusarium* infection in all vegetable crops as present in Table 5. The examination of the control plant roots and the *Fusarium* inoculated plants for a possible contamination with AMF was negative. In tomato, Bacto_Prof (+Fus) was the most effective product in root colonization (42%) followed by Endomyk_Prof (+Fus) (37%), Endomyk_Conc (+Fus) (30%) and then Endomyk_Basic (+Fus) with 33%. In pepper, Endomyk_Basic (+Fus) recorded the highest root colonization by 68% than other products, comparing the products with *Fusarium* infection and without *Fusarium*. It was clear that Endomyk_Prof (+Fus) performed better than other products, while in squash, Bacto_Prof (+Fus) was more efficient than other products by 65% followed by Endomyk_Conc (+Fus) (64%), Endomyk_Basic (+Fus) (60%) and Endomyk_Prof (+Fus) was less effective by 55%. However; Endomyk_Basic (+Fus) record the highest root colonization by 54% in eggplant, but Endomyk_Conc performed better than other products with *Fusarium* infection.

Treatment	*F. oxysporum*	Height (cm)	FSW (g)	DSW (g)	FRW (g)	Severity (cm)	Infection %
Bacto_Prof	+	34.6 ab	9.356 a	1.098 ab	4.866 a	1.2 a	3.50%
	-	35.9 a	9.916 a	1.192 a	4.726 a	0 c	0
Endomyk_Basic	+	35.1 a	9.464 a	1.07 ab	4.924 a	0.78 a	2.22%
	-	35.3 a	9.61 a	1.054 abc	4.922 a	0 c	0
Endomyk_Conc	+	35.1 a	9.45 a	1.106 ab	4.738 a	1.2 a	3.42%
	-	35.4 a	9.478 a	1.152 a	4.8 a	0 c	0
Endomyk_Prof	+	35.2 a	9.434 a	1.128 a	4.952 a	0.96 a	2.72%
	-	35.1 a	9.454 a	1.13 a	4.906 a	0 c	0
Control	+	30.4 c	6.264 b	0.866 c	3.276 b	1.76 b	5.80%
	-	32.4 bc	7.014 b	0.922 bc	4.224 ab	0 c	0

Values are average of five plants, values within each column followed by the same letter are not significantly different ($P<0.05$) according to LSD. (+) means inoculated with *F. oxysporum* while (-) means non-inoculated with *F. oxysporum*.

Table 3: Effect of *Fusarium oxysporum* with recommended dose of mycorrhizal products on growth of eggplant.

Treatment	*F. oxysporum*	Height (cm)	FSW (g)	DSW (g)	FRW (g)	Severity (cm)	Infection%
Bacto_Prof	+	45.7 bc	13.404 ab	3.506 a	4.534 a	4.3a	9.41%
	-	48.2 bc	13.792 ab	3.476 a	4.568 a	0c	0
Endomyk_Basic	+	53.3 a	14.562 a	3.406 a	5.038 a	5.4a	10.13%
	-	49 a	13.876 a	2.988 a	4.832 a	0c	0
Endomyk_Conc	+	45.8 abc	13.44 ab	3.284 a	4.656 a	5a	10.92%
	-	49.4 ab	13.874 a	3.55 a	4.9 a	0c	0
Endomyk_Prof	+	46.2 bc	13.314 ab	3.196 a	4.484 ab	5.4a	11.69%
	-	49.2 ab	13.87 a	3.102 a	5.05 a	0c	0
Control	+	42.9 cd	11.882 c	3.43 a	2.684 b	6.2b	14.45%
	-	45.6 d	12.476 bc	3.486 a	3.594 b	0c	0

Values are average of five plants, values within each column followed by the same letter are not significantly different ($P<0.05$) according to LSD. (+) means inoculated with *F. oxysporum* while (-) means non-inoculated with *F. oxysporum*.

Table 4: Effect of *Fusarium oxysporum* with the recommended dose of mycorrhizal products on squash growth.

Treatment	Infection	Tomato	Pepper	Squash	Eggplant
Bacto_Prof	+ Fus	42%	46%	65%	39%
	- Fus	31%	37%	52%	35%
Endomyk_Basic	+ Fus	33%	68%	60%	54%
	- Fus	30%	56%	54%	50%
Endomyk_Conc	+ Fus	30%	53%	64%	51%
	- Fus	25%	45%	52%	38%
Endomyk_Prof	+ Fus	37%	62%	55%	49%
	- Fus	28%	49%	51%	43%

(+ Fus) means inoculated with *F. oxysporum* while (- Fus) means non-inoculated with *F. oxysporum*.

Table 5: Effect of *Fusarium* infection on mycorrhizal root colonization of different products in different vegetable crops.

Discussion

All products reduced *Fusarium* infection by increasing the plant height and reducing the root percentage of infected plants. The competition for space and nutrients can change the morphology of root system, rhizosphere effect and the activation of plant defense mechanisms which are responsible for disease inhibition by VAM fungi [11]. Bacto_Prof recorded the highest in all parameters than other products with 42% root colonization in tomato experiment. Microconidia of *Fusarium* wilt, has been shown to be stimulated by tomato root exudates, but when colonized by AMF; it was clear that root exudates of plants inhibited the germination of *Fusarium* microconidia more than root exudates of non-mycorrhizal plants [12]. In pepper experiment; Endomyk_Basic was the most effective, but in VAM root colonization, Endomyk_Prof was the first place by 72% then Endomyk_Basic by 68%. Also Endomyk_Basic was the most effective product on both squash and eggplant.

Plant height, shoot fresh weight, shoot dry weight and root fresh weight were significantly different in mycorrhizal than non-mycorrhizal plants in all products, while number of plant flowers number, fruit number and fruit weight there, were no differences between mycorrhizal and non-mycorrhizal plants. *Glomus intraradices* increased plant yield, height and shoot fresh weight. However; pepper plants showed the least yield and height in comparison to tomato and eggplant [13], while in this experiment tomato showed more increase in height compared to pepper and eggplant, but pepper showed the highest mycorrhizal root colonization. The mycorrhizal tomato and eggplant plants had shown an increase in fresh, dry weight and mean plant height compared to non-mycorrhizal which was in agreement with the findings of Kargiannidis et al. [14]. Leaf area, shoot and root dry matter yields were higher in mycorrhizal tomato plants than in non-mycorrhizal [15]. Shoot weight of mycorrhizal tomato treated with or without *F. oxysporum* were more than non-mycorrhizal plants with *Fusarium* [9].

VAM can influence diseases caused by soil-borne pathogens affecting the root system. In general, mycorrhizal plants suffer less damage and the pathogen development will be inhibited [16,17]. *Fusarium oxysporum* is a soil-borne pathogen causing wilting in the vascular system of higher plants and severe yield losses for farmers. Management of *Fusarium* wilt is mainly through chemical soil fumigation (particularly methyl bromide) and resistant cultivars are the most cost-effective and environmentally safe methods of control. However, new races of pathogens overcoming host resistance can develop. The difficulty in controlling *Fusarium* wilt has stimulated the researches in biological control by micro-organisms such as arbuscular mycorrhizal fungi that are important members in the rhizosphere and considered as effective symbionts by protecting the plants from root rot pathogens. The arbuscular mycorrhizal fungi increase plant growth and enhanced plant tolerance to various stress factors. AMF may limit fungal root diseases by strengthening morphological traits of plants with some physiological and microbial modifications in the rhizosphere and by altering the chemical composition of plant tissues [11].

As soon as the VAM colonizes the plant roots the pathogen will be excluded. As a result, the most effective control is achieved when AMF colonization takes place before the pathogen attack the host cells. There are many factors involved in as; changes in root exudates which can cause changes in the rhizosphere microbial community, changes to root biochemistry connected with plant defense mechanisms. Changes to plant defense mechanisms, in another words induced resistance, resulted from a priming effect of the VAM colonization to respond faster to infection by pathogenic fungi [2].

There were no significant differences between mycorrhizal plants inoculated with *Fusarium oxysporum* and mycorrhizal plants alone, and this may be according to some reasons; as early planting date, delated the entrance of *Fusarium*, temperature or plant cultivar. The efficiency of VAM fungi can be influenced by plant pathogen, symbiotic fungi and environmental conditions [9]. While there were differences between mycorrhizal plants inoculated with *F. oxysporum* and plants inoculated with *F. oxysporum* alone. Pepper plants inoculated with *F. oxysporum* was not significantly different from plants inoculated with both VAM fungi and *F. oxysporum*. Mycorrhizal pepper and tomato plants inoculated with *F. oxysporum* showed higher fresh weight and plant height than plants inoculated with *F. oxysporum* only. *Fusarium* was more severe in non-mycorrhizal than in mycorrhizal tomato and pepper plants [16]. Pre-inoculation with VAM fungi was effective in delaying root infection with *Fusarium* [9].

The root colonization by the mycorrhizal products was not affected by the presence of *Fusarium* infection; the percentage of root colonization was higher in *Fusarium* infection in all examined plants. The number of *Fusarium* propagules was decreased significantly when plants were inoculated with *Glomus intraradices* and decreased root necrosis caused by *Fusarium* [18,19]. All mycorrhizal products were effective and significantly different than non-treated plants, and all mycorrhizal products were significantly inhibited *Fusarium* infection by enhanced the plant growth and increased root volume. Bacto_Prof increase plant growth and reduced *Fusarium* infection in tomato, while Endomyk_Basic was more efficient in pepper, squash and eggplant.

References

1. Kirk PM, Cannon PF, David JC, Stalpers J (2001) Ainsworth and Bisby's Dictionary of the Fungi (9thedn). Wallingford UK CAB International.
2. Gosling P, Hodge A, Goodlass G, Bending GD (2006) Arbuscular mycorrhizal fungi and organic farming. Agriculture, Ecosystems and Environment 113: 17-35.
3. Smith SE, Read DJ (1997) Mycorrhizal Symbiosis (2ndedn). Academic Press London.
4. Garmendia I, Goicoechea N, Aguirreolea J (2004) Effectiveness of three Glomus species in protecting pepper (Capsicum annuum L.) against verticillium wilt. Biological Control 31: 296-305.
5. Bethlenfalvay GJ, Barea JM (1994) Mycorrhizae in sustainable agriculture I Effects on seed yield and soil aggregation. American Journal of Alternative Agriculture 9: 157-161.
6. Al Raddad, Al Momany A (1993) Distribution of different Glomus species in rainfed areas in Jordan. Dirasat 20: 165-182.
7. Wetterauerr DG, Killorn RJ (1996) Fallow- and flooded-soil syndromes: effects on crop production. Journal of Production Agriculture 9: 39-41.
8. Caron M (1989) Potential use of mycorrhizal in control of soil-born diseases. Canadian Journal of Plant Pathology 11: 177-179.
9. Al Ameiri NS (1987) Interaction between vesicular arbuscular mycorrhizal fungi and Fusarium root rot of tomato. Master dissertation University of Jordan, Amman, Jordan.
10. Bierman B, Linderman R (1981) Quantifying vesicular-arbuscular mycorrhizae: proposed method towards standardization. New Phytologist Journal 87: 63-67.
11. Akkopru A, Demir S (2005) Biological Control of Fusarium Wilt in Tomato Caused by Fusarium oxysporum f sp lycopersici by AMF Glomus intraradices and some Rhizobacteria. Journal of Phytopathology 153: 544-550.
12. Steinkellner S, Hage-Ahmed K, Garcia-Garrido JM, Illana A, Ocampo JA (2012). A comparison of wild-type, old and modern tomato cultivars in the interaction with the arbuscular mycorrhizal fungus Glomus mosseae and the tomato pathogen Fusarium oxysporum f. sp. lycopersici. Mycorrhiza 22: 189-194.
13. Al Raddad, Al Momany A (1987) Effect of three vesicular arbuscular mycorrhizal isolates on growth of tomato, eggplant and pepper in a field soil. Dirasat 14: 161-168.
14. Kargiannidis N, Bletsos F, Stavropoulos N (2002) Effect of Verticillium wilt (Verticillium dahliae Kleb.) and mycorrhiza (Glomus mosseae) on root colonization, growth and nutrient uptake in tomato and eggplant seedlings. Scientia Horticulture 94: 145-156.
15. Al Karaki GN (2000) Growth of mycorrhizal tomato and minerals acquisition under salt stress. Mycorrhiza 10: 51-54.
16. Al Raddad, Al Momany A (1988) Effect of vesicular arbuscular mycorrhizae on Fusarium wilt of tomato and pepper. Alexandria Journal of Agricultural Research 33: 249-261.
17. Karajeh M, Al Raddad, Al Momany A (1999) Effect of VA Mycorrhizal fungus (Glomus mosseae Gerd and Trappe) on Verticillium dahliae Kleb. Dirasat 26: 338-341.
18. Caron M, Fortin JA, Richard C (1986) Effect of Glomus intraradices on infection by Fusarium oxysporum f.sp. radicis-lycopersici in tomatoes over a 12-week period. Canadian Journal of Botany 64: 552-556.
19. Caron M, Fortin JA, Richard C (1987) Effect of phosphorus concentration and Glomus intraradices on Fusarium crown and root rot of tomatoes. Phytopathology Journal 76: 942-946.

14

Black List of Unreported Pathogenic Bambusicolous Fungi Limiting the Production of Edible Bamboo

Louis Bengyella[1,2,3]*, Sayanika Devi Waikhom[1,2,3], Narayan Chandra Talukdar[1] and Pranab Roy[4]

[1]*Institute of Bioresources and Sustainable Development, Takyelpat, Imphal, Manipur 795001, India*
[2]*Department of Biotechnology, Haldia Institute of Technology, Haldia 721657, West Bengal, India*
[3]*Centre of Advanced Study in Life Sciences, Manipur University, Imphal, Manipur 795003, India*
[4]*Department of Biotechnology, Haldia Institute of Technology, Haldia-721657, West Bengal, India*

Abstract

Edible bamboo species are now domesticated and commercialized because of their nutraceutical values. The production of edible bamboo species are restrained by diseases caused by pathogenic bambusicolous fungi valued at 40% losses of the total $818.6 million generated annually in bamboo trade in North East India. Based on a systematic survey performed for 2 years in succession, only one Basidiomycota, a *Perenniporia* sp. was identified and validated by pathogenicity test. Ascomycota was the dominant and diverse group of pathogenic bambusicolous fungi. Some rDNA locus sequences failed to match sequences in the up-to-date databases and indicated novel species or genera. Divergence study based on rDNA locus showed that pathogenic bambusicolous fungi were located in the class of Ascomycetes, Sordariomycetes, Eurotiomycetes, Dothideomycetes and Basidiomycetes. The data demonstrated for the first time that *Fusarium*, *Cochliobolus*, *Daldinia*, *Leptosphaeria*, *Phoma*, *Neodeightonia*, *Lasiodiplodia*, *Aspergillus*, *Trichoderma*, *Peyronellaea*, *Perenniporia*, *Nigrospora* and *Hyporales* are potent pathogenic bambusicolous fungi genera restraining the production of edible bamboo *Dendrocalamus hamiltonii*.

Keywords: Fungal diversity; Phylogeny analysis; Pathogenicity test; *Trichoderma asperellum*; *Dendrocalamus hamiltonii;* rDNA

Introduction

Woody bamboo species are highly diverse and abundantly represented in Asian countries such as China, Japan and India etc. Raw bamboo products generate annual revenue of $818.60 million in North East India alone [1]. Bamboo is used in paper making, landscaping, soil conservation, handicrafts, construction, as well as food [2,3]. Nonetheless, it is predicted that half of the world woody bamboo species are in risk of extinction [4,5]. Because of the multipurpose usage and the risk of extinction, techniques for *in vitro* propagation and cultivation of endangered edible bamboo shoots had been developed [6,7].

Remarkably, bush fire, shifting cultivation, flowering boom followed by erratic death [3,8,9], pest and diseases are important factors accelerating the extinction of bamboo species. Although edible bamboo cultivation is plagued by these factors, low level production is exacerbated by harmful bambusicolous fungi. Bambusicolous means organismal life on bamboo [10]. Even though some bambusicolous fungi records are indexed (http://nt.ars-grin.gov/fungaldatabases), the list is not comprehensive for the following reasons: 1) The bamboo species hosting bambusicolous fungi are often not reported, 2) most bamboo species are in the wild and not domesticated for phytopathological scrutiny, and 3) the complex lifestyle of bamboo species which encompasses fast growth, giant height, often growing in difficult terrain limits surveillance and impedes insights on bamboo pathology.

Fungal diseases weaken the rate of growth and the quality of edible bamboo shoots. This is because bamboo shoots development depends on the health status of mother clump-rhizome and leaf canopy. To achieve the optimal production of edible bamboo, pathogenic fungi limiting cultivation must be identified. *Dendrocalamus hamiltonii* Nees et Arn. ex Munro is a sympodial commercial species, with erect and curve culms, and highly valued for its nutraceutical values [3,11]. It is richly distributed in North Eastern Himalayan region, India [12]. Young succulent bamboo shoots of *D. hamiltonii* are consumed fresh or fermented as vegetable; and preferred over other species because its fermented products retained good taste and low water content [13]. At present, there is no report on the diversity of pathogenic bambusicolous fungi of any edible bamboo species. To address this issue, landraces of edible bamboo species of *D. hamiltonii* were surveyed for a period of 2 years in succession for fungal diseases and pathogenicity test was used to validate the disease causing potential of the fungi. Herein, new pathogenic bambusicolous fungi and their phylogenetic link are established.

Materials and Methods

Study area, sampling and morphological identification of fungal pathogens

Landraces of edible bamboo species (*Dendrocalamus hamiltonii* GenBank® accession JX564903) were systematically surveyed in bamboo farms for 2 years in succession for the occurrence of fungal diseases during the month of July–August of 2011 to 2013. The farms are located in Imphal - East District, Manipur, India (Figure 1). The average age of bamboo clumps were 5-7 years old. The area often received an average rainfall of 1320 ± 3 mm and temperature of 29 ± 3°C during the months of July to August. Diseased plant tissue fragments (< 1 cm^2) from leaves, nodes and internodes were surface

***Corresponding author:** Louis Bengyella, Institute of Bioresources and Sustainable Development, Takyelpat, Imphal, Manipur 795001, India
E-mail: bengyellalouis@gmail.com

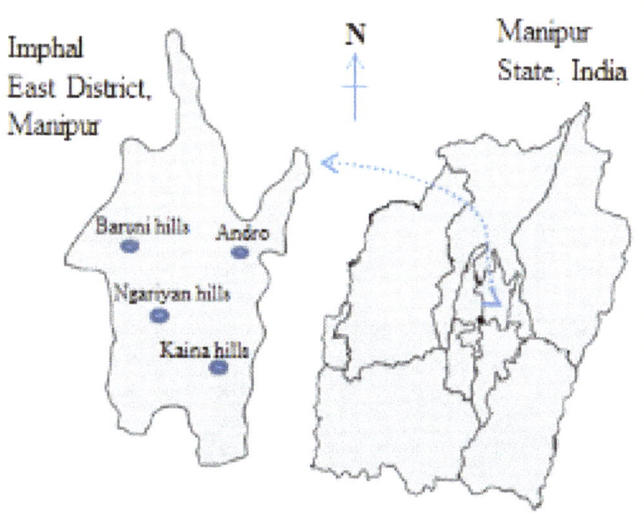

Figure 1: Locations of landraces population of *Dendrocalamus hamiltoniiin* Manipur, India where pathogenic bambusicolous fungi was explored.

sterilised in 0.1% sodium hypochlorite (5 min), 70% ethanol (2 min), and followed by washing in sterile water with three changes. The leaf pieces were plated on potato dextrose agar (PDA) medium (HiMedia®) fortified with 250 mg/L chloramphenicol and incubated at 25°C in the dark. Developed colonies were further purified on V8 agar medium for distinct morphological identification based on standard monographs taxonomic keys with the help of a microscope (Olympus BX61®, Japan).

DNA phylogeny

Sporulating fungi and non-sporulating fungi (that could not be identified morphologically) were characterised at the rDNA locus. Total genomic DNA was isolated from mycelium using UltraClean™ Microbial DNA isolation kits (MO Bio-Laboratories, Carlsbad, CA, USA) as described by the manufacturer. The integrity and quality of DNA was checked by agarose gel electrophoresis and absorbance measurements using a biospectophotometer (Shimadzu® BioSpec, Japan), respectively. rDNA locus comprising of partial sequence of 5.8S rRNA, complete internal transcribed region two (ITS2) and partial 28S rRNA region was amplified using the primer set (5′-tcctccgcttattgatatgc-3′, 5′-gcatcgatgaagaacgcagc-3′) [14] and the PCR conditions were as follows. PCR was performed in a 25 µl volume containing 2.5 µl of 10× DreamTaq buffer green, 1 µl of 2 mM dNTPs, 1 µl of 10 µM of forward and reverse primers each, 0.25 µl of DreamTaq® polymerase (ThermoScientific, UK) 1 µl of 10 ng DNA template and 18.5 µl nuclease free water. DNA template was denatured at 95°C for 3 min, followed by 35 cycles of 95°C for 30s, 55°C for 30s, 72°C for 48s and a final extension at 72°C for 5 min in a thermcycler (Bio-rad, C1000®). All products were profiled by electrophoresis on a 1% agarose gel and stained with ethidium bromide. The PCR products were purified and sequenced. Sequences were assigned to molecular species based on 98–100% sequence similarity threshold in the DNA database of Japan (DDBJ®) in accordance with standard monograph taxonomic keys. Multiple sequence alignment was performed in Muscle program [15] at default settings. Best substitution model parameters for phylogenetic inference were determined based on Akaike Information Criterion, corrected (AICc) and Bayesian Information Criterion (BIC). The maximum likelihood (ML) method was used for phylogenetic inference. All analysis was performed in MEGA 6.06 (updated v. 6140226) software [16]. The ML tree was statistically tested by 1000 bootstrap iterations.

Pathogenicity test

To validate Koch's postulates for the pathogenic bambusicolous fungi, pathogenicity test was performed as follows. Bamboo seeds of *D. hamiltonii* (GenBank® accession JX564903) were propagated in MS culture medium following previously established protocol [17] in a 20 cm long x 15 cm^3 diameter test tube. Following rooting, plants were progressively transferred to sterile soil (consisting of rice-straw vermin-compose-sand mixture (3:1)) in a 10 cm diameter pots under greenhouse conditions. Following the development of internodal culms with 15-20 true leaves, plants were sprayed with a suspension of 10^6 conidia/ml of each fungal pathogen under aseptic conditions. Each inoculated plant was enclosed with a plastic bag to create a near 100% humidity. Plants were observed every 12 h for the development of symptoms and pathogenicity test was performed three times. Only fungal pathogens which produced similar symptoms to those observed in the field are reported.

Results and Discussion

In Manipur, India, landraces of *D. hamiltonii* are densely populated in Imphal East District (Figure 1). This region often witness sporadic rainfall, foggy weather, and strong wind movement during July–August each year. *D. hamiltonii* is rich in nutraceutical values and highly demanded by consumers [3,11]. Because of the nutritional attributes and important population size of *D. hamiltonii* in Manipur, the study was focused on the fungal pathogens of this edible bamboo species.

A total of 32 bambusicolous pathogenic fungi identified and validated by Koch's postulates was deposited in DDBJ accessions (Table 1) and were used for phylogenetic reconstruction. Of the 32 fungal pathogens, 31 were Ascomycota distributed within the class of Dothideomycetes, Eurotiomycetes, Sordariomyctes and one was unclassified. Nonetheless, it has been shown that most fungi in these subclasses are pathogens [18,19]. Only one of the fungal pathogen (i.e. *Perenniporia* sp.) was Basidiomycetes (Table 1). Additionally, the pathogenic bambusicolous fungi belonged to the genera of *Fusarium, Cochliobolus, Daldinia, Leptosphaeria, Phoma, Neodeightonia, Lasiodiplodia, Aspergillus, Trichoderma, Peyronellaea, Perenniporia, Nigrospora* and *Hyporales*. It is estimated that there are over 630 Ascomycetes, 150 Basidiomycetes and 330 mitosporic taxa (100 coelomycetes and 230 hyphomycetes) infecting bamboo [10,20]. The finding in this study is in accordance with other data [10,20], that predominant bambusicolous fungi of bamboo are Ascomycetes (Table 1). Although *Hypocreaceae* is understood to be the common bambusicolous fungi [10], only one - *Hypocreales* sp. strain B101 was identified as a pathogenic bambusicolous via Koch's postulates (Table 1).

The combined sequences had an estimated transition/transversion bias ratio of 1.26. The Kimura 2-parameter [21] substitution model (+G, 5 categories, parameter = 3.50) produced the following nucleotide frequencies: A = 25.00%, T/U = 25.00%, C = 25.00%, G = 25.00% and sequence alignment is shown (Figure 2). The rate variations model that allowed for some sites to be evolutionarily invariable [+ I] was 4.71%. In the dataset, a total of 425 patterns were found out of 496 sites. Only 129 sites were without polymorphism (26.01%). From the sequence set, AICc = 2092.67, BIC = 2493.58 and the best substitution model used was T92 + G + I. Initial ribotype tree for the heuristic search was obtained by applying the Neighbor-Joining method to a matrix of pairwise distances estimated using the Maximum Composite Likelihood (MCL) approach.

Pathogens	DDBJ Accession	Strain	Tissue	Phylum	Class/Subclass	Collection date	Period of occurrence
Fusarium incarnatum	AB918015	B120	Leaf	Ascomycota		10-06-2012	May-June
Fusarium chlamydosporum	AB918016	B121	Internode	Ascomycota		18-07-2012	June-July
Fusarium camptoceras	AB918017	B122	Node	Ascomycota		07-07-2013	June-July
Fusarium proliferatum	AB918018	B124	Internode	Ascomycota		12-07-2013	June-July
Nigrospora oryzae	AB918019	B125	Leaf	Ascomycota		19-08-2013	July-September
Fusarium chlamydosporum	AB918020	B126	Leaf	Ascomycota		21-08-2012	July-August
Nigrospora sphaerica	AB918021	B127	Leaf	Ascomycota		15-03-2014	March-April
Fusarium oxysporum	AB918022	B129	Leaf	Ascomycota		03-05-2012	May-June
Chaetomium bostrychodes	AB918027	L3	Internode	Ascomycota		06-06-2012	June-July
Trichoderma reesei	AB918031	L9	Leaf	Ascomycota		15-07-2013	July-August
Fusarium proliferatum	AB918023	B130	Leaf	Ascomycota		04-07-2014	July-August
Trichoderma asperellum	AB918007	L7	Leaf	Ascomycota	Sordariomycetes	10-11-2012	October-November
Fusarium incarnatum	AB918010	B110	Leaf	Ascomycota		10-07-2012	July-August
Hypocreales sp	AB918034	B101	Leaf	Ascomycota		09-07-2012	June-July
Daldinia eschscholzii	AB918033	S1	Internode	Ascomycota		10-09-2013	August-September
Phoma plurivora	AB918009	B104	Leaf	Ascomycota		10-07-2013	June-July
Phoma herbarum	AB918006	L29	Leaf	Ascomycota		04-10-2013	October-November
Cochliobolus lunatus	AB918004	L26	Leaf	Ascomycota		09-06-2014	June-July
Lasiodiplodia theobromae	AB918000	L1	Node	Ascomycota		15-07-2012	June-July
Cochliobolus miyabeanus	AB918003	L17	Leaf	Ascomycota		13-08-2013	July-September
Peyronellaea glomerata	AB918011	B116	Leaf	Ascomycota		05-08-2012	August-November
Alternaria sp	AB918012	B117	Leaf	Ascomycota		07-10-2013	September-December
Leptosphaeria sacchari	AB918024	L10	Leaf	Ascomycota		09-08-2013	July-September
Lasiodiplodia theobromae	AB918025	L14	Leaf	Ascomycota		05-07-2012	July -August
Phoma herbarum	AB918026	L16	Internode	Ascomycota		06-08-2013	July -August
Lasiodiplodia theobromae	AB918028	L18	Leaf	Ascomycota	Dothideomycetes	07-07-2014	June-August
Cochliobolus miyabeanus	AB918032	LUN1	Leaf	Ascomycota		13-06-2012	July-October
Neodeightonia subglobosa	AB918035	S3	Leaf	Ascomycota		19-11-2013	October-November
Aspergillus fumigatus	AB918018	B118	Leaf	Ascomycota		03-08-2013	June-August
Aspergillus flavus	AB918002	L12	Leaf	Ascomycota	Eurotiomycetes	08-05-2012	
Aspergillus niger	AB918001	L11	leaf	Ascomycota		7-05-2013	
Ascomycetes sp	AB918014	B119	Leaf	Ascomycota	Unclassified	14-06-2013	June-August
Perenniporia sp.	AB918008	B100	Leaf	Basidiomycota	Basidiomycetes	02-06-2013	June-July

Table 1: Hit list of unreported pathogenic bambusicolous fungi limiting edible bamboo production.

The ML tree indicated that all *Fusarium* taxa formed a main node (at bootstrap value = 67%) and strongly supported by internal branches with bootstrap values > 98% (Figure 3). *Fusarium chlamydosporum* (2 isolates), *Fusarium proliferatum* (2 isolates), *Fusarium incarnatum* (2 isolates) were the most common *Fusarium* species (Figures 3 and 4). As shown (Figure 3), bambusicolous fungi population on edible bamboo *D. hamiltonii* is highly diversified. Generally, predominant group of fungi life in bamboo *D. hamiltonii* is the Ascomycota, estimated to fit in about 228 genera and 70 families [10]. In decreasing frequency of occurrences, *Hypocreaceae, Xylariaceae, Lasiosphaeriaceae, Clavicipitaceae, Phyllachoraceae, Lophiostomataceae, Diatrypaceae, hyaloscyhaceae, Paradiopsidaceae, Valsaceae* and *Pseudoperisporaceae* are reported families that successfully thrived on bamboo species [10].

Some fungi species were encountered only once or twice (Table 1), suggesting that the fungal community could change over time or natural fluctuation in the populations. Regardless of the 99% bootstrap values at the node associating Ascomycetes strain b119 and *Peyronellaea glomerata* strain b116 (Figure 3), we did not find similarity at the morphological level using standard monographs. Furthermore, Ascomycetes strain b119 did not match sequences in the databases at 100% threshold value. This may be an indication of the weakness in public DNA repositories to delineate all fungi. Within the surveillance period, dominant fungal genera were *Peyronellaea glomerata, Alternaria* sp., *Fusarium oxysporum, Aspergillus niger, Aspergillus fumigatus, Aspergillus flavus,* and *Cochliobolus lunatus* (Figures 3 and 4). *Fusarium chlamydosporum, Fusarium camptoceras, Fusarium oxysporum, Fusarium proliferatum* and *Fusarium incarnatum* were also identified (Table 1).

Aspergillus species have not been reported among the bambusicolous fungi in previous studies [10,22,23]. In this present study, *Aspergillus niger, Aspergillus flavus* and *Aspergillus fumigatus* (Figure 2a), *Neodeightonia subglobosa* (Figure 3b), *Trichoderma* species (Figure 3c and 3d), *F. incarnatum* (Figure 3e) were identified. All the *Aspergillus* species sporulated on *D. hamiltonii* during the infestation period of 72 h (Figure 5a-5c). *Trichoderma* species and *Aspergillus* species were recently shown to be pathogens of *Guadua* species, which are abundantly distributed in Ecuador, Chile and Peru (ftp://ftp.fao.org/docrep/fao/010/ah782e/AH782e00.pdf) only. This present study provide the first report of *Trichoderma* species (Figure 3c and 3d) and *Aspergillus* species causing diseases on edible bamboo *D. hamiltonii* (Figure 5a-5c). Although some *Aspergillus* spp. and *Trichoderma* spp. are used as biocontrol agent [24-26], they are important cellulase producers [27,28], which is an important factor for pathogenicity. On this basis, some *Aspergillus* spp. and *Trichoderma* spp. are opportunistic colonizers of economic importance [29-31].

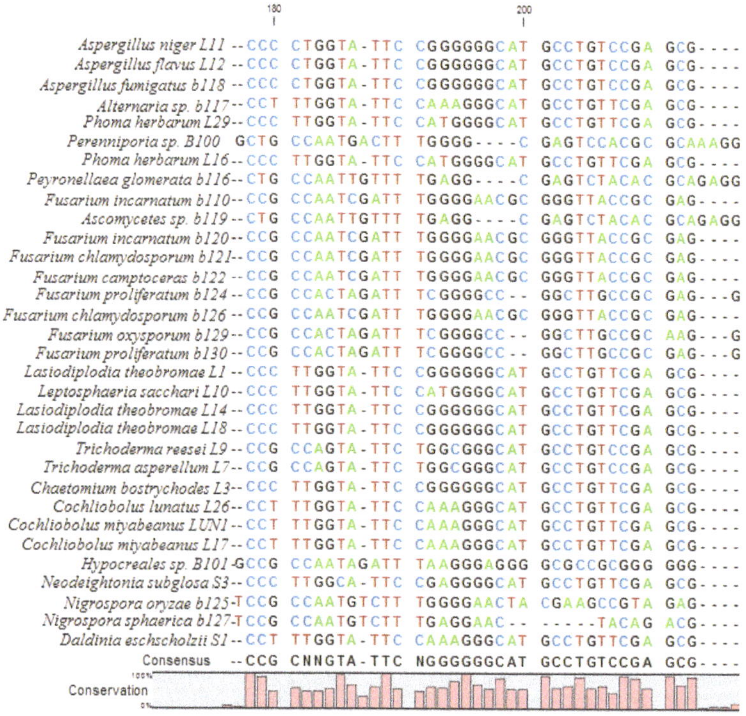

Figure 2: Multiple sequence alignment depicting the variations in bambusicolous fungi the alignment was performed in CLC workbench (Qiagen, Valencia, CA) and variable nucleotides are colored.

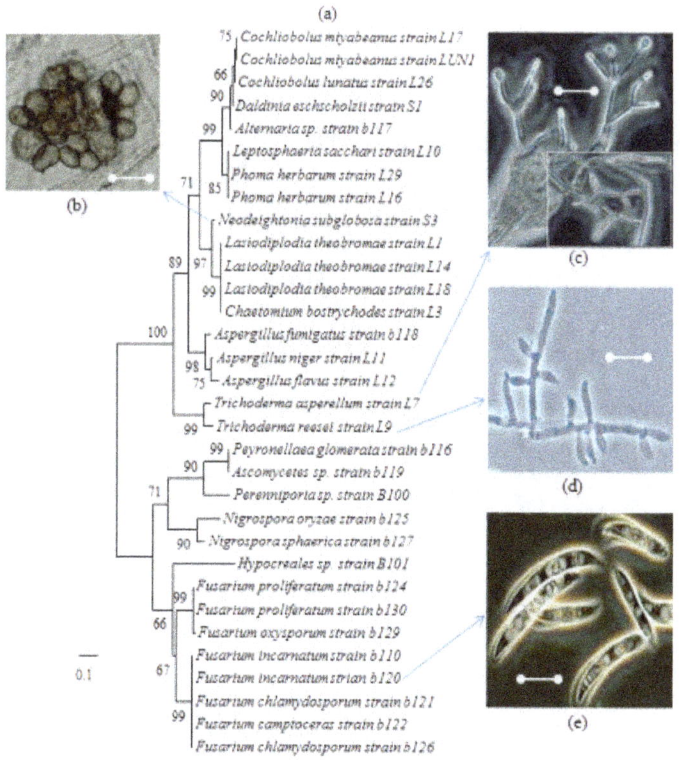

Figure 3: Taxonomical placement of unreported pathogenic bambusicolous fungi of edible bamboo. a: A maximum likelihood tree of highest log likelihood (-1116.47), associated taxa clustered together and supported with 1000 bootstrap reiterations. The ribotype tree is scaled, with branch lengths measured as the number of substitutions per site. b: Brown macroconidia of *Neodeightonia subglobosa*. **c:** Conidiophore of *Trichoderma asperellum* and close-up shows detail of hyphae branching. d: Conidiophore of *Trichoderma reesei*. e: Conidia of *Fusarium incarnatum*. All micrographs were acquired with Olympus DP70 camera (Olympus BX61, USA) at 1000× magnification and scale bars represent 15 μm.

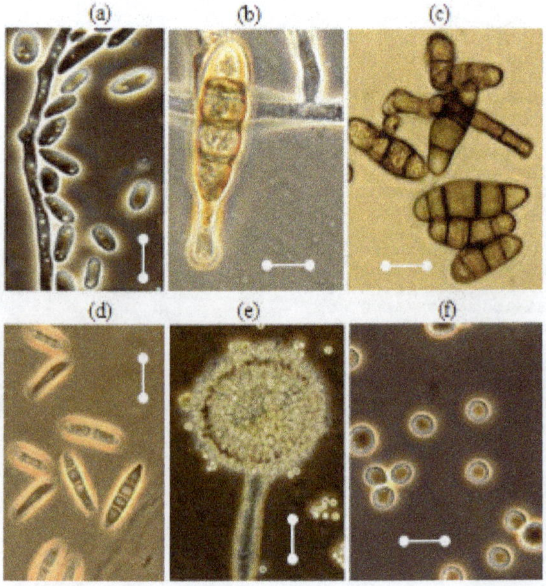

Figure 4: Micrographs of predominant fungi pathogens of edible bamboo *Dendrocalamus hamiltonii* cultured on V8 agar medium. a: *Peyronellaea glomerata* strain b116 showing details of hyphae, conidia and bar=30 μm. b: *Alternaria* sp. strain b117 showing details of hyphae, conidium and bar=10 μm. c: *Cochliobolus lunatus strain* L26 showing details of conidia and bar=20 μm. d: *Fusarium oxysporum* strain b129 and bar=25 μm. e: *Aspergillus flavus* strain L12 and bar=20 μm. f: *Hypocreales sp. strain B101* and scale bar=10 μm. Images were acquired with Olympus DP70 camera (Olympus BX61, USA) at 1000× magnification.

Figure 5: Pathogenicity test performed with plantlets of *D. hamiltonii* in test tube to verify Koch's postulates. a: Sporulating *Aspergillus niger* and colonization leaf tissue (400× magnification). b: Sporulating *Aspergillus fumigatus* and colonization of leaf tissue (400× magnification). c: Sporulating *Aspergillus flavus* and colonization of leaf tissue (400× magnification). d: Leaf rot disease caused by *Fusarium proliferatum*. e: Colonization marked by leaf rot caused by *Fusarium incarnatum* with evidence of fruiting bodies. e: Brown-to-black leaf lesion disease caused by *Cochliobolus lunatus*.

It was observed that all the *Fusarium* species caused rot disease of bamboo shoots, rot of growing culms, and rot of leaf tissues and damping-off of seed plantlets during pathogenicity test (Figure 5d and 5e). Noteworthy, this is the first report of *F. chlamydosporum*, *F. oxysporum*, *F. camptoceras*, *F. oxysporum*, *F. proliferatum* and *F. incarnatum* causing rot disease of bamboo in India. Under field conditions, *Fusarium* infected culms were bend and fallen. Also, *F. moniliforme* var. *intermedium* has been reported to be associated with rot of emerging culms in *B. bambos* [22]. Severe rot and blight diseases of bamboo have been observed in Bangladesh [32,33] and in India [22,34] caused by *Fusarium* species.

Recently in India, it was shown that *Fusarium semitectum* caused both blight and rot disease of *Bambusa tulda* [35]. Also, *F. oxysporum* and *F. chlamydosporum* have been reported in India on *Solanum tuberosum* L and *Capsicum annum* L, respectively [36,37]. *Cochliobolous* species caused foliar and sheath blight diseases, manifested by brownish oval-shaped and water-soaked lesions which became black as the bamboo leaf turned yellowish (Figure 5f). *Cochliobolus* species causes diseases on *Bambusa bambos* and *Dendrocalamus longispathus* [22], with similar characteristic symptoms to those described herein. Symptoms caused by *C. lunatus* in bamboo are similar to leaf spot disease of rice (*Oryza sativa*), wheat (*Triticum aestivum*), cassava (*Manihot esculenta*), sorghum (*Sorghum bicolor*) and potato (*Solanum tuberosum*) [38-42]. It was suggested that *C. lunatus* produced brown-to-black symptoms in many plant hosts because of its melaninated colonizing hyphae [42-44]. Nonetheless, other recurrent leaf spot diseases of bamboo are caused by many species of *Phyllachora* [44]. Interestingly, other studies [35,45,46] have reported new bambusicolous fungi causing a major threat to bamboo production (Table 2). The danger of all the reported bambusicolous pathogenic fungi is that, once bamboo shoots are infected in the field, fungal proliferation continues upto the market level and account to severe economic loses.

Blight and rot diseases of *B. tulda* caused by *Fusarium semitectum* [35].	
Bamboo rust disease of *B. vulgaris* caused by *Uredium* sp [45].	
Kweilingia rust of *B. vulgaris* caused by *Kwelingia divina* (syn. *Dasturella divina*) [45].	
Bamboo witches broom disease of *Phyllostachys bambusoides* caused by *Aciculosporium take* [46].	

*Permission for images was granted by Scot N, Matthew G, Tanaka E and Teron R.

Table 2: Some significant rare bamboo diseases recently communicated.

Conclusion

The study shows that poor pathological management of bambusicolous fungi is valued at 40% losses of the total $818.6 million generated annually in North East India. Until 2010, it was thought fungi belonging to the genera of *Kweilingia*, *Puccinia*, *Uredo*, *Phakospora*, *Stereostratum*, and *Tunicopsora* which caused bamboo rust diseases was the most predominant pathogenic bambusicolous fungi and distributed worldwide. In our study, two principal damages are often caused by these pathogenic bambusicolous fungi, *viz.*, 1) staining of bamboo shoots and 2) structural decay of bamboo shoots which leads to economic loses to all stakeholders in the commercial chain. Our data indicated that *Fusarium*, *Cochliobolus*, *Daldinia*, *Leptosphaeria*, *Phoma*, *Neodeightonia*, *Lasiodiplodia*, *Aspergillus*, *Trichoderma*, *Peyronellaea*, *Perenniporia*, *Nigrospora* and *Hyporales* are new pathogenic bambusicolous fungi genera limiting the production of *D. hamiltonii*. Given most bamboo species are endangered and threatened of extinction [4,5], further studies are required to understand the mechanism of bamboo invasion. The emergence of bambusicolous fungi reported on edible bamboo *D. hamiltonii* in this study illustrated the urgent need for developing a piecemeal control strategy [47].

Acknowledgements

This study was funded by the Academy of Sciences for Developing World (TWAS) and Department of Biotechnology (DBT), Government of India (Program No.3240223450). We thank Scot N, Matthew G, Tanaka E and Teron R for granting permission to use their photographs in Table 2.

References

1. Singh O (2008) Bamboo for sustainable India. Indian Forester, 134: 1193-1198.
2. Singh PK, Devi SP, Devi KK, Ningombam DS, Atokpam P (2010) Bambusa tulda Roxb. In Manipur State, India: Exploring the local values and commercial implications. Notulae Scientia Biologicae 2: 35-40.
3. Waikhom SD, Louis B, Sharma CK, Kumari P, Somkuwar BG, et al. (2013) Grappling the high altitude for safe edible bamboo shoots with rich nutritional attributes and escaping cyanogenic toxicity. Biomed Res Int 2013: 289285.
4. Walter KS, Gillett HJ (1998) IUCN red list of threatened species. IUCN-The World Conservation Union.
5. Hilton-Taylor C, Mittermeier RA (2000) IUCN red list of threatened species. IUCN-The World Conservation Union.
6. Mudoi KM, Siddhartha PS, Adrita G, Animesh G, Debashisha B, et al. (2013) Micropropagation of important bamboos: A Review. Afr J Biotechnol 12: 2770-2785.
7. Waikhom SD, Louis B (2014) An effective protocol for micropropagation of edible bamboo species (Bambusa tulda and Melocanna baccifera) through nodal culture. ScientificWorldJournal 2014: 345794.
8. Janzen DH (1976) Why bamboos wait so long to flower. Annu Rev Ecol Evol Syst 7: 347-391.
9. Waikhom SD, Louis B, Roy P, Singh MW, Bhardwaj PK, et al. (2014) Scanning electron microscopy of pollen structure throws light on resolving Bambusa-Dendrocalamus complex: bamboo flowering evidence. Plant Syst Evol 300: 1261-268.
10. Hyde KD, Zhou D, Dalisay T (2002) Bambusicolous fungi: A Review. Fungal Diver 9: 1-14.
11. Sooda S, Walia S, Gupta M, Soon A (2013) Nutritional characterization of shoots and other edible products of an edible bamboo - Dendrocalamus hamiltonii. Curr Res Nutr Food Sci 1: 169-176.
12. Bhatt BP, Singh K, Singh A (2005) Nutritional values of some commercial edible bamboo species of the North Eastern Himalayan region, India. J Bamboo Rattan 4: 111-124.
13. Waikhom SD, Ghosh S, Talukdar NC, Mandi SS (2012) Assessment of genetic diversity of landraces of Dendrocalamus hamiltonii using AFLP markers and association with biochemical traits. Genet Mol Res 11: 2107-2121.
14. White TJ, Bruns TD, Lee S, Taylor J (1990) Amplification and direct sequencing of fungal ribosomal RNA genes for phylogenetics. In: Innis MA, Gelfand DH, Sninsky JJ, White TJ (eds) PCR protocols, a guide to methods and applications. San Diego, California, USA, Academic Press, pp. 315-322.
15. Edgar RC (2004) MUSCLE: multiple sequence alignment with high accuracy and high throughput. Nucleic Acids Res 32: 1792-1797.
16. Tamura K, Stecher G, Peterson D, Filipski A, Kumar S (2013) MEGA6: Molecular Evolutionary Genetics Analysis version 6.0. Mol Biol Evol 30: 2725-2729.
17. Devi WS, Louis B, Sharma GJ (2012) In vitro seed germination and micropropagation of edible bamboo Dendrocalamus giganteus Munro using seeds. Biotechnol 11: 74-84.
18. Berbee ML (2001) The phylogeny of plant and animal pathogens in the Ascomycota. Physiol Mol Plant Pathol 59: 165-187.
19. Louis B, Waikhom SD, Singh MW, Talukdar NC, Roy P (2014) Diversity of ascomycetes at the potato interface: new devastating fungal pathogens posing threat to potato farming. Plant Pathol J 13: 18-27.
20. Umali TE, Quimio TH, Hyde KD (1999) Endophytic fungi in leaves of Bambusa tuldoides. Fungal Sci 14: 11-18.
21. Kimura M (1980) A simple method for estimating evolutionary rates of base substitutions through comparative studies of nucleotide sequences. J Mol Evol 16: 111-120.
22. Mohanan C (1994) Diseases of bamboos and rattans in Kerala. KFRI Research Report 98.
23. Morakotkarn D, Kawasaki H, Seki T (2007) Molecular diversity of bamboo-associated fungi isolated from Japan. FEMS Microbiol Lett 266: 10-19.
24. Itamar SM, Jane LF, Rosely SN (2006) Antagonism of Aspergillus terreus to Sclerotinia sclerotiorum. Brazilian J Microbiol 37: 417-419.
25. Suliman EA, Mohammed YO (2012) The activity of Aspergillus terreus as entomopathogenic fungi on different stages of Hyalomma anatolicum under experimental conditions. J Entomol 9: 343-351.
26. Hermosa MR, Grondona I, Iturriaga EA, Diaz-Minguez JM, Castro C, et al. (2000) Molecular characterization and identification of biocontrol isolates of Trichoderma spp. Appl Environ Microbiol 66: 1890-1898.
27. Zhao Z, Liu Z, Wang C, Xu J-R (2013) Correction: comparative analysis of fungal genomes reveals different plant cell wall degrading capacity in fungi. BMC Genomics 15: 6.
28. Jahromi MF, Liang JB, Rosfarizan M, Goh YM, Shokryazdan PY, et al. (2012) Efficiency of rice straw lignocelluloses degradability by Aspergillus terreus ATCC 74135 in solid state fermentation. Afr J Biotechnol 10: 4428-4435.
29. De Lucca AJ (2007) Harmful fungi in both agriculture and medicine. Rev Iberoam Micol 24: 3-13.
30. Louis B, Waikhom SD, Roy P, Bhardwaj PK, Singh MW, et al. (2014) Invasion of Solanum tuberosum L. by Aspergillus terreus: a microscopic and proteomics insight on pathogenicity. BMC Res Notes 7: 350.
31. Trabelsi S, Hariga D, Khaled S (2010) First case of Trichoderma longibrachiatum infection in a renal transplant recipient in Tunisia and review of the literature. Tunis Med 88: 52-57.
32. Gibson IAS (1975) Report on a visit to the republic of Bangladesh. Overseas Development Administration, London, U.K.
33. Rahman MA (1978) Isolation of fungi from blight affected bamboos in Bangladesh. Ban Bigyan Patrika 7: 42-49.
34. Jamaluddin BN, Gupta SC, Bohidar DVS (1992) Mortality of bamboo (Bambusa nutans Wall.) in coastal area of Orissa. J Trop Forestry 8: 252-261.
35. Gogoi J, Teron R, Tamuli AK (2013) Incidence of blight and rot diseases of Bambusa tulda Roxb. Groves in Dimapur district of Nagaland State. Int J Sci Nat 4: 478-482.
36. Kumar K, Bhagat S, Madhuri K, Birha A, Srivastava RC (2012) Occurrence of unreported fruit rot caused by Fusarium chlamydosporum on Capsicum annum in Bay Island, India. Vegetable Sci 39: 195-197.
37. Louis B, Roy P, Yekwa EL, Waikhom SD (2012) The farmers cry: Impact of heat stress on Fusarium oxysporum f.sp. dianthi, interaction with fungicides. Asian J Plant Pathol 6: 19-24.

38. Perfect JR, Schell WA (1996) The new fungal opportunists are coming. Clin Infect Dis 22 Suppl 2: S112-118.

39. Ahmad I, Iram S, Cullum J (2006) Genetic variability and aggressiveness in Curvularia lunata associated with rice-wheat cropping areas of Pakistan. Pak J Botany 38: 475-485.

40. Msikita W, Yaninek JS, Ahounou M, Baimey H, Fagbemissi R (1997) First report of Curvularia lunata associated with stem disease of cassava. Plant Dis 81: 112.

41. John EE, Louis KP (2006) Seed mycoflora for grain mold from natural infection in sorghum germplasm grown at Isabela, Puerto Rico and their association with kernel weight and germination. Plant Pathol J 5: 106-112.

42. Louis B, Roy P, Waikhom, SD, Talukdar NC (2013) Report of foliar necrosis of potato caused by Cochliobolus lunatus in India. Afr J Biotechnol 12: 833-835.

43. Louis B, Waikhom SD, Roy P, Bhardwaj PK, Singh MW, et al. (2014) Secretome weaponries of Cochliobolus lunatus interacting with potato leaf at different temperature regimes reveal a CL[xxxx]LHM-motif. BMC Genomics 15: 213.

44. Louis B, Waikhom SD, Roy P, Bhardwaj PK, Sharma CK, et al. (2014) Host-range dynamics of Cochliobolus lunatus: from a biocontrol agent to a severe environmental threat. Biomed Res Int 2014: 378372.

45. Pearce CA, Reddell P, Hyde KD (2000) Phyllachora shiraiana complex (Ascomycotina) on Bambusa arnhemica: a new record for Australia. Aus Plant Pathol 29: 205-210.

46. Scot Nelson, Matthew Goo (2011) Kweilingia rust of bamboo in Hawai'i. College of Tropical Agriculture and Human Resources, Plant Disease 74: 1-5.

47. Tanaka E (2009) Mechanisms of bamboo witches' broom symptom development caused by endophytic/epiphytic fungi. Plant Signal Behav 5: 415-418.

15

Entomopathogenic Effect of *Beauveria bassiana* (Bals.) and *Metarrhizium anisopliae* (Metschn.) on *Tuta absoluta* (Meyrick) (Lepidoptera: Gelechiidae) Larvae Under Laboratory and Glasshouse Conditions in Ethiopia

Tadele S* and Emana G

Addis Ababa University, College of Natural and Computational Sciences, Department of Zoological Sciences, Ethiopia

Abstract

Tomato leaf miner, *Tuta absolute (*Meyrick) is one of the major pest that infest tomato plant in all agro-ecological regions of the world where it present. Currently, the management strategies highly rely on chemical insecticides, which do not provide effective control and at the same time have environmental concern in addition to the residue left on the fruits. Hence, looking for alternative control measure is vital. Studies were conducted to determine the pathogenicity and virulence of three different concentrations of *Beauveria bassiana* and *Metarhizium anisopliae* against larvae of *T. absoluta* using the concentrations of 2.5×10^7, 2.5×10^8, and 2.5×10^9 conidia ml^{-1} under laboratory and glasshouse conditions. The experiments were carried out in the laboratory and glasshouse. Mortalities caused by *B. bassiana* isolate at the different concentrations ranged from 79.17% to 95.83% under laboratory and 73.0% to 84.04% under glasshouse, the highest mortality percentage being found at 2.5×10^9 conidia ml^{-1}. The isolate of *M. anisopliae* caused the highest mortality also at the highest concentration. The lowest lethal time for *B. bassiana* and *M. anisopliae*, were achieved by the concentration 2.5×10^9 (5.01 days) and 2.5×10^8 (5.21 days), respectively. The isolates of *B. bassiana* and *M. anisopliae*, at 2.5×10^9 conidia ml^{-1} are promising for use the integrated control of *T. absoluta* larvae.

Keywords: *Beauveria bassiana*; *Metarhizium anisopliae*; Efficacy; Conidia concentrations; Larval mortality; Virulence; Chemical insecticides

Introduction

Tomato leafminer, *Tuta absoluta* (Meyrick) is an oligophagous notorious pest of a number of economic crops including tomato. To overcome the problem of this pest, insecticides play a significant role globally. Tomato is a perishable commodity with a relatively short shelves life after harvest. This pest was initially reported in the central Rift Valley region of Ethiopia in 2012 [1]. Since the time of its initial detection, the pest has caused serious damages to tomato in invaded areas [2] and it is currently considered as a key threat to Ethiopian tomato production. If no control measures are taken, the pest can cause up to 80% to 100% yield losses by attacking leaves, flowers, stems and fruits [3]. Currently, chemical insecticides are heavily used by tomato growers against *T. absoluta*. However, the chemicals which are under use have negative impacts as the other chemical have. Hence, combination with other control methods like use of entomopathogen becomes imperative, as the continued use of chemical insecticides could harm non-target organisms [4] and the environment among others. The recommended waiting period which is required between application of conventional organophosphate pesticides group and consumption can hardly be afforded. Therefore, the current experiment was initiated to evaluate the efficacy of *M. anisopliae* and *B. bassiana* isolates against T. absoluta in the laboratory and glasshouse conditions

Materials and Methods

Description of the study area

The research was conducted under laboratory and glasshouse conditions at Ambo University glasshouse and plant Science laboratory. Ambo is far away from Addis Ababa 110 km and at geographical coordinate of 8°59`N latitude and 37.85°E longitude with an altitude of 2100 meter above sea level (m.a.s.l.) [5]. The mean daily temperatures were 22°C ± 2°C and 32°C ± 2°C for laboratory and glasshouse experiments, respectively.

Experimental design and materials used

The laboratory and glasshouse experiments were laid out in a Randomized Complete Block Design (RCBD) with three replications. Eight treatments were considered such treatments were *Beauveria bassiana* isolate at three different concentrations (2.5×10^7, 2.5×10^8 and 2.5×10^9 conidia ml^{-1}), similar concentrations were performed in *Metarhizium anisopliae* isolate. Chlorantraniliprole (Coragen 200 SC) as a standard check and untreated control was also considered for comparision.

Tomato cultivar known as "Coshoro" was brought from Melkasa Agricultural Research Center. The seeds were sown on the field for natural infestation of *T. absoluta*. Harboring *T. absoluta* larvae were collected from the fields of tomato and brought to the laboratory and glasshouse. The tomato leaf miner present on these collected tomato leaves were wrapped with wet cotton kept in plastic box (20×15 cm^2) in laboratory and glasshouse. After the emergency of adults rearing cages were prepared under glasshouse.

The insect was reared and maintained on tomato plants in the glasshouse until use. Leaves were examined under binocular microscope and *T. absoluta* larvae were counted. Spore suspensions were sprayed using a hand sprayer (1 liter of capacity). After treatment applications, the percent mortalities of the agents were observed at: 3, 5 and 7days in the laboratory and 3, 5, 7 and 10 days under glasshouse conditions.

***Corresponding author:** Shiberu T, Department of Zoological Sciences, Addis Ababa University, College of Natural and Computational Sciences, Ethiopia
E-mail: tshiberu@yahoo.com

Fungus culture and viability test

Isolates of *Beauveria bassiana* (PPRC-56) and *Metarhizium anisopliae* (PPRC-2) were obtained from Ambo Plant protection research center. These entomopathogenic fungi were cultured on potato dextrose Agar (PDA) medium containing 20 g glucose, 20 g starch, 20 g agar, and 1000 ml of distilled water in test tubes. The test tubes containing PDA medium was autoclaved at 121°C for 15-20 min and incubated at 27°C ± 1°C, 80% ± 5% relative humidity and photophase of 12 h for 15 days. The relative humidity was measured using Huger Hygrometer. The conidia were harvested by scraping the surface of 14-15 days old culture gently with inoculation needle. The mixture was stirred with a magnetic shaker for 10 min. The hyphal debris was removed by filtering the mixture through fine mesh sieve. The conidial concentration of final suspension was determined by direct count using Haemocytometer. Serial dilutions were prepared in distilled water containing 0.1% Tween-80 and preservedat 5°C until used.

Conidial viability was assessed according to Goettel and Inglis [6]. Three different concentrations were evaluated. The droplet of a diluted suspension was placed on a thin film of potato dextrose agar medium incubated at 27°C ± 1°C and 80% ± 5% relative humidity in the dark for 24 h. The conidia were stained with lacto-phenol cotton blue and germination was observed under microscope.

Mortality of *T. absoluta* under laboratory

The concentration of the stock suspension was adjusted to 2.5×10^7. 2.5×10^8 and 2.5×10^9 conidia ml^{-1} using an improved neubaour heamocytometer. To evaluate the efficiency of each of the fungal isolates on *T. absoluta*, 20 larvae were placed on a filter paper in 9 cm diameter petri-dish and 100 µl of the suspension was then spread. On similar trend the suspension was spread in glasshouse using hand sprayed was performed and after 3rd days of observation all counted larvae were collected from plants to brought in the laboratory to determined how many *T. absoluta* larvae were dead without being infested with fungal isolates. A control treatment was sprayed with only sterile distilled water as negative control.

Statistical analysis

The mean number of live larvae per plant or per leaf was tested for percent mortality. The data was subjected to analysis of variance (ANOVA) and the means were compared by least significant different (LSD) test at 0.05 levels, using SAS program version 9.1 [7]. Efficacy analysis was done based on data transformation to Arcsine $\sqrt{x+0.5}$ when necessary according to **Gomez and Gomez [8]**.

$$CM\ (\%) = \frac{[T(\%) - C(\%)]}{[100 - C(\%)]} X100$$

Where: CM (%) - Corrected mortality

T- Mortality in treated insects

C- Mortality in untreated insects

Isolate code	Host	Place of collection	Scientific Name	% Germination	Source
PPRC-56	*P. interrupta*	Berbere	*B. bassiana*	79	PPRC Ambo
PPRC-2	*P. interrupta*	Ashan	*M. anisopliae*	93	PPRC Ambo

Table 1: Pieces of information about indigenous entomopathogenic fungi in Ethiopia.

Treatments	Conc.	Mean percent mortality after treatment application		
		3 days	5 days	7 days
Beauveriabassiana **(PPRC-56)**	2.5×10^7	37.50^c	58.33^c	79.17^b
	2.5×10^8	70.83^{bc}	70.83^{bc}	83.33^{ab}
	2.5×10^9	79.17^{ab}	91.67^{ab}	95.83^a
Metarhiziumanisopliae **(PPRC-2)**	2.5×10^7	58.33^{bc}	58.33^{bc}	66.67^b
	2.5×10^8	58.33^{bc}	79.17^{bc}	79.17^{ab}
	2.5×10^9	66.67^{bc}	83.33^{abc}	87.50^{ab}
Chlorantraniliprole (Coragen 200 SC)		95.83^a	95.83^a	95.83^a
Control (water)		0.0^d	0.0^d	0.0^c

Note: Means with the same letter(s) in rows are not significantly different for each other. All treatment effects were highly significant at p<0.01 (DMRT)

Table 2: Mean percent mortality of *T. absoluta* treated with fungal isolates at different concentration over time under laboratory condition.

Results and Discussions

Under laboratory condition

The laboratory result also showed that percent mortality of *T. absoluta* larvae due to entomopatogenic fungi significant (P<0.01) differences among the concentrations of *B. bassiana* and *M. anisopliae* (Table 1). All concentrations of *B. bassiana* caused mortality of *T. absoluta* above 75% after treatment application of 7 days, indicating that 2.5×10^9 conidia ml^{-1} caused the highest mortality. For *M. anisopliae*, at the concentration of 2.5×10^9 conidia ml^{-1}, mortalities obtained with all concentrations were higher than 50%; however, the concentrations did differ statistically from each other after treatment application, and the highest mortality of *T. absoluta* larvae were observed with concentration 2.5×10^9 (87.5%) under laboratory condition (Table 2).

After 7th day of treatment application *B. bassiana* raveled that 79.17%, 83.33% and 95.83% mortality at 2.5×10^7, 2.5×10^8 and 2.5×10^9 concentrations, respectively. Similarly, *M. anisopliae* concentrations showed that 66.67%, 79.17% and 87.50% mortality at 2.5×10^7, 2.5×10^8 and 2.5×10^9 concentrations, respectively. There was a highly significant variation among the concentrations in causing mortality of *T. absoluta* larvae. The lowest mean percent mortality was caused by the *B. bassiana* at 3rd days of observation 37.50% which was not significantly different from *M. anisopliae* at 3rd days of 58.33 %. The highest mortality of *T. absoluta* was caused by *B. bassiana* 95.83% which did not significantly differ from the *M. anisopliae* which was 87.50% mortality. Based on the results of the virulence assays of *B. bassiana* and *M. anisopliae* had time taken by the three concentrations to caused percent mortality of *T. absoluta*. The effects of the concentrations varied significantly (P<0.01) with the lowest (3 days) recorded from concentration $2.5x10^7$ in *B. bassiana followed* by *M. anisopliae* (5 days) which recorded 58.33%. In the 7th day of the three concentrations the highest was recorded due to *B. bassiana* which was significantly (P<0.01) different from *M. anisopliae* concentrations.

The comparison among the different concentrations and treatments against *T. absoluta* the results indicated good performance and gradually increased from 3 to 7 days treatment application. The percent mortality according to Abbott formula [9], both agents at 2.5×10^9 conidial/ml gave statistically no significant (P<0.01) differences from the standard check (Coragen 200 SC) while highly significant different from untreated check after 3 days of treatment application (Table 2).

Concentration-response test

Percent mortality of *T. absoluta* larvae at different concentrations of *B. bassiana* and *M. anisopliae* shown in Table 2. There were no significant differences in mortality rates within each concentration except for the concentration of 2.5×10^9 conidia/ml in which the *B. bassiana* showed significantly higher mortality than *M. anisopliae*. The results of all concentrations except the first concentration 2.5×10^7 in

B. bassiana revealed the lowest at 3rd days of application but also highly significantly (P<0.01) among the concentrations requiring higher concentration (2.5 $10^9 \times 10^9$conidia ml^{-1}).

B. bassiana, strain presented the highest pathogenicity on *T. absoluta* larvae with 95.83% an average mortality, LC_{50}=2.5 × 10^9 conidia ml^{-1} and LT_{50}=5.01 days (Table 3). *M. anisopliae* strain was the most virulent on *T. absoluta* larvae presenting 87.50% mortality, LC_{50}=2.5 × 10^9 conidia ml^{-1} and LT_{50}=4.82 days. The LT_{90} values to *B. bassiana* strains on *T. absoluta* larvae ranged from 8.06 to 9.32 days, and for *M. anisopliae* strains on *T. absoluta larvae* ranged from 8.14 to 9.04 days (Table 3). The *M. anisopliae* strain presenting the lowest LC_{90} on *T. absoluta* larvae was 2.5 × 10^9 conidia ml^{-1}) and the highest LC_{90} was presented by *B. bassiana* 2.5 × 10^7 conidia ml^{-1}). Finally, for *T. absoluta* larvae the LC_{90} of both *B. bassiana and M. anisopliae* varied from 2.5 × 10^7 to 2.5 × 10^9 conidia ml^{-1} concentration (Table 4).

Under glasshouse conditions

The entomopathogenic fungal isolates were tested at three different concentrations for their percent mortality against *T. absoluta* in glasshouse to explore their potential to manage the pest population. Percent mortality of *T. absoluta* larvae were calculated for the different concentrations of the two isolates and showed increasing mortality with increasing spore concentration. Cumulative mortality of *T. absoluta* larvae over exposure period (3, 5, 7 and 10 days) was significantly (P<0.01) different for fungi isolates (Table 5). On the 3rd days of exposure maximum mortality 91.84 recorded from standard check, while the untreated control had 2.78% mortality. These were significantly different from all concentrations of the fungal isolates. Among the concentrations of entomopathogenic fungi maximum percent mortality was recorded at 2.5 × 10^9 conidial ml^{-1} of *B. bassiana* (84.04%) followed by *M. anisopliae* (76.31%) on 10th day after treatment application. At the highest concentration of conidial ml^{-1}, all *B. bcassiana* concentration gave the highest percent mortality (Table 5). The results indicated for pathogenicity of all the concentrations revealed that all of them are virulent, even three days after application causing significant mortality up to 64.05% when compared with untreated control.

Treatments		95% Confidence Limit		95% Confidence Limit	
	(Conidia ml^{-1})	LT_{50} (days)	Slope ± SE	LT_{90} (days)	Slope ± SE
B. bassiana **(PPRC-56)**	2.5 × 10^7	5.45	3.17 ± 0.52	9.32	2.28 ± 0.48
	2.5 × 10^8	5.21	3.80 ± 0.61	8.87	2.66 ± 0.37
	2.5 × 10^9	5.01	4.29 ± 0.82	8.06	3.06 ± 0.68
M. anisopliae **(PPRC-2)**	2.5 × 10^7	5.14	3.64 ± 0.56	9.04	2.86 ± 0.46
	2.5 × 10^8	5.02	3.63 ± 0.48	8.56	3.04 ± 0.58
	2.5 × 10^9	4.82	3.31 ± 0.64	8.14	3.31 ± 0.72

Table 3: Median lethal time (LT_{50}) and (LT_{90}) of *B. bassiana* and *M. anisopliae* against *T. absoluta*.

Treatments				95% Confidence Limit	
	(Conidia ml^{-1})	LC_{50}	Slope ± SE	LC_{90}	Slope ± SE
B. bassiana (PPRC-56)	2.5 × 10^7	4.23 ± 0.52	4.29 ± 0.82	9.68 ± 0.82	3.36 ± 0.41
	2.5 × 10^8	3.93 ± 0.61	3.80 ± 0.61	9.22 ± 0.61	2.64 ± 0.38
	2.5 × 10^9	3.50 ± 0.82	3.17 ± 0.52	8.46 ± 0.52	2.45 ± 0.28
M anisopliae (PPRC-2)	2.5 × 10^7	3.59 ± 0.56	3.31 ± 0.64	8.71 ± 0.64	2.47 ± 0.77
	2.5 × 10^8	3.26 ± 0.48	3.63 ± 0.48	8.13 ± 0.48	2.63 ± 0.58
	2.5 × 10^9	2.91 ± 0.64	3.64 ± 0.56	7.52 ± 0.56	3.54 ± 0.72

Table 4: Median lethal (LC_{50}) and (LC_{90}) of *B. bassiana* and *M. anisopliae* (100µ / larva) of *T. absoluta*.

Treatments	Conc.	Mean percent mortality after treatment application			
		3 days	5 days	7 days	10 days
Beauveriabassiana **(PPRC-56)**	2.5 × 10^7	43.85^c	57.57^{bc}	75.17^{ab}	81.64^{ab}
	2.5 × 10^8	56.27^{bc}	56.27^{bc}	76.62^{ab}	73.0^{abc}
	2.5 × 10^9	63.84^b	67.05^b	67.05^{bc}	84.04^{ab}
Metarhiziumanisopliae **(PPRC-2)**	2.5 × 10^7	38.76^c	42.93^c	53.37^c	53.37^d
	2.5 × 10^8	44.07^c	51.98^{bc}	61.49^{bc}	64.65^{cd}
	2.5 × 10^9	64.05^b	68.21^{bc}	71.98^{abc}	76.31^{bc}
Chlorantraniliprole (Coragen 200 SC)		91.84^a	91.84^a	91.84^a	91.84^a
Control		2.78^d	4.76^d	4.76^d	7.14^e
Note: Means with the same letter(s) in rows are not significantly different for each other. All treatment effects were highly significant at p<0.01 (DMRT)					

Table 5: Mean percent mortality of Entomopathogenic fungi at different concentration on larvae *T. absoluta* under glasshouse condition.

A positive relationship was recorded between mortality percentages and concentrations among the *B. bassiana* and *M. anisopliae* concentrations. Concurrently, with the increase in conidia concentration, a reduction in LT_{50} was observed. Concentrations of 2.5 × 10^9 from *B. bassiana*, at the concentrations 2.5 × 10^8 and 2.5 × 10^9 conidia ml^{-1}, presented the shortest lethal time (Table 5). These low values are probably associated to the presence of enzymes that aid in the process of penetration of the fungi [10].

The effect of entomopathogenic fungi were evaluated to determine the concentrations with high efficacy against larvae *T. absoluta* under laboratory and glasshouse conditions. Both fungal isolates were found to be pathogenic to *T. absoluta*. Though, there was a variation in their virulence against *T. absoluta*. The percent mortality for all the concentration gradually increased. The spore formation appeared on the larvae of *T. absoluta* took place after treatment exposure of the concentrations of the two isolates starting from the day three after treatment exposure, and thereby no hatched larvae were appeared in the concentrations of both isolates comparing the control treatment. The *M. anisopliae* in all concentrations were significantly less effective when compared with that of *B. bassiana* in terms of virulence. Virulence due to *B. bassiana* on 10th day was not significantly different from each other. This indicated that all *B. bassiana* concentrations were the best management option of *T. absoluta*. This finding confirms with earlier reports [11] who obtained high percent mortality during the evaluation time for *B. bassiana* and *M. anisopliae*.

The amount of conidia used should to attain a certain concentration and thus, achieving an efficacious penetration of the fungus on the insect cuticle and causing host death. Similar findings by Garcia *et al.* [12] were obtained, evaluating the insecticidal activity of *B. bassiana* strains and *M. anisopliae* on *Spodoptera frugiperda* and *Epilachna varivestis* larvae at six concentrations (10^4 to 10^9); *B. bassiana* strain was more virulent for *E. varivestis* larvae with a 93.3% mortality, LC_{50}=1.20 × 10^6 conidia ml^{-1} and LT_{50}=5.1 days. *B. bassiana* strain presented the highest mortality on *S. frugiperda* larvae (96.6%, LC_{50}=5.92 × 10^3 conidia ml^{-1} and LT_{50}=3.6 days). It was also reported by another authors differences among lethal times is a tool widely used in selecting strains, because it is interesting that the fungus quickly eliminate its host, as well [13]. These results are disagreed with Khalid et al. [14], evaluating the virulence of various strains of *B. bassiana* and *M. anisopliae* on *G. mellonella* larvae using 10^2, 10^3, 10^4, 10^5 and 10^6 conidia ml^{-1} concentration.

Thus, laboratory and glasshouse experiments suggested that *B. bassiana* and *M. anisopliae* have good effect on both egg and larvae of *T. absoluta*. Sabbour [15] also confirmed the effectiveness of both *B. bassiana* and *M. anisopliae* against larvae of *T. absoluta* under laboratory and greenhouse. The same results were obtained by Sabbour and Singer [16]; Sabbour and Abdel-Raheem [17]. These results agree with our findings and Cabello et al. [18] where stated that; the higher mortality of larvae under laboratory studies indicated *B. bassiana* could cause good larval mortality. At present, the knowhow of entomopathogenic fungi on *T. absoluta* was very limited because of very few studies that are available to indicate that the isolates causes the high mortality on other lepidopteran insects [19]. In this study it has been shown that all the fungal concentrations are effective against *T. absoluta*.

Our results confirmed that, the previous study of Shalaby et al. [11], they stated that when the second instar larvae fed on *M. anisopliae* the pathogen effect was evident by the 3rd day of evaluation after exposure in the concentration (10^7 and 10^8 conidia/ml). Dahliz et al. [20] have reported similar results with *Metarhizium*. Our result was confirmed the work of İnanl and Oldargc [21], they reported the studies conducted in Turkey, researchers compared the efficacy of *B. bassiana* and *M. anisopliae* on *T. absoluta* eggs and larvae; these two agents provided highly effective to control of *T. absoluta* larvae. Our results also indicated the potential of *B. bassiana* and *M. anisopliae* to control the larvae of *T. absoluta* in an integrated pest management programs. Neves and Alves [22] also noted, as more conidia penetrating, more toxins or enzymes are released, increasing the insect mortality. Though, the fungus action speed depends, besides the concentration, of the host species involved [23]. According Kleespies and Zimmermann [24], variation in virulence of entomopathogenic strains is a result of differences in the enzymes and toxins production in conidia germination speed, mechanical activity in the cuticle penetration, colonization capacity and cuticle chemical composition.

Conclusion and Recommendation

The most effective percent mortality of fungal isolates was found in *B. bassiana* followed by *M. anisopliae* at all concentrations. Both agents could be very well utilized as alternative to bio pesticides for the management of *T. absoluta*. It might be concluded that *B. bassiana* and *M. anisopliae* fungi present different capacity cause mortality of the insects, with the 2.5×10^9 conidian ml^{-1} *B. bassiana* strains as the most pathogenic for *T. absoluta*, as well as 2.5×10^9 conidian ml^{-1} *M. anisopliae* strains was also good virulence for *T. absoluta* and also presenting the lowest LC_{50} and LT_{50} values. Hence, insecticidal substances that have potential for use as alternative control measure. Therefore, further study on field conditions should be undertaken to evaluate effectiveness of experimental mycopesticide formulations in the management of *T. absoluta* under Tropical conditions in various economically important insect pests.

Acknowledgement

The authors would like to acknowledge Ambo University and Addis Ababa University for financial support. We greatly appreciate Ambo University, College of Agriculture and Veterinary Sciences, Department of Plant sciences, for allowing us to have access to their glasshouse to do the experiment. We also appreciate Mr. Fikadu Balcha for his technical assistance in data collection.

References

1. Gashawbeza A, Abiy F (2013) Occurrence and distribution of a new species of tomato fruit worm, *Tuta absoluta* Meyrick (Lepidoptera: Gelechiidae) in central rift valley of Ethiopia. Proceedings of the 4th Binneial Conference of Ethiopian Horticultural Science Society, Ethiopia.
2. Gashawbeza A (2015) Efficacy of selected insecticides against the South American tomato moth, *Tuta absoluta* Meyrick (Lepidoptera: Gelechiidae) on tomato in the central rift valley of Ethiopia. Afr Entomol 23: 410-417.
3. Öztemiz S (2012) The tomato leafminer (*Tuta absoluta* Meyrick (Lepidoptera: Gelechiidae) and it's biological control. KSU J Nat Sci 15: 47-57.
4. Landgren O, Kyle RA, Hoppin JA, Beane Freeman LE, et al. (2009) Pesticides exposure and risk of monoclonal gammopathy of undetermined significance in the agricultural Health study. Blood 113: 6386-6391.
5. Briggs P (2012) Ethiopia (Bradt Travel Guides) (6thedn), The Globe Pequot Press Inc., Guilford, Connecticut, USA.
6. Goettel MS, Inglis DG (1997) Fungi: Hyphomycetes. In: Lacey LA (Ed). Manual of techniques in insect pathology. Academic Press, London, UK.
7. SAS (2009). Statistical analysis system software. Ver. 9.1. SAS Institute Inc., Carry. NC.
8. Gomez KA, Gomez AA (1984) Statistical procedures for agricultural research. (2ndedn), John Wiley and Sons, Inc. New York, USA.
9. Abbott WS (1925) A method of computing the effectiveness of an insecticide. J Econ Entomol 18: 265-267.
10. St. Leger RJ, Durrands PK, Charnley AK, Cooper RM (1988) Role of extracellular chymoelastase in the virulence of *Metarhizium anisopliae* for *Manduca sexta*. J Invertebr Pathol 52: 285-293.
11. Shalaby HH, Faragalla FH, El-Saadany HM, Ibrahim AA (2013) Efficacy of three entomopathogenic agents for control the tomato borer, *Tuta absoluta* (Meyrick) (Lepidoptera: Gelechiidae). Nat Sci 11: 63-64.
12. Garcia C, Gonzalez M, Bautista M (2011) Pathogenicity of native entomopathogenic fungal isolates against Spodoptera frugiperda and *Epilachna varivestis*. Rev Colomb Entomol 37: 217-222.
13. Lohmeyer KH, Miller JA (2006) Pathogenicity of three formulations of entomopathogenic fungi for control of adult *Haematobia irritans* (Diptera: Muscidae). J Econ Entomol 99: 1943-1947.
14. Khalid AH, Mohamed AAA, Ahmed YA, Saad SE (2012) Pathogenicity of *Beauveria bassiana* and *Metarhizium anisopliae* against *Galleria mellonella*. Phytoparasitica 40: 117-126.
15. Sabbour MM (2014) Biocontrol of the Tomato Pinworm *Tuta absoluta*, Meyrick (Lepidoptera: Gelechiidae) in Egypt Middle East. J Agric Res 3: 499-503.
16. Sabbour MM, Singer SM (2014) Evaluations of two Metarhizium varieties against *Tuta absoluta* (Meyrick) (Lepidoptera: Gelechiidae) in Egypt. Int J Sci and Res 3: 9.
17. Sabbour MM, Abd-El-Raheem MA (2013) Repellent effects of Jatropha curcas, canola and Jojoba seed oil, against Callosobruchus maculates (F.) and Callosobruchus chinensis (L.). J Appl Sci Res 9: 4678-4682.
18. Cabello T, Granados JRG, Vila E, Polaszeck A (2009) Biological control of the South American tomato pinworm, *Tuta absoluta* (Lep.: Gelechiidae), with releases of *Trichogramma achaeae* (Hym.: *Trichogramma tidae*) in tomato greenhouses of Spain. Integrated control in protected crops. Mediterranean climate IOBC/wprs Bulletin 49: 225-230.
19. Kannan SK, Murugan K, Kumar AN, Ramasubramanian N, Mathiyazhagan P (2008) Adulticidal effect of fungal pathogen, *Metarhizium anisopliae* on malarial vector *Anopheles stephensi*. Afr J Biotechnol 7: 838-841.
20. Dahliz A (2014) Complex of natural enemies and control methods of the exotic invasive pest *Tuta absoluta* (Lepidoptera: Gelechiidae) in Southern Algeria" BP 17, Touggourt (Algeria).
21. Inanl C, Oldargc AK (2012) Effects of entomopathogenic fungi, *Beauveria bassiana* (Bals.) and *Metarhizium anisopliae* (Metsch.) on larvae and egg stages of *Tuta absoluta* (Meyrick) Lepidoptera: Gelechiidae. Ege Üniversitesi Ziraat Fakültesi Dergisi 49: 239-242.
22. Neves PJ, Alves SB (2000) Selection of *Beauveria bassiana* (Bals.) Vuill. and *Metarhizium anisopliae* (Metsch.) Sorok. Strains for control of *Cornitermes cumulans* (Kollar). Braz Arch Biol Technol 43: 319-328.
23. Sosa-Gómez DR, Moscardi F (1992) Epizootiology: Key of the problems for the microbial control with fungi. In: Symposium of biological control, 3, 1992, Águas de Lindóia. Anais. Jaguariúna: EMBRAPA-CNPDA. pp. 64-69.
24. Kleespies RG, Zimmermann G (1998) Effect of additives on the production, viability and virulence of blastospores of Metarhizium anisopliae. Biochem. Sci. Technol. 8(2):207-214.

16

Colletotrichum gloeosporioides: A True Endophyte of the Endangered Tree, *Cynometra travancorica* in the Western Ghats

Thulasi G Pillai[1]* and Jayaraj R[2]

[1]*Department of Forest Pathology, Kerala Forest Research Institute, Peechi, Thrissur-680751, India*
[2]*Divisions of Forest Ecology and Biodiversity Conservation, Kerala Forest Research Institute, Peechi, Thrissur-680751, India*

Abstract

Fungi form symbiotic associations with forest trees. True endophytes have been evolving with the host trees millions of years. Study was carried out to isolate and identify true endophytic fungi from the leaves of a rare tree species of *Fabaceae* family, *Cynometra travancorica*. Study was conducted in trees growing in four different forest areas of South Western Ghats in three different seasons, pre-monsoon, monsoon and post-monsoon. Twelve fungal cultures were obtained and one of the cultures was common in all the samples. Partial sequencing of genomic DNA was done for identification. The endophyte was identified as *Colletotrichum gloeosporioides*. This is the first report of *Colletotrichum gloeosporioides* as endophyte from *C. travancorica*. This organism can have an important role in the survival and protection of the tree against harsh environment and adverse conditions. Under favorable conditions, chances of the conversion of *C. gloeosporioides* as an endophyte to pathogen cannot be ruled out.

Keywords: Fungal endophyte; *Cynometra travancorica*; Western ghats; *Colletotrichum gloeosporioides*

Introduction

Plants form symbiotic associations with endophytic fungi which live inside healthy tissues as dormant microthalli. Studies conducted during the last 30 years have shown that presence of endophytic fungi is ubiquitous [1]. Endophytes can help the host plant by producing a surplus of substances that provide protection and survival. The metabolites from the endophytic fungi can have potential for use in modern medicine, agriculture, and industry [2]. Antibiotics, antimycotics, immunosuppressants, and anticancer compounds are only a few examples. Isolation of rohitukine, a chromane alkaloid possessing anti-cancer activity was reported from *Fusarium proliferatum* [3]. The diversity and quantity of endophytes depends on biotic, abiotic, and experimental factors like host species, type and phase disposition of plant organ, edaphic and climatic conditions, isolation procedure and number and size of samples [1]. Fungal endophytes in tropical forest trees are less studied, where abundance and diversity are great. The Western Ghats, especially the southern parts is one of the richest bio-geographic provinces of the Indian subcontinent *Cynometra travancorica* is an endangered tree species of the family Fabaceae [4]. The timber is useful and it is grown as avenue tree [5]. The tree is rare and endemic to Southern Western Ghats. The plant is rarely seen and narrowly distributed in evergreen and semi-evergreen forest of Southern Western Ghats. In the present study our aim was to isolate and identify the true fungal endophyte present in the leaves of *Cynometra travancorica*.

Materials and Methods

Sample collection

The material was collected from 4 different evergreen forest - Shendurney (8 8727 N, 77 1634 E), Siruvani (7°10 N, 10°55' E), Thamarassery Ghat (11°26' N, 75 53E) and Vellanimala (10° 25N and 76° 30E) of Kerala. The collection was made in 3 different season, pre monsoon, monsoon, post monsoon. Three samples per location per season were collected for the analysis.

Isolation of endophytes

Isolation of fungal endophyte was done as per standard procedures Kharwar et al. [6], with slight modifications. The leaves were surface sterilized with 75% ethanol for 90 seconds. The tissues were rinsed in sterile distilled water 3 times and allowed to surface dry in sterile conditions. Three trees were selected from a site. Five different media were used for fungal isolation.

- Potato dextrose agar (PDA)
- Sabouraud's dextrose agar
- PDA with Rose Bengal
- Water agar and
- Oat Meal Agar.

Oat Meal Agar was found to give optimum growth for fungal isolation (rapid growth and more number of colonies) and was selected for further study. Five segments of leaves with 0.5 cm diameter were evenly placed in each petri dishes containing Oat Meal Agar with streptopenicillin, to suppress bacterial growth and incubated for 20 days with 12 hrs. of light followed by 12 hrs. of dark cycles at room temperature. A total of 40 segments were analyzed. The tissues were monitored every day to check growth of endophytic fungal colonies. The hyphal tips when grew out from leaf segments were isolated, subcultured and stored at -4°C on Oat Meal Agar slants for preservation. The cultures were screened for the true endophytic fungi. The isolates were maintained by routine sub culturing. Identification was done by partial sequencing of genomic DNA.

***Corresponding author:** Thulasi G Pillai, Department of Forest Pathology, Kerala Forest Research Institute, Peechi, and Thrissur-680751, India
E-mail: thulasigpillai@gmail.com

Analysis of data

Frequency of occurrence of endophytes (%): Fungal population was quantified as frequency of occurrence. It is calculated as follows

[No of leaf discs colonised by a given fungus/Total number of explants observed] x 100 [7].

Colonisation rate

Colonization rate (CR) was calculated as the total number of leaf segments infected by fungi divided by the total number segments incubated [8]. Colonization rate was expressed as percentages, as widely used in the past endophyte studies, therefore, colonization rates can be used for comparative purposes.

[Total numbers of explants in a sample yielding 1 isolate or more/ Total number of leaf segments in that sample] x 100 [8].

Isolation rate: Isolation rates are used to demonstrate the degree of multiple colonization from the samples in different trials.

Isolation rate is determined by Frohlich et al. [9].

[Total number of isolates yielded by a given sample/Total number of explants in that sample].

Identification of fungal endophyte: Macro morphological characters were recorded. The culture failed to sporulate. The culture was further refined by partial sequencing of endophytic genome.

DNA Isolation: Genomic DNA was extracted from pure endophytic culture using Sigma Aldrich DNA extraction Kit. D1/D2 region of LSU (Large subunit 28S rDNA) gene was amplified by Polymerase Chain Reaction (PCR) from the isolated genomic DNA. DNA sequencing was carried out with PCR amplicon. D1/D2 region was amplified by PCR from fungal genomic DNA using universal Primers [10].

F- 5'-ACCCGCTGAACTTAAGC-3', R - 5'-GGTCCGT-GTTTCAAGACGG-3'. PCR was carried out in a final reaction volume of 25 μl with deionized water 17.0 μl, Taq buffer without $MgCl_2$ (10×) 2.5 μl, $MgCl_2$ (15 Mm) 1.0 μl, Forward Primer (10 pm/μl) 0.5 μl, Reverse Primer (10 pm/μl) 0.5 μl, dNTPs 1.0 μl, Taq DNA Polymerase (5 U/μl) 0.5 μl, Template DNA(20 ng/μl) 2.0 μl to a final volume of 25.0 μl. The PCR protocol designed for 30 cycles for the primers used were Initial denaturation 95°C 5 minutes, Denaturation 94°C 30 seconds, Annealing 48°C 30 seconds, Extension 72°C 45 seconds, Denaturation 94°C 30 seconds, 5 Final Elongation 72°C 10 minutes. Commercially available 100 bp ladder was used as standard molecular weight DNA. PCR-Product Electrophoresis was carried out. Amplified PCR product was purified using column purification as per manufacturer's guidelines, and further used for sequencing reaction (Figures 1 and 2) The concentration of the purified DNA was determined and was subjected to automated DNA sequencing on ABI3730xl Genetic Analyzer (Applied Bio systems, USA). The Sequencing was done at *UniBiosys* Biotech Research Labs, Kochi, and Kerala, India. The sequences after editing were then used for similarity searches using BLAST (Basic Local Alignment Search Tool) programme in the NCBI, USA GenBank (www.ncbi.nlm.nih.gov) DNA database for identifying the fungal strains.

Results

Isolation rate/colonization rate/frequency of occurrence of endophytes

About 12 pure cultures were obtained from the leaves of *Cynometra travancorica*. The leaf cultures constantly yielded one white fungus which was identified as *C. gloeosporioides*. The culture was identified as *C. gloeospoiroides*. The frequency, colonization rate and isolation rate were good (Table 1 and Figures 1-5).

Sequencing of D1/D2 region of LSU (Large Subunit 28S r DNA)

Sequenced data of D1/D2 region of LSU (Large Subunit 28S r DNA) after subjecting to BLAST with the nrd database of NCBI gene bank database revealed that the organism was *Colletotrichum gloeosporioides* with 100% similarity with accession number AB710144.1. Our accession number is KM823608.

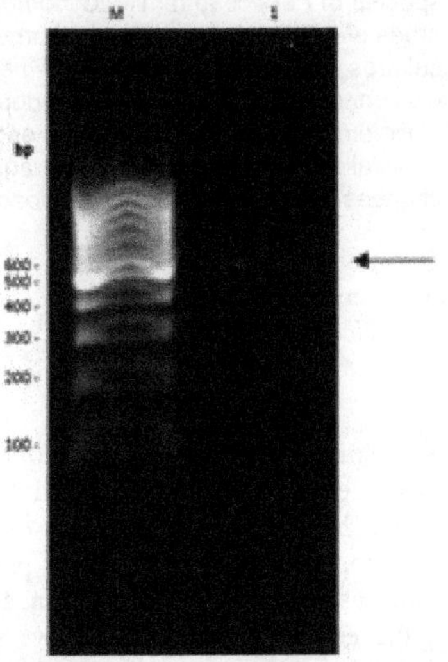

Figure 1: Agarose gel showing PCR amplification. M.DNA size marker of 100 bp, Lane one is amplified DNA from the fungus.

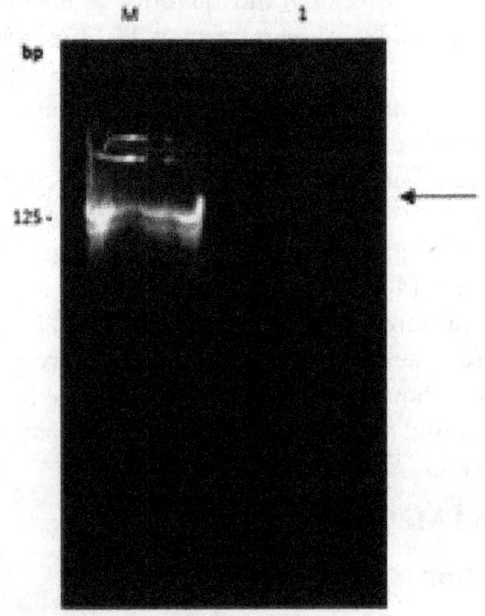

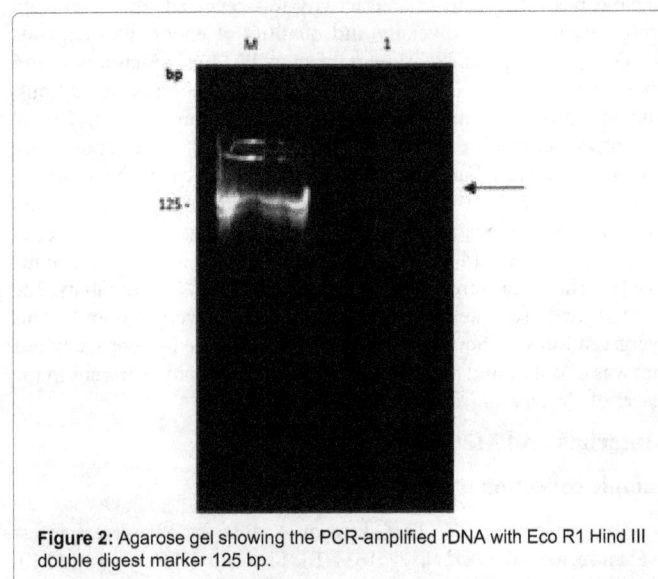

Figure 2: Agarose gel showing the PCR-amplified rDNA with Eco R1 Hind III double digest marker 125 bp.

Plant	Frequency of occurrence of endophytes (Value-%)	Colonisation rate (Value-%)	Isolation rate (Value-%)
C. travancorica leaves	70.6	86.6	80

Table 1: Isolation rate of the fungal endophytes from *C. travancorica*.

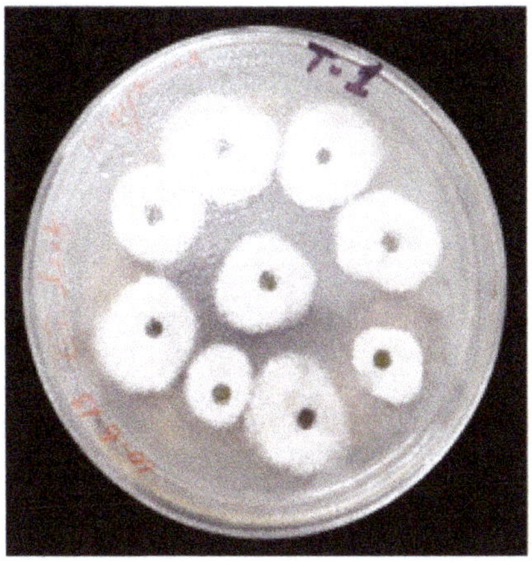

Figure 3: Isolation of funal endophytes obtained from the leaves of *Cynometra travancorica* on Oat meal agar. Colonisation of fungal and endophytes on leaf segments of *C. travancorica.*

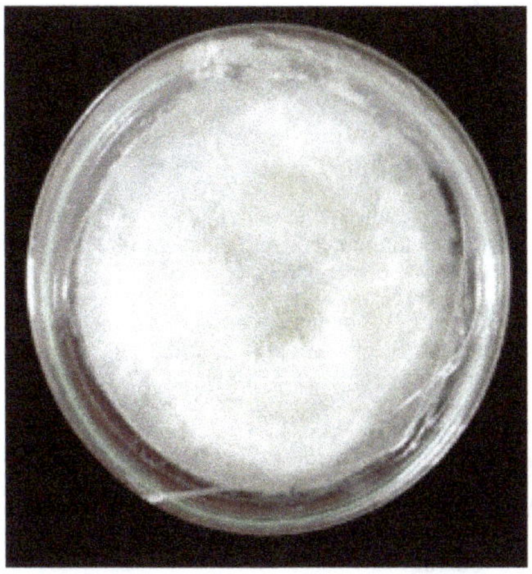

Figure 4: Pure culture of *C. gloeosporiodes.*

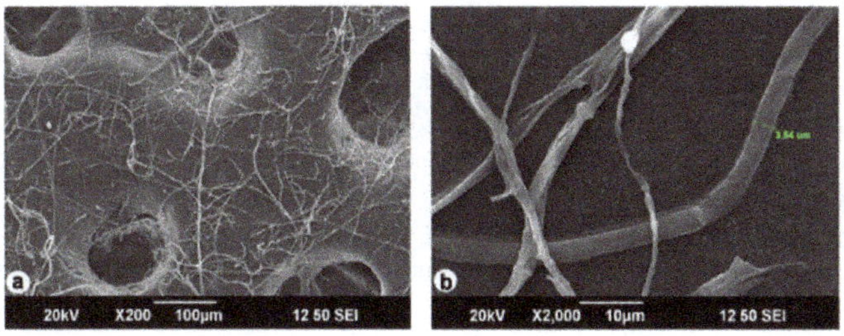

Figure 5: Scanning electron microscope image of mycelium of *C. gloeosporiodes* a. *Mycellium*, b. hyphae.

Sequence Data sample A FP

Sequence Data FP

GCAAGGATGCTGGCATAATGGTCATCAGCGA

Sequence Data RP

ATCCTTGCGGAGCGCGTACCTCAGCCCGCCGAAGGGTAT
TATGCAACAGGCTATAACACTCCCCGAAGAGAGCT
ACGTTCCTGAAGCTTTTGTCCCCGACAGCGAGCTGATGC
TGGCCTGAGCCGGCGAAGTGCCCCAGCCGCGAGAG
CTGGGTGATTCACCGGACGCAAGTCTGGTCACAAGCGCT
TCCCTTTTAACAATTTCACGTGCTGTTTAACCCTC
TTTTCAAAGTGCTTTTCATCTTTCGATCACTCTACTTGTG
CGCTATCGGTCTCTGGCCGGTATTTAGCTTTAGA
AGAAATATACCTCCCATTTTGAGCAGCATTCCCAAACTA
CTCGACTCGTCGAAGGAGCTTTACAAAGGCTTGGT
GTCCAACTGTACGGGGCTCTCACCCTCTCTGGCGTCCCG
TTCTAGGGAACTTGGAAGGCACCGCACCAAAAGCA
TCCTCTGCAAATTACAACTCGGGCCTAGGGCCAGATTTC
AAATTTGAGCTGTTGCCGCTTCACTCGCCGTTACT
GAGGCAATCCCTGTTGGTTTCTTTTCCTCCGCTTATTGAT

Consensus Sequence Data

ATCAATAAGCGGAGGAAAAGAAACCAACAGGGATTGCC
TCAGTAACGGCGAGTGAAGCGGCAACAGCTCAAATT
TGAAATCTGGCCCTAGGCCCGAGTTGTAATTTGCAGAGG
ATGCTTTTGGTGCGGTGCCTTCCAAGTTCCCTAGA
ACGGGACGCCAGAGAGGGTGAGAGCCCCGTACAGTTGG
ACACCAAGCCTTTGTAAAGCTCCTTCGACGAGTCGA
GTAGTTTGGGAATGCTGCTCAAAATGGGAGGTATATTTC
TTCTAAAGCTAAATACCGGCCAGAGACCGATAGCG
CACAAGTAGAGTGATCGAAAGATGAAAAGCACTTTGAAA
AGAGGGTTAAACAGCACGTGAAATTGTTAAAAGGG
AAGCGCTTGTGACCAGACTTGCGTCCGGTGAATCACCCA
GCTCTCGCGGCTGGGGCACTTCGCCGGCTCAGGCC
AGCATCAGCTCGCTGTCGGGGACAAAAGCTTCAGGAACG
TAGCTCTCTTCGGGGAGTGTTATAGCCTGTTGCAT
AATACCCTTCGGCGGGCTGAGGTACGCGCTCCGCAAGGA
TGCTGGCATAATGGTCATCAGCGA

Discussion

Endophytic fungal association with forest trees are always non-pathogenic [1]. The concept of "true endophyte" has been proved in our study. True endophytes are those fungi whose colonisation never results in visible disease symptoms [11]. Endophytic fungi on leaves of forest trees play ecological roles as accepted mutualists [12], latent pathogens [1] and saprobic decomposers after leaf death [13,14]. Out of the twelve cultures obtained, *Colletotrichum gloeosporioides* was isolated from *Cynometra travancorica* in all the three seasons studied and from all the four locations. An endophyte which is in association with a plant in all the conditions despite the climatic and geographical variations is true endophyte. Most of the endophytic cultures are sterile, making morphology based identification difficult. Earlier reports suggest that endophytes possess traits which are advantageous under extreme conditions. Support of these idea comes from the fact that some *Colletotrichum* species which are pathogenic on the main host species but symptomless endophytes on 'non -disease' host species, providing mutualistic benefits such as disease resistance, drought tolerance and growth enhancement. This differential behaviour may result from differences in fungal gene expression in response to the plant or differences in the ability of the plant to respond to the fungus. Mutualistic endophytes can be evolved from parasitic or pathogenic fungi [13].

Colletotrichum gloeosporioides was a pathogen for long period. The organism interacts with numerous plant species overtly as symptomatic pathogens and cryptically as asymptomatic pathogens. It is not clear whether these contrasting ecological modes are optional strategies expressed by individual *Colletotrichum* species or whether species ecology is explicitly pathogenic or endophytic. The ecological significance of endophytism is unclear. The genus *Colletotrichum* (Sordarimycetes, Ascomycota) comprises ~ 600 species [15]. As per ARS fungal databases, it attacks over 3,200 species of monocot and dicot plants. The infection strategy of this organism is a multistage hemibiotrophy [16]. *Colletotrichum* gloeospoirioides has been reported to produce taxol [17].

The establishment of endophytic association with plants are interesting, the initial steps being the same - recognition, germination and penetration and then a quiescence stage is developed. There may be some mechanism to avoid recognition also. For example a gene has been cloned from *Colletotrichum gloeosporioides* which is switched on during the initial phase of colonisation and switched off later during the neurotropic phase [18]. This gene encodes a glycoprotein that resembles plant cell wall proteins which is believed to coat the hyphae that the plant is unable to recognise as alien.

Colletotrichum are endophytes as well as disease agents of conifers [19-23] and ferns [24,25]. They are associated widely with both herbaceous and woody plants, though the latter appear mainly to contain colonies in fruits, leaves and other non-lignified tissues. The generic name *Colletotrichum* was introduced by Corda [26] for C. lineola, a species found associated with a member of the Apiaceae in the Czech Republic. The first applications of DNA sequence data to distinguish between *Colletotrichum* species were published by Mills et al. [27] and Sreenivasaprasad et al. [28], who identified sequence variation in the ITS1 region of nrDNA between six species of *Colletotrichum*, as well as detecting polymorphisms in the same region between strains of *C. gloeosporioides* from different hosts. The biotrophic life strategies adopted by *Colletotrichum* species may also contribute to their prominence as symptomless endophytes of living plant tissues [21,29-31]. Research into the molecular basis of host-parasite interactions in *Colletotrichum* is currently highly active and such approaches will dominate research in the future into the extent of host specificity exhibited by *Colletotrichum* species. Constant monitoring of the tree for disease symptoms are done to check whether the endophyte is transformed to pathogen.

No angiosperms that do not harbour endophytic *Colletotrichum* colonies are known so far. There are many reports about isolation of taxol, the anticancer drug from *Colletotrichum gloeosporioides* from different plants [18]. Our aim is to study the therapeutic potential of the fungi isolated from *C. travancorica*. The phytochemical evaluation is under study.

References

1. Sieber TN (2007) Endophytic fungi in forest trees:are they mutualist? Fungal Biology Reviews. 21: 75-89.
2. Strobel G, Daisy B, Castillo U, Harper J (2004) Natural products from endophytic microorganisms. J Nat Prod. 67: 257-68.
3. Kumara, Patel Mohana (2012) "Fusarium proliferatum, an endophytic fungus from Dysoxylum binectariferum Hook. f, produces rohitukine, a chromane alkaloid possessing anti-cancer activity." Antonie van Leeuwenhoek 2: 323-329.
4. IUCN (2011) IUCN Red list of Treatened species. IUCN, Gland.
5. Nayar MP, Sastry ARK (1988) Red data book on Indian Plants, Botanical Survey of India. Calcutta.
6. Kharwar RN, Verma VC, Strobel GA, Ezra D (2008) The endophytic fungal complex of Catharanthus roseus (L.) G. Don. Current Science 95: 228-233.
7. Photita W, Lumyong S, Lumyong P, Hyde KD (2001) Endophytic fungi of wild banana (Musa acuminate) at Doi Suthep Pui National Park, Thailand. Mycological Research, 105: 1508-1513.
8. Petrini O (1982) Notes on some species of Chloroscypha endophytic in Cupressaceae of Europe and North America. Sydowia, 35: 206-222.
9. Frohlick J, Hyde KD, Petrini O (2000) Endophytic fungi associated with palms. Mycological Research, 104: 1202-1212.
10. White TJ, Burns T, LEF S, Taylor J (1990) Amplification and direct sequencing of fungal ribosomal genes for phylogenetics. In: PCR protocols: A guide to Methods and Application. Academic Press. San Diego, USA. 315-322.
11. Mostert L, Crous DW, Petrini O (2000) Endophytic fungal association with shoots and leaves of Vitis Vinifera with specific reference to Phomopsis Viticola complex. Sydovia 52: 46-58.
12. Saikkonen K (2007) Forest structure and fungal endophytes. Fungal Biology Reviews 2: 67-74.
13. Osono T (2006) Role of phyllosphere fungi of forest trees in the development of decomposer fungal communities and decomposition processes of leaf litter. Canadian Journal of Microbiology 52: 701-716.
14. Promputtha I, Lumyong S, Dhanasekaren V, McKenzie EHC, Hyde et al (2007) A phylogenetic evaluation of whether endophytes become saprotrophs at host senescence. Microbial Ecology 53: 579-590.
15. Crous PW, Gams W, Stalpers JA, Robert V, Stegehuis G (2004) MycoBank: an online initiative to launch mycology into the 21st century. Stud Mycol 50: 19-22.
16. Perfect SE, Hughes HB, O'Connell RJ, Green JR (1999) Colletotrichum, a model genus for studies on pathology and fungal plant interactions. Fungal Genet Biol 27: 186-198.
17. Gangadevi V, Muthumary J (2008) Isolation of Colletotrichum gloeosporioides, a novel endophytic taxol-producing fungus from the leaves of a medicinal plant, Justicia gendarussa. Mycologia balcanica 5: 1-4.
18. Perfect SE, O'Connell R, Green EF, Doering-Saad C, Green JR (1998) Expression cloning of a fungal proline-rich glycoprotein specific to the biotrophic interface formed in the Colletotrichum–bean interface. Plant J; 15: 273-9.
19. Dingley JM, Gilmour JW (1972) Colletotrichum acutatum f.sp. pinea associated with terminal crook disease of Pinus spp. New Zealand Journal of Forestry Science 2: 192-201.
20. Wang YT, Lo HS, Wang PH (2008) Endophytic fungi from Taxus mairei in Taiwan: first report of Colletotrichum gloeosporioides as an endophyte of Taxus mairei.
21. Joshee S, Paulus BC, Park D, Johnston PR (2009) Diversity and distribution of fungal foliar endophytes in New Zealand Podocarpaceae. Mycological Research 113: 1003-1015.
22. Damm U, Cannon PF, Woudenberg JHC, Crous PW (2012) The Colletotrichum acutatum species complex. Studies in Mycology. 73: 37-113.
23. Leahy R, Schubert T, Strandberg J, Stamps B, Norman D (1995) Anthracnose of Leather leaf fern. Plant Pathology Circular 372: 4.
24. Mackenzie SJ, Peres NA, Barquero MP, Arauz LF, Timmer LW (2009) Host range and genetic relatedness of Colletotrichum acutatum isolates from fruit crops and leather leaf fern in Florida. Phytopathology. 99: 620-631.
25. Corda ACI (1831) Die Pilze Deutschlands. In: Deutschlands Flora in Abbildungen nach der Natur mit Beschreibungen (Sturm, J, ed.). Sturm, Nürnberg vol. 3, Abt. 12: 33-64, tab. 21-32.

26. Mills PR, Hodson A, Brown AE (1992) Molecular differentiation of Colletotrichum gloeosporioides isolates infecting tropical fruits. In: Bailey JA, Jeger MJ (eds) Colletotrichum: Biology, Pathology and Control CABI, Wallingford, UK, pp. 269-288.

27. Sreenivasaprasad S, Brown AE, Mills PR (1992) DNA sequence variation and interrelationship among Colletotrichum species causing strawberry anthracnose. Physiological and Molecular Plant Pathology. 41: 265-281. Botanical Studies 49: 39-43.

28. Lu G, Cannon PF, Reid A, Simmons CM (2004) Diversity and molecular relationships of endophytic Colletotrichum isolates from the Iwokrama Forest Reserve, Guyana. Mycological Research 108: 53-63.

29. Rojas EI, Rehner SA, Samuels GJ, Van Bael SA, Herre EA, et al. (2010) Colletotrichum gloeosporioides s.l. associated with Theobroma cacao and other plants in Panama: multilocus phylogenies distinguish pathogen and endophyte clades. Mycologia 102: 1318-1338.

30. Yuan ZL, Su ZZ, Mao LJ, Peng YQ Yang GM, et al. (2011) Distinctive endophytic fungal assemblage in stems of wild rice (Oryza granulata) in China with special reference to two species of Muscodor (Xylariaceae). Journal of Microbiology 49: 15-23.

31. O'Connell RJ, Thon MR, Hacquard S, Amyotte SG, Kleemann J, et al. (2012) Life-style transitions in plant pathogenic Colletotrichum fungi deciphered by genome and transcriptome analyses. Nature Genetics.

17

Induce Systemic Resistance against Root Rot and Wilt Diseases in Fodder Beet (*Beta vulgaris* L. var. *rapacea* Koch.) by Using Potassium Salts

Montaser F Abdel-Monaim*, Marwa AM Atwa and Kadry M Morsy

Plant Pathology Research Institute, Agriculture Research Center, Giza 12619, Egypt

Abstract

Rhizoctonia solani, Fusarium solani, F. oxysporum F. equiseti and *F. semitectum* were found to be associated with root rot and wilt symptoms of fodder beet plants collected from different fields in New Valley governorate, Egypt. All the obtained isolates were able to attack fodder beet plants (cv. Starmon) causing damping-off and root rot/wilt diseases. *R. solani* isolate FB1, *F. solani* isolate FB7 and *F. oxysporum* isolate FB11 were the more virulent ones in the pathogenicity tests. The efficacy of 4 different potassium salts for controlling damping-off, root rot and wilt diseases in fodder beet were evaluated *in vitro* and *in vivo*.

In vitro studies, all the tested potassium salts were significantly suppressed growth of the pathogenic fungi at different concentrations. $KHCO_3$ showed superior higher inhibitory effect on redial growth of the tested pathogenic fungi especially at higher concentration (20 mM).

Under green house and field conditions, all potassium salts significantly reduced damping-off and root rot/wilt severity and increased survival of plants. The reduction in damping-off and root rot/wilt increased with increasing of potassium salts concentration except potassium sulfate (K_2SO_4), while concentration 10 mM was more effective for reducing damping-off and root rot/wilt severity than 20 mM. K_2SiO_3 followed by K_2HPO_4 recorded highly protection against damping-off and root rot/wilt severity more than the other tested potassium salts. Under field conditions, all these potassium salts at different concentrations significantly submitted to various growth and yield parameters *viz.* root length, root diameters, fresh and dry weights compared with control during growing seasons 2013-14 and 2014-15. While, % dry maters was no significant in both growing seasons. The applied treatment K_2SiO_3 achieved the highest increase in all the mentioned parameters over the other entire three potassium salts during both growing seasons.

In physiological studies, activity of defense-related enzymes, including peroxidase (PO), polyphenol oxidase (PPO), phenylalanine ammonia lyase (PAL), and tyrosine ammonia lyase (TAL) and total phenols content were increased in inoculated plants with *R. solani, F. solani*, and *F. oxysporum* individually and treated with potassium salts compared with untreated plants. K_2SiO_3 at 20 mM showed the highest level of all oxidative enzymes activity and total phenols content followed by K_2HPO_4 and K_2SO_4 at 20 mM. Whereas, the least enzymes activity was recorded with $KHCO_3$ at 10 mM. These results suggested that these chemicals may be play an important role in controlling the fodder beet damping-off, root rot and wilt diseases; through they have induction of systemic resistance in fodder beet plants.

Keywords: Fodder beet; Root rot and wilt; Potassium salts; Induced resistance

Introduction

Fodder beet (*Beta vulgaris* L. var. *rapacea* Koch.) offers a higher yield potential than any other "arable" fodder crop. The roots have an excellent feed quality and they are very palatable to ruminant stock. The leaf can be utilized if required to boost the total fodder output even further. Fodder beet when grown under suitable conditions, can produce almost 20 t ha^{-1} dry matter yield compared with 13 ± 15 t DM/ha^{-1} from four harvests of grass. Approximately 75% of fodder beet dry matter is in the root component [1]. Including fodder beet in diet of cattle increases intake of dry matter that is quantitative and qualitative factors affecting intake of the basal diet [2,3].

Plant diseases caused by soil-borne plant pathogens considered the major problems in agricultural production throughout the world, reducing yield and quality of crops. Plant pathogens have caused an almost 20% reduction in the principal food and cash crops worldwide [4]. Root rot and wilt caused by soil-borne pathogenic fungi is one of the most serious diseases affected several cultivated plants in worldwide. It results in poor production, poor quality, poor milling returns and reduced agriculture income. This has a negative impact on the livelihood of farmers [5]. Fungal disease control is achieved through the use of fungicides which is hazardous and toxic to both people and domestic animals and leads to environmental pollution [6]. Therefore, a more balanced, cost effective and eco-friendly approach must be implemented and adopted farmers. In order to overcome such hazardous control strategies, scientists, researchers from all over the world paid more attention towards the development of alternative methods which are, by definition, safe in the environment, non-toxic to humans and animals and are rapidly biodegradable.

***Corresponding author:** Montaser F Abdel-Monaim, Plant Pathology Research Institute, Agriculture Research Center, Giza 12619, Egypt
E-mail: fowzy_2008@yahoo.com

The present research focuses on finding compounds that are safe to humans and the environment, *viz.* potassium salts are recorded by several investigators to have antimicrobial inhibitor effect as well as they play important role to induce plant resistance against damping-off, root rot and wilt diseases of fodder beet either *in vitro* or in *vivo* as well as its effective on plant growth and yield components in field.

Materials and Methods

Seeds and growth of plants

Fodder beet (*Beta vulgaris* L. var. *rapacea* Koch.) cultivar Starmon used in this study was obtained from the Forage Research Dep., Field Crops Research Institute, Agricultural Research Center, Giza, Egypt. Seeds were planted in plastic pots 30 cm diameter (2.4 kg soil), filled with a pasteurized mixture of soil and sand (4:1 w/w). Five seeds were sown in each pot and these pots were irrigated every three days.

Fungal isolation and pathogenicity tests

Samples of fodder beet plants showing root rot and wilt symptoms were collected from different farms of New Valley governorates. All samples were thoroughly washed with tap water several times, cut in small pieces and surface sterilized for 2 min in 2% sodium hypochlorite solution, then rinsed several times in sterilized distilled water and dried between sterilized filter papers. The surface sterilized samples were plated onto potato dextrose agar (PDA) medium and incubated at 25 ± 1°C. After 3-7 days incubation, the developed fungal colonies were purified by hyphal tip and/or single spore isolation techniques. The obtained fungal isolates were identified according to their cultural and microscopical characters as described by Booth [7] and Barnett and Hunter [8]. Subcultures of the obtained isolates were then kept on PDA slants and stored at 5°C for further studies.

Inoculum of the obtained isolates of soil borne pathogens was prepared on autoclaved barley medium (75 g washed dried barley grains, 100 g washed dried coarse sand and 75 ml tap water) in 500 ml glass bottles. Each bottle was inoculated with five discs (0.7 cm in diameter) of 4-day-old cultures of each isolates. Bottles were incubated at 25 ± 1°C for 15 days [9]. For each isolate, the contents of 20 bottles were thoroughly mixed in a plastic container and used as a source of inoculum. Soil and pots were sterilized with a 5% formalin solution for 15 min. Soil was covered with a polyethylene sheet for 7 days to retain the gas and left to dry for 2 weeks until all traces of formaldehyde disappeared. Pathogen inocula were added to the potted soil at a rate of 3% (w/w) and mixed thoroughly with the soil one week before planting. Three pots were used as replicates for each isolate (1-16) as well as control (uninfested soil). Fodder beet seeds of cv. Starmon were surface sterilized using 1% sodium hypochlorite for 2 min, rinsed in distilled water several times and sown in pots at rate 5 seeds pot^{-1}. These pots were irrigated every three days.

Assessment of disease severity

Percentage of damping-off was recorded after 35 days after planting. While severity of root rot and was determined 90 days after planting according to Abdou et al. [10] using a rating scale of 0-5 on the basis of root the discoloration or leaf yellowing as follows, 0=neither root discoloration nor leaf yellowing, 1=1-25% root discoloration or one leaf yellowed, 2=26-50% root discoloration or more than one leaf yellowed, 3=51-75% root discoloration plus one leaf wilted, 4=up to 76% root discoloration or more than one leaf wilted, and 5=completely dead plants. For each replicate a disease severity index (DS%) similar to that described by Liu *et al.* [11] was calculated as follows:

$$DSI = \frac{\Sigma d}{d\,max. \times n} \times 100$$

Whereas: d is the disease rating possible, d max is the maximum disease rating and n is the total number of plants examined in each replicate.

In vitro antifungal activity

The inhibitory effect of potassium salts *viz.* potassium phosphate dibasic (K_2HPO_4), potassium bicarbonate ($KHCO_3$), potassium sulfate (K_2SO_4), potassium silicate (K_2SiO_3) at different concentrations 5,10 and 20 mM (listed in Table 1) on the linear growth of *Rhizoctonia solani*, *Fusarium solani* and *F. oxysporum*, the fodder beet root rot and wilt pathogens, was evaluated. Tested solutions were added to conical flasks containing sterilized PDA medium before solidifying to obtain the proposed concentrations and shacked gently, then dispensed into sterilized Petri dishes (9-cm diameter). Petri dishes were individually inoculated with equal disks (7-mm-diam.), taken from 7-day-old cultures of tested fungi. The Petri dishes containing the PDA medium inoculated with the tested pathogens alone served as control. All plates were incubated at 25 ± 1°C. Each treatment was represented by 3 plates as replicates. Linear growth of tested fungi was measured when the control plates (medium free of potassium salts) reached full growth and the average growth diameter was calculated. Mycelial growth inhibition was calculated by using the formula:

Mycelial growth inhibition (%)=100 (C-T/C)

Where C=growth in control and T=growth in treatment.

Evaluation effect of potassium salts on damping-off, root rot and wilt diseases under greenhouse conditions

The fungal inoculum of *R. solani* (isolate FB1), *F. solani* (isolate FB7) and *F. oxysporum* (isolate FB 11) were prepared as described before in pathogenicity test. Plastic pots (30 cm diameter) were packed with sterilized sandy clay soil infested with fungal inocula at the rate 3% (w/w), seven days before planting. Fodder beet cv. disinfested seeds were soaked in the solution of each potassium salts (Table 1) for 12 hr. [12], then sown in infested pots at rate 5 seeds pot^{-1}. Also, in control treatment, fodder beet seeds soaked in water for the same time and seeding in infested soil with the pathogen at the same rate. Three pots were used per treatment as a replicates. Percentages of damping-off, root rot and wilt severity were recorded after 35 and 90 days from planting, respectively.

Evaluation effect of potassium salts on damping-off, root rot and wilt diseases and under greenhouse conditions

This experimental, factorial block design experiment was conducted at sowing date of 1st November of two successive growing seasons 2013/14 and 2014/15 in a field naturally infected with the causal organisms of root rot and wilt diseases of fodder beet located at the experimental farm of Kharga Agric. Station, New Valley Governorate. The main plots were potassium salts tested, sub plots were concentrations. Healthy fodder beet seeds were soaked in the solutions of the potassium salts for 12 hrs.

Materials	Chemical Composition	Molecular Weight	Used Concentration
Potassium phosphate dibasic	K_2HPO_4	174.18 g/mol	5,10,20 mM
Potassium bicabonate	$KHCO_3$	100.12 g/mol	5,10,20 mM
Potassium sulfate	K_2SO_4	174.26 g/mol	5,10,20 mM
Potassium silicate	K_2SiO_3	154.28 g/mol	5,10,20 mM

Table 1: Chemical formula of potassium salts.

[12]. A plot was 3 × 3.5 m with five rows; 50 cm row spacing, seeds were sown in hills (2 seeds hill-1 and 25 cm apart). In the control treatment, fodder beet seeds were soaked in water for the same time and sown with the same method. Fertilizers application at the rate of recommended doses. The crop was irrigated at 12-15 days intervals. Hand thinning to one plant per hill after 5 weeks from planting [3]. Percentages of damping-off and root rot/wilt disease index were calculated after 35 and 120 days from planting, respectively. At harvesting, 10 plants from the central ridges were pulled up to determine the following growth traits and forage yield.

1. Root length (cm)=distance between the beginning of the root to an end.
2. Root diameter (cm)= Circumference of circle when the maximum width of root divided on 2.14.
3. Fresh and dry weights of roots (ton/fed.).
4. Dry maters (%)=Dry weight of roots/Fresh weight of roots × 100

Biochemical changes associated with induced resistance

Activities of peroxidase (PO), polyphenol oxidase (PPO), phenylalanine ammonia lyase (PAL) and tyrosine ammonia lyase (TAL) and total phenols content was studied in tissue extracts of fodder beet plants surviving treatment with K_2HPO_4, $KHCO_3$ and K_2SiO_3 at 20 mM and K_2SO_4 at 10 mM, as well as in untreated seeds. All treatments were grown in soil infested with *R. solani*, *F. solani*, *F. oxysporum* as individually. One gram of plant tissue was homogenized in 10 mL of ice-cold 50 mM potassium phosphate buffer (pH 6.8) containing 1M NaCl, 1% polyvinylpyrrolidone, 1 mM EDTA, and 10 mM β-mercaptoethanol [13]. After filtration through cheesecloth, the homogenates were centrifuged at 8,000 rpm at 4°C for 25 min. The supernatants (crude enzyme extract) were stored at −20°C or immediately used for determination of PO, PPO, PAL and TAL activities and total protein. For the determination of enzyme activities, each treatment consisted in four replicates (three plants/ replicate) and two spectrophotometric readings were taken per replicate using a Milton Roy 1201 Spectrophotometer (PEMED', Denver, CO, USA). The bioassay experiments were repeated twice.

PO activity

PO activity was determined directly using a spectrophotometrical method [14] using guaiacol as common substrate. The reaction mixture consisted of 0.2 mL crude enzyme extract and 1.40 mL of a solution containing guaiacol, hydrogen peroxide (H_2O_2) and sodium phosphate buffer (0.2 mL 1% guaiacol+0.2 mL 1% H_2O_2 +1 mL 10 mM potassium phosphate buffer). The mixture was incubated at 25°C for 5 min and the initial rate of increase in absorbance was measured over 1 min at 470 nm. Activity was expressed as units of PO/mg protein [15].

PPO activity

The activity of PPO was determined by adding 50 μL of the crude extract to 3 mL of a solution containing 100 mM potassium phosphate buffer, pH 6.5 and 25 mM pyrocatechol. The increase of absorbance at 410 nm during 10 min at 30°C, was measured [16]. One PPO unit was expressed as the variation of absorbance at 410 nm per mg soluble protein per min.

PAL activity

PAL activity was determined following a previously-described direct spectrophotometric method [17]. Two hundred microlitres of the crude enzyme extract previously dialyzed overnight with 100 mM Tris-HCl buffer, pH 8.8, were mixed to obtain a solution containing 200 μL 40 mM phenylalanine, 20 μL 50 mM β-mercaptoethanol, and 480 μL 100 mM Tris-HCl buffer, pH 8.8. After incubation at 30°C for 1 hr, the reaction stopped by adding 100 μL 6 N HCl. Absorbance at 290 nm was measured and the amount of trans-cinnamic acid formed was evaluated by comparison with a standard curve (0.1~2 mg/mL trans cinnamic acid) and expressed as units of PAL/min/mg protein.

TAL activity

TAL activity was determined using the same method used for PAL except L-tyrosine was used instead of phenylalanine.

Protein concentration

Total protein content of the samples was quantified according to the method described by Bradford [18].

Determination of phenolic compounds

To assess phenolic content, 1 g fresh plant sample was homogenized in 10 mL 80% methanol and agitated for 15 min at 70°C. One milliliter of the extract was added to 5 mL of distilled water and 250 μL of 1 N Folin-Ciocalteau reagent and the solution was kept at 25°C. The absorbance was measured with a spectrophotometer at 725 nm. Catechol was used as a standard. The amount of phenolic content was expressed as phenol equivalents in mg/g fresh tissue [19].

Statistical analysis

All experiments were performed twice. Analyses of variance were carried out using MSTAT-C program version 2.10 [20] (1991). Least significant difference (LSD) was employed to test for significant difference between treatments at $P \leq 0.05$ [21].

Results

Isolation, purification and identification of the cauasl fungi

Isolation trials from naturally rotted roots of fodder beet plants coallected from many field grown in New Valley governorate yeilded sexteen fungal isolates. The obtained isoltes were purified using single spore method and/or hyphal tip technique. The purified fungi were identified as *Rhizoctonia solani* (5 isolates), *F. solani* (4 isolates), *Fusarium equiseti* (1 isolates), *F. oxysporum* (4 isolates), *F. semitectum* (2 isolates). These fungi were maintained as pure cultures on ager slants kept in refrigerator at 5 °C until using in further studies.

Pathogenicity tests

Pathogenicity tests of the isolated fungi (Figure 1) reveal that all the isolates were pathogenic to fodder beet plants cv. Starmon with deferent degrees caused damping-off and root rot/wilt symptoms. In this respect *R. solani* isolate FB1 caused the highest damping-off (60%) followed by *R. solani* isolate FB2 and *F. solani* isolate FB7 (53.33%) while *F. equiseti* isolate FB10 caused the lowest damping-off (6.67%). On the other hand, *F. oxysporum* isolate FB11 recorded the highest severity of wilt disease (60.33%) followed by *R. solani* isolate FB3 (45.36% root rot). Generally, *R. solani* isolate FB1, *F. solani* isolate FB7 and *F. oxysporum* FB11 were the highest pathogenic fungi isolated from fodder beet while recorded the lowest survival plants (9.65, 14.11 and 13%, respectively) since these isolates were used in following studies *in vitro* and/or *in vivo*. While, both *F. semitectum* isolates showed the lowest damping-off and root rot severity therefore they were neglected from the following studies.

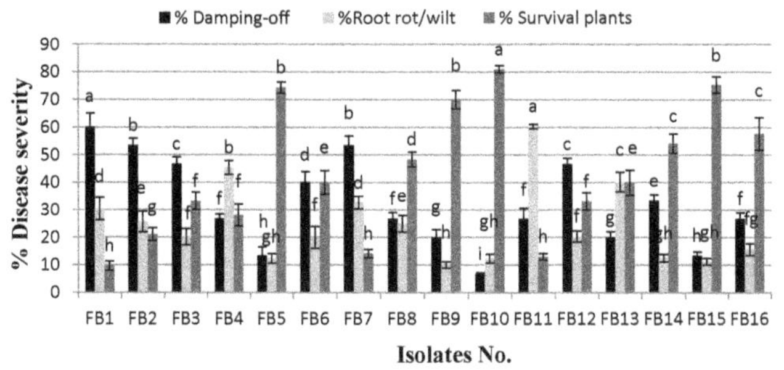

Figure 1: Pathogenicity tests of soil borne fungi isolated from fodder beet roots under the greenhouse conditions. Mean ± SDs per isolate are shown. Different letters indicate significant differences among treatments within the same color column according to least significant difference test (P ≤ 0.05). Percentages of damping-off were recorded 35 days after planting, while root rot/wilt disease severity was determined 90 days after planting.

Potassium salts	Concen. (mM)	% Inhibition		
		Rhizoctonia solani	*Fusarium solani*	*F. oxysporum*
K_2HPO_4	5	25.36	33.26	36.25
	10	32.25	41.25	44.14
	20	36.47	46.25	50.14
	Mean	31.36	40.25	43.51
$KHCO_3$	5	32.25	36.35	41.29
	10	41.25	46.54	55.14
	20	50.12	52.36	62.14
	Mean	41.21	45.08	52.86
K_2SO_4	5	32.36	35.14	36.36
	10	38.53	42.15	47.24
	20	44.23	47.05	58.71
	Mean	38.37	41.45	47.44
K_2SiO_3	5	31.25	36.96	38.96
	10	36.96	44.72	48.75
	20	43.59	50.21	58.96
	Mean	37.27	43.96	48.89
LSD at 0.05 for:				
Potassium salts (A)=		2.59		
Concentrations (B)=		3.47		
Pathogenic fungi (C)=		2.25		
Interaction (A × B × C)=		7.59		

Table 2: Effect of different concentrations of potassium salts on mycelial growth of *R. solani*, *F. solani* and *F. oxysporum in vitro.*

Effect of potassium salts on the redial growth of pathogenic fungi

Data in Table 2 show that all concentrations of the tested potassium salts resulted in a significantly suppressed redial growth of the tested pathogenic fungi (*R. solani, F. solani, F. oxysporum*) compared with the check treatment (control). The growth inhibition (%) of the tested fungi was increased with the increasing of concentrations of all tested substances. $KHCO_3$ showed superior higher inhibitory effect on redial growth of the tested pathogenic fungi especially at higher concentration (20 mM). In this regard, the recorded reduction in the growth of *R. solani, F. solani, F. oxysporum* was 50.12, 52.36 and 62.14%, respectively. On the other hand, the growth of *F. oxysporum* followed by *F. solani* showed the most affective then *R. solani.*

Effect of potassium salts of damping-off and root rot/wilt under greenhouse conditions

Fodder beet seeds soaking in tested potassium salts reduced significantly damping-off and root rot/wilt caused by *R. solani, F. solani* and *F. oxysporum* compared with control (Table 3). The reduction in damping-off and root rot/wilt increased with increasing of potassium salts concentration except K_2SO_4 while concentration 10 mM was more effective for reducing damping-off and root rot/wilt severity than 20 mM. K_2SiO_3 followed by potassium phosphate dibasic (K_2HPO_4) recorded highly protection against damping-off and root rot/wilt severity more than the other tested potassium salts. K_2SiO_3 and K_2HPO_4 at 20 mM recorded the highest reduction of damping–off caused by *R. solani* (6.67, 13.33%), *F. solani* (6.67, 6.67) and *F. oxysporum* (6.67 and 13.33%) compared with 66.67, 40, 26.67% in control, respectively. Similar results were obtained with root rot /wilt incidence caused by *R. solani* and *F. solani* and *F. oxysporum* while, fodder seeds treated with K_2SiO_3 and K_2HPO_4 at 20 mM reduced root rot/wilt severity from 22.14, 33.29 and 56.39% in control to 4.59, 5.67, 10.58 in case of K_2SiO_3 and 6.36, 8.52 and 6.54 in case of K_2HPO_4, respectively.

Effect of potassium salts of damping-off and root rot/wilt under field conditions

Data present in Table 4 show that all tested concentrations of potassium salts significantly reduced damping-off and root rot/wilt diseases under nutrition infection in field during growing seasons (2013-14 and 2014-15) compared with control. The reduction in damping-off and root rot/wilt increased with increasing of potassium salts concentration except K_2SO_4 while concentration 10 mM was more effective for reducing damping-off and root rot/wilts severity than 20

Potassium salts	Concen. (g/L)	*Rhizoctonia solani*		*Fusarium solani*		*F. oxysporum*	
		% Damping-off	% Root rot	% Damping-off	% Root rot	% Damping-off	% Wilt
K_2HPO_4	5	33.33	15.67	20.00	13.26	13.67	13.67
	10	20.00	10.34	13.33	10.24	13.67	8.36
	20	13.33	6.36	6.67	8.52	13.67	6.54
	Mean	22.22	10.79	13.33	10.67	13.67	9.52
$KHCO_3$	5	40.00	18.36	33.33	30.24	20.00	42.36
	10	26.67	12.36	20.00	22.42	20.00	33.14
	20	20.00	8.45	13.33	19.36	13.67	25.42
	Mean	28.89	13.06	22.22	24.01	17.89	33.64
K_2SO_4	5	40.00	19.36	33.33	25.67	20.00	44.67
	10	20.00	10.36	26.67	13.37	13.67	25.34
	20	26.67	15.35	33.33	19.49	20.00	35.52
	Mean	28.89	15.02	31.11	19.51	17.89	35.18
K_2SiO_3	5	26.67	10.36	13.33	10.24	13.67	20.23
	10	20.00	7.24	13.33	7.69	13.67	13.25
	20	6.67	4.59	6.67	5.67	6.67	10.58
	Mean	17.78	7.40	11.11	7.87	9.00	14.69
Control		66.67	22.14	40.00	33.29	26.67	56.39
LSD at 0.05 for:		Damping-off			Root rot/wilt		
Potassium salts (A)=		3.29			2.87		
Concentrations (B)=		4.85			4.80		
Pathogenic fungi (C)=		3.47			3.54		
Interaction (A×B×C)=		10.48			8.76		

Table 3: Effect of fodder beet seeds treatment with potassium salts on damping-off, root rot/wilt diseases under greenhouse conditions.

Potassium salts	Concen. (g/L)	Season 2013-2014 % Damping-off	Season 2013-2014 % Root rot/wilt	Season 2014-2015 % Damping-off	Season 2014-2015 % Root rot/wilt
K_2HPO_4	5	15.24	15.24	12.35	12.24
	10	12.35	10.32	10.33	8.25
	20	10.33	7.36	8.21	6.36
	Mean	12.64	10.97	10.30	8.95
$KHCO_3$	5	25.36	19.35	24.14	18.52
	10	20.55	15.34	18.25	12.36
	20	14.86	16.25	13.24	14.96
	Mean	20.26	16.98	18.54	15.28
K_2SO_4	5	20.14	18.69	17.67	17.25
	10	14.25	10.24	10.25	9.58
	20	16.36	12.36	13.24	12.36
	Mean	16.92	13.76	13.72	13.06
K_2SiO_3	5	10.25	12.36	8.56	10.25
	10	7.36	6.36	6.25	5.36
	20	5.28	5.45	5.28	5.56
	Mean	7.63	8.06	6.70	7.06
Control		35.26	25.36	30.25	26.35
LSD at 0.05 for:					
Potassium salts (A)=		2.65	2.44	2.47	2.31
Concentrations (B)=		3.01	2.09	2.85	2.59
Pathogenic fungi (C)=		2.69	2.14	2.54	2.51
Interaction (A×B×C)=		7.48	6.51	7.01	6.78

Table 4: Effect of fodder beet seeds treatment with potassium salts on damping-off, root rot/wilt diseases during 2013/14 and 2014/15 growing seasons under field conditions.

mM in both growing seasons. K_2SiO_3 was more effective for controlling damping-off and root rot/wilt severity than the other tested potassium salts especially in case of higher concentration (20 mM). While K_2SiO_3 at 20 mM reduced damping off from 35.26 and 30.25% to 5.28, 5.28% and reduced root rot/wilt from 25.36 and 26.35% in control to 5.45 and 5.36 in both growing seasons, respectivily. On the other hand, $KHCO_3$ and K_2SO_4 were the lowest efficient in reducing damping-off and root rot/wilt in both growing seasons.

The effect of potassium salts on fodder beet vigor and yield under field conditions

Fodder beet seed soaking in any of these potassium salts at different concentrations were significantly submitted to various growths and yield parameters *viz.* root length, root diameters, fresh and dry weights, except % dry maters was no significant, compared with control during growing seasons 2013-14 and 2014-15 (Tables 5 and 6). The enhancement in growth and yield parameters were increased by increasing potassium salts concentration except K_2SO_4 at concentration 10 mM was more effective for increasing plant growth and yield parameters than 20 mM in both growing seasons. The applied treatment K_2SiO_3 achieved the highest increase in all the mentioned parameters over the other entire three potassium salts during both growing seasons. The average root length of untreated seeds (control) was 29.26 and 31.02 cm/root in control; it recorded 51.42 and 51.68 cm/root in K_2SiO_3 treatment at 20 mM in both seasons, respectively. The diameter of root was 24.64 and 27.17 cm recorded in K_2SiO_3 at 20 mM compared to 12.33 and 13.45 cm in control. Also, fresh weight increased from 38.56 and 39.14 ton per fed. in control to 69.17 and 70.67 when applied K_2SiO_3 at 20 mM treatment. The dry weight increased from 4.49 and 5.32 ton per fed. In control to 9.71 and 10.06 in treated with K_2SiO_3 at 20 mM. The percentage of dry mater increased from 11.46 and 12.42 in control to 14.03 and 14.24% in seed treated with K_2SiO_3 at 20 mM in both growing seasons respectively. On the other hand, fodder seeds treated with $KHCO_3$ recorded the lowest increased of various growths and yield parameters in both growing seasons.

Potassium salts	Concen. (gm/L)	Root length	Root Diam.	Fresh weight (ton/fed.)	Dry weight (ton/fed.)	% Dry maters
K_2HPO_4	1	41.63	18.29	53.10	7.50	14.12
	2	43.90	21.23	55.84	7.96	14.25
	4	45.07	23.18	59.66	8.75	14.66
	Mean	43.53	20.90	56.20	8.07	14.35
$KHCO_3$	1	33.96	14.23	43.26	6.02	13.92
	2	35.42	15.49	44.96	6.3	14.01
	4	36.58	17.01	46.32	6.59	14.23
	Mean	35.32	15.58	44.85	6.30	14.05
K_2SO_4	0.50	35.69	16.36	44.96	6.28	13.97
	1.0	39.53	20.14	47.25	6.69	14.16
	2.0	37.59	19.24	46.18	6.59	14.27
	Mean	37.60	18.58	46.13	6.52	14.13
K_2SiO_3	0.50	46.24	20.86	62.59	8.84	14.13
	1.0	48.95	23.88	66.64	9.40	14.10
	2.0	51.42	24.64	69.17	9.71	14.03
	Mean	48.87	23.13	66.13	9.32	14.09
Control		29.26	12.33	38.56	4.49	11.46
LSD at 0.05 for:						
Potassium salts (A)=		4.55	2.64	3.97	1.25	ns
Concentrations (B)=		3.96	2.55	3.47	1.19	ns
Interaction (A×B)=		7.99	4.95	6.85	2.47	ns

Table 5: Effect of fodder beet seeds treated with potassium salts on growth and yield parameters during 2013/14 growing season under field conditions.

Potassium salts	Concen. (gm/L)	Root length	Root Diam.	Fresh weight (ton/fed.)	Dry weight (ton/fed.)	% Dry maters
K_2HPO_4	1	43.87	19.48	54.31	7.60	13.99
	2	46.14	21.90	57.42	8.14	14.17
	4	48.01	23.08	59.23	8.54	14.42
	Mean	46.01	21.48	56.99	8.09	14.19
$KHCO_3$	1	35.49	15.69	44.23	6.23	14.09
	2	37.25	16.58	46.32	6.49	14.01
	4	39.85	17.02	47.25	6.69	14.16
	Mean	37.53	16.43	45.93	6.47	14.09
K_2SO_4	0.50	36.96	17.42	45.02	6.33	14.06
	1.0	40.10	21.02	46.36	6.54	14.11
	2.0	39.63	19.96	48.57	6.86	14.12
	Mean	38.90	19.47	46.65	6.58	14.10
K_2SiO_3	0.50	47.47	21.60	62.52	8.78	14.04
	1.0	50.91	25.56	65.95	9.4	14.25
	2.0	51.86	27.17	70.67	10.06	14.24
	Mean	50.08	24.77	66.38	9.41	14.18
Control		31.02	13.45	39.14	5.32	12.42
LSD at 0.05 for:						
Potassium salts (A)=		4.36	2.69	4.96	1.35	ns
Concentrations (B)=		3.45	2.58	3.64	1.15	ns
Interaction (A × B)=		7.56	4.59	8.02	2.48	ns

Table 6: Effect of fodder beet seeds treated with potassium salts on growth and yield parameters during 2014/15 growing season under field conditions.

Biochemical changes associated with inducers PO, PPO, PAL and PAT activities

The effect of potassium salts *viz.* K_2HPO_4, KHCO3, K_2SO_4 and K_2SiO_3 as inducer chemicals on the activities of PO, PPO, PAL and TAL of the fodder beet plants grown in soil infested with *R. solani, F. solani, F. oxysporum* separately was studied. The data are presented in Figures 2-5 show that all tested potassium salts increased the activity of PO, PPO, PAL and TAL in the fodder beet compared with untreated plants (control). K_2SiO_3 at 20 mM showed the highest level of all oxidative enzymes activity followed by K_2HPO_4 at 20 mM and K_2SO_4 at 20 mM. Whereas, the least enzymes activity was recorded with $KHCO_3$ at 10 mM. On the other hand, fodder beet plants inoculated with *R. solani* were recorded the highly level of PO, PAL, TAL enzymes more than plants inoculated with *F. solani* or *F. oxysporum* either in treated and untreated fodder beet plants. While, PPO enzyme was more activity in case of fodder plants inoculated with *F. oxysporum* than the other tested fungi.

Effect of potassium salts on total phenols content

Data present in Figure 6 indicate that total phenolic compounds were higher in fodder beet plants treated with all the tested potassium salts than those of untreated infected control. The higher total phenolic contents were recorded plants treated with K_2SiO_3 at 20 mM followed by K_2HPO_4 at 20 mM. While, the lowest content of total phenolic compounds was recorded in plants treated with K_2SO_4 at 10 mM. on the other hand, fodder beet plants inoculated with *R. solani* gave highly content of phenolic compounds than plants inoculated with *F. solani* or *F. oxysporum* either in treated plants with potassium salts or untreated.

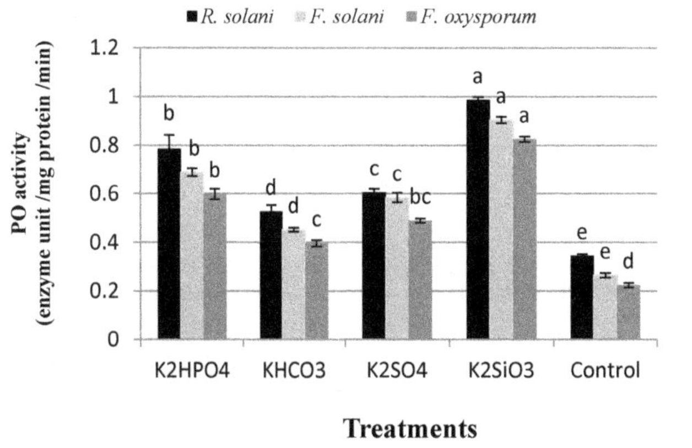

Figure 2: Effect of potassium salts on peroxidase activity (PO) in inoculated fodder beet plants. Mean ± SDs for nine plants per treatment are shown. Different letters indicate significant differences between treatments according to LSD test ($P \leq 0.05$).

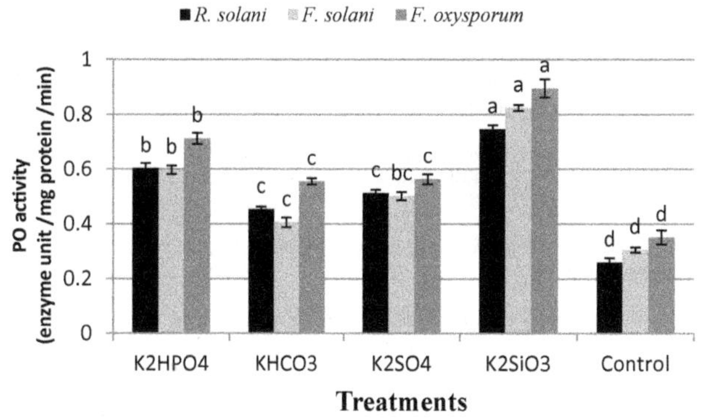

Figure 3: Effect of potassium salts on polyphenol oxidase activity (PPO) in inoculated fodder beet plants. Mean ± SDs for nine plants per treatment are shown. Different letters indicate significant differences between treatments according to LSD test ($P \leq 0.05$).

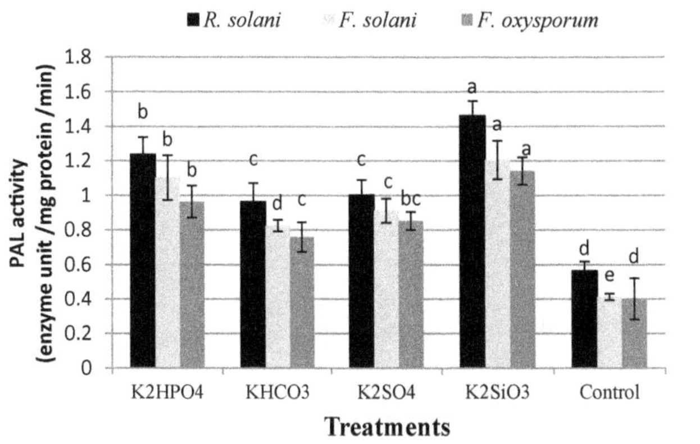

Figure 4: Effect of potassium salts on phenylalanine ammonia lyase activity (PAL) in inoculated fodder beet plants. Mean ± SDs for nine plants per treatment are shown. Different letters indicate significant differences between treatments according to LSD test ($P \leq 0.05$).

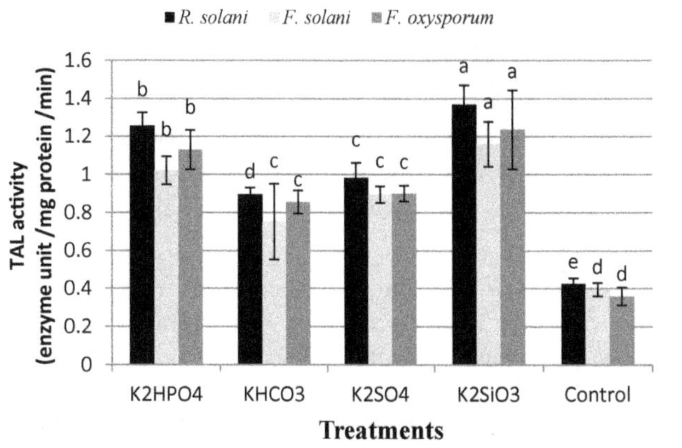

Figure 5: Effect of potassium salts on tyrosine ammonia lyase activity (TAL) in inoculated fodder beet plants. Mean ± SDs for nine plants per treatment are shown. Different letters indicate significant differences between treatments according to LSD test ($P \leq 0.05$).

Discussion

Plant diseases caused by soil-borne plant pathogens considered the major problems in agricultural production throughout the world, reducing yield and quality of crops. Plant pathogens have caused an almost 20% reduction in the principal food and cash crops worldwide [4,22].

Control of soil borne pathogens with chemicals is difficult because of their ecological behavior, their extremely broad host range and the high survival rate of resistant forms such as sclerotia and chlamediospores under different environmental conditions [23]. In recent years, public demands to reduce pesticide use, stimulated by greater awareness of environmental and health issues as well as the development of fungicide resistant strains of pathogens, have created

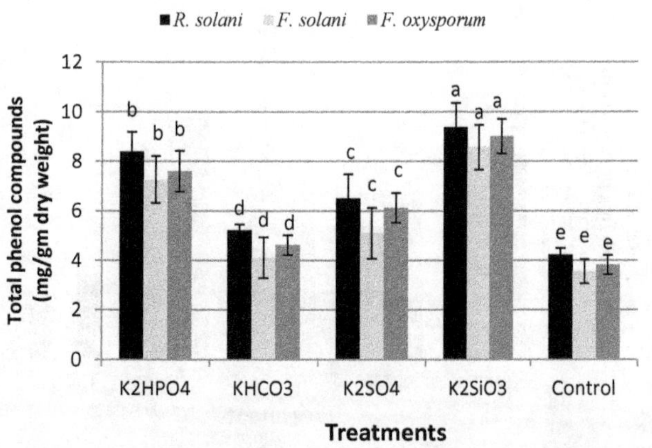

Figure 6: Effect of potassium salts on total phenol content (TPC) in inoculated fodder beet plants. Mean ± SDs for nine plants per treatment are shown. Different letters indicate significant differences between treatments according to LSD test ($P \leq 0.05$).

the need to find alternatives to pesticides. Natural substances such as some potassium salts may be used to achieve this aim. The main advantages of using potassium salts compared with fungicides include their relatively low mammalian toxicity, a broad spectrum of modes of action and relatively low cost [24]. They also have wide spectrum antimicrobial properties. They have been shown to be effective growth inhibitors of some soil borne fungal pathogens [25,26].

In the present study, it was planning to investigate the possibility of minimizing the infection with damping-off, root rot and wilt diseases caused by *R. solani*, *F. solani*, *F. oxysporum* of fodder beet using some potassium salts *viz.* K_2HPO_4, $KHCO_3$, K_2SO_4 and K_2SiO_3 at 5, 10, 20 mM as resistance inducer. The obtained data revealed that all potassium salts caused significantly suppressed redial growth of the tested pathogenic fungi *in vitro*. The growth suppression was increased by increasing of potassium salts concentrations and $KHCO_3$ showed superior higher inhibitory effect on redial growth of the tested pathogenic fungi especially at higher concentration (20 mM). Also, all tested potassium salts significantly reduced damping-off, root rot and wilt diseases incidence either under artificial infection in greenhouse of natural infection in field, compared with the control treatment. In general, K_2SiO_3 at the high concentration (20 mM) was more efficient in reducing disease incidence followed by K_2HPO_4 at 20 mM. Moreover, K_2SO_3 and $KHCO_3$ was the lowest affected one by the investigated diseases.

Also, under field conditions, all potassium salts improved growth and yield components of fodder beet in both growing seasons. The applied treatment K_2SiO_3 achieved the highest increase in all the mentioned parameters (root length, root diameters, fresh and dry weights and % dry maters) over the other entire three potassium salts during both growing seasons.

On the other hand, potassium salts due to many biochemical changes in fodder beet plants either in inoculated or un-inoculated plants with tested pathogens. While, all tested potassium salts increased activity of peroxidase, polyphenol oxidase, phenyalalanine ammonia lyase and tyrosine ammonia lyase as well as total phenolic compounds. K_2SiO_3 was recorded the highest activity of all enzymes and total phenolic compounds followed by K_2HPO_4, while $KHCO_3$ and K_2SO_4 recorded the lowest ones in this respect.

Many investigations reported the use of potassium salts as a chemical agent for induction of plant resistance [27,28]. Furthermore, there has been considerable interest in the use of potassium bicarbonate and potassium phosphate, potassium sulfate and potassium silicate for controlling various fungal diseases in plants [29-32].

Potassium is a mobile element with multiple functions in the plant. It acts as a counter-ion for anion transport, regulates stomatal aperture and the water potential of plant cells, affects cell wall plasticity, as well as other roles [33]. It promotes wound healing and decreases frost injury [34]. Potassium deficiency has been found to be linked to diseases in a number of temperate crops [34] and a high K supply can improve resistance of plants to fungal and bacterial pathogens [35,36]. The mechanism of resistance in some disease-resistant genotypes might be related to a greater efficiency in K uptake [37]. The potassium bicarbonate causes the collapse of hyphal walls and shrinkage of conidia, [38].

In general, potassium application improves plant health and vigour, making infection less likely or enabling a quick recover [39]. Potassium probably exerts its greatest effects on disease through specific metabolic functions that alter compatibility relationships of the host-parasite environment and increases the production of disease inhibitory compounds, such as phenols, phytoalexins and auxins around infection sites of resistant plants [40,41].

In conclusion, the present study provides further evidence that may facilitate applying simple non-toxic chemicals as potassium salts for controlling damping-off, root rot and wilt diseases in fodder beet. Their low cost, low toxicity to the man and environmental pollution make them ideal seed soaking for disease control under field conditions and increased root yield and dry mater.

References

1. DAF (Department of Agriculture and Food) (1998) Root, fodder crop, pulse and oilseed varieties. Irish recommended list. Government Stationary Office, Dublin, p. 17.

2. Turk M (2010) Effects of fertilization on root yield and quality of fodder beet (*Beta vulgaris* var. crassa Mansf.). Bulgarian Journal of Agricultural Science 16: 212-219.

3. Abdel -Naby ZM, Shafie WWM, Sallam AM, El-Nahrawy SM, Abdel-Ghawad MF (2014) Evaluation of seven fodder beet genotypes under different Egyptian ecological conditions using regression, cluster models and variance measures of stability. Int J Curr Microbiol App Sci 3: 1086-1102.

4. Oerke EC, Dehne HW, Schonbeck F, Weber A (1994) Crop Production and Crop Protection – Estimated Losses in Major Food and Cash Crops, Elsevier Science, Amsterdam, pp 808.

5. El-Mohamedy RSR, Abdel-Kader MM, Abd-El-Kareem F, El-Mougy NS (2013) Essential oils, inorganic acids and potassium salts as control measures against the growth of tomato root rot pathogens in vitro. Journal of Agricultural Technology 9: 1507-1520.

6. De Waard A, Georgopoulos SG, Hollomon DW, Ishii H, Leroux P, et al. (1993) Chemical control of plant diseases: problems and prospects. Ann Rev Phytopathol 31: 403-423.

7. Booth C (1985) The genus Fusarium. Surrey: Commonwealth Mycological Institute.

8. Barnett HL, Hunter BB (1986) Illustrated Genera of Imperfect Fungi. 4 th Ed., Macmillan Publishing Co., New York.

9. Abo-Elyousr KAM, Mohammed H (2009) Biological Control of Fusarium Wilt in Tomato by Plant Growth-Promoting Yeasts and Rhizobacteria. The Plant Pathology J 25: 199-204.

10. Abdou El-S, Abd-Alla HM, Galal AA (2001) Survey of sesame root rot/wilt disease in Minia and their possible control by ascorbic and salicylic acids. Assuit J Agric Sci 32: 135-152.

11. Liu L, Kloepper JW, Tuzun S (1995) Introduction of systemic resistance in cucumber against fusarium wilt by plant growth-promoting rhizobacteria. Phytopathol 85: 695-698.

12. Somda I, Leth V, Sereme P (2007) Antifungal effect of *Cymbopogon citratus*, *Eucalyptus camaldulensis* and *Azadirachta indica* oil extracts on sorghum seed-borne fungi. Asian J Plant Sci 6:1182-1189.

13. Biles CL, Martyn RD (1993) Peroxidase, polyphenoloxidase, and shikimate dehydrogenase isozymes in relation to tissue type, maturity and pathogen induction of watermelon seedlings. Plant Physiol Biochem 31: 499-506.

14. Hammerschmidt R, Nuckles EM, Ku J (1982) Association of enhanced peroxidase activity with induced systemic resistance of cucumber to *Colletotrichum lagenarium*. Physiol Plant Pathol 20: 73-82.

15. Urbanek H, Kuzniak-Gebarowska E, Herka H (1991) Elicitation of defense responses in bean leaves by *Botrytis cinerea* polygalacturonase. Acta Physiol Plant13: 43-50.

16. Gauillard F, Richard-Forget F, Nicolas J (1993) New spectrophotometric assay for polyphenol oxidase activity. Anal Biochem 215: 59-65.

17. Cavalcanti FR, Resende ML, Carvalho CP, Silveira JA, Oliveira JT (2007) An aqueous suspension of *Crinipellis perniciosa* mycelium activates tomato defense responses against *Xanthomonas vesicatoria.* Crop Prot 6: 729-38.

18. Bradford MM (1976) A rapid and sensitive method for the quantitation of microgram quantities of protein utilizing the principle of protein-dye binding. Anal Biochem 72: 248-254.

19. Saikia R, Yadav M, Varghese S, Singh BP, Gogoi DK, et al. (2006) Role of riboflavin in induced resistance against Fusarium wilt and charcoal rot diseases of chickpea. Plant Pathol J 22: 339-47.

20. MSTAT-C (1991) A Software Program for the Design, Management and Analysis of Agronomic Research Experiments. Michigan State University, pp. 400.

21. Gomez KA, Gomez AA (1984) Statistical procedures for agricultural research. New York: Wiley Interscience Publication p. 678.

22. Villajuan-Abgona R, Kageyama K, Hyakumachi M (1996) Biological control of Rhizoctonia damping-off of cucumber by non-pathogenic binucleate Rhizoctonia. Euro J Plant Pathol 102: 227- 235.

23. Yangui T, Rhouma A, Triki MA, Gargouri K, Bouzid J (2008) Control of damping-off caused by *Rhizoctonia solani* and *Fusarium solani* using olive mill waste water and some of its indigenous bacterial strains. Crop Protection 27: 189-197.

24. Olivier C, Halseth DE, Mizubuti SG, Loria R (1998) Post-harvest application of organic and inorganic salts for suppression of silver scurf on potato tubers. Plant Dis 82: 213-217.

25. Arslan U, Kadir I, Vardar C, Karabulut OA (2009) Evaluation of antifungal activity of food additives against soilborne phytopathogenic fungi. World J Microbiol Biotechnol 25: 537-543.

26. Ordonez C, Alarcón A, Ferrera R, Hernández LV (2009) In vitro antifungal effects of potassium bicarbonate on *Trichoderma* sp. and *Sclerotinia sclerotiorum*. Mycoscience 50: 380-387.

27. Stromberge A, Brishammar S (1991) Induction of systemic resistance in potato (*Solanum tuberosum* L.) plants to late blight by local treatment with *Phytophthora infestans*, *Phytophthora cryptogea* or dipotassium phosphate. Potato Res 34: 219-225.

28. Yurina TP, Karavaev VA, Solntsev MK (1993) Characteristics of metabolism in two cucumber cultivars with different resistance to powdery mildew. Russian Plant Physiol 40: 197-202.

29. Karabulut OA, Smilanick JL, Gabler FM, Mansour M, Droby S (2003) Near-harvest applications of *Metschnikowia fructicola*, ethanol, and sodium bicarbonate to control postharvest diseases of grape in central California. Plant Dis 87: 1384-1389.

30. Smilanick JL, Mansour MF, Sorenson D (2006) Pre- and postharvest treatments to control green mold of citrus fruit during ethylene degreening. Plant Dis 90: 89-96.

31. Ragab MMM, Ashour AMA, Abdel-Kader MM, El-Mohamady R, Abdel-Aziz A (2012) In vitro evaluation of some fungicides alternatives against *Fusarium oxysporum* the causal of wilt disease of pepper (*Capsicum annum* L.). International J of Agric and Forestry 2: 70-77.

32. Safari S, Soleimani MJ, Zafari D (2012) Effects of silicon pretreatment on the activities of defense-related enzymes in cucumber inoculated with *Fusarium oxysporum*. Adv Environ Biol 6: 4001-4007.

33. Rice RW (2007) "The Physiological Role of Minerals in Plants." In Mineral Nutrition and Plant Disease, edited by L. E. Datnoff, Wade H. Elmer and D. M. Huber, 9-30. St. Paul, Minn: American Phytopathological Society, 2007.

34. Palti J (1981) Cultural Practices and Infectious Crop Diseases. Vol. 9. New York: Springer-Verlag.

35. Marschner H (1995) Mineral nutrition of higher plants, 2ndedn. Academic, London.

36. Perrenoud S (1990) Potassium and Plant Health. IPI Research Topics No. 3. 2nd Ed. International Potash Institute, Bern, Switzerland.

37. Prabhu M, Veeraragavathatham D, Srinivasa K (2003) Effect of nitrogen and phosphorous on growth and yield of brinjal. South-Indian-Hort 51: 152-156

38. Ziv O, Zitter TA (1992) Effects of bicarbonates and film-forming polymers on cucurbit foliar diseases. Plant Dis 76: 513-517.

39. Perrenoud S (1993) Fertilizing for High Yield Potato. IPI Bulletin 8. 2nded. International Potash Institute, Basel, Switzerland.

40. Legrand M, Kauffmann S, Geoffroy P, Fritig B (1987) Biological function of pathogenesis-related proteins: Four tobacco pathogenesis-related proteins are chitinases. Proc Natl Acad Sci USA 84: 6750-6754.

41. Abd-El-Kareem F, El-Mougy Nehal S, El-Gamal NG, Fotouh YO (2004) Induction of resistance in squash plants against powdery mildew and Alternaria leaf spot diseases using chemical inducers as protective or therapeutic treatments. Egypt J Phytopathol 32: 65-76.

Screening of Sugarcane Varieties/Lines against Whip Smut Disease in Relation to Epidemiological Factors

Sarmad Mansoor*, Asmlam Khan M and Nasir Ahmed Khan

Department of Plant Pathology, University of Agriculture, Faisalabad, Pakistan

Abstract

Whip smut caused by (*Ustilago scitaminea*) is an important fungal disease, which is widely distributed all over the world causing huge losses in sugarcane crop. Sugarcane crop basically required humid and hot climate for its development which is also favorable for the different diseases in sugarcane. Epidemiological factors play an important role for the development and the management of different diseases. They are also used as disease prediction models. Epidemiological factors are very important for the development and spread of pathogen causing smut of sugarcane. Out of fifteen promising varieties/ lines, eight were found resistant (S2006-US-469, S2006-US-272, S2005-US-54, S2008-AUS-130, S2006-US-658, S2008-AUS-190, S2008-AUS-107, S2009-SA-169), six were moderately susceptible(S2008-M-34, S2008-AUS-133, S2003-US-127, S2003-US-704, S2008-Fd-19, S2008-AUS-87) and one (S2003-US-618 had susceptible reaction against the disease. There was positive correlation of relative humidity with disease incidence and negative correlation of maximum and minimum temperature with disease incidence.

Keywords: Sugarcane; Smut disease; Epidemiological factors

Introduction

Sugarcane (*Saccharum officinarum* L.) (Punjabi: Ganna, Urdu: Naishkar, Kamad) belongs to family *Poaceae* and crop is grown under 30° south to 30° north latitude with climatic conditions ranging from sub-tropical to tropical regions [1]. In Pakistan, Sugarcane is cultivated on a range of one million hectare. Sugarcane growing zones of Pakistan fall between 24° N latitude in Sindh to 34° N latitude in KPK. Pakistan is at fifth position with respect to Sugarcane production in the world. The most important thing is that the sugar industry shares of Pakistan economy about 1.9% of GDP [2]. There are numerous restraints, including diseases such as Whip Smut, Red Rot, Pokkah Boeng, Red Stripe, Rust and Sugarcane Mosaic and Brown stripe [3,4]. In 1877 the smut of sugarcane was first time reported in (Natal) South Africa. Whip Smut is extremely critical disease of Sugarcane in Pakistan wherever the crop is grown. Whip smut is caused by *Ustilago scitaminea*, which belongs to the phylum Basidiomycota [5] occurs in a few physiological races [6,7]. The temperature ranges between (25-30°C) supports the disease' development. The smut of sugarcane is prevalent in all the world countries where the sugarcane crop is cultivated. Use of susceptible varieties show more losses because of intensive cultivation, secondary infection, and poor management practices [8]. 52-73% yield losses occur in ratoons crops [9]. Sandhu, [10] specified yield losses of 70.7% to 75.3%. Total crop failure is possible if susceptible varieties are used and conditions are favorable for disease development [11]. It can cause major losses as well as juice quality losses. 3-7% sucrose content of infected variety is reduced [12]. Disease incidence increase was found to be linked with increasing age of the crop and varietal susceptibility. After 120 days of planting the appearance of the apical whips was found. When the second flush of whip was produced, it produces very large quantity of teliospores and these spores effect the lateral and terminal buds of rapidly growing crop. The emergence of the third level of whips and the infection caused by this level is supposed to be very serious in the epidemiology of whip smut disease [13].

Smut inoculation techniques in sugarcane plantlets and examined the chance of screening for smut resistance at the plantlet period. Injury paste technique was found the extreme event of whip smut production, followed by paste; on the other hand, soaking method had the minimum occurrence of smut [14]. The susceptible varieties show significant losses due to poor management practices, secondary infection and intensive cultivation. The most suitable and economical process to control the disease is the use of resistant varieties. The resistant germplasm of sugarcane plays a leading role for assessment of resistant varieties through breeding program [15]. Disease development is dependent on the environmental conditions and the resistance of the sugarcane varieties grown. The most recognizable diagnostic feature of a smut infected plant is the emergence of a "smut whip" [16]. According to Sreeramulu [17], the day time dispersal of spores is maximum. The maximum dispersal of spores takes place at 24 to 27°C and 50 to 60% R.H. Crop age and cycle at the time of infection appear to be important [18]. Resistance of a variety retain only for a few years. A variety resistant previously pertaining race may become susceptible to the new physiological race with change in climatic conditions. Pre-release evaluation of varieties / lines, is therefore, important in relation to epidemiological factors. The objective of my present research to screen the sugarcane clones for the smut tolerance and to study the influence of epidemiological factors on the occurrence of smut disease in sugarcane. This research work was based on the hypothesis, through evaluation of sugarcane varieties in relation to epidemiological factors may be helpful for management of whip smut disease. Use of resistant lines/varieties along with proper management practices and study of epidemiological factors will be helpful to reduce the losses caused by whip smut.

Materials and Methods

Fifteen (15) varieties/ lines were grown in field area of Ayub Agriculture Research Institute (AARI), Sugarcane Research Institute

***Corresponding author:** Sarmad Mansoor, Department of Plant Pathology, University of Agriculture, Faisalabad, Pakistan
E-mail: sarmadmansoor09@gmail.com

(SRI), Faisalabad during 2015-2016. The varieties were planted under RCBD design with three replication. The plot size was kept as 2.4 m width and 3 m length [19]. The varieties sown were cut into small pieces (setts). The length of one sett was about 45 cm with 3 buds present on it. Forty eight (48) setts of each variety were taken for three row plantation (with 16 setts in one row). These setts were dipped in spore suspension for 30 min prior to plantation. The fungal spores entered into the cane setts which were used to evaluate the disease incidence. Plantation of sugarcane inoculated setts was done in February 2015 in three meter long plot under RCBD design with three replications / repeats at sugarcane experimental area, Sugarcane Research Institute (SRI), Faisalabad in clay loam soil. Thus each treatment was comprised of 48 smut-inoculated setts per variety [20]. Data for the number of smutted tillers was collected with a regular interval of 30 days. The data was collected by counting the number of smutted tillers in each variety. The layout plan for the sugarcane varieties was made in such a way that there were 15 varieties, 3 replications and 3 factors (maximum temperature, minimum temperature and relative humidity) were studied with respect to disease incidence in each variety each month. Data was collected monthly from June to December 2015. To collect the data, number of smutted tillers and total number of tillers were counted.

Meteorological data

Meteorological data for temperature and humidity were collected from the meteorological department, AARI, Faisalabad. Meteorological data was in the form of computerized spread sheet on which day to day information for maximum temperature, minimum temperature and relative humidity was listed. Meteorological data was calculated to conclude mean values of maximum temperature, minimum temperature and relative humidity for the whole month.

Statistical analysis

The analysis of the information was done based on the percentage of infected strains of the last observation and these were processed using the statistical parametric analysis for randomized blocks.

Correlation and regression analyses with epidemiological factors to determine the relationship between epidemiological factors and disease incidence. The prediction equation used was

$$Y = a + b1x1 + b2x2 + b3x3$$

Where, Y = Predicted disease incidence; a = Intercept; b1 - b3 = Regression coefficients; X1= Average maximum temperature (°C), X2= Average minimum temperature (°C), X3-= Average relative humidity (%), R2 = coefficient of determination.

Recording smut incidence

The trial was closely monitored for appearance of first smut whips and recorded monthly intervals until the trial was completed (Table 1).

Smut reaction

Due to its vegetative mode of propagation sugarcane is prone to infect by systemic pathogens. Among, smut disease caused by *Ustilago scieminae* is a dreadful disease of sugarcane and is endemic in most of the tropical regions. The most eco-friendly means to contain the pathogen through the use of resistant varieties/lines. In the present investigation, fifteen sugarcane promising clones were evaluated for their resistance against whip smut pathogen under field conditions. It was concluded that, out of fifteen promising lines/varieties eight were found resistant, six moderately susceptible and one had susceptible reaction against the disease (Table 2). The resistance/susceptibility of the variety were determined by bud morphological characters. In the most resistant varieties the germplasm adapted sub apical position in the bud whereas the susceptible varieties the position was apical. The position was considered to be associated with the tendency of the bud to sprout which makes it more vulnerable to the entry of promycelium and hence more prone to infection. Hence, bud scales acted as morphological barrier and restricted smut pathogens. Source of resistant against whip smut available in sugarcane clones and it can be further manipulated through breeding program for evolution of new high yielding sugarcane varieties [15,21].

Results of screening has shown that out of 15 varieties, Eight (8) varieties were found resistant (S2006-US-469, S2006-US-272, S2005-US-54, S2008-AUS-130, S2006-US-658, S2008-AUS-190, S2008-AUS-107, S2009-SA-169), six (6) moderately susceptible (S2008-M-34, S2008-AUS-133, S2003-US-127, S2003-US-704, S2008-Fd-19, S2008-AUS-87), and one (1) susceptible (S2003-US-618). The varieties which were somewhat resistant were suppressed by the invasion of pathogen as invading plants showed poor growth with large number of thin canes.

Epidemiological factors on smut incidence

Temperature (max. temperature and min. temperature), relative humidity are important factors in smut epidemiology.

Characterization of environmental conditions conductive for whip smut of sugarcane disease development on seven varieties

Seven varieties of sugarcane (S2003-US-618, S2008-M-34, S2008-AUS-133, S2003-US-127, S2003-US-704, S2008-Fd-19, S2008-AUS-87) showed significant correlation with temperature (maximum and minimum) and relative humidity. These varieties were employed to characterize the critical ranges of environmental conditions (maximum and minimum temperature and relative humidity) conductive for the

Response		Disease incidence (%)
Resistant	R	0-5
Moderately Resistant	MR	5.1-15
Moderately Susceptible	MS	15.1-30
Susceptible	S	Above 30

Table 1: Smut description, rating and infection were done as explained by Rao et al.

Varieties	D.I (%)	Response
S-2003-US-618	46.43	S
S-2008-M-34	25.02	MS
S-2006-US-469	0	R
S-2006-US-272	0	R
S-2005-US-54	0	R
S-2008-AUS-133	18.45	MS
S-2008-AUS-130	0	R
S2003-US-127	21.80	MS
S-2006-US-658	0	R
S-2008-AUS-190	0	R
S-2003-US-704	16.58	MS
S-2008-Fd-19	24.42	MS
S-2008-AUS-107	0	R
S-2008-AUS-87	19.34	MS
S-2009-SA-169	0	R

Table 2: Evaluation of sugarcane clones to smut (%) incidence.

whip smut of sugarcane disease development. The results demonstrate that at maximum temperature of 38.75ºC the variety S2008-AUS-133 showed the minimum disease incidence of 3.28% (Figure 1). While at the minimum temperature of 25.5ºC the variety S2003-US-618 showed the maximum disease incidence of 46.52% (Figure 2). In case of relative humidity at 70.5% relative humidity the maximum disease incidence of 46.52% was recorded in the variety S2003-US-618 while at the 48.5% relative humidity the minimum disease incidence of 3.28% was recorded in the variety S2008-AUS-133 (Figure 3).

These results clearly demonstrated that the maximum temperature and minimum temperature were negatively correlated with the whip smut disease incidence, while relative humidity was positively correlated with the whip smut disease incidence as shown in Table 3.

Discussion

Whip smut of sugarcane (*U. scitaminea*) is very destructive disease in all sugarcane grown areas of the world. It usually causes losses from germination to maturity of the crop. There was a need to highlight resistant lines among different clones of sugarcane. To fulfill this need, research on screening of different varieties was done on the basis of disease rating scale [22]. Conditions are critically important in the development and spread of the pathogen causing smut of sugarcane. Some of these can be utilized to form the basis of disease prediction model. They may vary in their combinations in different agro climatic zones and influence not only the pathogen but also the host. The present findings are in accordance with Sreeramulu et al. [17] reporting that there is definite diurnal and seasonal rhythms in the spore incidence, the day time dispersal of spores is maximum. The maximum dispersal of spores takes place at 24 to 27°C and 60 to 70% R.H. The difference in diseases severity may be attributed to the environmental conditions. Factors such as maximum temperature, minimum temperature and relative humidity were studied with special reference to the varietal reactions of different varieties. It was observed that all the factors maximum temperature, minimum temperature and relative humidity had statistically significant correlation with varieties. Disease severity was maximum at temperature range 25-27°C. With the decrease in temperature 38-25ºC from June 2015 to December

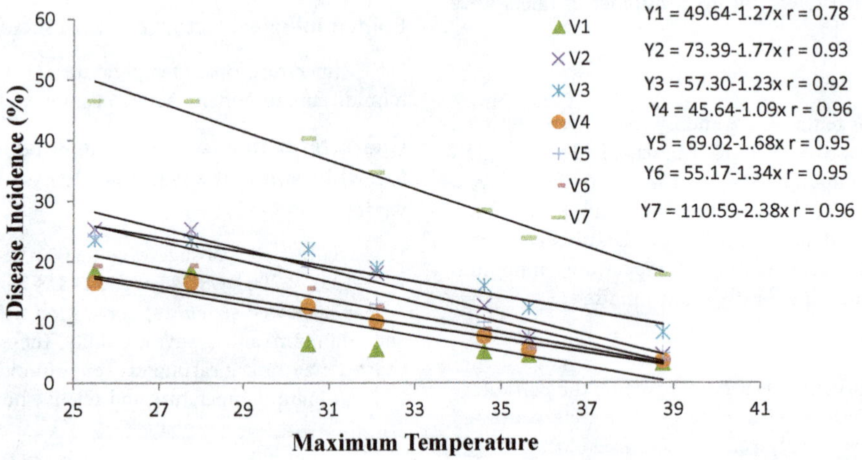

Figure 1: Correlation of maximum temperature with whip smut disease incidence on different varieties.

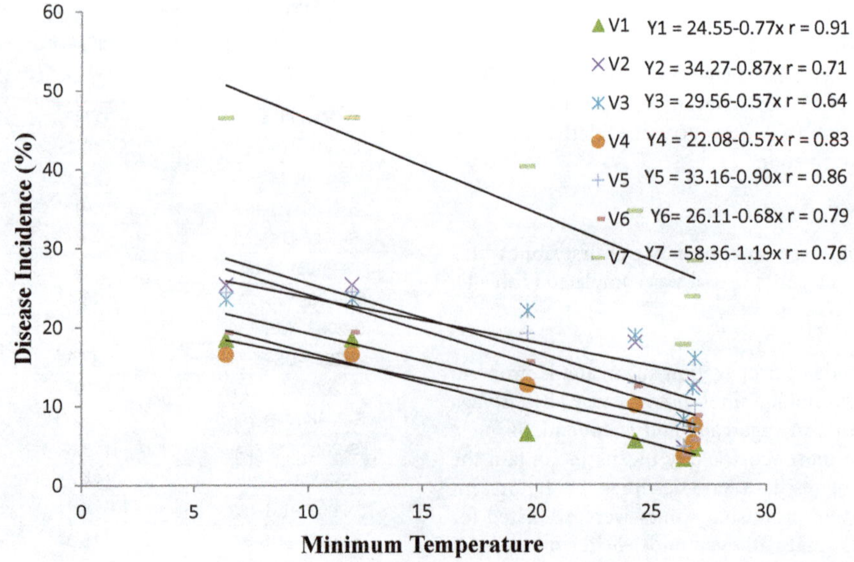

Figure 2: Correlation of minimum temperature with whip smut disease incidence on different varieties.

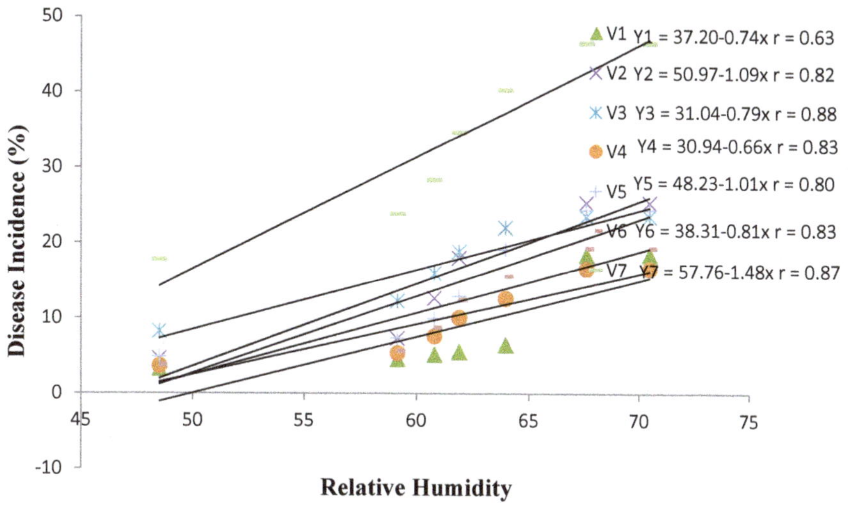

Figure 3: Correlation of relative humidity with whip smut disease incidence on different varieties.

Sr. No.	Varieties	Max Temp. (°C)	Min Temp. (°C)	RH (%)
1	S2008-AUS-133	-0.89** 0.01	-0.96** 0.00	0.79* 0.03
2	S2008-M-34	-0.97** 0.00	-0.85* 0.02	0.91** 0.00
3	S2003-US-127	-0.99** 0.00	-0.80* 0.03	0.94** 0.00
4	S2003-US-704	-0.98** 0.00	-0.92** 0.00	0.92** 0.00
5	S2008-Fd-19	-0.98** 0.00	-0.93** 0.00	0.90** 0.01
6	S2008-AUS-87	-0.98** 0.00	-0.89** 0.01	0.91** 0.00
7	S2003-US-618	-0.98** 0.00	-0.88** 0.01	0.94** 0.00

Upper value indicate Pearson's correlation coefficient, while the lower value indicate level of significance at 1% (0.00-0.01) and at 5% (0.02-0.03) probability.
** = Highly Significant; * = Significant; No Sign = Non Significant

Table 3: Correlation of environmental factors with whip smut of sugarcane disease on different varieties.

2015, disease incidence or severity was increased. On the other hand, it had been observed that disease severity was maximum at relative humidity range 65-70%. With increase in relative humidity 65-70%, overall disease incidence was increased whereas Singh and Budhraja, [23] reported that disease incidence was maximum at optimum temperature of 28°C as this temperature favours the maximum growth of smut pathogen (*Ustilago scitaminea*). The smut spores are killed instantaneously at 62°C but can survive more than three day in ice [24,25]. Whereas it has also been reported that high temperature 25-30°C is the most favourable temperature for the development of whip smut disease. So our results regarding disease incidence match with the previous investigations. Disease incidence was maximum in variety S2003-US-618 at temperature 27.75°C and relative humidity 70%. So this variety was the most susceptible among all the varieties in every month from June to December. Besides this, the variety S2003-US-704 had shown minimum disease incidence among seven varieties in which disease was appeared.

Conclusion

It can be concluded that the intensity of sugarcane smut incidence highly influenced by the epidemiological factors. The prevalence of optimum temperature during the crop stage of germination to tillering, increased the setts and soil borne teliospores germination subsequently it may give rise to infection hyphae which are capable of infecting sugarcane bud. In addition, the temperature has an enhanced effect on the release and dispersal of smut spores in the air. Fifteen sugarcane varieties / lines were screened out to find the resistant lines. When these fifteen varieties /lines were compared on the basis of recommended scale under natural conditions, Eight (8) varieties or lines (S2006-US-469, S2006-US-272, S2005-US-54, S2008-AUS-130, S2006-US-658, S2008-AUS-190, S2008-AUS-107, S2009-SA-169) were graded as resistant, (S2008-M-34, S2008-AUS-133, S2003-US-127, S2003-US-704, S2008-Fd-19, S2008-AUS-87) were found as moderately susceptible and (S2003-US-618) was found susceptible. Maximum Disease Incidence was observed at (25-27°C) and at R.H (65-70%) and minimum disease incidence was observed at (38.75°C) and at R.H (48.5%). Environmental conditions especially maximum temperature, minimum temperature and relative humidity, which showed that maximum and minimum temperature and relative humidity had great influence on the incidence of whip smut disease of sugarcane. There is a negative correlation between maximum and minimum temperature and disease incidence whereas the correlation between relative humidity and disease incidence was recorded as positive.

Acknowledgment

This research is collaboration between the University of Agriculture

Faisalabad, Pakistan and Ayyub Agriculture Research Institute (AARI) Faisalabad, Pakistan. The author gratefully acknowledged the support of his supervisor Prof. Dr Muhammad Aslam Khan and Co- supervisor Dr. Muhammad Abdul Shakoor for their innovative ideas to complete this research.

References

1. Rao TCR, Bhagyalakshmi KV, Rao JT (1979) Indian atlas of sugarcane. Sugarcane Breeding Institute, Coimbatore, India.
2. Shaukat MI (2009) A comprehensive studies on sugarcane. University of Agriculture, Faisalabad, Pakistan.
3. Chattha AA, Afzal M, Iqbal MA, Ahmad F, Chattha MU (2004) CPF-243, an early maturing, high yielding and high sugar variety 6: 25-27.
4. Luthra JC, Suttar A, Sandhu SS (1940) Experiments on the control of smut of sugarcane. Proceedings in Indian Academy of Sciences Sec B 12: 118-128.
5. Rott P, Bailey A, Comstock JC, Croft BJ (2000) Whip smut of Sugarcane. A guide to sugarcane diseases. CIRAD and ISSCT Publishing Co. Amsterdam, Netherland pp: 339-341.
6. Grishan MP (2001) An international project on genetic variability within sugarcane smut. Proc Int Soc Sugar Technol 24: 459-461.
7. Agnihotri VP (1983) Smut of Sugarcane. Diseases of Sugarcane and Sugarbeet. Oxford and IBH Publishing Co. 66 Janpath, New Dehli, India pp: 65-86.
8. Whittle AM (1982) Yield loss in sugar-cane due to Culmicolous smut infection. Trop Agric Trinidad 59: 239-242.
9. Mohan RNV, Praksam P (1956) Studies on sugarcane smut. Proc Intern Soc Sugarcane Technol 17: 1048-1057.
10. Sandhu SA, Bhatti DS, Rattan BK (1969) Extent of losses caused by smut (Ustilago scitaminea Syd.). Jour Res (PAU) 6: 341-344.
11. Lee-Lovick GL (1978) Smut of Sugarcane Ustilago scitaminea. Plant Pathol 57: 181-188.
12. Sandhu SS, Mehan VK, Ram RS, Shani SS, Sharma JR (1975) Screening of promising sugarcane varieties for resistance to smut by Ustilago scitaminea Syd. in the Punjab. Indian Sugar 25: 423-426.
13. Bergamin A, Amorim L, Cardoso CON, Da Silva WM, Sanguino A, et al. (1989) Epidemiology of sugarcane smut in Brazil. Sugarcane pp: 211-216.
14. Olweny CO, Ngugi K, Nzioki H, Githiri SM (2008) Evaluation of smut inoculation techniques in sugarcane seedlings. Sugar Tech 10: 341-345.
15. Begum F, Talukdar MI, Iqbal M (2007) Performance of various promising lines for resistance to sugarcane smut (Ustilago scitaminea Sydow). Pak Sugar J 22: 16-18.
16. Comstock JC (2000) "Smut." In a guide to sugarcane diseases (Ed. Philippe Rott, Roger, A. Bailey, C. J. Comstock, J. B. Croft, and A. Salem Saumtally) Montpellier, France pp: 181-185.
17. Sreeramulu T (1973) Aero-mycological observations and their implications in the epidemiology of some diseases of Sugarcane. Indian Natn Sci Acad Bull 46: 506-510.
18. Ferreira SA, Comstock JC (1989) Diagnosis of whip smut of sugarcane. Indian Sugar J 11: 35-39.
19. Bock KR (1964) Studies on sugarcane smut (Ustilago scitaminea) in Kenya. Trans Brit mycol SOC 3: 403-417.
20. Islam MA, Miah MNA, Rahman MA, Kader MA, Karim KMR (2009) Performance of Sugarcane with Different Planting Methods and Intercrops in Old Himalayan Piedmont Plain Soils. Int J Sustain Crop Prod 4: 55-57.
21. Sabalpara AN, Vishnav MU (2002) Screening of sugarcane clone/ varieties for resistance to smut caused by Ustilago scitaminea. Ind Sugar J 12: 507-509.
22. Rao GP, Triphathi DNP, Upadhaya UC, Singh RDR, Singh RR (1996) New promising red rot and smut resistant sugarcane varieties for eastern Utterpardesh. Ind Sugar J 14: 261-263.
23. Singh K, Budhrajam TR (1964) The role of bud scales as barriers against smut infection. Proceedings Bien Conf Sugarcane Res Dev 5: 687-690.
24. Appalanarasayya P (1964) Some physiological studies on sugarcane smut (Ustilago scitaminea). Indian Phytopath 17: 284-287.
25. Saxena SK, Khan AM (1963) Effect of temperature on spore germination. Jour Indian Bot Soc 42: 195-203.

In vitro Studies on Branch Canker Pathogen (*Macrophoma* sp.) Infecting Tea

Mareeswaran J*, Nepolean P, Jayanthi R, Premkumar Samuel Asir R and Radhakrishnan B
Plant Pathology Division, UPASI Tea Research Institute, Valparai-642127 Coimbatore District, Tamil Nadu, India

Abstract

Branch canker is the main stem disease of *Camellia* sp. caused by *Macrophoma* sp. In this study, branch canker pathogen was isolated, bought to pure culture and maintained in potato dextrose agar medium (PDA). A total number of 150 bacterial and 40 fungal strains were isolated from different agro climatic zone of south India, which are region specific and native strains (resembling *Pseudomonas* spp. *Bacillus* spp. and *Trichoderma* spp.). Among the total number of bacterial and fungal isolates, 6 bacterial and 3 *Trichoderma* spp. Showed antagonistic effect against the branch canker pathogen. The study clearly indicates that *Bacillus* spp. *Pseudomonas* spp. followed by *Trichoderma* spp. showed higher antagonistic potential against the test pathogen. The study also includes that, the selected botanical fungicides, neem kernel extract, garlic extract, *Aloe vera*, Tulsi and Expel (Botanical fungicides) at different concentration were carried out against *Macrophoma* sp. Results showed that, commercially available botanical fungicide (Expel) is effective to control the growth of branch canker pathogen compare then other chemical and botanical fungicides. The commonly used fungicides in tea plantation such as Hexaconazole (Contof 5E), Tebuconazole (Folicur) and Tridemorph (Calixin) were evaluated against *Macrophoma* sp. under *in vitro* conditions. The results indicated that Tebuconazole all the three concentrations at 1.78 ppm was found to be the most effective in suppressing the growth of branch canker pathogen. The results concluded that biocontrol agents (*Bacillus* spp. *Pseudomonas* spp and *Trichoderma* spp.), botanical fungicide (Expel) and chemical fungicide (Tebuconazole) are very effective to control the branch canker pathogen under *in vitro* conditions.

Keywords: Biocontrol agents; Botanical and chemical fungicides; *Camellia* sp.; *Macrophoma* sp.

Introduction

Tea, an evergreen plant is one of the most popular, non-alcoholic beverages consumed by nearly half the world population. Tea is produced from the young shoots of the commercially cultivated tea plant (*Camellia* sp.). India is the one of the largest producer and consumer of tea in the world with an area of 5.75 lacks/ha under tea cultivation. Tea is attacked by number of pests and diseases which are the major limiting factors in crop productivity. The first comprehensive account on the pests and diseases of tea was presented by Watt [1]. Majority of tea pathogens are of fungal origin and more than 300 species of fungi are reported to affect different parts of the tea plant [2-4]. Mann and Hutchinson [5] recorded various diseases and that was substantiated by Petch [6]. Sarmah [7] described all parasitic and non-parasitic/physiological diseases. Among the stem diseases of tea, branch canker caused by *Macrophoma theicola* is a predominant stem inhabiting fungal disease which has been reported from Ceylon. Branch canker, *Macrophoma theicola* occurs in drought susceptible areas where soil is poor. In Kangra valley, Himachal Pradesh this disease was observed after rainy season, whereas the occurrence of the disease was very rare in Darjeeling [7]. *M. theicola* has been observed to cause twig die-back of mature tea in Taiwan [8]. In general, tea bush affected by sun-scorch is prone to this disease. The diseased patches on the branches appear as slightly sunken lesions surrounded by a ring of callus growth [7]. The affected branches are killed slowly by the invading fungus until the disease spreads to the collar when upper portion of the plant dies. In mild infestations, the canker is callused over completely within a few months, but the fungus may renew its growth forming concentric cankers under adverse conditions. Fructifications are produced on the dead bark during wet weather conditions. To control the disease; the affected branches should be cut out to clean healthy wood. Plants should be protected from sun-scorch and pruning should be avoided during dry weather. The crop loss due to this disease depends upon pathogen and the geographical area [9]. In Taiwan, around 40% of the tea bushes were killed by twig dieback and in south-east Asian countries, root rot disease was responsible for major crop loss [4,10]. Low yield due to incidence of collar and branch canker caused by *Phomposis theae* and *Macrophoma theicola* was reported from central Africa [11]. It has been difficult to control branch canker as it grows with the saprophytic fungi on the plant stem. Being a related anamorph genera of *Botryosphaeria* along with *Botryodiplodia, Diplodia, Fusicoccum, Lasiodiplodia, Macrophomposis* and *Sphaeropsis,* it was difficult to separate it from others as its morphological features were poorly defined [12]. The present study involves the isolation, morphological identification and the effect of different chemical and botanical fungicides, bio-control agents on *Macrophoma* sp.

Materials and Methods

Sample collection

Survey was conducted in major tea growing areas of south India (The Anamallais, Central Tranvancore, High Range, Wayanad, Coonoor and Koppa) to collect soil samples in order to isolate biocontrol agents and branch canker fungal pathogens.

***Corresponding author:** Mareeswaran J, Plant Pathology Division, UPASI Tea Research Institute, Valparai-642127 Coimbatore District, Tamil Nadu, India
E-mail: jmareeswaran11@gmail.com

Isolation of branch canker pathogen

The infected stem portions were collected. The samples were washed in distilled water and were cut into small piece. Surface sterilized with 0.1% mercuric chloride for few seconds followed by sterile distilled watering, 2-3 times. After surface sterilization the infected portions were blotted on sterile filter paper and then inoculated on water agar plates amended with streptomycin (50 mg/lit). Plates were incubated for 3 to 5 days. The grown mycelial tips from water agar plates were aseptically transferred to potato dextrose agar medium (PDA). Pure cultures were obtained from the primary plates by colonies initiated from single spores or from hyphal tips. Single-spore cultures were made by preparing a suspension of spores in distilled sterile water and spreading it over water agar plates. Single germinated spores were removed on a small amount of agar with a transfer needle to a PDA medium. Distinct hyphal tips were cut from the well grown water agar plate and then sub-cultured repeatedly on PDA to obtain pure culture of the fungus.

Isolation of bio-control agents from soil

Soil samples at 0"- 9" depth were collected from three tea growing districts, High range Munnar, Central Travancore, Koppa and The Annamallais for isolation of biocontrol agents (*Trichoderma* spp. *Bacillus* spp. and *Pseudomonas* spp.). The Biocontrol agents were isolated by standard serial dilution plating techniques, sub cultured, brought to purity and stored in slants at 4°C. The cultures were identified using standard bacteriological techniques.

Screening for antagonism

The isolated bacterial and fungal strains (*Trichoderma* spp. *Bacillus* spp. and *Pseudomonas* spp.) were screened for their antagonistic potential against the pathogen, following dual culture technique [13]. The mycelial plug of four day old, actively growing *Macrophoma sp.* was ground and spread uniformly on PDA medium plate with the help of a sterilized spatula. These plates were then spot inoculated within 24 h culture of isolated bacterial strains. Plates were incubated at 30 ± 2°C for 3-5 days. The antagonism was graded by measuring the zone of inhibition produced around the bacterial strains. The grading was given as (-) no antagonism, (+) those showing inhibition zone of <0.5 cm, (++) with 0.5 cm to 1 cm and (+++) those with >1.0 cm. In the case of *Trichoderma, in vitro* screening was done by placing a mycelial plug of 4 days old culture of both pathogens and the antagonist. Time for the first contact between the antagonist and the pathogen and the advancement of the antagonist on the pathogen colony was noted and the efficient strains were short listed. A control plate was maintained for comparison. Radial growth of the pathogenic fungi was measured after comparison with control. Percent inhibition was calculated by the formula given by Bell et al. [14]

Compatibility of pathogen against chemical fungicides

Three systemic fungicides, hexaconazole (Contof 5E), tebuconazole (Folicur) and tridemorph (Calixin) which are commercially being used in the tea fields for the control of various fungal pathogens were selected for the compatibility study of the branch canker pathogen. Three dosages of the systemic chemical fungicide were tested. RD-recommended dosage (3.57 ppm), LR- lower recommended dosage (1.78 ppm) and HR- higher recommended dosage (5.35 ppm). These dosages were mixed with PDA and poured in sterile petri plates and allowed to solidify. Small blocks of the pure culture of the pathogen were cut using sterile cork borer (7 mm) and inoculated onto the solid medium. A control plate, devoid of the chemical fungicide was made as reference. The growth of the pathogen was observed for 10 days (3rd, 5th, 7th and 10th days) and recorded.

Compatibility of pathogen towards botanical fungicides

Neem kernel extract: Dry neem seeds (approximately 100 g) were ground using a mortar and pestle. The powder was tied in a sterile muslin cloth and soaked in 250 ml sterile distilled water and left to stand overnight at room temperature. The extract was filtered using Whatmann filter paper No 1. The filtrate was added to the PDA medium at different concentrations (5%, 7.5% and 10%) to find out the effective dosage at which the pathogen cannot survive. The plates along with the extract, at various doses were inoculated by placing small block of the pure culture (7 mm). A control plate devoid of fungicide was maintained as the reference. The growth is observed for 10 days (at 3rd, 5th, 7th and 10th days) measured and recorded.

Garlic: 30g of garlic was made to paste using mortar and pestle and mixed with 30 ml of sterile distilled water. The extract was filtered using muslin cloth and the extract was added to the PDA medium at different concentrations (1%, 2.5% and 5%) to find the percent inhibition at various dosage of garlic extract. The plates were inoculated the growth was measured and recorded as mentioned as earlier for neem kernel extract.

Tulsi: Tulsi leaves were cleaned with sterile distilled water and ground with 5 ml of 95% ethanol using mortar and pestle. The ground paste was centrifuged at 5000 rpm for 5 min. The collected extract was used to prepare the disc.

Disc preparation: Discs were prepared with Whatmann No. 1 filter paper. The extract was added to 95% ethanol at various concentrations (5%, 7.5% and 10%) and the prepared discs were immersed in it. The control discs were prepared by soaking the discs to 95% ethanol. The discs were kept in hot air oven at 45°C and left overnight to dry. The PDA plates were swabbed with the pure culture over the entire surface of the plate. This procedure was repeated twice and the plate was rotated 60° each time to ensure an even distribution of the culture. The appropriate discs were placed (with plant extracts) evenly (no closer than 24 mm from centre to centre) on the surface of the agar plate either by using sterile forceps or the dispensing apparatus. After 7 days of incubation, each plate was examined and measured for the diameters of the zones of complete inhibition. The zones were measured to the nearest mm using a ruler.

Aloe vera: Gel portion of the leaf was separated using a sterile blade and ground using a mortar and pestle. The ground gel was filtered and the extract was collected. The collected extract was mixed up to 95% ethanol at various concentrations (5%, 7.5%, 10% and 100%). The control discs were prepared using in 95% ethanol and dried in hot air oven at 45°C overnight. The disc preparation and the procedure were same as that of tulsi.

Expel: The commercially available botanical fungicide Expel which is being widely used in the tea field was evaluated at low dose- 1.5 ppm, recommended dose- 3 ppm and high dose- 4.5 ppm. The procedure was same as that of chemical fungicide method.

Results and Discussion

Survey was conducted in major tea growing areas of south India to collect the soil samples and disease specimens to isolate bio-control agents and branch canker fungal pathogen. A total of four branch canker pathogen and biocontrol were isolated from different tea growing district like the Anamallais (MT APF1), Central Tranvancore

(MT HE 02), Coonoor (MT C2 03) and Koppa (MT KH 04) also specific same areas (The Anamallais - 2 *Bacillus* spp. and 2 - *Trichoderma* spp. Central Travancore - 1 *Bacillus* sp., Koppa – 2 *Pseudomonas* spp. and The Nilgiris - 1 *Pseudomonas* sp and 1- *Trichoderma* sp.) were showed bacterial and fungal biocontrol agents. The branch canker pathogen was morphologically, spore characteristically identified used as standard reference book image Petch [6] and confirmed as *Macrophoma* sp (Figure 1). A total number of 150 bacterial and 40 fungal isolate (resembling *Pseudomonas* spp. *Bacillus* spp. and *Trichoderma* spp.), were isolated and screened six bacterial and three *Trichoderma* spp. showed higher antagonistic effect against branch canker pathogen (Table 1). The antagonism was graded by recording the zone of inhibition produced around the bacterial strains. From this study it was concluded that *Bacillus* spp. *Pseudomonas* spp. followed by *Trichoderma* spp. were more inhibitory effect against branch canker pathogen (Figures 2 and 3). Three fungicides, hexaconazole (Contof 5E), tebuconazole (Folicur) and tridemorph (Calixin) were evaluated against *Macrophoma* sp. under *in vitro* condition. The results indicated that, Tebuconazole all the three concentrations at 1.78 ppm was found to be the most effective in suppressing the growth of pathogen followed by hexaconazole and tridemorph. Hexaconazole at 1.78 ppm and tridemorph at 3.57 ppm were found to be optimum for the control of pathogen growth (Table 2). The study revealed that Tebuconazole completely inhibited the growth of branch canker pathogen compared to that other two fungicides. There was absolutely no growth in the fungicide amended plates even at a lower concentration (Figure 4). Among the botanical fungicides tested, expel showed the highest percentage of inhibition against *Macrophoma* sp under *in vitro* condition (Table 3). While Tulsi, Neem kernel, *Aloe vera* and garlic extract had no growth effect of *Macrophoma* sp. (Table 4). In this study, commercially available botanical fungicide (Expel) is effective to control the branch canker disease without any residual effect and maintaining the soil structure, bush health when compared to the

Figure 1: Microscopic view of *Macrophoma* sp. A single pycno spore is magnified through 40x which was isolated from Branch canker infected stem obtained from the Nilgiris.

Tea growing districts	Number of bacterial isolates	Number of *Trichoderma* spp. isolates	No. of antagonist against *Macrophoma* sp.	
			Bacterial strains (*Bacillus* spp. and *Pseudomonas* spp.)	*Trichoderma* spp.
1. The Anamallais	25	15	2 (2 cm)*	2
2. The Nilgiris	50	5	1 (>1 cm)	1
3. Central Travancore	50	15	1 (>1 cm)	-
4. Koppa	25	5	2 (1-2 cm)*	-
Total	150	40	6	3

Table 1: List of biocontrol bacterial and fungal strains isolated from various tea growing areas *Values in parentheses indicate inhibition zone produced by bacterial antagonist.

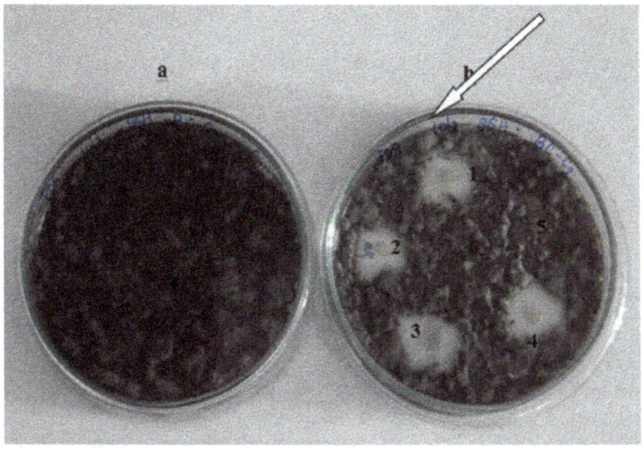

Figure 2: Control plate (a) *Macrophoma* sp. spreaded PDA plate is free from bacterial antagonist and (b) *Bacillus* spp. and *Pseudomonas* spp. inhibited the growth of *Macrophoma* sp. (Arrow indicates the zone of inhibition between fungal pathogen & bacterial antagonist).

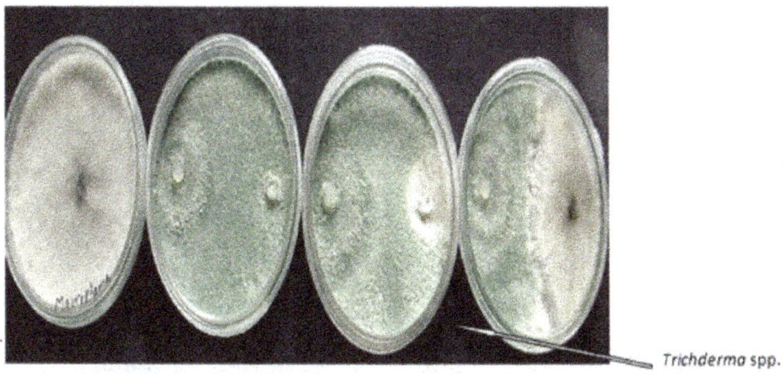

Figure 3: *Trichderma* spp. against *Macrophoma* sp. pathogen.

Fungicides		Fungicide concentration (ppm)	Mean radial growth (cm)	% inhibition of growth (%)
1.Hexaconazole	RD	3.57	0.00	100
	LR	1.78	7.58	75.8
	HR	5.35	0.00	100
2.Tebuconazole	RD	3.57	0.00	100
	LR	1.78	0.00	100
	HR	5.35	0.00	100
3.Tridemorph	RD	3.57	0.00	100
	LR	1.78	0.62	6.24
	HR	5.35	0.00	100
Control plate	UT	-	9.00	0

Table 2: *In vitro* efficacy of different fungicides on *Macrophoma* sp. *On 10th day. Values in the parentheses indicate percent inhibition of the pathogen. RD: Recommended Dosage; LR: Lower Recommended Dosage; HR: Higher Recommended Dosage and UT: Untreated control.

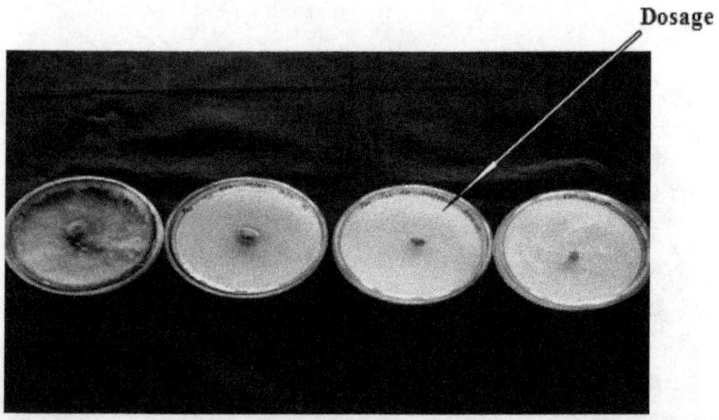

Figure 4: Tebuconozole inhibitied the growth of *Macrophoma* sp. with three different concentration. UT-Untreated control, LD: Lower Recommended Dosage, RD: Recommended Dosage, HR: Higher Recommended Dosage.

Isolates	3rd day				5th day				10th day			
Dosage	U	LD	RD	HD	U	LD	RD	HD	U	LD	RD	HD
MT APF1	8.50	-	-	-	9.00	-	-	-	9.00	74.4	-	-
MT HE 02	9.00	85.0	86.6	87.7	9.00	64.4	67.7	85.0	9.00	60.3	62.3	76.1
MT C2 03	3.50	48.2	54.9	-	4.55	47.4	54.0	69.2	8.10	40.0	52.8	65.8
MT KH 04	9.00	86.6	88.8	89.9	9.00	83.3	87.7	87.7	9.00	75.4	82.4	84.4

Table 3: Effect of botanical fungicide (Expel) on *in vitro* growth of Macrophoma spp. Means of 5 replicates and four different isolates. *(-) no growth *the values indicate percentage inhibition *the values of untreated are indicated in centimeters. U: Untreated, LD: Low Dose, RD: Recommended Dose and HD: High Dose.

Fungicides	Fungicide Concentration (ppm)	Mean radial growth	% inhibition of growth (%)
1. *Aloe vera*	5%	-	-
	10%	-	-
	100%	-	-
2. Tulsi	5%	-	-
	10%	-	-
	100%	-	-
Control plate UT	-	+	+

Table 4: *In vitro* efficacy of different botanical fungicides on Macrophoma spp. (Means of 5 replicates and 4 different isolates). Garlic and Neem Kernel were noticed same results (-) samples showed no growth. *On 10th day. Values in the parentheses indicate percent inhibition of the pathogen. (+) samples showed growth (-) samples showed no growth.

other chemical fungicides. Same result recorded with Nepolean et al. [15] expel botanical fungicide and bacterial biocontrol agents were showed good results against wood rot pathogen. Long term application of PGPR resulted in reduced disease incidence in field grown tea plants. When fungicide or biocontrol agents were incorporated their efficiency in controlling the disease was also improved. Continuous application of PGPR helped the plants to build up natural resistance to the disease. Silva et al. [16] reported reduction of fungicide application number of rounds (50%) in tomato plants treated with *Bacillus cereus*, which provided protection against multiple diseases. In recent studies on antagonistic potential of biocontrol agents against tea pathogens, *Hypoxylon* sp. and *Pestalotiopsis* sp. were tested under *in vitro* level and the results indicated that *Pseudomonas* sp. and *Trichoderma* sp. exhibited superior antagonistic potential against the grey blight and wood rot pathogens [17]. The study clearly indicated that each 3 strains of *Bacillus* and *Pseudomonas* that showed higher antagonism against branch canker pathogen. *Trichoderma* spp. isolated from such a region showed effective antagonism against *Macrophoma* sp. and *Bacillus* spp. provided excellent control of the branch canker disease. Similar results were reported by Nandakumar et al. [18], Vivekananthan et al. [19], Vidhyasekaran and Muthamilan [20] and Ramamoorthy et al. [21], for the control of various fungal pathogens. When groundnut plants were sprayed with *P. fluorescencs*, increased activity of PAL was observed and correlated with the lesser disease incidence [22]. In the present study, *Bacillus* spp. and *Pseudomonas* spp. followed by *Trichoderma* spp. showed more inhibitory effect against *Macrophoma* sp. under *in vitro* condition. Standard fungicides and biological control agents provided satisfactory control of the disease under the field conditions without any residual effects on tea. In this result accordance with Premkumar and Baby [23] have published the latest recommendations on the control of blister blight and grey blight in tea and also Karthika and Muraleedharan [24] supported that, fungicides residues were lost during the shoot expansion time and the 10th day, the level of residues on tea shoots are definitely lower than the limits of residue effect. Hence upon the climatic factor, i.e., due to such as mainly growth dilution, rainfall elution, thermal degradation and photodegradation. Both the fungal and bacterial biocontrol agents provided superior control for the integrated management of grey blight disease. Jo and willson [25] found that the exogenous application sof carbon and nitrogen sources increased the population of biocontrol agent, *P. syringae* in the phyllosphere and increased the biocontrol efficacy. The present study revived that the potential of the selected chemical fungicides (hexaconazole, tebuconazole and tridemorph). To sum up, the present investigation proved beyond doubt that various botanical fungicides like (Expel) neem kernel extract, garlic extract, aloe vera, tulsi and expel were experimented. It was found that expel showed the highest percentage of inhibition against *Macrophoma* sp. while Tulsi, Neem kernel, Garlic extract, and *Aloe vera* had no growth effect of test pathogen.

Conclusion

The study indicated that biocontrol agents (*Bacillus* spp. *Pseudomonas* spp. and *Trichoderma* spp.), botanical fungicide (Expel) and chemical fungicide (Tebuconazole) are very effective to control the branch canker pathogen under *in vitro* conditions. There was absolutely no growth in the fungicide amended plates even at a lower concentration. From this study, it was critically evaluated that *Bacillus* spp. and *Pseudomonas* spp. followed by *Trichoderma* spp. botanical fungicide (Expel) and chemical fungicide (tebuconazole) strengthens the integrated disease management of branch canker disease in tea.

Acknowledgements

The authors are thankful to UPASI Tea Research Institute, Valparai, Coimbatore dist. for their constant encouragement. We wish to extend our heartfelt thanks to Tea Board of India, Kolkata, for the financial support.

References

1. Watt G (1898) The pests and blights of tea plants. Cal. pp. 443-459.
2. Agnihothrudu V (1964) A world list of fungi reported on tea (Camellia spp.). J Madras Univ 34: 155-271.
3. Agnihothrudu V (1967) Some probable traumatic parasites on tea and rubber in southern India. Indian Phytopath 20: 196-198.
4. Chen ZM, Chen XF (1990) The diagnosis of tea disease and their control. Shanghai Sci. And Tech. Publ. Shanghai, China p. 275.
5. Mann HH, Hutchinson CH (1904) Red rust, (2ndedn) I.T.A. Bull. p. 26.
6. Petch T (1923) Diseases of the tea bush. Macmillan and Co. Ltd, London. p. 220.

7. Sarmah KC (1960) Diseases of tea and associated crops in north east India. Indian Tea Association, Scientific Department, Tocklai Experimental Station Memorandum No 26, p. 68.

8. Arulpragasam PV (1992) Disease control in Asia, In: Wilson KC, Clifford MN (eds) Tea cultivation to consumption. Chapman and Hall, London pp. 353-373.

9. Hajra GN (2001) Tea cultivation, comprehensive treatise. International book Distributing Company, Lucknow. p. 518.

10. Muraleedharan N, Chen ZM (1997) Pests and diseases of tea and their management. Journal Plant Crops 25: 15-43.

11. Rattan PS, Sobrak A (1976) Incidence of Phomposis stem and branch canker. Annual report. The Tea research Foundation of Central Africa. Malawi.

12. Crous PW, Palm ME (1999) Reassessment of the anamorph genera Botryodipoldia, Dothiorella and Fusicoccum. Sydowia 52: 167-175.

13. Rajendiran R, Jegadeeshkumar D, Sureshkumar BT, Nisha T (2010) In vitro assessment of antagonistic activity of Trichoderma viride against post harvest pathogens. Journal of Agric Technol 6: 31-35.

14. Bell DK, Wells HD, Markham CR (1982) In vitro antagonism of Trichoderma Species against six fungal plant pathogens. Phytopathol 72: 379-382.

15. Nepolean P, Balamurugan A, Jayanthi R, Mareeswaran J, Premkumar R (2014) Bio efficacy of certain chemical and biofungicides against wood rot pathogen. Journal of Plantation Crops 42: 341-347.

16. Silva HSA, Romeiro RS, Carrer Filho R, Pereora JLA, Mounteer A (2004) Induction of systemic resistance by Bacillus cereus against tomato foliar diseases under field conditions. J Phytopathology 152: 371-375.

17. Vidhya Pallavi R, Nepolean P, Balamurugan A, Pradeepa N, Kuberan T, et al. (2010) In-vitro studies on antagonistic potential of biocontrol agents against tea pathogens, Hypoxylon sp. and Pestalotiopsis sp. Rubber Research Institute of India, Kottayam. pp. 172-173.

18. Nandakumar R, Babu S, Viswanathana R, Raguchander T, Samiyappan R (2001) Induction of systemic resistance in rice against sheath blight disease by plant growth promoting rhizobacteria. Soil Biol Biochem 33: 603-612.

19. Vivekananthan R, Ravi M, Ramanathan A, Samiyappan R (2004) Lytic enzymes induced by Pseudomonas fluorescens and other biocontrol organisms mediate defence against the anthracnose pathogen in mango. World J. Microbiol. Biotechnol 20: 235-244.

20. Vidhyasekaran R, Muthamilan M (1995) Development of formulations of Pseudomonas fluorescens for control of chickpea wilt. Plant Dis 79: 782-786.

21. Ramamoorthy V, Viswanthan R, Raguchander T, Prakasam V, Samiyappan R (2001) Induction of systemic resistance by plant growth promoting rhizobacteria in crop plants against pest and diseases. Crop Prot 20: 1-11.

22. Meena B, Radhajeyalakshmi R, Marimuthu T, Vidhyasekaran P, Doraisamy S, et al. (2000) Induction of Pathogenesis related proteins, phenolics and phenylalanine ammonia lyase in groundnut by Pseudomonas fluorescens. J Plant Dis Prot 107: 514-527.

23. Premkumar R, Baby UI (2005) Blister blight control a review of current recommendations. Journal of Plantation Crops 101: 26-34.

24. Karthika C, Muraleedharan NN (2009) Contribution of leaf growth on the disappearance of fungicides used on tea under South Indian agroclimatic conditions. J Zhejiang Univ Sci B 10: 422-426.

25. Ji P, Wilson M (2003) Enhancement of population size of a biological control agent and efficacy in control of bacterial speck of tomato through salicylate and ammonium sulfate amendments. Appl Environ Microbiol 69: 1290-1294.

Natural Products to Control Postharvest Gray Mold of Tomato Fruit-Possible Mechanisms

Firas Ali Ahmed, Brent S Sipes and Anne M Alvarez*

Department of Plant and Environmental Protection Sciences, 3190 Maile Way University of Hawaii at Manoa, Honolulu, HI, 96822, USA

Abstract

Grey mold is the one of most important postharvest disease of tomato fruit worldwide. Treatments with edible, natural products are needed to reduce losses and contribute to food sustainability. Based on the hypothesis that inhibition of *Botrytis* spore germination will significantly reduce postharvest losses, botanicals were tested for their effects on conidia. Ten *Botrytis* isolates from rotting tomato fruit collected at seventeen different sites in Hawaii were identified morphologically and confirmed by DNA sequence analysis. Effects of plant extracts on spore germination were assessed at several dilutions. Leaves of candidate biocontrol plants were frozen at -20°C and plant fluids were sterilized by passing through a 0.22 µm millipore. The effect of plant extracts on germination of *Botrytis* spores (10^6/ml) in sterile malt extract broth was evaluated in multi-well microplates using preparations ranging from 10 to 40%. Changes in absorbance measured at 6, 12, and 24 hours after inoculation were analyzed with Gen5 software. *Capsicum chinense* cultivars Datil and *C. annuum* Carnival completely inhibited fungal germination at all evaluation times. Extracts from *Capsicum* species were superior to all other extracts tested. Treatments with 40% extracts increased the generation of intercellular reactive oxygen species in the fungal conidia. Plasma membrane damage was shown with fluorescence microscopy when extract-treated conidia were stained with propidium iodide. Loss of integrity in the spore plasma membrane appears to account for the inhibition of *Botrytis* spore germination. Extracts from the two pepper cultivars, Datil and Carnival, showed promise as pre-harvest sprays in the greenhouse and as edible coatings on tomato fruit postharvest to reduce grey mold.

Keywords: Postharvest diseases; *Botrytis cinerea*; Natural products; Reactive oxygen species; Membrane damage

Introduction

Tomato (*Solanum lycopersicum* Mill) is one of the most important vegetables produced globally. Fresh tomato production is highest among all vegetable crops in the U.S and the U.S. is a main producer of tomato in the world [1]. However, postharvest diseases of fruit and vegetables cause major losses in food production. Approximately 23-25% and 31-38% of harvested fruit are lost to postharvest spoilage in the USA and the world respectively [2]. Gray mold, caused by *Botrytis cinerea,* is considered one of the most destructive postharvest diseases of tomato in both field and greenhouse production where the temperature and humidity are conducive to fungal development [3]. Infection of tomatoes by *B. cinerea* causes major economic losses at the pre- and post-harvest stages [4]. Currently, synthetic fungicides are commonly used to control postharvest infections of *B. cinerea* [5].

Increasing international concerns for public health, excessive use of pesticides and development of fungal strains resistant to fungicides have resulted in the introduction of regulations and policies to limit use of synthetic pesticides over the past decades [6,7]. Alternative economically feasible approaches are needed to reduce postharvest gray mold of tomato fruit. Toward this end, antimicrobial characteristics of various plant extracts have been described for use in plant protection [8]. Many applications of substances, such as chitosan, extracts of *Azadirachta indica* seed, essential oils, medicinal plants, and mineral nutrients such as selenium, and borate have been investigated for control of postharvest fruit spoilage of fruit and several have been successfully applied [9-12].

Several plants in the family Solanaceae have antimicrobial and antifungal properties [13-15]. Pepper plants (*Capsicum* sp.) also have a wide range of uses, including pharmaceutical, natural pigment agents, cosmetics, and as ornamentals [16]. Phytochemical analysis of *Capsicum* species demonstrated that the presence of tannins, alkaloids, steroids, glycosides, flavonoids, phenols and terpenoid, provided a wide range of antimicrobial activities [17,18].

Waltheria indica is a woody plant belonging to the family Malvaceae. *Waltheria* is widespread in tropical and subtropical regions including Hawaii [19], South America [20], and South Africa [21]. Its antibacterial and antifungal activity has been documented in several studies [22,23]. The phytochemical investigation from the crude extract of *W. indica* indicated the presence of several chemical compounds such as alkaloids, flavonoids, tannins, sterols, terpenes, and saponins [24]. Scant information is available regarding the antimicrobial activity of pepper and Waltheria extracts on postharvest fungi or their modes of action.

In the present study, leaf extracts of *Capsicum chinense* (Datil), *C. annuum* (Carnival), *C. frutescens* (Hawaiian chili pepper), and *Waltheria indica* (sleepy morning) were tested as potential natural antifungal agents using spore germination of *B.cinerea* as a measure. The role of intercellular reactive oxygen species (ROS) and the integrity of the plasma membranes in relationship to germination of *B. cinerea* spores were also examined.

Material and Methods

Fungal isolates

Fungi were isolated from diseased tomato fruit collected from 37 markets located throughout the island of Oahu, Hawaii. Small sections (1 mm^2) were excised from a range of lesions, incubated on water agar. Single hyphal tips were transferred to V-8 agar in 9 cm petri dishes

***Corresponding author:** Anne M Alvarez, Department of Plant and Environmental Protection Sciences, 3190 Maile Way University of Hawaii at Manoa, Honolulu, HI, 96822, USA, E-mail : alvarez@hawaii.edu and incubated for 10 to 14 d at 20°C with a 12-h photoperiod [25]. The isolates were stored at 2°C in sterile soil for further study. Fresh cultures were established as needed by plating 0.2 g soil on V-8 petri plates. Ten isolates of *B. cinerea* were selected for further study.

Pathogenicity tests

Tests were conducted to determine the most virulent isolates of *B. cinerea* on tomato fruit (common market, cherry and grape tomatoes), detached leaves and whole plants. Fruit were inoculated with mycelium of 10-day-old-cultures of the selected isolates of *B.cinerea* using 0.2-10 µl pipette tips. On the detached leaves, 6-mm fungal plugs were placed on 4-week-old surface sterilized tomato leaves placed in 9-cm-d petri dishes including wet filter paper. The petri plates were held for 72 h at 23°C in the dark. On tomato plants, spore suspensions of 1×10^5 spores/ml from 10-day-old cultures of each isolate of *B. cinerea* were prepared using a hemocytometer. Fungal spore suspensions were sprayed on 4 week-old tomato plants (Favorite, F1) growing in 10-cm-d plastic pots. The plants were placed in plastic bags, sealed, and held for 24 h at room temperature. Plastic bags were removed and plants returned to the greenhouse for 14 days. For disease assessment of tomato fruit and detached leaves, the lesion diameter of inoculated fruit and leaves were measured after 72 h at 23°C. On tomato plants, the disease severity was scored on diseased plant utilizing a 0-5 scale.

The experiments were set up as Complete Randomized Design (CRD) with four replications. Data were analyzed using (SAS 9.2 V.USA) and means were compared by Duncan's multiple range tests. Differences at $p<0.05$ were considered significant. All experiments were repeated three times.

DNA extraction

Fungal DNA was extracted from freshly collected mycelium of 10-day-old cultures using the Microbial DNA Isolation Kit (MO BIO, Laboratories, Inc.) according to manufactures' instructions. The concentration of extracted DNA was measured with a Nano Drop 1000 spectrophotometer (Thermo scientific, V 3.7., USA) and equilibrated using distilled water. Extracted DNA was stored at 4°C until used.

PCR amplification

The ITS region of the fungal isolates was amplified with the primer pair ITS3 (5-GCA TCG ATG AAG AAC GCA GC-3) and ITS4 (5-TCC TCC GCT TAT TGA TAT GC-3) [26]. PCR was performed with a REDTaq DNA polymerase (Sigma, St. Louis, MO). A 20 µl reaction mix contained 1 µl of ITS3 primer, 1 µl ITS4 primer, 10 µl REDTaq, 4 µl ddH_2O, and 4 µl of template. PCR was carried out using a Real-time PCR detection system (BIO-RAD) with initial denaturation for 3 min at 95°C, followed by 36 cycles of denaturation for 30 s at 94°C, primer annealing for 30 s at 55°C, and extension for 1 min at 72°C. A final extension for 5 min at 72°C was performed [27]. Each PCR reaction was run with a negative control (no DNA). The PCR products were electrophoresed on 1.5% agarose gels, stained with 0.4 µg/ml ethidium bromide, and bands visualized with a UV illuminator. DNA concentration estimates were calculated by comparing the fluorescent intensities of the product with DNA standards on a 2-log ladder (New England Biolabs, Inc.).

Sequence analysis

PCR product was cleaned using ExoSAP-1T (Affymetrix, Inc., USA). A mix of 5 µl of post-PCR reaction and 2 µl ExoSAP-IT reagents were combined. The mix was incubated at 37°C for 15 min following by incubation at 80°C for 15 min. Each purified template was sequenced on both strands using two flanking primers (ITS3- ITS4) (ASGPB Lab). The sequences of ITS 3 and 4 regions of the tested isolates were edited in order to generate a consensus sequence from forward and reverse sequences in the amplicon using sequence assembly software (DNA BASER). A consensus sequence was analyzed by NCBI BLAST database seeking fungal identities.

Plant extraction and multi-well plates

Leaf extracts were made from *C. chinense*, *C. annuum*, and *C. frutescens* obtained from Waimanalo Research Station and *W. indica* obtained from H1/University Ave (Native Hawaiian Plant Demonstration Site). Plant materials were extracted by following method described by [28] with some modification. Fresh plant material was collected in resealable plastic bags and placed in a freezer for a minimum of 12 h at -20°C. As plant material was tested, it was withdrawn from the freezer and allowed to thaw for a minimum of 20 min. Freezing and thawing fractured the plant cells. The plastic bag was tilted so that the fluid collected in one corner, and extracted plant fluids collected in plastic weight boats. The extract was filter-sterilized by passing through a 0.22 µm Millipore filter. Filtered 10, 20, 30, 40% extract solutions were prepared by adding 1, 2, 3, and 4 ml of sterile extract with 9, 8, 7,and 6 ml respectively of a *B.cinerea* spore suspension (1×10^5 conidia/ml) in sterile malt extract broth.

Fifty µl of the plant extract and spore suspension mixture was pipetted into each well of a row of a 96 multi-well plate. A nontreated spore suspension was added to one row of each plate for comparison, and a blank row was included as an instrument check. Each well was sealed with a sheet of parafilm to prevent cross-contamination by volatiles. After 6, 12, 24 h incubation at 25°C, the density of fungal growth in the wells was measured with an Epoch microplates reader (BioTek Instrument, Inc.). Absorbency/optical density was measured for each well using a 492-mm filter. Background readings from the average of the nontreated samples were subtracted by the Gene5 software program to provide net fungal growth in tested samples.

Reactive oxygen species (ROS)

A method using fluorescent probe 2,7-dichlorodihydrofluorescein diacetate(DCFH-DA)was used with slight modifications to assess the intercellular level of ROS in *B. cinerea* [12]. Spores of *B. cinerea* were cultured in malt extract broth supplanted with 40% *C. chinense* or *C.annuum* leaf extract and incubated for 2, 4, and 6 h at 23°C. The spores of 1×10^5 per ml were collected and washed with 10 mM potassium phosphate buffer (pH 7.0) and incubated for 1 h in the same buffer containing 10 µM DCFH-DA (dissolved in dimethyl sulfoxide).After washing twice with potassium phosphate buffer, spores were examined under a Zeiss microscope (Axioscop.A1, USA) using a fluorescein 2, 7- dichlorodihydro-specific filter. At least 100 spores were examined for each treatment with three replicates.

Membrane integrity

A propidium iodide (PI) staining method was used to detect the membrane integrity of *B. cinerea* conidia [29]. Spores of *B. cinerea* were cultured in malt extract broth medium supplemented with 40% of *C. chinense C.annuum* leaf extract and incubated 2, 4, and 6 h at 23°C. The spores of 1×10^5 per ml were collected and stained with 10 µg/ml PI for 5 min at 30°C. The spores were centrifuged and washed twice with 10 mM potassium phosphate buffer (pH 7.0) to remove residual dye. The spores were observed under a Zeiss microscope (Axioscop.A1, USA) and red stained conidia were recorded. At least 100 spores were examined for each treatment with three replicates.

Statistical analysis

All statistical analyses were performed using SAS software version 9.2 (SAS Institute Inc., Cary, NC, USA). Data were subjected to one-way analysis of variance (ANOVA). Where appropriate, means were compared suing Duncan's Multiple Range Test. Differences at $P<0.05$ were considered significant.

Results

Ten *Botrytis* isolates were made from 33 pathogenic-fungal genera recovered from infected tomato in a previous survey. Based on the morphological characteristics of isolates, microscopic observations of their conidiophores and conidia, all were identified as *Botrytis* spp.

Molecular identification

A PCR product with expected size 370 bp was amplified successfully for all fungal isolates (Figure 1). A NCBI BLAST web (http://blast.ncbi.nlm.nih.gov/Blast.cgi) for the ITS3-ITS4 region to determine fungal identity and found that sequence maximum identity of >98% with other *B. cinerea* entries. In addition, the BLAST results of *Botrytis* isolates B03 and B09 matched ≥99% to *B. cinerea* base pairs.

Pathogenicity tests

The ten isolates of *Botrytis* spp. were highly variable with respect to disease on tomato fruit, detached leaves, and whole plant. The lesion diameter ranged from 70, 32, and 25 to 16, 10, 10 mm^2 on common market, cherry, and grape tomato, respectively; lesion size ranged from 28 to 14 mm on detached leaves; and the severity scale ranged from 1 to 5 on whole plants, showing wide pathogenic variability. On tomato fruit, grey mold development differed on common market, cherry and grape tomatoes (Table 1). All isolates were pathogenic on the original host from which they were isolated. The largest lesion diameters on inoculated fruit were caused by *B. cinerea* B03, and were significantly in comparison to other isolates and the control (Table 1 and Figure 2). Necrotic lesions developed on detached leaves after 3 d for all isolates. Lesion diameters differed among Botrytis isolates with lesion size of B03 and B09 greater ($p<0.05$) compared with other isolates (Figure 3). On whole tomato plants, isolate B03 was the most virulent among all the isolates 14 days after inoculation (Figure 4). B08 and B09 were less virulent on tomato plants. Other isolates varied in their disease severity. Since *B. cinerea* isolate B03 was the most virulent isolate; it was used for further study

Inhibitory effect of plant extracts on spore germination

Plant extracts showing the greatest antifungal activity were those from *C. chinense* and *C. annuum*. Spores treated with these extracts at 30 and 40% concentrations showed significant differences compared to spores treated with broth (control) for all measurement of absorbance. Treated spores with these extracts registered under 0.1 absorbance in fungal growth after 6, 12, and 24 h. No significant differences were observed with the control were observed (Figure 5). An interaction between main treatments and all concentrations was not detected, indicating that the extracts effectively reduced *Botrytis* spore germination in all concentrations. The 40% *C. chinense* and *C. annum* extracts completely inhibited spore germination and fungal growth of *B. cinerea* after 24 h compared with the control (Figure 5). The extracts of *W. indica* and *C. frutescens* had intermediate antifungal activity against spore germination of *B. cinerea* at the 30 and 40% concentrations. The absorbance was less than 0.1 after 6 and 12 h but increased to more than 0.3 with 24 h of incubation (Figure 5). No effects were observed on spore germination with any of the plant extracts at the 10 or 20% concentrations (Figure 5). All plant extracts showed OD greater than 0.3 at concentrations of 10 and 20% after 24 h (Figure 5). Mycelia growth did not occur when spores were treated with extract concentrations of 40% in *C. chinense* or *C. annuum* after 24 h. However, growth was less in spores treated with the same concentration of *W. indica* and *C. frutescens* extracts compared with the positive and negative control (Figure 5).

Reactive oxygen species

Reactive oxygen species generation was recorded when *B. cinerea* was exposed to 40% extracts of *C. chinense* cv. Datil and *C. annuum* cv. Carnival for 0, 2, 4, and 6 h at 23°C (Figure 6). With an increased incubation period, the number of stained spores increased. The highest percentage of stained spores 92% and 66% for *C. chinense* and *C. annuum* extracts respectively. ROS of spores treated with the 40% extracts were greater compared with control treatment where 6% ROS was observed (Figure 6). A majority of spores in the control treatment were not stained by DCFH-DA, implying poor ROS production at this time, indicating that an increasing amount of oxidizing molecules was produced in spores exposed to the extracts of *C. chinense* and *C. annuum*.

Plasma membrane integrity

With an increased incubation period, damage to cell membranes increased (Figure 7). The highest percentage of damaged plasma

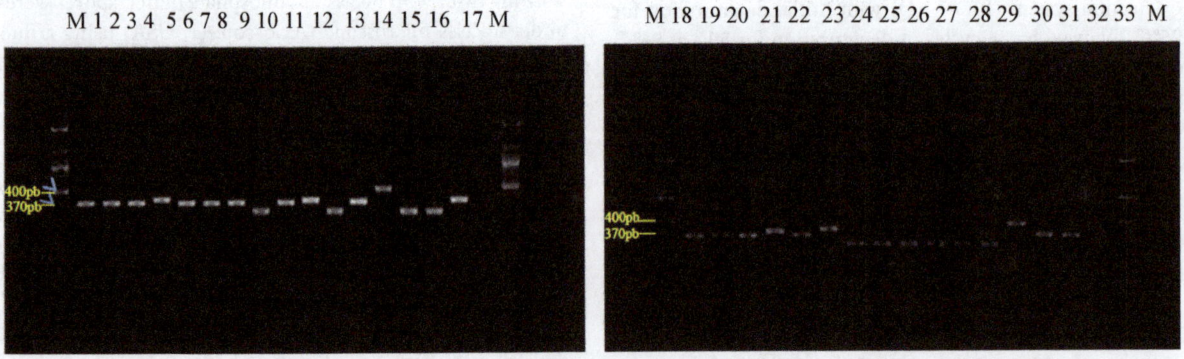

Figure 1: An amplified PCR products following gel electrophoresis using DNA from isolated fungi. M = Marker (0.1-10.0 kb); 1-3 = *Botrytis* sp.; 4 = *Rhizopus* sp.; 5-6 = *Botrytis cinerea*; 7 = *Botrytis* sp.; 8 = *Mucor* sp.; 9-10 = *Fusarium* sp.; 11 = *Penicillium*. 12 = *Stemphylium* sp.; 13=*Colletotrichium* sp.; 14-15: = *Geotrichium* sp.; 16 = *Phoma* sp.; 17 = NC (No DNA); 18-20 = *Botrytis* sp.; 21 = *Alternaria* sp.; 22 =*Botrytis* sp.; 23 = *Fusarium* sp.; 24-29=*Geotrichium* sp. 30 = *Phoma* sp.; 31-32 = *Mucor* sp., 33 = NC (No DNA).

	Colony diameter (mm) following inoculation onto fruit of three tomato varieties		
Isolate ID	Common market	Cherry	Grape tomato
B01	68.00^{a}	27.00^{b}	20.00^{b}
B02	30.00^{c}	22.00^{c}	10.00^{c}
B03	70.00^{a}	32.00^{a}	25.00^{a}
B04	30.00^{c}	12.00^{de}	12.00^{c}
B05	30.00^{c}	16.00^{d}	10.00^{c}
B06	16.00^{d}	14.00^{de}	13.00^{c}
B07	59.00^{b}	22.00^{c}	18.00^{b}
B08	60.00^{b}	22.00^{c}	20.00^{b}
B09	61.00^{b}	25.00^{bc}	20.00^{b}
B10	59.00	10.00^{e}	10.00^{c}
Control	1.00^{e}	1.00^{f}	1.00^{d}

Mean values followed by different letters within a column are significantly different according to Duncan's multiple Range Tests ($p \leq 0.05$).

Table 1: Pathogenicity of ten *Botrytis* isolates on common market, cherry and grape tomato after 72 h incubation at 23°C.

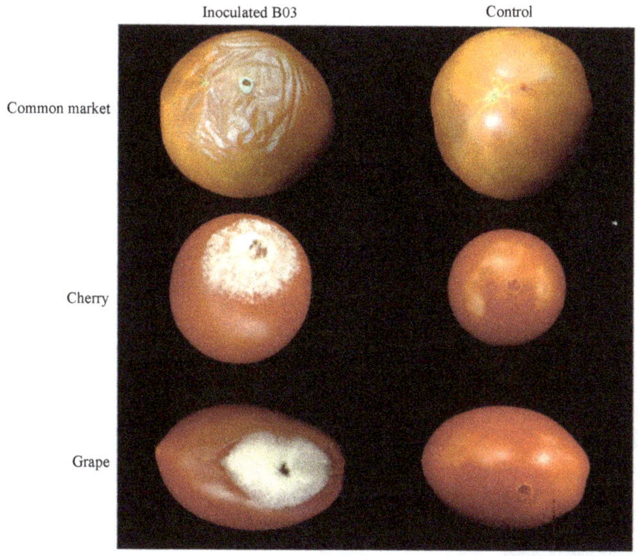

Figure 2: Lesions on tomato fruit inoculated with *Botrytis cinerea* B03.

membrane of stained spores was 89 and 51% for *C. chinense* and *C. annuum* extracts respectively following 6 h incubation. The control treatment had only 30% damage ($p<0.05$) (Figure 7). No difference was observed among treatments and control after 0 and 2 h incubation, indicating that an increasing of exposed period of spores to the extracts was needed to affect cell membrane integrity (Figure 7).

Discussion

The highly variable severity of symptoms caused by *Botrytis* isolates on tomato fruit, detached leaves, and whole plant suggest the existence of wide pathogenic variability among the isolates Pathogenic behavior of isolates is frequently as a result of deviation in cell wall degrades enzyme (CWDEs) activities and existence of oxalic acid and secretion pathogenicity factors [30,31]. High concentration of oxalic acid is associated with high level of pathogen virulence [32,33]. Some studies demonstrated that variation of *Botrytis* isolates in secreted polygalacturonases and pectin lyases contributed to the pathogenicity range of pathogens [34]. Other studies showed pathogenic behavior of isolates is associated with dispersal of a transposable element that contributes to pathogenicity and disease severity [35,36].

Many natural fungicides occur in plants and can serve as safe alternatives to synthetic fungicides. Some compounds could be applied directly after formulated or serve as a pattern for artificial compounds. Pesticides resistance developed is a major problem in efficiency using of fungicide against *Botrytis cinerea*, such as Fenhexamid [37,38].

Most of plant species utilized in this experiment are edible and traditionally use in cured recipes [24,39]. Among them, *Capsicum chinense* (Datil) and *C. annuum* (Carnival) extract showed high antifungal activity followed by *C. frutescens* and *Waltheria indica*. Capsicum species extract were demonstrated to have a great antifungal activity against *Botrytis cinerea* [28]. The antifungal activity of Capsicum extract could be attributed by capsaicin, phenolic compounds, flavonoids, steroids, alkaloids, and tanins [17,40]. Although these extracts have been recognized possess of antifungal compound, they have not been developed into commercial product for postharvest treatments because industries utilize synthetic products as easier method. *C. chinense* and *C.annuum* extracts inhibited spore germination of *B.cinerea*. Capsicum extracts have potential as propitious antifungal agents as well as alternatives to synthetic fungicides for postharvest fungal infections. Capsicum extracts could be used as pre- and/or postharvest applications for protection of tomato fruit from gray mold disease.

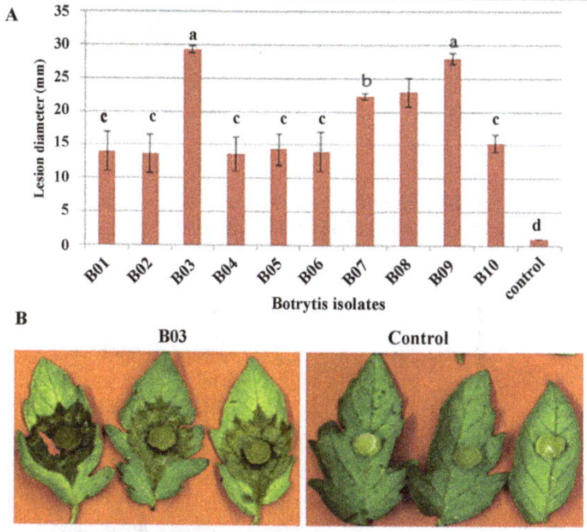

Figure 3: (A) Lesion diameter of ten *Botrytis* isolates 72h after inoculation of detached tomato. Vertical bars indicate standard error (±SE). Bars with the same letters are not significantly different according to the Duncan's multiple range test ($p \leq 0.05$). (B) Leaf lesions caused by *Botrytis cinerea* (B03) 72 h after plug inoculation.

A

Botrytis isolates	Disease severity
Control (ddH2O)	0
B01	3
B02	1
B03	5
B04	2
B05	2
B06	2
B07	3
B08	4
B09	4
B10	3

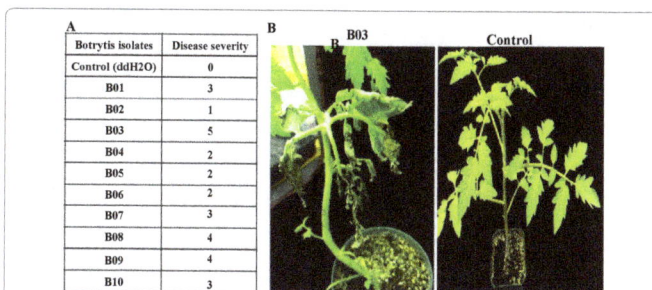

Figure 4: (A) Disease severity rating of *Botrytis* isolates following spray inoculation onto leaves of tomato (variety Favorite) (B) Symptoms caused by *Botrytis cinerea* (B03) two weeks after inoculation; Disease severity rating: (0) no lesions (healthy); (1) small number of lesions (almost healthy); (2) small number of lesions without died leaves; (3) small number of lesions with 1-5 dead leaves; (4) large number of lesions with 6-10 dead leaves; (5) large number of lesions with 11-20 dead leaves.

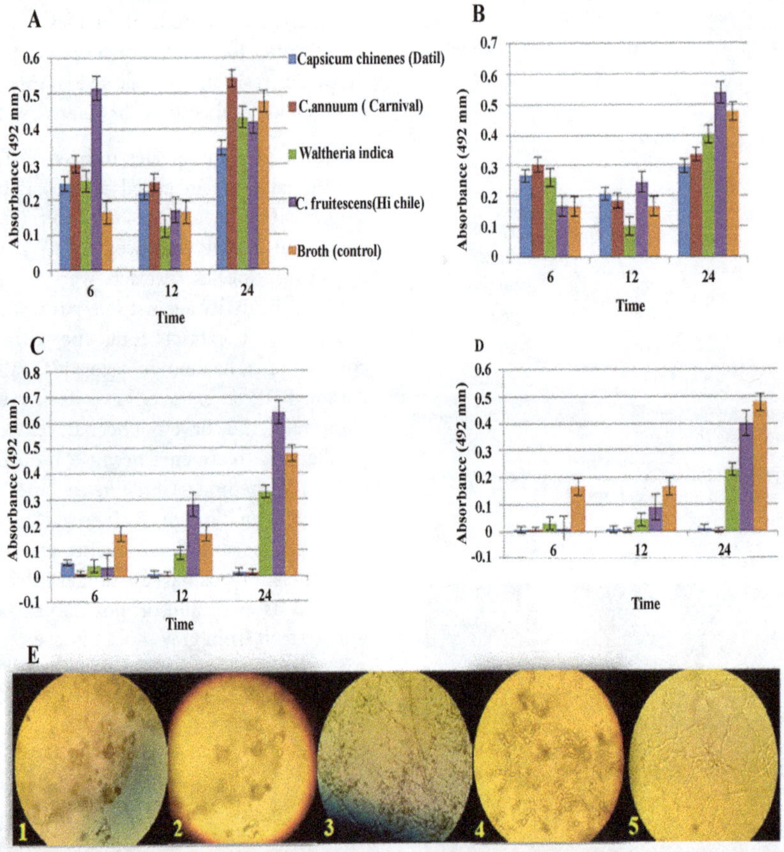

Figure 5: Effect of plant extracts on spore germination of *Botrytis cinerea* measured by absorbance of spore suspensions (492 mm) at 6, 12, 24 h. (A) 10% (B) 20% (C) 30% (D) 40% (E) Micrographs of spores treated with 40% extracts after 24h of (1) *Capsicum chinense* (Datil); (2) *C.annuum* (Carnival); (3) *C. frutescens* (Hawaii Chile pepper); (4) *Waltheria indica*; (5); malt extract broth (control). Vertical bars indicate standard error (±SE).

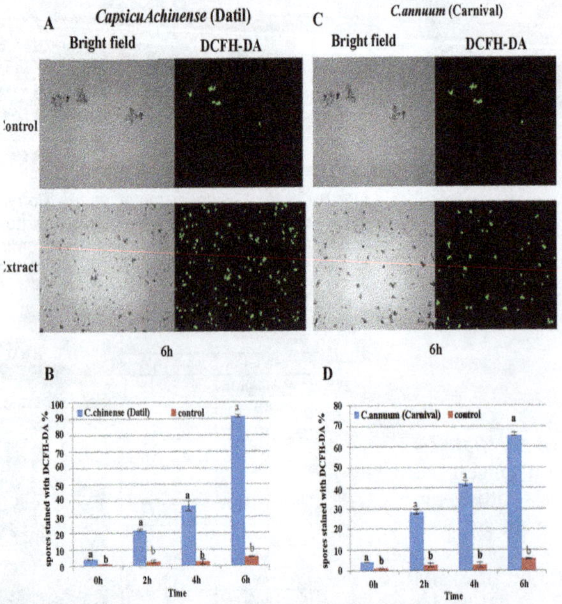

Figure 6: Effect of *Capsicum chinense* cv. Datil (A) and *C.annuum* cv. Carnival (B) extracts at 40% on production of reactive oxygen species in spores of *Botrytis cinerea*. (C) and (D) the percentage of spores stained with DCFH-DA. Spores incubated for 0, 2, 4, and 6h. Vertical bars indicate standard error (±SE). Columns followed by different letter are significantly different according to the Duncan's multiple range test ($p<0.05$).

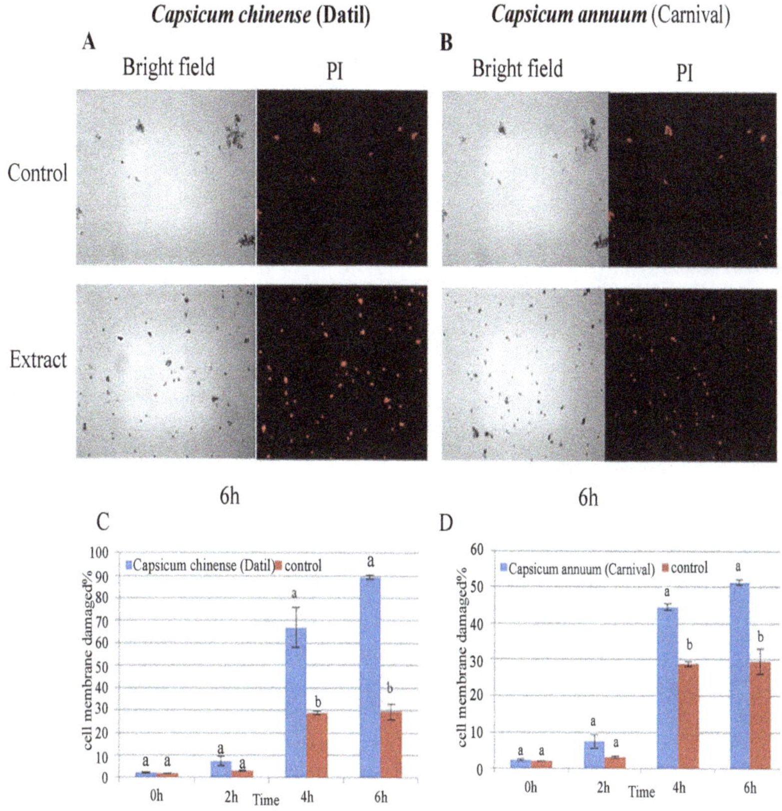

Figure 7: Effect of *Capsicum chinense* cv. Datil (A) and *C. annuum* cv. Carnival (B) extracts at 40% on cell membrane in spores of *Botrytis cinerea*. The percentage of cell membrane damage of spores treated with a 40% extract of *C. chinense* cv. Datil (C) or *C. annuum* cv. Carnival (D) and stained with propidium iodide. Spores were incubated for 0, 2, 4, and 6h. Vertical bars indicate standard error (±SE). Bars with the same letter are not different according to the Duncan's multiple range test ($p<0.05$).

The mechanism of action associated with extracts of Capsicum seems to be related to production of ROS. Cellular dysfunction or cell death is a consequence of oxidative damage to cellular components caused by ROS [12,29,41]. ROS was induced by the Capsicum extract and inhibited spore germination of *B.cinerea*. PI staining implied that the accumulation of ROS in spore cells led cell membrane damage or reduced the intact membrane. Moreover, acknowledging of the mode of action of Capsicum extracts on their effects on *B.cinerea* would be useful for management of pre and postharvest application against gray mold. Consequently, further studies are needed on Capsicum extracts not only fungal pathogenic fungi but also on pathogenic bacteria that can also cause postharvest diseases on fruit and vegetables such as *Pectobacterium* sp.

Conclusion

Botrytis isolates differed in their virulence that was shown during the pathogenicity test. Also, we conclude that *C. chinense* cv. Datil and *C. annuum* cv. Carnival extracts at concentrations of 40% inhibited spore germination of *B.cinerea*. In addition, Capsicum extracts induce ROS generation in fungal spores, leading to oxidation and cell membrane damage. The oxidative damage could be the mechanism of Capsicum extracts against *B.cinerea*. These two natural products could be alternative to synthetic fungicide to control postharvest gray mold disease in tomato.

Acknowledgments

This research was supported by Higher Committee for Educational Development in Iraq and the University of Hawaii Foundation Program on Plant Disease Control.

References

1. USDA-ERS (2012a) Vegetables and Pulses Data.
2. Kantor LS, Lipton K, Manchester A, Oliveira V (1997) Estimating and addressing America's food losses. Food reviw 20: 3-11.
3. Elad Y, Williamson B, Tudzynski P, Delen N (2007) Botrytis: Biology, Pathology and Control Springer, The Netherlands.
4. Soylu EM, Kurt S, Soylu S (2010) In vitro and in vivo antifungal activities of the essential oils of various plants against tomato grey mould disease agent *Botrytis cinerea*. Int J Food Microbiol 143: 183-189.
5. Mari M, Francesco AD, Bertolini P (2014) Control of fruit postharvest diseases: old issues and innovative approaches. Stewart postharvest review 1: 1-4.
6. Holmes GJ, Eckert JW (1999) Sensitivity of *Penicillium digitatum* and *P. italicum* to Postharvest Citrus Fungicides in California. Phytopathology 89: 716-721.
7. Liu J, Sui Y, Wisniewski M, Droby S, Liu Y (2013) Review: Utilization of antagonistic yeasts to manage postharvest fungal diseases of fruit. Int J Food Microbiol 167: 153-160.
8. Del Campo J, Amiot MJ, Nguyen-The C (2000) Antimicrobial effect of rosemary extracts. J Food Prot 63: 1359-1368.
9. Bautista-Baños S, Hernandez-Lauzardo AN, Velazquez-Del Valle MG, Hernández-López M, Barka EA, et al (2006) Chitosan as a potential natural compound to control pre and postharvest diseases of horticultural commodities. Crop Prot 25:108-118.
10. Emma O, Theresa N, Ucheoma O (2013) Control of postharvest bacterial diseases of tomato in Abia State, South Eastern Nigeria. Biol Agri Healthc 3:24-25.

11. Ibrahim FA, Al-Ebady N (2014) Evaluation of antifungal activity of some plant extracts and their applicability in extending the shelf life of stored tomato fruits. J Food Process Technol 6: 1-6.

12. Wu Z, Yin X, Bañuelos GS, Lin ZQ, Zhu Z, et al. (2016) Effect of Selenium on Control of Postharvest Gray Mold of Tomato Fruit and the Possible Mechanisms Involved. Front Microbiol 6: 1441.

13. Bosland PW (1996) Capsicums: Innovative uses of an ancient crop. Progress in new crops. ASHS Press, Arlington, VA pp: 479-487.

14. Dorantes L, Araujo J, Carmona A, Hernandez-Sanchez H (2008) Effect of capsicum extracts and cinnamic acid on the growth of some important bacteria in dairy products. Food Engineering: Integrated Approaches, (Springer).

15. Omolo MA, Zen-Zi W, Amanda KM, Jennifer HC, Nina LC, et al (2014) Antimicrobial properties of Chili peppers. J Infect Dis Ther 2: 145.

16. Brito-Argáez L, Moguel-Salazar F, Zamudio F, González-Estrada T, Islas-Flores I (2009) Characterization of a Capsicum chinense seed peptide fraction with broad antibacterial activity. Asian J Biochem 3: 77-87.

17. Gayathri N, Sekar T, Gopalakrishnan M (2016) Phytochemical screening and antimicrobial activity of Capsicum chinense Jacq. Int J Adv in Pharmac 1: 12-20.

18. Vinayaka K, Prashith-Kekuda TR, Nandini KC, Rakshitha MN, Ramya, M, et al (2010) Potent insecticidal activity of fruits and leaves of *Capsicum frutescens*. Der Pharm Lett 2: 172-176.

19. Abbott IA, Shimazu C (1985) The geographic origin of the plants most commonly used for medicine by Hawaiians. J Ethnopharmacol 14: 213-222.

20. Olajuyigbe O, Babalola A, Afolayan A (2011) Antibacterial and phytochemical screening of crude ethanolic extracts of *Waltheria indica* Linn. Afr J Microbiol Res 22: 3760-3764.

21. Mathabe MC, Nikolova RV, Lall N, Nyazema NZ (2006) Antibacterial activities of medicinal plants used for the treatment of diarrhoea in Limpopo Province, South Africa. J Ethnopharmacol 105: 286-293.

22. Almagboul A, Bashir A, Salih A, Farouk A, Khalid S (1988) Antimicrobial activity of certain Sudanese plants used in folkloric medicine. Screening for antibacterial activity (V). Fitoterapia 59: 57-62.

23. Garba S, Salihu L, Ahmed M (2012) Antioxidant and antimicrobial activities of ethanol and n-hexane extracts of *Waltheria indica* and *Mucona pruriens*. J Pharm Sci Innov 5: 5-8.

24. Zongo F, Ribuot C, Boumendjel A, Guissou I (2013) Botany, traditional uses, phytochemistry and pharmacology of *Waltheria indica* L. (syn. Waltheria americana): a review. J Ethnopharmacol 148: 14-26.

25. Carisse O, Heyden HV (2014) Relationship of Airborne *Botrytis cinerea* Conidium Concentration to Tomato Flower and Stem Infections: A Threshold for De-leafing Operations. Plant Dis 1: 137-142.

26. Nikolcheva LG, Cockshutt AM, Bärlocher F (2003) Determining diversity of freshwater fungi on decaying leaves: comparison of traditional and molecular approaches. Appl Environ Microbiol 69: 2548-2554.

27. Evans TN, Watson G, Rees GN, Seviour RJ (2014) Comparing activated sludge fungal community population diversity using denaturing gradient gel electrophoresis and terminal restriction fragment length polymorphism. Antonie Van Leeuwenhoek 105: 559-569.

28. Wilson CL, Solar JM, El Ghaouth A, Wisniewski ME (1997) Rapid evalution of plant extracts and essential oils for antifungal activity against *Botrytis cinerea*. Plant Dis 2: 204-210.

29. Wu ZI, Ban JS, Li M, Bin YX, Qing LZ, et al (2014) Inhibitory effect of selenium against *Penicillium expansum* and its possible mechanisms of action. Curr Microbiol 69: 192-201.

30. Bellincampi D, Cervone F, Lionetti V (2014) Plant cell wall dynamics and wall-related susceptibility in plant-pathogen interactions. Front Plant Sci 5: 228.

31. Kubicek CP, Starr TL, Glass NL (2014) Plant cell wall-degrading enzymes and their secretion in plant-pathogenic fungi. Annu Rev Phytopathol 52: 427-451.

32. Mbengue M, Navaud O, Peyraud R, Barascud M, Badet T, et al. (2016) Emerging trends in molecular interactions between plants and the broad host range fungal pathogens *Botrytis cinerea* and *Sclerotinia sclerotiorum*. Front Plant Sci 422: 1-9.

33. Kumari S, Tayal P, Sharma E, Kapoor R (2014) Analyses of genetic and pathogenic variability among *Botrytis cinerea* isolates. Microbiol Res 169: 862-872.

34. Cotoras M, Silva E (2005) Differences in the initial events of infection of *Botrytis cinerea* strains isolated from tomato and grape. Mycologia 97: 485-492.

35. Samuel S, Veloukas T, Papavasileio A, Karaoglanidis GS (2012) Differences in frequency of transposable elements presence in *Botrytis cinerea* populations from several hosts in Greece. Plant Dis 9: 1286-1290.

36. Fournier E, Giraud T, Albertini C, Brygoo Y (2005) Partition of the *Botrytis cinerea* complex in France using multiple gene genealogies. Mycologia 97: 1251-1267.

37. Angelini RMM, Rotolo C, Masiello M, Gerin D, Pollastro S, et al. (2014) Occurrence of fungicide resistance in populations of *Botryotinia fuckeliana* (*Botrytis cinerea*) on table grape and strawberry in southern Italy. Pest Manag Sci 70: 1785-1796.

38. Hahn M (2014) The rising threat of fungicide resistance in plant pathogenic fungi: Botrytis as a case study. J Chem Biol 7: 133-141.

39. Everitt JH, Drawe DL, Lonard RI (1999) Field guide to the broad-leaved herbaceous plants of South Texas used by livestock and wildlife. (Texas Tech University Press, USA) p: 244.

40. Nascimento PL, Nascimento TC, Ramos NS, Silva GR, Gomes JE, et al (2014) Quantification, antioxidant and antimicrobial activity of phenolics isolated from different extracts of *Capsicum frutescens* (Pimenta Malagueta). Molecules 19: 5434-5447.

41. Shi X, Li B, Qin G, Tian S (2012) Mechanism of antifungal action of borate against *Colletotrichium gloeosporioides* related to mitochondrial degradation in spores. Postharvest Biol Technol 67: 138-143.

Race Diversity of *Pyrenophora tritici-repentis* in South Dakota and Response of Predominant Wheat Cultivars to Tan Spot

Abdullah S, Sehgal SK and Ali S*

Department of Agronomy, Horticulture, and Plant Science (AHPS), South Dakota State University, Brookings, USA

Abstract

The fungus *Pyrenophora tritici-repentis* (*Ptr*) causing tan spot (TS) is an important pathogen of wheat in the US Northern-Great-Plains. Knowledge of physiological variation in the pathogen population is essential in the development of durable TS resistant cultivars. Eight *Ptr* races have been identified based on three host selective toxins (Ptr ToxA/Ptr ToxB/Ptr ToxC), which are associated with necrosis and chlorosis symptoms. The information about *Ptr* race structure and reaction of wheat cultivars grown in SD to tan spot is scarcely available. In this study, 569 isolates of *Ptr* collected from wheat were genotyped for *Ptr ToxA* and *Ptr ToxB* genes and a subset of 134 isolates were evaluated for their race identity on a wheat differential set. *Ptr ToxA* and *Ptr ToxB* genes were amplified in 89.6% and 0.4% isolates, respectively. The remaining 57 (10%) isolates lacked both toxins genes. The characterization of 134 isolates exhibited diverse race structure with 74.6%, 18.7%, 1.49% , and <1% isolates categorized as race 1, 4, 5, and 2, respectively. Another six (4.5%) isolates behaved like race 2 but lacked *Ptr ToxA* gene, hence could not fit under the currently known eight races. Our results determine the diversity of *Ptr* population that exists in SD and establish the presence of race 5 in SD for the first time. Since races 1 and 5 are most prevalent in the region, we screened 45 most predominant wheat cultivars against these races and *Ptr ToxA*. We observed eleven cultivars resistant or moderately resistant to both races, however, seven spring wheat cultivars showed susceptibility to both races 1 and 5. Continued cultivation of wheat cultivars susceptible to both races could play a role in the establishment and development of new races. Continuous germplasm enhancement and periodically monitoring *Ptr* population can help in better TS management.

Keywords: Tan spot; *Drechslera tritici-repentis*; Wheat; Host-selective toxins; Ptr ToxA; Ptr ToxB; Race 1; Race 5

Introduction

Tan spot caused by an ascomycete fungus *Pyrenophora tritici-repentis* (Died.) Drechs. (anamorph: *Drechslera tritici-repentis* Shoe.) is an important foliar disease of wheat in the US Northern Great Plains (NGP) [1]. In addition to wheat, a primary and economical host, the fungus has a wide host range that includes cereals such as barley, oat, rye and many non-cereal grasses [2-4]. The fungus produces oval-shaped tan necrotic spots with a chlorotic halo and a small black spot in the center of the leaves of susceptible cultivars. Tan spot can cause a significant yield loss up to 50%, attributed to low 1000 KWT, shriveled kernels, decrease in leaf area for photosynthesis, and number of grains/spike [5-7]. The fungus overwinters from one growing season to the next on infested wheat residue primarily in the form of sexual fruiting bodies "Pseudothecia" in the region [8]. These fruiting bodies produce ascospores under cool and wet conditions in spring, which serve as a primary source of inoculum for disease initiation. Further, these sexual fruiting bodies provide ample opportunities for the fungal sexual recombination that could lead to change in the pathogen virulence, especially when two distinct races reside on the residue of a cultivar susceptible to multiple races. Virulence variation in *P. tritici-repentis* has been observed based on an isolate's ability to produce necrosis and/or chlorosis symptoms on appropriate wheat differential genotype. So far, the isolates from wheat and alternative host plants have been grouped into eight races [2,9-11] and race 1 was the most prevalent race observed in Africa, Asia, Europe, North and South America [2,9,11-15]. Further, *P. tritici-repentis* produces three host-selective toxins, *Ptr ToxA*, *Ptr ToxB,* and *Ptr ToxC*, which are associated with necrosis and chlorosis symptoms in toxins sensitive wheat genotypes [11,16-18]. At present gene-specific molecular markers are available for *Ptr ToxA* (associated with necrosis symptom) and *Ptr ToxB* (chlorosis) and can be used to determine if an isolate carries these genes. However, presence or absence of *Ptr ToxA* and *Ptr ToxB* genes in the isolate allows us to potentially categories it into group 1 (races 1, 2, 5, 7): which carries one or both of these genes or group 2 (race 3 or 4): which lacks these genes or group 3: as new race(s) [11]. Lack of availability of race 3 (*Ptr ToxC*) and race 4 specific molecular markers still warrants for phenotyping the isolates on tan spot wheat differential set to confirm the race identity [11].

Wheat is the second most important crop in SD planted on 2.7 million acres with the total production of 103.16 million bushels [19]. Wheat productivity is impacted by various pests and diseases, including TS in the US NGP. TS alone can cause 5% to 29% yield reduction in South Dakota depending on the inoculum level and cultivar susceptibility under suitable environment for disease development [5]. Excessive use of fungicide could make the pathogen fungicide resistant, reduce its utility and ultimately impact crop productivity [20,21]. In this scenario, development of TS resistant cultivars seems to be the best approach for disease management, especially for a marginally profitable crop like wheat.

A comprehensive knowledge of physiological variation in the pathogen population and reaction of commonly grown cultivars in the region is essential for developing TS management strategies including durable disease resistant cultivars. Limited information is available on *P. tritici-repentis* race structure prevalent in wheat and reaction of wheat cultivars to TS in South Dakota. The objectives of this study were to 1) characterize race structure of *P. tritici-repentis* isolates recovered from wheat in South Dakota and 2) evaluate the current status of TS

***Corresponding author:** Ali S, Department of Agronomy, Horticulture, and Plant Science (AHPS), South Dakota State University, Brookings, SD 57007, USA
E-mail: shaukat.ali@sdstate.edu

resistance in wheat cultivars widely grown in South Dakota and the region. The information obtained in this study would help the regional wheat breeders and pathologists in the development of TS management strategies.

Materials and Methods

Leaf samples collection and recovery of *P. tritici-repentis* isolates

Leaf samples with tan spot symptoms were collected from 55 locations during the 2012 (n=18), 2013 (n=19), and 2014 (n=18) growing seasons in South Dakota. About 15 to 20 leaves were randomly collected from each location. The samples were collected from both South Dakota State University Experimental Research Stations and commercial field plots when the crop was mostly at milk-stage (Feekes 10.54-11.1). The leaves were cut into about 2 cm long segments with at least one TS lesion/segment, placed in paper bags (one sample/bag), and stored in a refrigerator until used. *P. tritici-repentis* isolates were recovered from the leaf segments following the method described by Ali and Francl [22]. In short, 40-50 segments randomly selected of each sample were placed in 9 cm Petri plates with three layers of moist filter paper. Plates were incubated under fluorescent light for 24 h at room temperature (22°C ± 1°C) and 24 h in dark at 16°C for spores formation. The segments were examined under a stereoscope for *P. tritici-repentis* conidia. Thereafter 10-15 single conidia were collected with a flamed steel needle and transferred individually onto V8PDA plates [23]. The cultures were grown for 5-6 days in the dark and then stored at -20°C following the procedure of Jodhal and Francl [24] until they were characterized for their race structure. In total 569 isolates, including 8 isolates from Nebraska, were recovered in 2012 (n=138), 2013 (n=176) and 2014 (n=255) over 3-year period.

Genotypic characterization of *P. tritici-repentis* isolates for *Ptr ToxA* and *Ptr ToxB* genes

Fresh cultures of all 569 *P. tritici-repentis* isolates were obtained by growing them from their stored cultures at -20°C individually on V8PDA medium for 5 days in the dark. Each isolate's mycelia were removed from the agar surface with a sterile scalpel and placed in a 2 ml micro-centrifuge tube, and dried overnight in a water bath at 37°C. Thereafter, the mycelia were ground into a fine powder using a first prep machine Retsch MM 301 (Retsch., Clifton, NJ). DNA from all the isolates was recovered by following the method of Moreno et al. [25]. The DNA concentration was adjusted to 25 ng/µl using a Nano drop (Counterpane Inc. Tacoma, WA) and run on a 0.8% agarose gel to verify the DNA quality. The genotype of the *P. tritici-repentis* isolates for *Ptr ToxA* and *Ptr ToxB* genes was determined using the *Ptr ToxA* and *Ptr ToxB* genes specific primers developed by Andrie et al. [26]. The conformity of the isolates of *P. tritici-repentis* was further verified by using two *P. tritici-repentis* mating type genes (*MAT1-1* and/or *MAT1-2*) specific primers [27]. PCR reactions were performed in total 20 µl volume (2 µl genomic DNA @25 ng/µl, 0.8 µl of each primer (10 mM), 0.5 µl dNTP (200 µM), 2 µl 10x thermophol buffer, 0.2 µl 10 U/ml Taq DNA Polymerase and 13.7 µl of molecular biology grade water). PCR reaction was conducted in a S-1000 thermal cycler (Bio-Rad, Hercules, CA) using amplification steps of 94°C for 1 min, followed by 30 cycles of 94°C for 45 sec, 55°C for 30 sec and 72°C for 1 min with a final extension of 72°C for 7 min. We pooled PCR products from housekeeping genes (*MAT1-1 or MAT1-2*) with Ptr ToxA or Ptr ToxB specific PCR products as a positive amplification control for each isolate. The amplified products were electrophoresed on 1.5% agarose gel and scored with reference to 1 Kb ladder (New England Biolabs, Ipswich, MA). *P. tritici-repentis* isolates, Pti2 (race 1) and DW7 (race 5) were used as positive controls for *Ptr ToxA* and *Ptr ToxB* genes, respectively.

Phenotypic characterization of *P. tritici-repentis* isolates

Fresh cultures of randomly selected 134 of 569 isolates genotyped for *Ptr ToxA* and *Ptr ToxB* genes were obtained by following the procedure described by Ali and Francl [22]. In short, all 134 isolates cultures were initiated by plating their frozen dry plugs on fresh V8PDA Petri plates. The plugs of each isolate were placed in 3 plates (one plug/plate) and were incubated in dark for 5 days. The plates were flooded with distilled sterile water (DSW) and the mycelial growth was flattened with a sterile test tube. The excess water was decanted from the plates and then placed under fluorescent light for 24 h at room temperature (22°C ± 1°C) and for 24 h in dark at 16°C for spore production. The conidia were harvested with sterile disposable plastic inoculating loops (Fisher Scientific, Asheville, NC) by adding about 30 ml of DSW into each plate. The final spore concentration was adjusted to 2500 spores/ml prior to inoculations. *P. tritici-repentis* isolates, Pti2 (race 1) and DW7 (race 5) were included as positive checks for validation of inoculation procedure and race identification. Two-week-old seedlings of all 4 differentials genotypes Glenlea, 6B365, 6B662, and Salamouni [11] were raised in 3 × 9 cm plastic containers (Stuewe & Sons, Inc. Tangent, OR) filled with Sunshine Mix 1 (Agawam, MA) with three replications inoculated individually with all 134 isolates by spraying their spore suspension with a handheld CO_2 pressurized sprayer (Preval, Coal City, IL). Three seedlings/container were maintained throughout the experimentation. The seedlings were placed in a humidity chamber at 100% humidity for 24 h for enhancing the chances of fungal infection. Thereafter, the seedlings were moved to a greenhouse bench at South Dakota State University (SDSU) for seven days for symptom development. The isolates were grouped under appropriate race based on their ability to produce necrosis and chlorosis symptoms on the differentials genotype [11].

Evaluation of wheat cultivars for their reaction to tan spot using race 1 and 5 and *Ptr ToxA*

Two-week-old seedlings of 41 (hard red spring=29 and hard red winter=12) wheat cultivars which have been planted as popular varieties in South Dakota (Tables 1a and 1b) were raised in plastic containers as described under section A. In addition, four spring wheat advanced lines were included in the experiment. All cultivars seed was kindly provided by SDSU spring and winter wheat breeding programs. The purpose of the cultivars evaluation was to know if any of the cultivars are susceptible to both race 1 and 5 or race 5 that may play any role in the establishment of *P. tritici-repentis* race 5. Tan spot wheat differential genotypes Glenlea, 6B365, 6B662, and Salamouni were included in the experiment as checks for inoculation procedure validation and disease rating. Seedlings of all 45 wheat genotypes were inoculated individually using race 1 and race 5 spore's suspension. Nine seedlings (3 seedlings/container) of each cultivar were inoculated and the experiment was repeated once. The inoculated seedlings were rated for their reaction based on the 1-5 rating scale [23].

All 45 wheat genotypes were also evaluated for Ptr ToxA reaction by following the procedure of Faris et al. [28]. Briefly, three fully expanded second leaves of three two-week-old seedlings of all 45 genotypes was infiltrated with Ptr ToxA (10 µg/ml), using a needleless syringe and infiltrated leaf area was marked by a non-toxic permanent marker. The seedlings were rated as sensitive/insensitive based on presence/absence of necrosis, respectively. Wheat differential genotypes Glenlea (Ptr ToxA sensitive) and Salamouni (insensitive) were included in the experiment as positive controls. The experiment was repeated once. Ptr ToxA was kindly provided by Dr. S. Mienhardt, Department of Plant Pathology, North Dakota State University, Fargo, North Dakota.

Number	Cultivar/Line	*P. tritici-repentis* races				Ptr ToxA
		Race 1		Race 5		
		Lesion type	Reaction	Lesion type	Reaction	
1	Advance	1.8	MR	2.3	MR	+
2	Barlow	1.5	R	1.9	MR	-
3	Breaker	1.0	R	2.0	MR	+
4	Brick	1.0	R	2.2	MR	+
5	Briggs	2.0	MR	2.6	MR	+
6	Chris	1.0	R	4.3	S	-
7	Elgin-ND	1.1	R	3.8	MS	+
8	Faller	1.0	R	3.6	MS	-
9	Forefront	3.7	MS	4.4	S	+
10	LCS Albany	1.1	R	1.4	R	+
11	LCS Breakaway	4.0	S	4.0	S	-
12	LCS Powerplay	1.2	R	1.8	MR	+
13	Linkert	2.0	MR	4.1	S	-
14	Mott	3.5	MS	1.8	MR	-
15	Norden	3.7	MS	4.0	S	+
16	Oxen	3.1	MS	1.1	R	+
17	Prosper	2.0	MR	3.6	MS	-
18	RB07	1.7	MR	1.3	R	+
19	Rollag	1.1	R	3.9	S	-
20	Russ	3.9	S	4.0	S	+
21	Sabin	1.3	R	1.0	R	+
22	Samson	1.1	R	1.2	R	-
23	SD4189	3.8	MS	4.0	S	+
24	SD4215	1.9	MR	1.0	R	+
25	Select	4.0	S	3.8	MS	+
26	SY Soren	1.0	R	1.0	R	-
27	SY Rowyn	3.0	MS	1.2	R	+
28	Transverse	3.3	MS	2.1	MR	+
29	Vantage	1.7	MR	1.3	R	-
30	Velva	1.2	R	3.8	MS	-
31	WB Mayville	4.0	S	1.1	R	+
32	SD4148	1.6	MR	3	MS	+
33	SD4011	1.1	R	2.1	MR	+

Table 1a: Reaction of hard red spring wheat cultivars/lines to *P. tritici-repentis* race 1, race 5, and Ptr ToxA (Tan spot).

Number	Cultivar/Line	*P. tritici-repentis* races				Ptr ToxA
		Race 1		Race 5		
		Lesion type	Reaction	Lesion type	Reaction	
1	Alice	3.0	MS	1.1	R	+
2	Expedition	3.3	MS	1.7	MR	+
3	Freeman	1.6	MR	1.0	R	-
4	Grainfield	3.0	MS	1.0	R	+
5	Ideal	4.0	S	1.0	R	+
6	Lyman	1.9	MR	1.0	R	+
7	Millennium	2.8	MR	1.2	R	+
8	Overland	3.2	MS	1.2	R	+
9	Redfield	2.1	MR	1.0	R	-
10	SY Wolf	1.0	R	1.0	R	-
11	WB Matlock	3.8	MS	1.3	R	+
12	Wesley	2.2	MR	1.0	R	-

Table 1b: Reaction of hard red winter wheat cultivars to *P. tritici-repentis* race 1, race 5, and Ptr ToxA (Tan spot).

Results and Discussion

Molecular evaluation of *P. tritici-repentis* isolates for *Ptr ToxA* and *Ptr ToxB* genes

We confirmed all 569 isolates were *P. tritici-repentis* by mating type (*MAT1-1 or MAT1-2*) primers. Further, amplification of *Ptr ToxA* (585 bp) and *Ptr ToxB* (295 bp) genes specific PCR products from isolates Pti2 (race 1) and DW7 (race 5) validated the success of PCR assay (Figure 1). All isolates were then scored for presence or absence of the *Ptr ToxA* and *Ptr ToxB* genes and the isolates were divided into three groups and that include: 1) isolates carrying *Ptr ToxA* gene, 2) isolates carrying *Ptr ToxB* gene, and 3) the isolates without *Ptr ToxA* and *Ptr ToxB* genes (Figure

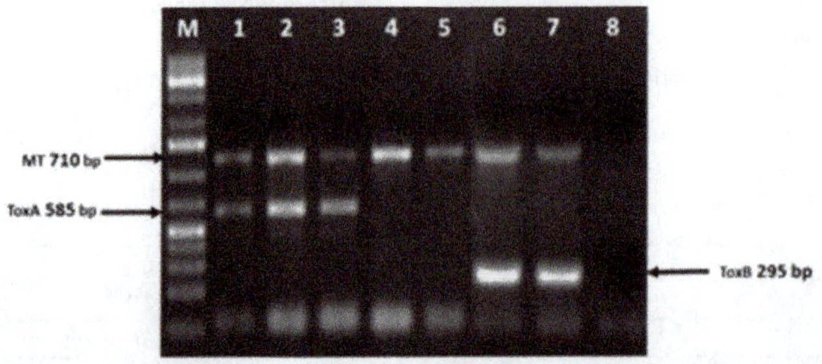

Figure 1: Characterization *P. tritici-repentis* isolates based on *ToxA*, *ToxB* primer amplification and Mating type (*MT*) genes. Lane 1-3: *Ptr ToxA* gene from unknown isolates SD13-101, SD12-S26-SN27-P2, and Pti2 (race1 positive control); Lane 4-5: isolates 12-S11-SN11-P7.2 and 12-S11-SN11-P7.3 without *ToxA/ToxB*; Lane 6-7: isolates SD13-103 (unknown) and DW-7 (race 5 positive control) amplified *Ptr ToxB* and Lane 8: water (as negative control).

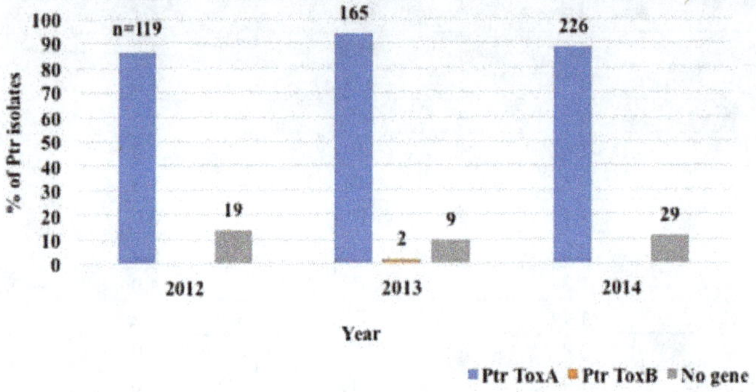

Figure 2: Genotypic characterization of *P. tritici repentis* isolates for *Ptr ToxA* and *Ptr ToxB* genes prevalent in wheat in South Dakota in 2012-2014.

2). The isolates (n=502, 89.5%) belonging to group 1 can potentially be race 1, 2, and 8 as these races carry *Ptr ToxA* gene, however the lack of *Ptr ToxB* gene in all these isolates eliminates the possibility of race 8 [11]. In addition, all eight isolates from Nebraska also amplified *Ptr ToxA* gene and were placed in group 1. Further, the frequency of isolates with *Ptr ToxA* gene was higher in all three years, 2012 (n=119, 86.2%), 2013 (n=165, 93.8%), and 2014 (n=226, 88.3%) compared to isolates with *Ptr ToxB* gene (Figure 2). The presence of *Ptr ToxA* gene at high frequency in the evaluated isolates shows predominance of race 1 in South Dakota which is similar to *P. tritici-repentis* race structure reported from different countries [2,13,14,29]. Only two (0.4%) isolates that amplified *Ptr ToxB* gene (group 2) could possibly be of either race 5 (*Ptr ToxB*) or race 6 (*Ptr ToxB* and *Ptr ToxC*) depending on the phenotypic reaction of these isolates (chlorosis on differential genotype 6B662- race 5 or chlorosis on both genotypes 6B662 and 6B365-race 6). Ours results suggest that though only two isolates with *Ptr ToxB* gene were observed from one location in 2013, there is prevalence of either race 5 or race 6 in South Dakota. Though earlier we have reported race 5 in North Dakota [30], this is the first time we observed race 5 in South Dakota suggesting that the race is expanding and this could aggravate the challenges in tan spot management in the region.

Lack of molecular markers for *Ptr ToxC* gene that distinguishes race 1 (*Ptr ToxA* and *Ptr ToxC*) from 2 (*Ptr ToxA*); race 3 (*Ptr ToxC*) from race 4 (carries neither of the three toxins genes); and race 5 (*Ptr ToxB*) from 6 (*Ptr ToxB* and *Ptr ToxC*) is a bottleneck in determining the isolate's absolute race.

We categorized 57 (10%) isolates into group 3 that were devoid of both *Ptr ToxA* and *Ptr ToxB* genes and potentially be of race 3 and 4 [11], or a new unknown virulent race similar to those reported in previous studies [9,26,31,32]. The isolates without *Ptr ToxA* and *PtrToxB* genes were observed in all three years 2012, (19 isolates, 13.8%), 2013 (9 isolates, 5.1%) and in 2014 (29 isolates, 11.7%) but they varied in their proportion (Figure 2). Our results indicate that majority (n=502) of the isolates collected over the three-year period from South Dakota carry *Ptr ToxA* gene that validate reports indicating prevalence of *Ptr ToxA* carrying isolates in abundance in South Dakota [12,31,32]. Availability of *P. tritici-repentis* whole genome information [33] may help in accelerating the development of molecular markers for race 3 and 4 and minimize the dependency on phenotyping which is a time-consuming procedure and requires a well-trained person for observing the isolates phenotypic reaction. However, presently the exact race identification of the isolates is only possible through their phenotyping on the differential set [11,34].

Phenotypic evaluation of *P. tritici-repentis* isolates for race identification

A subset of randomly chosen 134 of 569 *P. tritici-repentis* isolates studied for *Ptr ToxA* and *Ptr ToxB* genes were further characterized for their race structure on tan spot wheat differential set. The 134 isolates were grouped into races 1, 2, 4, and 5 (Figure 3) based on phenotypic reaction on the differential set indicating the prevalence of diverse population of *P. tritici-repentis* in wheat in South Dakota. As expected Pti2 (race 1) produced necrosis and chlorosis and DW7 (race 5) produced chlorosis on appropriate differential genotypes [11] validating the inoculation procedure. Among these four races, the majority of the isolates from 2012 (40%), 2013 (84%), and 2014 (80%) were identified as race 1. However, during 2012 in addition to race 1, a significant proportion of isolates were identified as an unknown race (Figure 3). Prevalence of race 1 at high frequency in the state can be expected as most of the cultivars grown in the South Dakota exhibit some level of susceptibility to tan spot race 1 (Tables 1a and 1b). Also, our phenotyping results conform with the previous studies that reported race 1 as the most prevalent race in many countries including

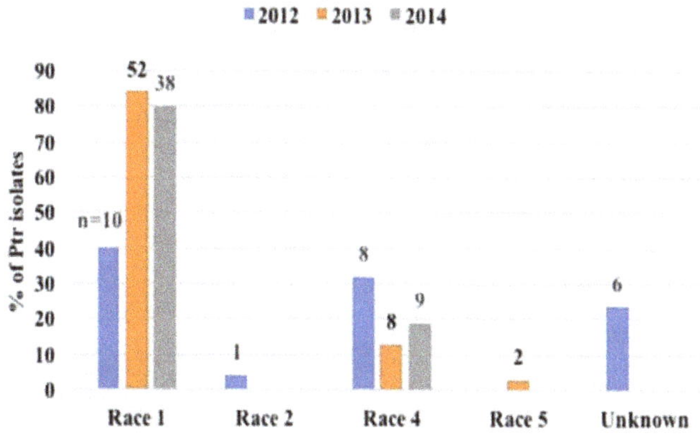

Figure 3: Race characterization of *P. tritici-repentis* isolates prevalent on wheat in South Dakota in 2012-2014.

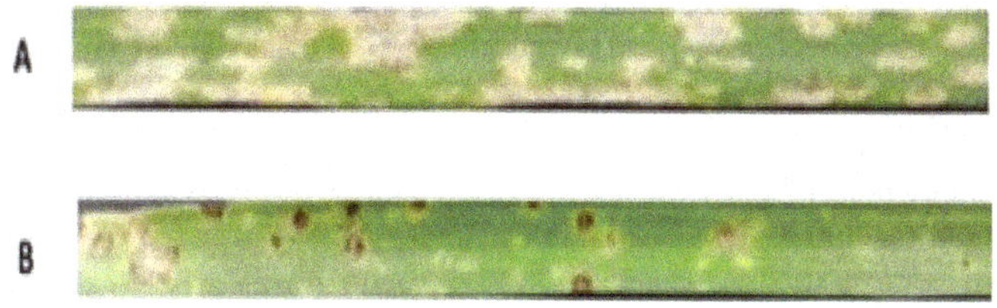

Figure 4: Reaction of Ptr in wheat. (A) Isolate SD12-S11-P7.2 (Lacks in *Ptr ToxA* gene) produced necrosis on Glenlea (B) no chlorosis on 6B365.

the US [2,9,11,14,29,35]. One reason for dominance of race 1 over other races observed in various countries may be deployment of tan spot susceptible cultivars to race 1 on a larger acreage, as documented from two independent studies where screening with multiple *P. tritici-repentis* races identified majority of genotypes were susceptible to race 1 [36,37]. However, predominance of race 1 does not completely eliminate the possibility of other races and their prevalence at low frequency has been reported [2,10,11,13,29].

Further, 25 (18.65%) isolates were identified as race 4 and were the second most prevalent race in all three years, 2012 (32%), 2013 (13%), and 2014 (19%) (Figure 3). We observed a higher proportion of race 4 (13% to 32%) in our study as compared to around 5% reported in some earlier studies [2,10,13,29]. A higher prevalence of *P. tritici-repentis* race 4 has also been reported from the fungal alternative hosts like non-cereal grasses, smooth bromegrass, intermediate wheatgrass, western wheatgrass etc. [2] and rye [38], in the US NGP. In rye, 78% of 103 isolates characterized for their race structure were identified as race 4 [38]. Recovery of race 4 in a slight higher frequency in wheat in this study could be the race 4 isolates transferring from rye and establishing as a weak pathogen on wheat. With rye being mostly planted adjacent to wheat field plots in the NGP, it is highly likely that race 4 isolates from rye may survive on wheat crop.

Only 1.4% of the 134 isolates we phenotypically evaluated on differential set were race 5 (Figure 3). Although race 5 was detected at very low frequency and only in 2013, however, this is the first report of prevalence of race 5 in SD. Our leaf sampling was random in all three years, however, no detection of race 5 in 2012 and 2014 could be due to the majority of leaf samples likely collected from race 5 resistant cultivars. Only one isolate of the 134 isolates characterized was identified as race 2. We categorized 31 isolates that lacked *Ptr ToxA* and *Ptr ToxB* genes to be either of race 3 or race 4, however, phenotypic characterization showed six isolates did not fit under any of the 8 races. These isolates produced necrosis (Figure 4) like race 2, however, lacked Ptr ToxA so were categorized as unknown races. Recovery of these types of *P. tritici-repentis* isolates is not surprising as they have been noted in other independent studies as well [9,15,26,31]. Further, the occurrence of necrosis producing isolates without *Ptr ToxA* gene on the *Ptr ToxA* sensitive wheat genotype indicates presence of additional toxin(s) responsible for necrosis symptoms. These isolates further need to be investigated for their mechanism of host-pathogen interaction.

Reaction of wheat cultivars to tan spot (*P. tritici-repentis* race 1, race 5, and *Ptr ToxA*)

In this study, we evaluated 45 wheat cultivars (hard red spring=33 and hard red winter=12) including four advanced lines to tan spot using race 1 and race 5 to determine the potentially prevalent races in the region based on their susceptibility. Wheat cultivars belonging to both spring and winter wheat classes exhibited diverse reaction, ranging from susceptible to resistant, to both races (Tables 1a and 1b). In general, a higher number of spring wheat cultivars exhibited susceptibility to race 1 and race 5 as compared to winter wheat cultivars. Seven of the 33 spring wheat cultivars/lines, Forefront, LCS Breakaway, Norden, Russ, Select, and SD4189 exhibited susceptibility to both races 1 and 5 (Table 1a). Fourteen cultivars (42%) showed susceptible to moderately susceptible reaction to race 5, whereas 10 cultivars (30%) were susceptible to race 1. Wheat cultivars susceptible to both race 1 and 5 could play an important role in the development of new virulent race(s) such as race 6 (a combination of *Ptr ToxB* and *Ptr ToxC*), race 7 (*Ptr ToxA* and *Ptr ToxB*), and race 8 (*Ptr ToxA, Ptr ToxB, and Ptr ToxC*) as they are the combination of race 1 (*Ptr ToxA* and *Ptr ToxC*) and race 5 (*Ptr ToxB*). This is highly possible to occur, as the fungus *P. tritici-repentis* is residue-borne in nature and present year-round on wheat residue due to minimum tillage practices in South Dakota and the region. Presence of two different races (1 and 5) in the same leaf and/or in different plants within the same field provides ample opportunities to the fungus for sexual reproduction and lead to evolution of a new race.

In contrast to spring wheat, about 50% (n=6) winter wheat cultivars exhibited susceptibility to tan spot race 1 and developed lesion type 3-3.8 (moderately susceptible) except Ideal (lesion type 4; susceptible). All winter wheat cultivars were moderately resistant to resistant to race 5 as they developed lesion type 1-2 (Table 1b). The majority of the winter wheat cultivars exhibited resistance to tan spot (race 5) which minimizes their role in the race 5 establishment in the state.

These results indicate that spring wheat cultivars that are susceptible to both races may be a potential source for survival of both races 1 and 5 in the region. Though race 5 was detected from only one location in 2013, monitoring the fungal isolates on wheat cultivars susceptible to both race 1 and 5 or only race 5 may provide better information on the race 5 prevalence in South Dakota.

All 45 wheat cultivars/lines varied in their reaction to Ptr ToxA as they developed either necrosis (sensitive) or no necrosis (insensitive). We found a very weak correlation (r=0.24) between the susceptibility-sensitivity and resistance-insensitivity to the toxin in spring wheat, however a good correlation (r=0.72) was observed in winter wheat (Tables 1a and 1b). Nine spring wheat cultivars were susceptible; whereas two cultivars were susceptible to race 1 but insensitive to the toxin. Also, 10 cultivars were resistant to race 1 and insensitive to the toxin; whereas 12 cultivars showed resistance to race 1 but sensitivity to Ptr ToxA (Table 1a). For example, spring wheat cultivar 'LCS Breakway' showed susceptibility (lesion type 4) to race 1 (Ptr ToxA producers) but exhibited insensitivity to the toxin; whereas cultivar 'Select' developed susceptible reaction (lesion type 4) to race 1 and exhibited sensitivity to the toxin. Six winter wheat cultivars that developed a susceptible reaction to race 1 were also developed necrosis (sensitive) to Ptr ToxA. Four cultivars showed resistance to race 1 were also insensitive to the toxin; whereas two winter wheat cultivars exhibited a resistant reaction to race 1 but were sensitive to Ptr ToxA. Our results are in accordance with previous studies where all 4 possible combinations of tan spot susceptibility and Ptr ToxA sensitivity (susceptibility-sensitivity, susceptibility-insensitivity; resistant-sensitivity, and resistant-insensitivity) were observed in wheat genotypes [9,16,39,40].

The results of this study indicate that a diverse population of *P. tritici-repentis* exists on wheat in South Dakota with race 1 being the most prevalent. In addition, we observed race 5 for the first time in South Dakota and identified wheat cultivars susceptible to race 1 and 5. Planting of such cultivars on a large acreage coupled with favorable climatic conditions might play an important role in the development of more virulent races. Further identification of isolates lacking *Ptr ToxA* gene can help in understanding the mechanism of pathogenicity/virulence and help in development of better TS management strategies. Our results suggest that screening of germplasm against both races 1 and 5 should be performed prior to using them in the development of tan spot resistant cultivars and regular monitoring of the pathogen population in the region can lead to better management of leaf spot diseases like tan spot.

References

1. Hosford RM (1982) Tan spot - Developing knowledge 1902-1981. In: Hosford RM (Ed). Tan spot of wheat and related diseases workshop. Fargo ND: North Dakota Agricultural Experimental Station, North Dakota State University, Fargo, N.D., USA.
2. Ali S, Francl LJ (2003) Population race structure of *Pyrenophora tritici*-repentis prevalent on wheat and non-cereal grasses. Plant Dis 87: 418-422.
3. Krupinsky JM (1992) Grass hosts of *Pyrenophora tritici* repentis. Plant Dis 76: 92-95.
4. Sprague R (1950) Diseases of cereals and grasses in North America. Ronald Press Co., New York, USA.
5. Buchenau GW, Smolik JD, Ferguson MW, Cholick FA (1983) Wheat diseases in South Dakota. Ann Wheat Newsletter 29: 147.
6. Shabeer A, Bockus WA (1988) Tan spot effects on yield and yield components relative to growth stage in winter wheat. Plant Dis 72: 599-602.
7. Sharp EL, Sally BK, McNeal FH (1976) Effect of Pyrenophora wheat leaf blight on the thousand-kernel weight of 30 spring wheat cultivars. Plant Dis 60: 135-138.
8. De Wolf ED, Effertz RJ, Ali S, Francl LJ (1998) Vistas of tan spot research. Can J Plant Pathol 20: 349-370.
9. Ali S, Gurung S, Adhikari TB (2010) Identification and characterization of novel isolates of *Pyrenophora tritici*-repentis from Arkansas. Plant Dis 94: 229-235.
10. Lamari L, Bernier CC (1989b) Virulence of isolates of *Pyrenophora tritici*-repentis on eleven wheat cultivars and cytology of the differential host reaction. Can J Plant Path 11: 284-290.
11. Lamari L, Srelkove SE, Yahyaoui A, Orabi J, Smith RB (2003) The identification of two new races of *Pyrenophora tritici*-repentis from the host center of diversity confirms a one to one relationship in tan spot of wheat. Phytopathology 93: 391-396.
12. Aboukhaddour R, Turkington TK, Strelkov SE (2013) Race structure of *Pyrenophora triciti*-repentis (Tan spot of wheat) in Alberta, Canada. Can J Plant Pathol 35(2): 256-268.
13. Benslimane H, Lamari L, Benbelkacem A, Sayoud R, Bouznad Z (2011) Distribution of races of *Pyrenophora tritici*-repentis in Algeria and identification of a new virulence type. Phytopathol Mediterr 50: 203-211.
14. Gamba FM, Strelkov SE, Lamari L (2012) Virulence of P*yrenophora tritici*-repentis in the southern cone region of South America. Can J Plant Pathol 34: 545-550.
15. Postnikova EN, Khasanov BA (1998) Tan spot in Central Asia. Helminthosporium Blights of Wheat: Spot Blotch and Tan Spot. In: Duveiller E, Dubin HJ, Reeves J, McNab A (Eds). CIMMYT (International Maize and Wheat Improvement Center), Mexico, D. F., Mexico.
16. Abdullah S, Sehgal SK, Jin Y, Turnipseed B, Ali S (2017) Insights into tan spot and stem rust resistance and susceptibility by studying the pre-green revolution global collection of wheat. Plant Pathol J 33: 125-132.
17. Ciuffetti LM, Manning VA, Pandelova I, Betts MF, Martinez JP (2010) Host-selective toxins, Ptr ToxA, and Ptr ToxB, as necrotrophic effectors in the *Pyrenophora tritici*-repentis-wheat interaction. New Phytologist 187: 911-919.
18. Friesen TL, Rasmussen JB, Kwon CY, Ali S, Francl LJ, et al. (2002) Reaction of Ptr ToxA-insensitive wheat mutants to *Pyrenophora tritici*-repentis race 1. Phytopathology 92: 38-42.
19. National agricultural statistics service (NASS) (2016) Crop production (August 2016), Agricultural Statistics Board, United States Department of Agriculture (USDA).
20. FRAC (Fungicide Resistance Action Committee) (2002) http://www.frac.info/frac/index.htm (18 December 2006)
21. Sierotzki H, Frey R, Wullschleger J, Palermo S, Karlin S, et al. (2007) Cytochrome b gene sequence and structure of *Pyrenophora teres* and *Pyrenophora tritici*-repentis and implication for QoI resistance. Pest Manag Sci 63: 225-233.
22. Ali S, Francl LJ (2001) Recovery of *Pyrenophora tritici*-repentis from barley and reaction of 12 cultivars to five races and two host-specific toxins. Plant Dis 85: 580-584.
23. Lamari L, Bernier CC (1989) Evaluation of wheat for reaction to tan spot (*Pyrenophora tritici*-repentis) based on lesion type. Can J Plant Pathol 11: 49-56.
24. Jordahl JG, Francl LJ (1992) Increase and storage of cultures of *Pyrenophora tritici*-repentis. Advances in tan spot research. In: Francl LJ, Krupinsky JM, McMullen MP (Eds). North Dakota Agric. Exp. Stn., Fargo.
25. Moreno MV, Stenglein SA, Blatti PA, Perollo AE (2008) Pathogenic and molecular variability among isolates of *Pyrenophora tritici*-repentis causal agent of tan spot of wheat in Argentina. Eur J Plant Pathol 122: 239-252.
26. Andrie RM, Pandelova I, Ciuffetti LM (2007) A combination of phenotypic and genotypic characterization *Pyrenophora tritici*-repentis race identification. Phytopathology 97: 694-701.

27. Lepoint P, Renard ME, Legrève A, Duveiller E, Maraite H (2010) Genetic diversity of the mating type and toxin production genes in *Pyrenophora tritici*-repentis. Phytopathology 100: 474-483.

28. Faris JD, Anderson J, Francl LJ, Jordahl JG (1996) Chromosomal location of a gene conditioning insensitivity in wheat to a necrosis-inducing culture filtrate from *Pyrenophora tritici*-repentis. Phytopathology 86: 459-463.

29. Sarova J, Hanzalova A, Bartos P (2005) Races of *Pyrenophora tritici*-repentis in the Czech Republic. Acta Agrobotanica 58:73-78.

30. Ali S, Francl LJ, DeWolf ED (1999) First report of *Pyrenophora tritici*-repentis race 5 from North America Plant Dis 83: 591.

31. Abdullah S, Sehgal SK, Ali S, Zilvinas L, Mariana I, et al. (2017) Characterization of *Pyrenophora tritici*-repentis (tan spot of wheat) races in Baltic States and Romania. Plant Pathol J 33: 133-139.

32. Mironenko NV, Baranova OA, Kovalenko NM, Mikhalova LA (2015) Frequency of *ToxA* gene in North Caucasian and North-West Russian populations of *Pyrenophora tritici*-repentis. Mikol Fitopatol 49: 325-329.

33. Friesen TL, Ali S, Klein KK, Rasmussen JB (2005) Population genetic analysis of a global collection of *Pyrenophora tritici*-repentis, causal agent of tan spot of wheat. Phytopathology 95: 1144-1150.

34. Birren B, Lander E, Galagan J, Nusbaum C, Devon K et al. (2007) Pyrenophora genome project. Broad Institute of Harvard and MIT.

35. Engle JS, Madden LV, Lipps PE (2006) Distribution and pathogenic characterization of *Pyrenophora tritici*-repentis and *Stagonospora nodorum* in Ohio. Phytopathology 96: 1355-1362.

36. Ali S, Singh PK, McMullen PM, Mergoum M, Adhikari TB (2008) Identification of new sources of resistance to multiple leaf spotting diseases in wheat germplasm. Euphytica 159: 167-179.

37. Singh PK, Mergoum M, Ali S, Adhikari TB, Elias EM, et al. (2006) Evaluation of elite wheat germplasm for resistance to tan spot. Plant Dis 90: 1320-1325.

38. Abdullah S, Sehgal SK, Glover KD, Ali S (2017) Reaction of global collection of rye (*Secale cereale* L.) to tan spot and *Pyrenophora tritici-repentis* races in South Dakota. Plant Pathol J.

39. Noriel AJ, Xiaochum S, Willimam B, Guihua B (2011) Resistance to tan spot and insensitivity to Ptr ToxA in wheat. Crop Sci 51: 1059-1067.

40. Oliver RP, Lord M, Rybak K, Faris JD, Solomon PS (2008) Emergence of tan spot disease caused by toxigenic *Pyrenophora tritici*-repentis in Australia is not associated with increased deployment of toxin sensitive cultivars. Phytopathology 98: 488-491.

Integrated Management of Faba Bean Black Root Rot (*Fusarium solani*) through Varietal Resistance, Drainage and Adjustment of Planting Time

Belay Habtegebriel[1]* and Anteneh Boydom[2]

[1]*Ethiopian Institute of Agricultural Research, Plant Protection Research Center, P.O. Box 37, Ambo, Ethiopia*
[2]*Ethiopian Institute of Agricultural Research, Holleta Agricultural Research Center, Holleta, Ethiopia*

Abstract

Black root rot, caused by *Fusarium solani*, is one of the most important diseases of faba bean causing up to 70% on farm yield loss in severe conditions in Ethiopia. Integrated management is the most promising alternative to control the disease. Three faba bean varieties, two drainage methods and three planting dates were evaluated under high disease inoculum pressure on a sick plot for two consecutive cropping seasons in a 3×2×3 factorial experiment. Results showed that the resistant variety Wayu was least affected by the disease (18.86% dead plants at harvest) when sown early on a flat bed. The susceptible variety Kassa was highly affected by the disease (89%) on a raised bed. All the three varieties performed well on raised beds (41.16%) than on flat beds (51.29% dead plants). Early or late planting resulted in significantly lesser overall percentage of dead plants (38.85%) and (44.23%) respectively as compared to optimum planting date (55.59%). Significant interactions were observed between variety and drainage method (P=0.003, F = 6.94, df= 2) which resulted in the least percentage of dead plants (21% and 20% on a flat and raised bed respectively) of variety Wayu compared to the moderately resistant variety Wolki (69% on flat and 36% on raised) and the susceptible variety Kassa (63% flat and 67% raised). The yield (g/plot)also varied significantly with variety wolki giving the highest yield (856 g/plot) followed by variety Wayu (883 g/plot). It is concluded that all the three factors are important for management of the disease but emphasis should be given to varietal resistance and use of raised beds. The two varieties Wolki and Wayu are recommended with raised beds for higher yield and variety improvement programs.

Keywords: Faba bean; Root rot; *Fusarium solani;* Drainage; Varietal resistance

Introduction

Root rots are the most important diseases of faba bean (*Vicia faba* L.) which are caused by *Fusarium solani (*black root rot), *Rhizoctonia solani* (wet root rot) and *Sclorotium rolfsi* (collar rot) in most growing areas [1-3]. In particular field, grown beans are highly destructed by the black root rot pathogen *F. solani* (Mart) Appel and Wollenw [4]. *Fusarium solani* is among the most commonly isolated soil borne pathogens causing root rot [5]. The disease occurs at early stage in the growing season causing seedling death [6]. In Ethiopia also, black root rot disease is a major biotic stress in faba bean growing areas [7,8] causing up to 70% yield loss on farmers' fields in severe conditions [9,10]. Up to 84% yield, losses have been reported on other pulses such as the common beans due to root rot caused by *Fusarium solani* [11]. It is the second most important disease of faba bean and when favorable conditions prevail and severe infections occur, faba bean black root rot (BRR) can cause complete crop loss [6]. Some studies have shown up to 45% yield loss on farmers' fields [12].

Black root rot almost entirely occurs in black clay soils, which are characterized by water logging that predisposes the plant to the disease [6]. Severe rotting causes black discoloration of the roots followed by death of the plant [12,13]. Other symptoms include elongated reddish lesions on primary roots, longitudinal cracks on the outer root and destruction of the tap root [14]. Optimum soil temperature for the development of *F.solani* is 25°C [15].

Management of root rots is a difficult task as most pathogens live near the rhizospher and survive for a long period by forming resistant structures [16]. Chemical control of faba bean root rot is neither efficient nor economical. Management options are mostly agronomic practices such as crop rotation, good soil drainage and use of disease free or fungicide treated seeds that may help reduce losses and there are no adequate control measures for *Fusarium* rots in the field [17]. Use of broad bed furrows (BBF) and resistant varieties have also been suggested for the management of the disease [18]. However, the disease remains difficult to control especially in black clay soils.

The objective of this study was therefore to establish the roles of managing faba bean black root rot through the integrated use of varietal resistance, soil drainage and adjustment of planting time.

Materials and Methods

Three faba bean varieties viz. a resistant variety Wayu, a moderately resistant variety Wolki and a susceptible check variety Kassa were used for the experiments. Two drainage methods (flat bed and raised bed) and three planting times (early, optimum and late) were used. Each treatment contained a variety a drainage method and a planting time. There were a total of 18 treatments (Table 1) which were set up in a 3×2×3 factorial arrangement. The experimental design was randomized complete block design (RCBD) with three replications. Plot size was 4.8 m^2 and the distance between blocks and plots was 1.5 m and 1 m respectively. The spacing between rows and plants was 40 cm and 5 cm respectively. The experiments were carried out in a well-developed sick plot containing high amount of root rot inoculum at Ambo Plant Protection Research Center (APPRC). The experiments were carried out in two consecutive main cropping seasons (2009 and 2010).

***Corresponding author:** Belay Habtegebriel, Ethiopian Institute of Agricultural Research, Plant Protection Research Center, P.O. Box 37, Ambo, Ethiopia
E-mail: belayhw@yahoo.com

Treatment	Variety	Drainage	Time of planting
T1	Wolki	Flat	Optimum (June 26)
T2	Wolki	Flat	Early (June 11)
T3	Wolki	Flat	Late (July 3)
T4	Wayu	Raised	Optimum (June 26)
T5	Wayu	Raised	Early (June 11)
T6	Wayu	Raised	Late (July 3)
T7	Kassa	Flat	Optimum (June 26)
T8	Kassa	Flat	Early (June 11)
T9	Kassa	Flat	Late (July 3)
T10	Wolki	Raised	Optimum (June 26)
T11	Wolki	Raised	Early (June 11)
T12	Wolki	Raised	Late (July 3)
T13	Wayu	Flat	Optimum (June 26)
T14	Wayu	Flat	Early (June 11)
T15	Wayu	Flat	Late (July 3)
T16	Kassa	Raised	Optimum (June 26)
T17	Kassa	Raised	Early (June 11)
T18	Kassa	Raised	Late (July 3)

Table 1: Treatment combinations used for field experiments on management of faba bean black root rot using variety, drainage and time of planting.

Data collection was done at emergence, seedling, podding, maturity and harvesting stages by rouging out dead plants. Data analysis and statistical comparison was conducted using the ANOVA procedure of the SAS software version 9.2. followed by mean separation.

Results

In the 2009 cropping season, the percentage of dead faba bean plants grown under high disease pressure from black root rot at harvest ranged from 36.17 for variety Wolki sown late on flat bed to 84.87 for Kassa sown early on a flat bed. Although the disease appeared and caused the plants to die, there were no significant differences among all the individual factors and treatments including interaction effects at all the growth stages (Table 2). In the following cropping season (2010) however, significant variations were observed at different growth stages. The percentage of dead plants at harvest ranged from the least (18.86%) for the resistant variety Wayu sown early on a flatbed to the maximum (89.03%) for the susceptible variety Kassa sown early on a raised bed. The percentage of dead plants varied significantly among the individual factors i.e. varieties, drainage methods and planting time (Table 3). All the three varieties performed well on raised beds (41.16% dead plants) than on flat beds (51.29% dead plants). Early or late planting resulted in significantly lesser overall percentage of dead plants (38.85%) and (44.23%) respectively as compared to optimum planting (55.59%). There was no significant difference between planting early or late in the season.

However, as shown in Figure 1, interaction effects were significant only between varieties and drainage methods (P=0.003, df=2, F=6.94). The least percentages of dead plants in this interaction were 19.88% followed by 21.13% recorded from variety Wayu sown on raised and flat beds respectively. The susceptible variety Kassa showed the highest overall percentage of dead plants (65.21%) while the resistant variety Wayu exhibited the least (20.50%) over all dead plants in this interaction.

In the 2009 crop season, the yield (g/plot) did not vary significantly except for the overall effect of varieties (P = 0.0464 F= 3.35, df = 2) (Table 4) and the three way interactions among variety, planting time and drainage method (P= 0.0251, F= 2.78 df= 6) (Table 5). Variety Wolki sown late on flat beds (855.87 g/plot) gave the highest yield followed by variety Wayu sown at optimum date on flat beds (882.73 g/plot). The least yield was obtained from variety Wolki sown on a flat bed at optimum time (54.03 g/plot).

In the 2010 crop season, yield was significantly affected by the

Treatment	Before emergence	Seedling	Podding	Maturity	Harvesting
T1	36.0	62.9	66.7	75.2	83.1
T2	40.1	65.6	65.6	65.6	68.2
T3	27.6	30.3	30.7	30.7	36.2
T4	35.7	68.9	68.9	69.3	74.3
T5	21.5	58.1	58.1	58.1	66.7
T6	28.1	41.7	41.7	41.7	43.0
T7	16.9	65.1	65.1	65.1	72.8
T8	35.5	68.0	68.0	72.1	84.9
T9	21.1	59.6	59.6	59.6	68.6
T10	29.2	64.9	64.9	64.9	67.3
T11	21.3	65.6	65.6	65.6	65.6
T12	29.2	70.4	70.4	70.4	74.6
T13	25.9	39.3	39.5	39.7	43.0
T14	26.8	51.3	51.3	51.3	54.4
T15	22.1	91.7	91.7	91.7	91.7
T16	21.9	61.6	61.6	61.6	71.1
T17	28.3	59.0	59.0	59.0	59.2
T18	42.1	65.8	65.8	65.8	76.5
DMRT (0.05%)	ns	ns	ns		ns

Table 2: Overall means of percentage of dead faba bean plants due to black root rot at different growth stages in 2009 crop season.

	Variety				
	Kassa	Wayu	Wolki	F	P
Mean	65.21**a**	20.5 **c**	52.96 **b**	36.80	<0.0001
	Drainage				
	Flat	Raised		5.32	0.027
Mean	51.29 **a**	41.16 **b**			
	Planting time				
	Early	Optimum	Late		
Mean	38.85 **b**	55.59 **a**	44.23**b**	5.04	0.011

Table 3: One way table of means of percentage of dead plants due to black root rot for three faba bean varieties, two drainage methods and three planting times at harvesting in 2010 crop season.

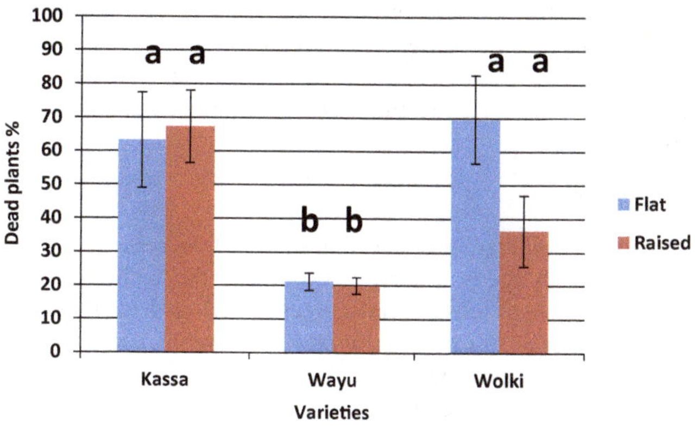

Figure 1: Variety X drainage interaction means ± standard error of percentage of dead plants due to black root rot for three faba bean varieties and two drainage methods at harvesting in 2010 crop season (P=0.003, F = 6.94, df= 2, LSD 5%).

Crop season	Variety					
		Kassa	Wayu	Wolki	F	P
2009	Mean	188.9 **b**	**497.4 a**	321.8.**a b**	3.35	<0.0464
	Variety					
2010		Kassa	Wayu	Wolki	F	P
	Mean	359.88 **c**	**844.51 a**	319.43. **b**	23.14	<0.0001

Table 4: One way table of means of yield (g/plot) for three faba bean varieties grown under disease pressure from black root rot in 2009 and 2010 crop seasons.

	Varieties						F	P
	Kassa		Wayu		Wolki			
Drainage method	Raised	Flat	Raised	Flat	Raised	Flat		
Planting time								
Early	200.10	62.57	267.80	615.20	289.96	258.10	2.78	0.025
Late	209.47	396.13	752.43	82.96	127.33	855.87		
Optimum	182.97	82.05	383.03	882.73	345.30	54.03		

Table 5: Three way table of overall mean yield (g/plot) of variety X drainage X planting time interaction of three faba bean varieties grown under black root rot disease pressure at harvesting in the 2009 crop season.

Crop growth stage	Crop season	Variety					
			Kassa	Wayu	Wolki	F	P
Before Emergence	2009	Mean	3.18 b	16.52 a	2.12 b	67.4	<0.0001
		Variety					
			Kassa	Wayu	Wolki	F	P
Seedling	2010	Mean	10.38b	17.51 a	6.29 b	13.06	<0.0001
		Drainage					P
Seedling	2010		Flat	Raised		5.15	0.029
		Mean	13.45 a	9.33 b			

Table 6: One-way table of means of percentage of dead plants of three faba bean varieties grown under disease pressure from black root rot before emergence and at seedling stage in 2009 and 2010 crop seasons.

overall main effect of varieties (P = <0.0001, F = 23.14, df = 2). There were no other individual or interaction effects of the factors on the yield. The highest yield (844.51 g/plot) was obtained from variety Wayu (Table 4).

The overall percentage of seeds which died before emergence significantly varied among the three varieties in the 2009 field experiments with no interaction effects. Highest pre-emergence seed death was observed on variety wayu (16.52%) (Table 6). Similarly, the overall percentage of dead plants at the seedling stage significantly varied among the three varieties and between flat and raised seed beds in the 2010 crop season. Except these, no significant variations were observed in the rest of the crop stages both in 2009 and 2010 crop seasons (Table 6).

Discussion

Since controlling black root rot disease of faba bean by chemical means is unlikely, the use of alternative control methods is indispensible. In this study, use of resistant varieties (either the moderately resistant variety wolki or the resistant variety wayu) in combination with raised beds for adequate drainage showed significant reduction in the number of dead plants in a sick plot. Use of good cultural practices such as use of resistant varieties, adjustment of planting time and adequate drainage is known to reduce root rot incidence and have been suggested by many authors [17,18]. Similarly, this study has shown that varietal resistance and drainage methods play an important role in reducing disease pressure from black root rot and increasing yield of faba bean in vertisols where the disease is most serious.

In the 2009 field experiments, no significant differences were observed among the different factors at all the growth stages except at the pre-emergence stage. This may be attributed to the environmental conditions which favored the high disease pressure. Moreover, Fusarium root rot resistance, being quantitative, is strongly affected by the environment [19], but in the second year, variations became visible. The resistance of the varieties also varied depending on the year and the crop growth stage. For example variety wayu exhibited significantly lesser resistance (higher number of dead seeds) before emergence and at seedling stage in 2009 and 2010 respectively, while it performed well once emerged compared to the susceptible variety kassa which showed significantly higher emergence percent and less number of dead plants at seedling stage. Faba bean root rot can infect seeds prior to emergence or shortly after emergence resulting in damping off and death of seedlings which also results in uneven plant stands [11,20]. Faba bean root rot causes reduction in yield and efficiency of nitrogen fixation [20] which may weaken the plant's resistance. The significant interaction between varieties and drainage methods resulted in significantly lower percentage of dead plants for the variety wayu in 2010. Similarly, planting of resistant varieties on raised beds (ridges) has been proven to be useful in water logged soils [21]. This indicates the need for the use of more than one cultural practice. Combining two or more cultural practices have been reported to result in additive and positive interactions in root rot control [22]. In addition to cultural practices, application of DAP and farm yard manures (FYM) have been reported to improve tolerance to root rot indicating the need for integrated disease management which is most preferred for management of root rots [11]. In general however, in developing countries, where up to 100% yield loss is recorded from susceptible cultivars, use of resistant varieties is the most viable measure for root rot caused by *F. solani* [23]. For example, in Ethiopia, use of variety wayu and broad bed and furrow (BBF) for improving drainage has reduced the incidence of black root rot resulting in increased yield [19] Wayu is one of the varieties released in 2002 for root rot resistance in Ethiopia with another variety Selale to perform well under waterlogged conditions [24]. This study has further confirmed that wayu can perform even better when used in combination with raised beds and when planted early before the soils become waterlogged.

In conclusion, this study has demonstrated that all the three factors (varietal resistance, drainage method and sowing dates) are important for effective management of the disease. But emphasis should be given to varietal resistance and use of raised beds especially on vertisols. The two varieties Wolki and Wayu are recommended with raised beds for higher yield and variety improvement programs focusing on resistance to black root rot of faba bean caused by *Fusarium solani*. Black root rot disease is known to be a major problem in areas such as north shoa. Further studies on farmers' fields are recommended over locations where the disease is a problem every year such as north Shoa of Ethiopia.

Acknowledgement

The authors would like to thank Dr. Getaneh Woldeab and Ato Birhanu Bekele who handled the early stages of the experiments. Compliments also go to Ato Tesfaye Hailu and Ato Beddasso Jebessa for technical assistance during the experiments. The research was funded by the Ethiopian Institute of Agricultural Research (EIAR) regular budget.

References

1. Madkour MA, Abou-Taleb EM, Oka-sha AM (1983) Aceton inhibition of *Rhizoctonia solani* growth. Phytopathol Z 107: 111-116.

2. Abdel-Rehim MA, Abou-Taleb EM, Al-Mounofe OA, Raffat FM, Tohamy A (1987) The efficacy of seed treatment with calcium compounds on controlling damp-ing-off of certain vegetable crops. Alexandria J Agric Res 32: 333-334.
3. Celar F (2000) Cucurbit disease. Sodobno Kmetijstvo 33: 162-165.
4. Akrami M, Ibrahimov AS, Zafari DM, Valizadeh E (2009) Control of Fusarium Rot of bean by combina-tion of *Trichoderma harzianum* and *Trichoderma asperillum* in green house conditions. Agricultural Journal 4: 121-123.
5. Helsper JPFG, Van Norelydi´x A, Burgermeyer K, Hoogendijk JM (1994) Effect of the absence of condensed tannins in faba beans (*Vicia faba*) on resistance to foot rot, ascochyta blight and chocolate spot. J Agric Sci 123: 349-355.
6. Tadesse N, Ahmed S, Gorfu D, Beshir T, Fininsa C, et al. (2008) Review of Research on Diseases of Food Legumes. In: Abraham Tadesse (ed.). Increasing Crop Production through Improved Plant Protection Volume I. Plant Protection Society of Ethiopia (PPSE), 19-22 December 2006. Addis Ababa, Ethiopia. PPSE and EIAR, Addis Ababa, Ethiopia pp: 598.
7. Beshir T (1995) Development of wilt/root rot resis-tant cultivars in faba bean. In: Eshetu B.,Abdurahman A. and Aynekulu Y.(eds.). Proceeding of the third annual conference of crop protection society of Ethiopia May 18-19, 1995 CPSE, Addis Ababa.
8. Beshir T (1999) Evaluation of the potential of Tri-choderma viride as biological control agent of root rot diseases *Fusarium solani* of faba bean. Pest management journal of Ethiopia 3: 91-94.
9. Stewart RD, Yirgu D (1967) Index of plant diseases in Ethiopia. Experimnetal sation Bull. No 30, HISU. College of Agriculture, Debre-zeit p: 95.
10. Assefa H, Gorfu D (1985) A review of food legume disease research in Ethiopia. In: Tsedeke Abate (ed.). A review of crop protection research in Ethiopia. Addis Ababa Ethiopia pp: 345-500.
11. Otsyula RM, Ajanga SI, Buruchara RA, Wortmann CS (1998) Development of bean root rot control strategy for western Kenya. African crop science Journal 6: 61-67.
12. PPRC (1996) Progress report for the period 1995/96. Ambo.
13. Dereje G, Tesfaye B (1994) In: Cool Season Food Legumes of Ethiopia (Asfaw Telaye, Faba bean diseases in Ethiopia Geletu Bejiga, Mohan C. Saxena and Mahmoud B. Solh, eds.). Proceedings of the First National Cool season Food Legumes Review Conference, 16-20 December 1993, Addis Ababa, Ethiopia pp: 328-345.
14. CIAT (Centro International de Agricultura Tropica) (1987) Principal diseases of beans in Africa. Study Guid. Series:04EB-06.13. Cali, Colombia.
15. Negussie T (1989) Effect of temperature on the growth of important wilt and root rot fungi of chickpea. Page 24 in Abstracts of The 14th Annual Meeting of the Ethiopian phytopathological Committee, 30-31 March, 1989. Ambo, Ethiopia.
16. El-Mougy NS, Abdel-Kader MM (2009) Seed and Soil treatments as Integrated Control Measure against Faba Bean Root Rot Pathogens. Plant Pathology Bulletin 18: 75-87.
17. Goellner K, Loehrer M, Langenbach C, Conrath U, Koch E, et al. (2010) Phakopsora pachyrhizi, the causal agent of Asian soybean rust. Mol Plant Pathol 11: 169-177.
18. International Center for Agricultural Research in the Dry Areas (ICARDA) (2006) NVRSRP/IFAD Project – Progress Report 2004-2005.
19. Schneider KA, Grafton KF, Kelly JD (2001) QTL analysis of resistance to Fusarium root rot in bean. Crop Science 41: 535-542.
20. Conner RL, Chang KF, Hwang SF, McLaren DL, Bruce D (2009) Gossen Impact and Control of Fusarium Root Rot in Faba Bean. Agriculture and Agri-Food Canada.
21. Bruchara RA, Rusuku G (1992) Root rots in the great lakes region. Pro. Of the Pan African Bean Pathology Working group Meeting, Thika, Kenya pp: 26-30.
22. CIAT (1992) (Centro International de Agricultura Tropica) Principal diseases of beans in Africa. CIAT Annual Report. Bean Programme, Cali, Colombia p: 385.
23. Mukankusi CM, Melis RJ, Derera J, Buruchara RA, Mark D (2011) A screening technique for resistance to Fusarium root rot of common bean. African Journal of Plant Science 5: 152-161.
24. NAIA (National Agricultural Input Authority) (2003) Crop variety register, Ethiopia.

Soil-borne and Compost-borne *Aspergillus* Species for Biologically Controlling Post-harvest Diseases of Potatoes Incited by *Fusarium sambucinum* and *Phytophthora erythroseptica*

Rania Aydi Ben Abdallah[1,2*], Hayfa Jabnoun-Khiareddine[2], Boutheina Mejdoub-Trabelsi[2,3] and Mejda Daami-Remadi[2]

[1]*National Agronomic Institute of Tunisia, 1082, Tunis, University of Carthage, Tunisia*
[2]*UR13AGR09- Integrated Horticultural Production in the Tunisian Centre-East, Regional Center of Research on Horticulture and Organic Agriculture, University of Sousse, 4042, Chott-Mariem, Tunisia*
[3]*Higher Agronomic Institute of Chott-Mariem, University of Sousse, 4042, Chott-Mariem, Tunisia*

Abstract

Nine isolates of *Aspergillus* spp., isolated from soil and compost were tested *in vitro* and *in vivo* for their antifungal activity against *Fusarium sambucinum* and *Phytophthora erythroseptica*, the causal agents of the Fusarium dry rot and pink rot of potato tubers. Tested using the dual culture method, the pathogen growth of *F. sambucinum* and *P. erythroseptica* was inhibited by 27 to 68% and 16 to 25% by all *Aspergillus* species, respectively. The highest inhibitory activity against both pathogens was induced by the isolate CH12 of *A. niger*. A significant reduction of the mycelial growth of both pathogens tested using the inverse double culture method involves the presence of volatile antifungal metabolites. Their effectiveness was also evaluated as tuber treatment prior to inoculation with the pathogens. The highest effectiveness in reducing Fusarium dry rot severity was recorded on tubers treated with the isolate CH12 of *A. niger*. This study also revealed that the efficacy of *Aspergillus* spp. as biocontrol agents may be enhanced by varying the timing of their application. In fact, the lesion diameter of dry and pink rots was reduced by 54-70 and 52% with preventive application, respectively. However, this parameter decreased by 21-48 and 47% when the *Aspergillus* spp. were applied simultaneously with pathogens, respectively. Similarly, diseases' severity, estimated based on average penetration of *F. sambucinum* and *P. erythroseptica*, was reduced by 57-77 and 55% with preventive treatments and by 29-68 and 44% with simultaneous application, respectively. This study reveals that *Aspergillus* spp., isolated from compost and soil, exhibits an interesting antifungal activity toward *F. sambucinum* and *P. erythroseptica* and may represent a potential source of biopesticide. Testing of their culture filtrates, their organic extracts and their toxicity may give additional information on their safe use as biocontrol agents.

Keywords: *Aspergillus* spp; Biological control; Dry rot; *Fusarium sambucinum*; Mycelial growth; *Phytophthora erythroseptica*; Pink rot

Introduction

Potato (*Solanum tuberosum* L.) is one of the most important vegetable crops in the world [1-3]. In Tunisia, it is one of the strategic crops occupying about 16% of all Tunisian cultivated areas [4]. However, potato production is threatened by several fungal diseases resulting in considerable yield losses [5,6]. These diseases can affect potato at any growth stage or even during storage, where potato tubers may exhibit diverse types of rots. Dry rot induced by different *Fusarium* species, Pythium leak caused by *Pythium* spp. and pink rot incited by *Phytophthora erythroseptica* are responsible for important tuber storage losses in the world and in Tunisia [7-13].

Significant losses in storage with estimates of up to 100% have been reported both in developed and developing countries when disease management is neglected [14]. Losses associated with dry rot have been estimated to range from 6 to 25%, and occasionally losses as great as 60% have been reported during long-term storage [15]. The primary sources of inoculum are contaminated or infected seed tubers and infested soil [5]. Tuber infection by *F. sambucinum* occurs through wounds produced during planting, harvesting or transport. Dry rot causes a dry and crumbly decay with abundant white, yellow or carmine-coloured mycelium depending on *Fusarium* species [15]. For pink rot, the oomycete, *P. erythroseptica* infects potato tubers through stolons or lenticels via zoospores and through cracks and cuts made during harvest and handling operations [5,16-17]. The infected tuber flesh becomes soft, spongy and watery with a light brown color. A clear liquid came out from a cut tuber and the surface acquires a pink coloration which turns brown and darkens after a few hours [18]. Thus, the health of seed tubers, the management practices during the growing period, the harvesting and handling practices and the environmental conditions maintained throughout storage are key factors affecting tuber infection by these pathogens [19].

Currently, the primary control for these diseases in storage facilities includes elimination of infected tubers prior to storage and storage management using forced air ventilation, and controlled temperature and humidity feedback systems [20]. In fact, there is a shortage of postharvest fungicides to completely manage these pathogens [21]. The control of potato dry rot has been achieved, for many years, by postharvest application of thiabendazole, a Benzimidazole fungicide [5]. However, *F. sambucinum* resistant to this fungicide and other benzimidazole fungicides were discovered in many parts of the world, leading to reduced effectiveness in controlling dry rot [15]. For pink

***Corresponding author:** Rania Aydi Ben Abdallah, National Agronomic Institute of Tunisia, 1082, Tunis, University of Carthage, Tunisia, E-mail: raniaaydi@yahoo.fr

rot, chemical control is limited to a few compounds. In fact, fungicides containing metalaxyl and mefenoxam were used effectively to control pink rot in the 1990's. However, metalaxyl-resistant isolates of *P. erythroseptica* are now widespread and this may lead to failure of these chemicals to control pink rot [22]. Phosphorous acid and many other disinfectants are currently registered for postharvest management of pink rot. However, use of these products does not completely control storage pathogens [23].

Furthermore, all commonly grown potato cultivars are susceptible to potato pink and dry rots. Therefore, lack of available post-harvest fungicides and disease-resistant cultivars has prompted the search for new and efficient alternative methods as seed tuber and/or postharvest treatments to reduce incidence and severity of dry and pink rots.

Recent studies indicate that generally recognized as safe (GRAS) compounds and microbial antagonists could eventually be integrated into dry rot management strategies [24]. In fact, many biocontrol agents have been explored to control potato postharvest diseases such as *Pseudomonas* spp., *Enterobacter* spp., *Trichoderma* spp, *Aspergillus* spp. which reduced the severity of dry and pink rots caused by *F. sambucinum* and *Phytophthora* spp., respectively [25-27]. *Aspergillus niger, A. flavus, A. terreus, A. fumigatus* have been shown to control potato postharvest diseases caused by *Pythuim ultimum* [28-30], *Fusarium sambucinum* [31] and *Phytophthora* sp. [32]. These *Aspergillus* species used as biological control agents against many fungal pathogens such as *F. oxysporum, Sclerotinia sclerotiorum, Pythium* spp. act through mycoparasitism, competition, mycelial lysis and antibiosis via the synthesis of volatile and/or non-volatile metabolites [11,33-34]. *Aspergillus* species are ubiquitous in most agricultural soils and they generally produce a variety of secondary metabolites exhibiting inhibitory effects on several soil-borne microorganisms [35]. Moreover, many *Aspergillus* species are considered among the potential thermotolerant antagonists surviving after soil solarisation such as *A. terreus, A. ochraceus* and *A. fumigatus* [36]. They are also among the most abundant in composts [37-39] and they play a significant role in the composting process as well as in suppressing the growth of pathogenic fungi [40-42]. In recent studies, many atoxigenic *Aspergillus* spp. strains, mostly isolated from soil, were used as effective biocontrol agents against *Aspergillus* toxigenic strains [43]. In fact, naturally occurring populations of atoxigenic strains are considered as potentially important reservoirs for the selection of strongest biocompetitors [43]. Furthermore, Frisvad and Larsen [44] mentioned that the genus *Aspergillus* is rich in species and these species are able to produce a large number of extrolites, including secondary metabolites, bioactive peptides/proteins, lectins, enzymes, hydrophobins and aegerolysins.

Therefore, the objective of the present study was to evaluate the *in vitro* and *in vivo* antifungal potential of *Aspergillus* species (*A. niger, A. terreus, A. flavus, Aspergillus* sp.) isolated from compost, solarised and non-solarised soils to control *F. sambucinum* and *P. erythroseptica*. The timing of their application as tuber treatment was also tested to optimize their suppressive effects toward potato Fusarium dry rot and pink rot.

Materials and Methods

Plant material

Apparently healthy and undamaged potato tubers (*Solanum tuberosum* L.) cv. Spunta were used in this study. This cultivar was known by its susceptibility to Fusarium dry rot and pink rot [7,11].

Pathogens

The isolates of *F. sambucinum* and *P. erythroseptica* used in this study were obtained from potato tubers showing typical symptoms of dry rot and pink rot, respectively. They were cultured on Potato Dextrose Agar (PDA) and incubated at 25°C for seven days before use.

Biocontrol agents

Four *Aspergillus* species namely *A. niger, A. terreus, A. flavus* and *Aspergillus* sp., isolated from soil or compost, were used in this study (Table 1). They were cultured on PDA medium and incubated at 25°C for seven days before being used in the bioassays.

Assessment of the *in vitro* antifungal activity of *Aspergillus* spp. against *Fusarium sambucinum* and *Phytophthora erythroseptica*

The antagonism of *Aspergillus* spp. isolates against *F. sambucinum* and *P. erythroseptica* was tested *in vitro* using the dual culture method and the inverse double technique on PDA medium. The first technique consists in depositing equidistantly two agar plugs (diameter 6 mm) colonized by the pathogen (removed from a 7-days-old culture at 25°C) or the antagonist (removed from a 7-day-old culture at 25°C) in the same Petri dish containing PDA medium supplemented with streptomycin sulfate (300 mg/L) [45].

The second method consists in transplanting the antagonist and the pathogen in two separate Petri dish containing PDA medium supplemented with streptomycin sulfate (300 mg/L). Thereafter, a 6 mm agar plug of the antagonist, removed from the margin of an actively growing culture, was placed in the centre of the bottom dish whereas the agar plug (diameter 6 mm) colonized by the pathogen (removed from a 7-days-old culture at 25°C) was placed in the centre of the top Petri dish. Both dishes were sealed with parafilm layers to prevent loss of volatile substances [46]. The two pathogens (*F. sambucinum* and *P. erythroseptica*) were thus exposed to the influence of the volatile substances released from isolates of *Aspergillus* spp. tested. Control cultures are challenged by the pathogen only in the PDA medium without transplanting antagonists, for the two confrontation methods.

The diameter of the pathogen colony (treated and control) was measured after 7 and 4 days of incubation at 20°C for *F. sambucinum* and *P. erythroseptica*, respectively. The mycelial growth inhibition percentage was calculated according to the following formula:

$I\,\% = [(C2-C1)/C2] \times 100$ with C2: Mean diameter of the control colony and C1: Mean pathogen colony diameter in the presence of the antagonist [33].

Assessment of the *in vivo* antifungal activity of *Aspergillus* spp. against *Fusarium sambucinum* and *Phytophthora erythroseptica*

Apparently healthy potato (cv. Spunta) tubers were used in this study. They were washed under running tap water to remove excess soil, dipped into 10% sodium hypochlorite solution for 5 min, rinsed twice with sterile distilled water (5 min each) and air-dried.

Tubers were wounded at two sites along the tuber longitudinal axis by a disinfected Pasteur pipette occasioning wounds of 6 mm in diameter and in depth, which serve as infection sites. Tuber inoculation was made by deposing in each wound an agar plug (6 mm diameter) colonized by the pathogen removed from a 7-day-old culture at 25°C.

Conidial suspensions (10^6 CFU/mL) of the biocontrol agents, tested individually, were applied by injecting 100 µl in each wound either simultaneously with the pathogen or 24 h before. The positive control was inoculated with the pathogen only whereas the negative control was uninoculated and untreated with any of the antagonists tested.

All tubers were incubated at high relative humidity for 21 days at 20°C and for 7 days at 25°C for *F. sambucinum* and *P. erythroseptica*, respectively. After this incubation period, tubers were cut longitudinally via sites of inoculation and lesion diameter, width (l) and depth (p) of the occasioned rot were measured. The penetration of the pathogen was calculated using the following formula:

$P\ (mm)=[(l/2)+(P-6)]/2$ [47].

The rate of reduction for the lesion diameter of dry and pink rots and the penetration of the pathogens were calculated using the following formula:

$I\ \%=[(C2-C1)/C2]*100$ with C2: lesion diameter of the rot or penetration of the pathogen in the untreated control and C1: lesion diameter of the rot or penetration of the pathogen in the presence of the antagonist [33].

Statistical analysis

Data were subjected to a one-way analysis of variance (ANOVA) using SPSS 16.0. For the *in vitro* tests, each individual treatment was repeated three times and the essays were analyzed in a completely randomized model. The *in vivo* essay was analyzed in a completely randomized factorial model with two factors (antagonistic treatments and timing of application) and each individual treatment was repeated six times. Means were separated using Student-Newman-Keul's (SNK) test (at $P \leq 0.05$).

Results

Effect of *Aspergillus* spp. on the mycelial growth of *Fusarium sambucinum* and *Phytophthora erythroseptica*

Analysis of variance showed a highly significant (at $P \leq 0.05$) inhibitory effect of the antagonists tested on the average colony diameter of *F. sambucinum* and *P. erythroseptica*, recorded after 7 and 4 days of incubation, respectively. Tested using the dual culture method, *F. sambucinum* and *P. erythroseptica* growth at 20°C were inhibited by 27 to 68% and by 16 to 25% by all *Aspergillus* isolates, respectively. The highest inhibitory activity against pathogens, as compared to the untreated control, was induced by the isolate CH12 of *A. niger* (Figures 1 and 2). An important inhibition of *F. sambucinum* by 37, 32 and 35% was also induced by the isolates CH1 and MC2 of *A. niger* and MC5 of *A. flavus*, respectively. The isolates CH2 and MC8 of *A. terreus*, CH8, CH4 and CH3 of *Aspergillus* sp. reduced the colony diameter of *F. sambucinum* by 31, 30, 29, 28 and 27%, respectively. *A. niger* was also active against *P. erythroseptica*, causing an inhibition of 24 and 22%, by the isolates MC2 and CH1, respectively. An inhibition of *P. erythroseptica* mycelial growth by 20 to 21% was achieved using the isolates CH2 of *A. terreus*, MC5 of *A. flavus*, and MC8 of *A. terreus*. The isolates of *Aspergillus* sp. (CH8, CH4 and CH3) were less active against *P. erythroseptica* with an inhibition of 19, 18 and 16%, respectively (Figure 1).

Tested using the inverse double culture method, a significant reduction of *F. sambucinum* mycelial growth, by 27.84 and 27.45% compared to the untreated control was induced by the isolates CH2 of *A. terreus* and MC5 of *A. flavus*, respectively (Figures 3a and 4). All *Aspergillus* spp. tested had significantly inhibited *P. erythroseptica* by distance culture method from 9.24 to 23.09% compared to the untreated control. The highest reduction of *P. erythroseptica* mycelial growth was achieved using the isolates CH1, CH12, MC2 of *A. niger* and MC5 of *A. flavus* (Figures 3b and 4). An inhibition varying from 15.70 to 17.78% was caused by *Aspergillus* sp. isolates (CH3, CH4 and CH8) and MC8 of *A. terreus*. The isolate CH2 of *A. terreus* showed the lowest inhibition of *P. erythroseptica* at distance compared to the untreated control (Figure 3b).

Effect of the application timings of *Aspergillus* spp. on potato dry rot and pink rot severity

Results analyzed by ANOVA revealed a significant variation in Fusarium dry rot severity, recorded after 21 days of incubation at 20°C, depending upon treatments tested, timings of their application and their interaction. The lesion diameter of dry rot was reduced by 54 to 70%, as compared to the untreated control, with preventive application. However, this parameter decrease varied from 21 to 48% only when the *Aspergillus* spp. were applied simultaneously with the pathogen (Figures 5a and 8). Disease severity, estimated based on average penetration of *F. sambucinum*, was reduced by 57 to 77% with preventive treatments compared with 29 to 68% obtained with simultaneous application (Figures 5b and 8).

The highest effectiveness of *Aspergillus* spp. in reducing Fusarium dry rot severity, as compared to the inoculated and untreated control tubers, was recorded on potato tubers treated with the isolate CH12

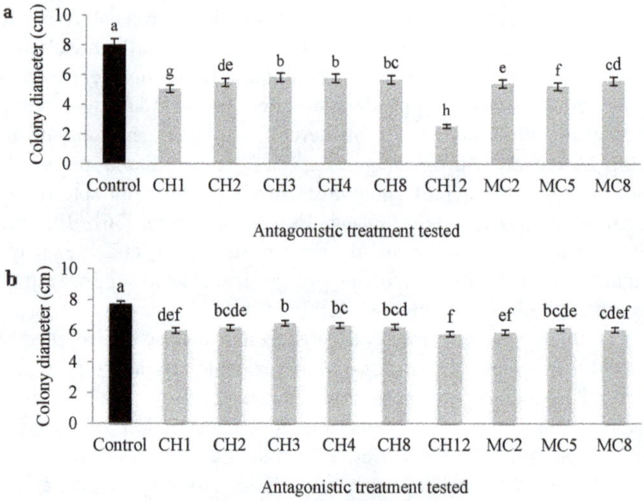

Figure 1: Effect of *Aspergillus* spp. on the mycelial growth of *Fusarium sambucinum* (a) and *Phytophthora erythroseptica* (b) noted after 7 and 4 days of incubation at 20°C, respectively, using the dual culture method. CH1, CH12, MC2: *A. niger*, CH2, MC8: *A. terreus*; CH3, CH4, CH8: *Aspergillus* sp.; MC5: *A. flavus*. Bars affected with the same letter are not significantly different according to Student-Newman-Keul's test at $P \leq 0.05$.

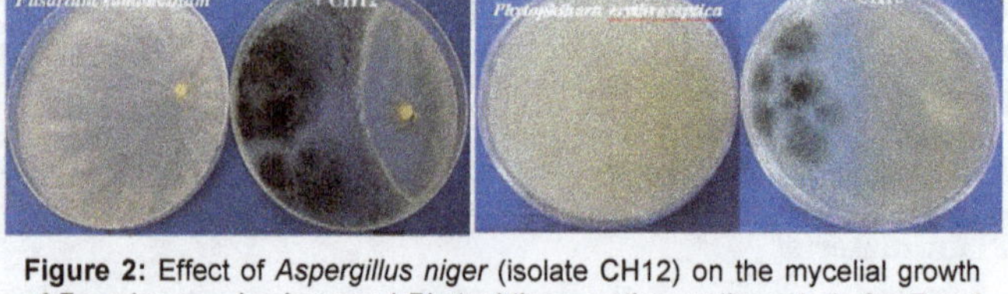

Figure 2: Effect of *Aspergillus niger* (isolate CH12) on the mycelial growth of *Fusarium sambucinum* and *Phytophthora erythroseptica* noted after 7 and 4 days of incubation at 20°C, respectively, as compared to the untreated controls.

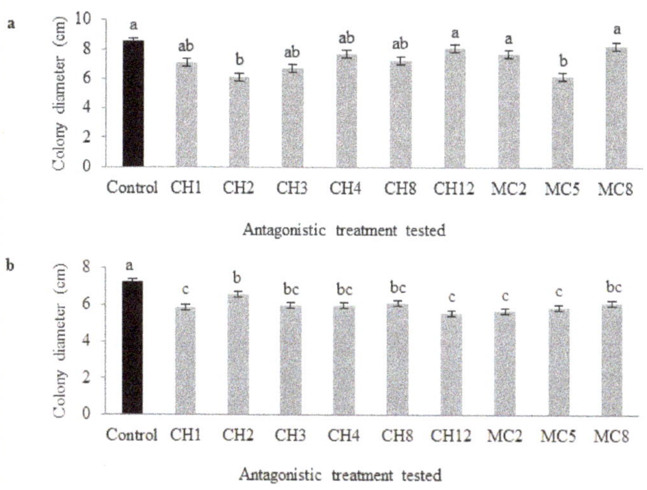

Figure 3: Effect of *Aspergillus* spp. on the mycelial growth of *Fusarium sambucinum* (a) and *Phytophthora erythroseptica* (b) noted after 7 and 4 days of incubation at 20°C, respectively, tested using the inverse double culture method. CH1, CH12, MC2: *A. niger*, CH2, MC8: *A. terreus*; CH3, CH4, CH8: *Aspergillus* sp.; MC5: *A. flavus*. Bars affected with the same letter are not significantly different according to Student-Newman-Keul's test at $P \leq 0.05$.

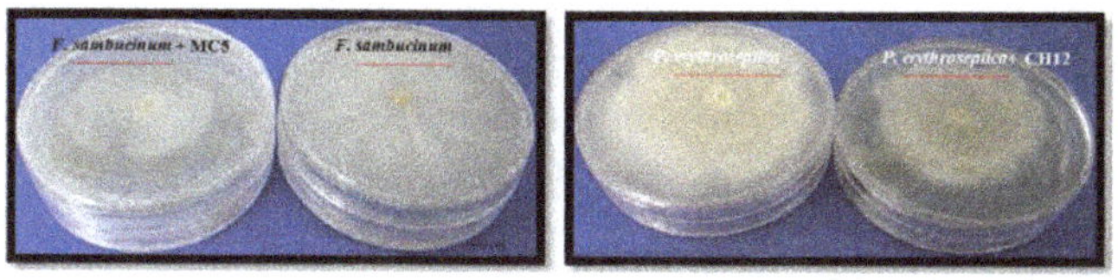

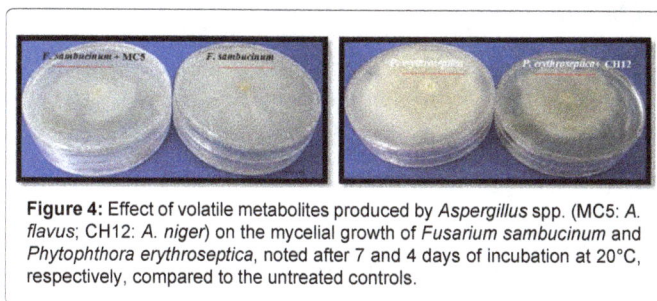

Figure 4: Effect of volatile metabolites produced by *Aspergillus* spp. (MC5: *A. flavus*; CH12: *A. niger*) on the mycelial growth of *Fusarium sambucinum* and *Phytophthora erythroseptica*, noted after 7 and 4 days of incubation at 20°C, respectively, compared to the untreated controls.

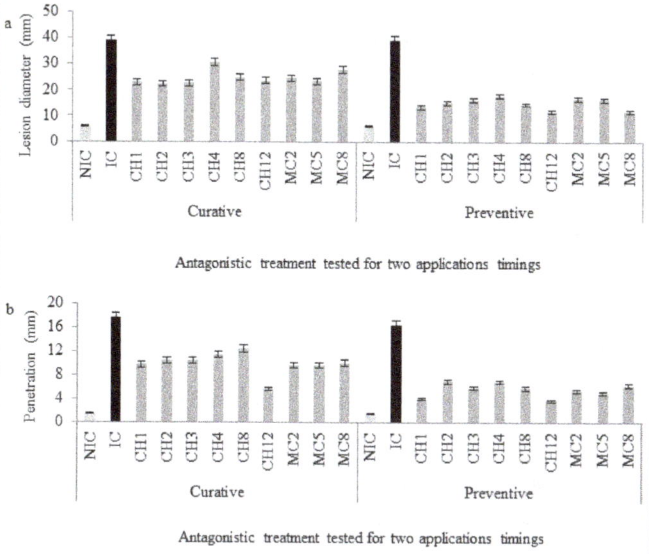

Figure 5: Effect of the application timing of *Aspergillus* spp.-based treatments on potato tubers on the lesion diameter (a) and the average penetration (b) of *Fusarium sambucinum* noted after 21 days of incubation at 20°C. CH1, CH12, MC2: *A. niger*, CH2, MC8: *A. terreus*; CH3, CH4, CH8: *Aspergillus* sp.; MC5: *A. flavus*. NIC: Uninoculated and untreated control; IC: Inoculated and untreated control. LSD (Timings of application × Antagonistic treatment tested): 5.8 mm (a) or 2.28 mm (b) at $P \leq 0.05$.

of *A. niger* in both application timings. This isolate reduced the lesion diameter by 55% and the average penetration of the pathogen by 73%, as compared to the inoculated and untreated control tubers (Figures 5 and 9).

Analysis of variance showed a significant (at $P \leq 0.05$) variation in pink rot severity, recorded after 7 days of incubation at 25°C, depending upon treatments tested and the timings of their application. The interaction between these two factors did not show significant variation on the reduction of pink rot severity. Whatever the timing of their application, a reduction of the lesion diameter of pink rot varied between 83 and 88% with all biological treatments tested compared to the inoculated and untreated control (Figure 6a). All isolates of *Aspergillus* spp. tested had also limited the average penetration of *P. erythroseptica* by about 82 to 91% compared to the inoculated and untreated control (Figure 6b).

A significant decrease in pink rot severity was also achieved using the antagonists tested preventively. The lesion diameter of pink rot was reduced by 52% with preventive application as compared to 47% when the *Aspergillus* spp. were applied simultaneously with the pathogen (Figure 7a). All antagonists tested and applied 24 h prior to inoculation with *P. erythroseptica* led to 55% decrease in the average penetration of the pathogen compared to 44% obtained with simultaneous application (Figures 7b).

Discussion

In recent years, intense research efforts have been devoted to the development of antagonistic microorganisms to control potato diseases such as *Penicillium*, *Trichoderma*, *Aspergillus*, *Gliocladium*, etc [11,48]. In fact, several studies have promoted the antagonistic activity of *Aspergillus* species which have been identified in compost- or residue-amended substrates, in solarised and non-solarised soils, in the rhizosphere, in decaying plant material, in stored grains, etc. Furthermore, some of them are essential component of many compost

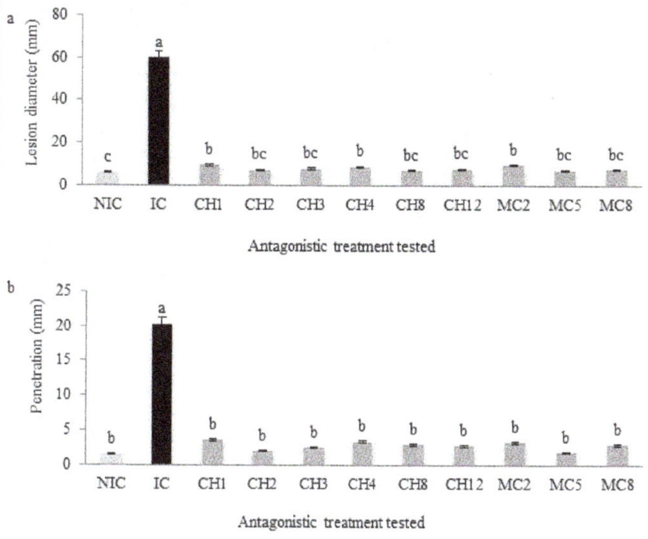

Figure 6: Effect of *Aspergillus* spp. based-treatments of potato tubers on the lesion diameter (a) and the average penetration of *Phytophthora erythroseptica* (b), noted after 7 days of incubation at 25°C. CH1, CH12, MC2: *A. niger*, CH2, MC8: *A. terreus*; CH3, CH4, CH8: *Aspergillus* sp.; MC5: *A. flavus*. NIC: Uninoculated and untreated control; IC: Inoculated and untreated control. Bars affected with the same letter are not significantly different according to Student-Newman-Keul's test at $P \leq 0.05$.

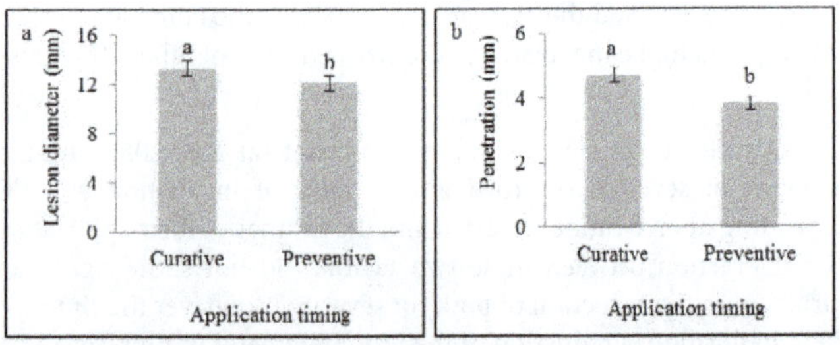

Figure 7: Effect of the application timing of *Aspergillus* spp.-based treatments on potato tubers on the lesion diameter (a) and the average penetration (b) of *Phytophthora erythroseptica*, noted after 7 days of incubation at 25°C. Bars affected with the same letter are not significantly different according to Student-Newman-Keul's test at $P \leq 0.05$.

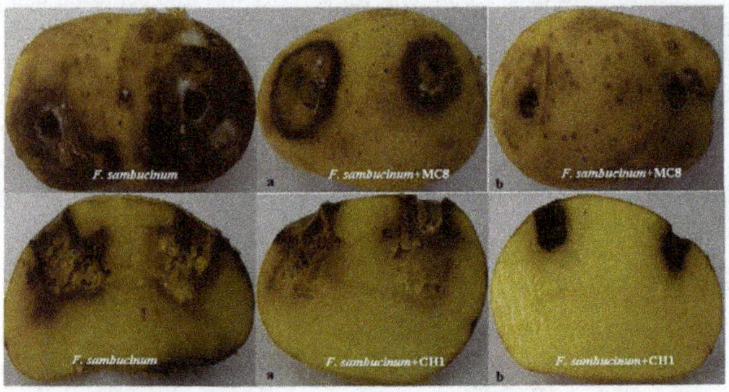

Figure 8: Effect of *Aspergillus* spp. (MC8: *A. terreus*; CH1: *A. niger*) applied simultaneously with pathogens (a) or 24 h before (b) on the severity of dry rot, noted after 21 ays of incubation at 20°C, as compared to the inoculated and untreated controls.

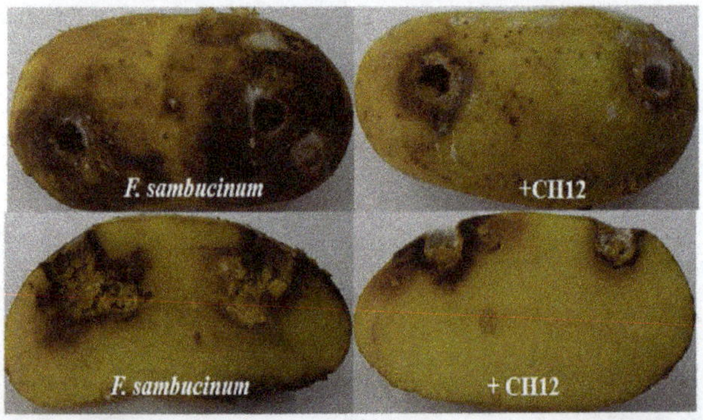

Figure 9: Reduction of Fusarium dry rot severity induced by the isolate CH12 of *Aspergillus niger* as compared to the inoculated and untreated control.

microbiota [28,49-52]. Moreover, *Aspergillus* species have been reported as endophytes with antifungal activity [53-55] and able to produce several metabolites such as phenolic and bioactive flavonoid compounds.

In the current study, the use of naturally occurring *Aspergillus* species, isolated from compost, solarised and non-solarised soils, to control potato dry and pink rots caused respectively by *F. sambucinum* and *P. erythroseptica*, was evaluated *in vitro* and *in vivo*.

Tested using dual culture method, all the nine isolates of *Aspergillus* spp. tested, had inhibited the mycelial growth of *F. sambucinum* and *P. erythroseptica* by 27 to 68% and 16 to 25%, respectively. These findings confirm previous studies showing that native compost and soil-borne *Aspergillus* spp. significantly reduced the *in vitro* growth of *Pythium* spp., causing potato leak as well as that of many *Fusarium* species responsible for potato dry rot in Tunisia [11,29]. Furthermore, results revealed the inhibitory effects of *Aspergillus* species and isolates regardless of their origin of isolation. These findings are in accordance with several previous studies showing the potential antagonistic effects of *Aspergillus* isolated from many substrates. In fact, Barocio-Ceja et al. [56] showed that *Aspergillus* sp. "VC11" isolated from chicken-manure vermicompost inhibited the growth of *F. oxysporum* by 27%, that of *F. subglutinans* by 24% and that of *Rhizoctonia* sp. by 25%. Similarly, Israel and Lodha [57] reported the hyperparasitism of *A. versicolor* isolated from heated (naturally or solarized) cruciferous residue-amended soils against *F. oxysporum cumini*. Moreover, Adebola and Amadi [51] found that three *Aspergillus* species, namely *A. niger*, *A. fumigatus* et *A. repens*, isolated from the rhizosphere of black pod-infected cocoa trees showed ability to inhibit growth of *Phytophthora palmivora in vitro* and that the inhibition zones produced might be due to the production of antifungal metabolites by the test antagonists.

Our *in vitro* results showed that the highest inhibitory activity against both pathogens was induced by *A. niger*. These findings are in accordance with Dwivedi and Enespa [58] who found that among the biocontrol agents tested, *A. niger* isolated from tomato field was most effective and completely inhibited the mycelial growth of *F. solani* and *F. oxysporum* f. sp. *lycopersici*. In the same sense, Tiwari et al. [59] showed that *A. niger* inhibit the growth of all the fungal species tested which belong to ten genera of wood decay fungi (*Trametes*, *Stereum*, *Pycnoporus*, *Phellinus*, *Lenzites*, *Phellinus*, *Earliella Gloeophyllum*, *Flavodon* and *Daedalea*), from 29.2 to 66.7%. In addition, Venkatasubbaiah and Safeeulla [60] found that *A. niger*, isolated from the rhizosphere of coffee seedlings, was antagonistic to *Rhizoctonia solani in vitro* and exhibited hyperparasitism against this collar rot pathogen in dual culture experiments. The compost inhabiting *A. niger* was also able to inhibit the growth of soil-borne pathogens including *Macrophomina phaseolina*, *Pythium aphanidermatum* and *Rhizoctonia solani* [61]. In addition, besides the inhibition of the mycelial growth, an atoxigenic *A. niger* FS10 isolated from Chinese fermented soybean was found to be able to inhibit spore germination and also to degrade the aflatoxin B1 of toxigenic *A. flavus*, and to reduce its mutagenicity and toxicity [62].

The suppressive effects of the microbial agents used in this study against potato pathogens may involve several mechanisms of action. In fact, to antagonize *F. sambucinum* and *P. erythroseptica*, *Aspergillus* isolates had probably deployed mycoparasitism, involving direct contact between the tested antagonist and the pathogens, (ii) production of antibiotic-type secondary metabolites, which spread through the medium, and (iii) competition for nutrients and space [63,64]. In fact, the first study showing that *A. terreus* can parasitize the sclerotia of *Sclerotinia sclerotiorum* under both laboratory and field conditions was reported by Melo et al. [50]. Recently, the work of Hu et al. [65] clearly demonstrates that the mycoparasite characteristics of *Aspergillus* sp. ASP- 4 may be enhanced and related to its ability to produce a range of extracellular enzymes, such as chitinases and other antifungal extrolites which help colonization of *S. sclerotiorum* sclerotia; as well as thought to be the mode of action of *A. terreus* (*Aspergillus* section Terrei). In addition, chitinases have been shown to be produced by several *Aspergillus* spp. [65]. In fact, antifungal activity of crude and purified chitinase produced by *A. niger* LOCK 62 was observed against *F. culmorum*, *F. solani*, and *R. solani*. The growth inhibition of *F. culmorum* was the strongest both by crude and purified enzymes (70 and 60%, respectively) whereas the growth of *F. solani* was strongly inhibited by crude chitinase (73%) [66]. Recently, Frisvad and Larsen [44] mentioned that *Aspergillus* species are usually very efficient specialized metabolite producers which produces a wide array of small molecule extrolites (secondary metabolites or specialized metabolites), but also other bioactive molecules. Zhang et al. [53] have also isolated an endophytic *A. clavatonanicus* from a twig of Chinese *Taxus mairei* able to produce clavatol and patulin which exhibited inhibitory activity *in vitro* against several plant pathogenic fungi, i.e., *Botrytis cinerea*, *Didymella bryoniae*, *F. oxysporum* f. sp. *cucumerinum*, *R. solani*, and *P. ultimum*. Recently, Patil et al. [54] have identified an isolate of *A. flavus*, as an endophytic fungus from Indian medicinal plant *Aegle marmelos*, able to produce phenolic and bioactive metabolite, characterized to be a flavonoid rutin with excellent biological activities.

Tested using the inverse double culture method, *Aspergillus* spp. isolates exhibited antifungal activity toward *F. sambucinum* and *P. erythroseptica*. The reduction of mycelial growth of both pathogens by *Aspergillus* spp. isolates involved the presence of volatile metabolites produced by each active antagonist tested. In the same sens, Gao et al. [67] revealed the presence of nine volatile metabolites produced by *Aspergillus* spp. (*A. fumigatus*, *A. versicolor*, *A. sydowi*, *A. flavus*, *A. niger*). In addition, Israel and Lodha [57] reported that, apart from releasing antibiotic substances, *A. versicolor* also produces volatile metabolites such as various hydrocarbons, alcohols, ketones, ethers, esters and sulphur-containing compounds. Moreover, volatile metabolites, emitted by *A. niger*, had been shown to inhibit the growth of *Colletotrichum gloeosporioides* [68]. Recently, Thakur and Harsh [69] demonstrates that *A. niger*, isolated from the phylloplane of healthy *Piper longum* plants, possess higher antagonistic efficacy in inhibiting the mycelial growth of *C. gloeosporioides in vitro* by producing volatile metabolites.

Our results make clear that the inhibitory effect of *Aspergillus* spp. against *F. sambucinum* and *P. erythroseptica in vitro* varied depending on species and isolates of the same antagonist. In fact, populations of *A. flavus* in agricultural fields consist of complex communities that exhibit considerable genetic diversity based on phylogenetic and vegetative compatibility group analyses [70,71]. Natural communities of *A. flavus* consist of individuals that vary widely in ability to produce aflatoxin [72]. In the same sense, Khan and Anwer [73] mentioned that the effectiveness of soil aggregate isolates of *A. niger* against *R. solani* varies with the isolate. In fact, the HCN-producing isolates AnC2 and AnR3, belonging to group I, produced greater amounts of siderophore and solubilized phosphorus than the other isolates and caused the greatest inhibition of colonization by *R. solani* in dual culture. They also suppressed the root rot on eggplant and the soil population of *R. solani* in pot soil [73].

The dry and pink rots are serious post-harvest potato diseases that cause important storage losses in Tunisia [7,9]. In the present study, four species of *Aspergillus* were used as antagonistic agents against the causative agents of these two tuber diseases. Results indicate that all the *Aspergillus* spp. tested significantly reduced the lesion diameter of the dry and pink rots as well as the average penetration of *F. sambucinum* and *P. erythroseptica*, whatever the timing of their application. In fact, the antagonistic potential of *Aspergillus* sp. against both diseases confirmed previous local reports on biocontrol of soil-borne pathogens with *Aspergillus* sp. isolated from compost teas [11,28-29]. Furthermore, efficacy of *Aspergillus* species in controlling fungal diseases has been proved. In fact, *Aspergillus niger* has a fair capacity to suppress plant pathogens and to increase the yield of the plants it colonizes [74,73]. In this sens, Khan and Anwer [73] reported that soil application of *A. niger* aggregate isolates, especially AnC2 and AnR3, suppressed *R. solani* infection and also significantly increased eggplant yield. The production of NH_3, HCN and siderophore may have contributed to the suppression of *R. solani*. Hu et al. [65] mentioned also that their results strongly suggest that *Aspergillus* sp. ASP-4 has huge potential to be a successful biocontrol agent for Sclerotinia stem rot of oilseed rape. In fact, this antagonist significantly inhibits *S. sclerotiorum* by parasitism of the sclerotia, resulting in a marked decrease in the formation of apothecia in the field. In addition, Dwivedi and Enespa [58] reported that *Aspergillus* species were the most effective antagonists followed by *Penicillium* spp. and *Trichoderma* spp. for controlling the wilt of tomato and brinjal crops. In addition, the use of these bio-agents are not only safe for the farmers and consumers, but also eco-friendly, cost effective, easy to produce and easy to apply the formulations [58]. Furthermore, results obtained by Wang et al. [75] indicated that the ASD strain, *A. flavipes*, could be useful as a potential biocontrol agent against the soil-borne fungus *Phytophthora capsici* and suggest that the increased activity of defence-related enzymes might be part of the mechanism of *A. flavipes* in controlling *P. capsici*. In other studies, Chuang et al. [76] reported that, based on pot experiments, significant increases in plant (*Brassica chinensis* Linn.) dry weight and N and P contents were observed with the addition of isolate 6A of *A. niger*, which has the ability to solubilise the native soil phosphate, which is not available to plant.

The present study also highlighted the importance of the timing of antagonist's application on their effectiveness in dry and pink rots control. It was clearly evidenced that these compost and soil-borne fungi are able to significantly reduce rot development and severity when applied preventively following tuber wounding. In fact, when applied 24 h before inoculation with the pathogen, the *Aspergillus* isolates tested reduced more effectively the severity of Fusarium dry rot and pink rot. Similarly, Aydi et al. [29] recorded a better antagonistic effect of *Aspergillus* spp. when applied 24 h before inoculation with *Pythium ultimum* on potato tubers. Similar findings were reported by Daami-Remadi et al. [28] who used *Aspergillus* spp., *Penicillium* sp. and *Trichoderma* sp. as antagonistic agents against *P. ultimum* and *P. aphanidermatum*. Aydi-Ben Abdallah et al. [32] demonstrated that an important decrease in pink rot severity was achieved using preventive treatments of potato tubers with the culture filtrates and organic extracts from *Aspergillus* spp. In the same sense, applied preventively (24, 48 and 72 h before inoculation with the pathogen), dry rot severity incited by *F. sambucinum* was also limited using *Bacillus* spp. isolates [10] suggesting that this factor (i.e. treatment timing) may have a direct effect on disease control efficiency. In fact, Mohale et al. [77] reported that the timing of application of atoxigenic *Aspergillus* strains relative to the onset of host tissue colonisation by toxigenic strains, the innate competitive ability of the atoxigenic strains in the presence of their toxigenic counterparts are all important for the successful implementation of control strategies.

Conclusion

Aspergillus spp., isolated from soil and compost, exhibited an interesting antifungal activity against *F. sambucinum* and *P. erythroseptica in vitro*, using the dual culture and the inverse double culture methods, showing that these antagonists are able to release volatile and non-volatile metabolites. Thus, these *Aspergillus* species may represent a potential source of biologically active compounds and constitute a promising alternative to manage these increasingly important diseases.

Our studies demonstrated that *Aspergillus* species are potential native biological control agents against both potato diseases. They have shown great promise in their ability to reduce potato dry and pink rots. Their effectiveness was proved under extremely favourable conditions for *F. sambucinum* and *P. erythroseptica* development, such as use of susceptible potato cultivar (cv. Spunta), tuber wounding before inoculation, incubation under high moisture and disease conducive temperature, altogether have normally favoured pathogen expression and consequently, rot development.

Our results suggest that *Aspergillus* spp. are successful biocontrol agents for potato tuber rots in storage. These *Aspergillus* spp. may be further stimulated or enhanced by soil amendments and fertilizers, which result in the release of volatile chemicals that are toxic to potato pathogens (e.g., ammonia and glucosinolates), but can stimulate the growth and proliferation of *Aspergillus* species and other soil-borne mycoparasites [78,79].

Acknowledgements

This work was funded by the Ministry of Higher Education and Scientific Research of Tunisia through the funding allocated to the research unit UR13AGR09-Integrated Horticultural Production in the Tunisian Centre-East. My sincere gratitude goes to all the staff of the Regional Centre of Research in Horticulture and Organic Agriculture (CRRHAB) for their welcome and pleasant working conditions.

References

1. Fuglie KO (1994) The demand for potatoes in Tunisia: Are they a cereal substitute. Eur Rev Agric Econ 21: 277-286.
2. Wang B, Yin Z, Feng C, Shi X, Li Y, et al. (2008) Cryopreservation of potato shoot tips. Fruit Veg Cereal Sci Biotechnol 2: 46-53.
3. Öztürk E, Kavurmaci Z, Kara K, Polat T (2010) The effect of different nitrogen and phosphorus rates on some quality of potato. Potato Res 53: 309-312.
4. Djébali N, Tarhouni B (2010) Field study of the relative susceptibility of eleven potato (*Solanum tuberosum* L.) varieties and the efficacy of two fungicides against *Rhizoctonia solani* attack. Crop Prot 29: 998-1002.
5. Secor GA, Gudmestad NC (1999) Managing fungal diseases of potato. Can J Plant Pathol 21: 213-221.
6. Taylor RJ, Pasche JS, Gudmestad NC (2008) Susceptibility of eight potato cultivars to tuber infection by *Phytophthora erythroseptica* and *Pythium ultimum* and its relationship to mefenoxam-mediated control of pink rot and leak. Ann Appl Biol 152: 189-199.
7. Triki MA, Priou S, El Mahjoub M (1996) Activités inhibitrices in vitro de quelques substances chimiques et souches antagonistes de *Trichoderma* sp. vis-à-vis des Erwinia spp. des *Fusarium* spp. et de *Phytophthora erythroseptica* agents des pourritures des tubercules de pomme de terre. Annales de l'INAT 69: 171-184.
8. Salas B, Secor GA, Taylor RJ, Gudmestad NC (2003) Assessment of resistance of tubers of potato cultivars to *Phytophthora erythroseptica* and *Pythium ultimum*. Plant Dis 87: 91-97.
9. Daami-Remadi M, El Mahjoub M (1996) Fusariose de la pomme de terre en Tunisie-III; Comportement des variétés de pomme de terre vis-à-vis des souches locales de *Fusarium*. Annales de l'INRAT 69: 113-130.
10. Chérif M, Omri N, Hajlaoui MR, Mhamdi M, Boubaker A (2001) Effect of some fungicides on *Fusarium roseum* var. *sambucinum* causing potato tuber dry rot and *Trichoderma* antagonists. Annales de l'INRAT 74: 131-149.
11. Daami-Remadi M, Jabnoun-Khiareddine H, Ayed F, Hibar K, Znaïdi IEA, et al. (2006) In vitro and in vivo evaluation of individually compost fungi for potato *Fusarium* dry rot biocontrol. J Biol Sci 6: 572-580.
12. Priou S, Triki MA, El Mahjoub M, Fahem M (1997) Assessing potato cultivars in Tunisia for susceptibility to leak caused by *Pythium aphanidermatum*. Potato Res 40: 399-406.
13. Triki MA, Priou S, EL Mahjoub M (2001) Effects of soil solarization on soil-borne populations of *Pythium aphanidermatum* and *Fusarium* solani and on the potato crop in Tunisia. Potato Res 44: 271-279.
14. Wale A, Platt HW, Cattlin N (2008) Diseases, pests and disorders of potatoes. Taylor & Francis, CRS Press.
15. Secor GA, Salas B (2001) *Fusarium* dry rot and *Fusarium* wilt. In: 2nd Compendium of potato diseases. American Phytopathological Society Press St. Paul MN.
16. Salas B, Stack RW, Secor GA, Gudmestad NC (2000) The effect of wounding, temperature and inoculum density on the development of pink rot of potatoes caused by *Phytophthora erythroseptica*. Plant Dis 84: 1327-1333.
17. Taylor RJ, Salas B, Gudmestad NC (2004) Differences in etiology affect mefenoxam efficacy and the control of pink rot and leak tuber diseases of potato. Plant Dis 88: 301-307.
18. Platt B (2008) Maladies de la pomme de terre causées par des oomycètes. Cahier Agriculture 17: 361-367.
19. Gachango E, Kirk W, Schafer R (2012) Effects of in-season crop-protection combined with postharvest applied fungicide on suppression of potato storage diseases caused by oomycete pathogens. Crop Prot 41: 42-48.
20. Knowles NR, Plissey ES (2008) Maintaining tuber health during harvest, storage and post-storage handling. In: Potato Health Management. USA: APS Press St Paul MN.
21. Olsen N, Kleinkopf G, Woodell L (2003) Efficacy of chlorine dioxide for disease control on stored potatoes. Am J Potato Res 80: 387-395.
22. Taylor RJ, Pasche JS, Gudmestad NC (2006) Biological significance of mefenoxam resistance in *Phytophthora erythroseptica* and its implications for the management of pink rot of potato. Plant Dis 90: 927-934.
23. Miller JS, Olsen N, Woodell L, Porter LD, Calayson S (2006) Postharvest applications of zoxamide and phosphate for control of potato tuber rots caused by oomycetes at harvest. Am J Potato Res 83: 269-278.
24. Bojanowski A, Avis TJ, Pelletier S, Tweddell RJ (2013) Management of potato dry rot. Postharvest Biol Technol 10: 99-109.
25. Schisler DA, Slininger PJ, Bothast RJ (1997) Effects of antagonist cell concentration and two-strain mixtures on biological control of *fusarium* dry rot of potatoes. Phytopathology 87: 177-183.
26. Etebarian HR, Scott ES, Wicks TJ (2000) *Trichoderma* harzianum T39 and T. virens DAR 74290 as potential biological control agents for *Phytophthora erythroseptica*. Eur J Plant Pathol 106: 329-337.
27. McLeod A, Labuschagne N, Kotzé JM (1995) Evaluation of *Trichoderma* for biological avocado root rot in bark medium artificially infested with *Phytophthora cinnamomi*. South African Avocado Growers' Association Yearbook 18: 32-37.
28. Daami-Remadi M, Dkhili I, Jabnoun-Khiareddine H, El Mahjoub M (2012) Biological control of potato leak with antagonistic fungi isolated from compost teas and solarized and non-solarized soil. In Daami-Remadi M. (Ed) Potato Pathology. Pest Technol 6 (Special Issue 1): 32-40.
29. Aydi R, Hassine M, Jabnoun-Khiareddine H, Ben Jannet H, Daami-Remadi M (2013a) Valorization of *Aspergillus* spp. as biocontrol agents against Pythium and optimization of their antagonistic potential in vitro and in vivo. Tunisian J Med Plants Nat Prod 9: 70-82.
30. Aydi-Ben Abdallah R, Hassine M, Jabnoun-Khiareddine H, Haouala R, Daami-Remadi M (2014b) Antifungal activity of culture filtrates and organic extracts of *Aspergillus* spp. against *Pythium ultimum*. Tunisian J Plant Prot 9: 17-30.
31. Aydi R, Hassine M, Jabnoun-Khiareddine H, Ben Jannet H, Daami-Remadi M, (2014a) Study of the antifungal potential of *Aspergillus* spp. and their culture filtrates and organic extracts against *Fusarium sambucinum*. Tunisian J Med Plants Nat Prod 11: 15-29.
32. Aydi-Ben Abdallah R, Hassine M, Jabnoun-Khiareddine H, Ben Jannet H, Daami-Remadi M (2013b) Etude du pouvoir antifongique in vitro et in vivo des filtrats de culture et des extraits organiques des *Aspergillus* spp. contre *Phytophthora* sp. Microbiologie & Hygiène Alimentaire 25: 15-25.
33. Kaewchai S, Soytong K (2010) Application of biofongicides against *Rigidoporus microporus* causing white root disease of rubber trees. J Agri Technol 6: 349-363.
34. Bhattacharyya PN, Jha DK (2011) Optimization of cultural conditions affecting

growth and improved bioactive metabolite production by a subsurface *Aspergillus* strain TSF 146. Int J Appl Biol Pharm Technol 2: 133-143.

35. Siddiqui IA, Shaukat SS, Khan A (2004) Differential impact of some *Aspergillus* species on *Meloidogyne javanica* biocontrol by *Pseudomonas fluorescens* strain CHA0. Lett Appl Microbiol 39: 74-83.

36. Tjamos EC, Paplomatas EJ (1987) Effect of soil solarization on the survival of fungal antagonists of *Verticillium dahliae*. EPPO Bull: 17: 645-653.

37. El-Masry MH, Khalil AI, Hassouma MS, Ibrahim HAA (2002) In situ an in vitro suppressive effect of agricultural composts and their water extracts on some phytopathogenic fungi. World J Microbiol Biotechnol 18: 551-558.

38. Molla AH, Fakhru'l-Razi A, Hanafi MM, Abd-Aziz S, Alam MZ (2002) Potential non-phytopathogenic filamentous fungi for bioconversion of domestic wastewater sludge. J Environ Sci Health A Tox Hazard Subst Environ Eng 37: 1495-1507.

39. Naidu Y, Meon S, Kadir J, Siddiqui Y (2010) Microbial starter for the enhancement of biological activity of compost tea. Int J Agric Biol 12: 51-56.

40. Raimbult M (1998) General and Microbiological aspects of solid substrate fermentation. Electronic J Biotechnol 1: 2-15.

41. Muhammad S, Amusa NA (2003) In vitro inhibition of growth of some seedling blight inducing pathogen by compost-inhabiting microbes. Afr J Biotechnol 2: 161-164.

42. Suárez-Estrella F, Vargas-Garcia C, Lopez MJ, Capel C, Moreno J (2007) Antagonistic activity of bacteria and fungi from horticultural compost against *Fusarium oxysporum* f. sp. melonis. Crop Prot 26: 46-53.

43. Degola F, Berni E, Restivo FM (2011) Laboratory tests for assessing efficacy of atoxigenic *Aspergillus flavus* strains as biocontrol agents. Int J Food Microbiol 146: 235-243.

44. Frisvad JC, Larsen TO (2015) Chemodiversity in the genus *Aspergillus*. Appl Microbiol Biotechnol 99: 7859-7877.

45. Dennis C, Webster J (1971) Antagonistic properties of species-groups of *Trichoderma*, production of non-volatile antibiotics. Trans Brit Mycol Soc 57: 25-39.

46. Hibar K, Daami-Remadi M, Khiareddine H, El Mahjoub M (2005) Effet inhibiteur in vitro et in vivo du *Trichoderma harzianum* sur *Fusarium oxysporum* f. sp. *radicis-lycopersici*. Biotechnol Agron Soc Environ 9: 163-171.

47. Lapwood DH, Read PJ, Spokes J (1984) Methods for assessing the susceptibility of potato tubers of different cultivars to rotting by *Erwinia carotovora* subsp. atroseptica and carotovora. Plant Pathol 33: 13-20.

48. El-Kot Gan (2008) Biological control of black scurf and dry rot of potato. Egypt J Phytopathol 36: 45-56.

49. Hoitink HAJ, Inbar Y, Boehm MJ (1991) Status of compost-amended potting mixes naturally suppressive to soilborne diseases of floricultural crops. Plant Dis 75: 869-873.

50. Melo IS, Faull JL, Nascimento RS (2006) Antagonism of *Aspergillus terreus* to *Sclerotiana sclerotiorum*. Braz J Microbiol 37: 417-419.

51. Adebola MO, Amadi J,E (2010) Screening three *Aspergillus* species for antagonistic activities against the cocoa black pod organism (*Phytophthora palmivora*). Agric Biol J N Am 1: 362-365.

52. Vibha (2010) Effect of fungal metabolites and amendments on mycelium growth of *Rhizoctonia solani*. J Plant Prot Res 50: 93-97.

53. Zhang CL, Zheng BQ, Lao JP, Mao LJ, Chen SY, et al. (2008) Clavatol and patulin formation as the antagonistic principle of *Aspergillus clavatonanicus*, an endophytic fungus of Taxus mairei. Appl Microbiol Biotechnol 78: 833-840.

54. Patil MP, Patil RH, Maheshwari VL (2015) Biological Activities and Identification of Bioactive Metabolite from Endophytic *Aspergillus flavus* L7 Isolated from Aegle marmelos. Curr Microbiol 71: 39-48.

55. Soltani J, Hosseyni Moghaddam MS (2015) Fungal endophyte diversity and bioactivity in the Mediterranean cypress Cupressus sempervirens. Curr Microbiol 70: 580-586.

56. Barocio-Ceja NB, Ceja-Torres LF, Morales-García JL, Silva-Rojas HV, Flores-Magallón R, et al. (2013) In vitro biocontrol of tomato pathogens using antagonists isolated from chicken-manure vermicompost. Int J Exp Bot 82: 15-22.

57. Israel S, Lodha S (2005) Biological control of *Fusarium* oxysporum f. sp. cumini with *Aspergillus* versicolor. Phytopathol Mediterr 44: 3-11.

58. Dwivedi SK, Enespa (2013) In vitro efficacy of some fungal antagonists against *Fusarium* solani and *Fusarium* oxysporum f. sp. lycopersici causing brinjal and tomato wilt. International Journal of Biological and Pharmaceutical Research 4: 46-52.

59. Tiwari CK, Paribar J, Verma RK (2011) Potential of *Aspergillus niger* and *Trichoderma viride* as biocontrol agents of wood decay fungi. J Indian Acad Wood Sci 8: 169-172.

60. Venkatasubbaih P, Safeeulla KM (1984) *Aspergillus niger* for biological control of *Rhizoctonia solani* on coffee seedling. Trop Pest Manag 30: 401-406.

61. Ramzan N, Noreen N, Shahzad S (2014) Inhibition of in vitro growth of soil-borne pathogens by compost-inhabiting indigenous bacteria and fungi. Pak J Bot 46: 1093-1099.

62. Xu D, Wang H, Zhang Y, Yang Z, Sun X (2013) Inhibition of non-toxigenic *Aspergillus niger* FS10 isolated from Chinese fermented soybean on growth and aflatoxin B1 production by *Aspergillus flavus*. Food Control 32: 359-365.

63. Hoitink H, Boehm M (1999) Biocontrol within the context Of soil microbial communities: A Substrate-Dependent Phenomenon. Annu Rev Phytopathol 37: 427-446.

64. Sabet KK, Saber MM, El-Naggar MAA, El-Mougy NS, El-Deeb HM, et al. (2013) Using commercial compost as control measures against cucumber root-rot disease. J Mycol 2013: 1-13.

65. Hu X, Webster G, Xie L, Yu C, Li Y, Liao X (2013) A new mycoparasite, *Aspergillus* sp. ASP-4, parasitizes the sclerotia of *Sclerotinia sclerotiorum*. Crop Prot 54: 15-22.

66. Brzezinska MS, Jankiewicz U (2012) Production of antifungal chitinase by *Aspergillus niger* LOCK 62 and its potential role in the biological control. Curr Microbiol 65: 666-672.

67. Gao P1, Korley F, Martin J, Chen BT (2002) Determination of unique microbial volatile organic compounds produced by five *Aspergillus* species commonly found in problem buildings. AIHA J (Fairfax, Va) 63: 135-140.

68. Pandey RR, Arora DK, Dubbey RC (1993) Antagonistic interaction between fungal pathogens and phylloplane fungi of guava. Mycopathologia 124: 31-39.

69. Thakur S, Harsh SNK (2014) In vitro potential of volatile metabolites of phylloplane fungi of piper longum as biocontrol agent against plant pathogen. International Journal of Science and Nature 5: 33-36.

70. Barros GG, Torres AM, Rodriguez MI, Chulze SN (2006) Genetic diversity within *Aspergillus flavus* strains isolated from peanut-cropped soils in Argentina. Soil Biol Biochem 38: 145-152.

71. Mehl HL, Cotty PJ (2010) Variation in competitive ability among isolates of *Aspergillus flavus* from different vegetative compatibility groups during maize infection. Phytopathology 100: 150-159.

72. Atehnkeng J, Ojiambo PS, Cotty PJ, Bandyopadhyay R (2014) Field efficacy of a mixture of atoxigenic *Aspergillus flavus* Link: Fr vegetative compatibility groups in preventing aflatoxin contamination in maize (*Zea mays* L.). Biol Control 72: 62-70.

73. Khan MR, Anwer MA (2007) Molecular and biochemical characterization of soil isolates of *Aspergillus niger* aggregate and an assessment of their antagonism against *Rhizoctonia solani*. Phytopathol Mediterr 46: 304-315.

74. Khan MR, Anwar MA, Khan SM, Khan MM (2006) An evaluation of isolates of *Aspergillus niger* against *Rhizoctonia solani* in vitro and in vivo on eggplant. Tests of Agrochemicals and Cultivars 27: 31-32.

75. Wang WH, Zhao X, Liu C, Liu L, Yu S, et al. (2015) Effects of the biocontrol agent *Aspergillus* flavipes on the soil microflora and soil enzymes in the rooting zone of pepper plants infected with *Phytophthora capsici*. J Phytopathol 163: 513-521.

76. Chuang CC, KuoYL, Chao CC, Chao WL (2007) Solubilization of inorganic phosphates and plant growth promotion by *Aspergillus niger*. Biol Fertil Soils 43: 575-584.

77. Mohale S, Medina A, Magan N (2013) Effect of environmental factors on in vitro and in situ interactions between atoxigenic and toxigenic *Aspergillus flavus* strains and control of aflatoxin contamination of maize. Biocontrol Sci Technol 23: 776-793.

78. Huang Z, White DG, Payne GA (1997) Corn Seed Proteins Inhibitory to *Aspergillus flavus* and Aflatoxin Biosynthesis. Phytopathology 87: 622-627.

79. Yang Y, Jin D, Wang G, Wang S, Jia X, et al. (2011) Competitive biosorption of Acid Blue 25 and Acid Red 337 onto unmodified and CDAB-modified biomass of *Aspergillus* oryzae. Bioresour Technol 102: 7429-7436.

24

Periodic Existence of Mycorrhizal Fungi in Roots and Non-root Dissident Portions of Some Bulbous Plants

Muhammad Ali*, Muhammad Adnan and Mehra Azam

Institute of Agricultural Sciences,University of the Punjab, Lahore-54590, Pakistan

Abstract

Decaying and senescing scale-like leaves and roots were collected regularly from Botanical Garden, University of the Punjab, Lahore for a period of four months, with an interval of fifteen days. Roots and other portions of both the plants when processed and examined revealed the occurrence of AM fungal structures. However, AM structures were totally missing in *Allium cepa*, and roots of *Amaryllis vittata*. Thick hyphal mats with clusters and clumps of spores were often seen in the decaying scale-leaves. The vesicles in these portions were large sized and thick walled. As regard the seasonal variations, the side by side of hyphal, arbuscular and vesicular infections varied among the samples collected throughout the season. Seasonal variations in Glomalean spore dynamics were observed with respect to number while glomeromycetous species richness varied in the rhizosphere soil of the three plants. The recent research was conducted to evaluate the configuration of occurrence of arbuscular mycorrhizal fungal structures in decaying scale like leaves and root system of three bulbous plants i.e., *Allium cepa*, *Amaryllis vittata* and *Zaphyranthes citrina*.

Keywords: Arbuscular mycorrhizae; Scale-leaves; *Allium cepa; Amaryllis vittata; Zaphyranthes citrine*; Species richness

Introduction

Arbuscular Mycorrhizae (AM) is mutually beneficial relationship amongst fungi and roots of the higher plants. Colonization of roots by mycorrhiza has been revealed to recover development and yield of numerous field crops including leguminous crops, cereals, vegetables and oil crops [1-5]. Mycorrhizal associations increase plant growth and productivity by increasing nutrient element uptake [6] and improving resistance to abiotic [7,8] and biotic [9] stress factors. Plants often benefit from the presence of mycorrhizal associates via a variety of mechanisms including improvement of soil structure, mobilization of essential minerals, enhancement of desiccation resistance, and protection from pathogens and herbivores [10]. Recently some workers have reported that AM also increase the crop tolerance against allelopathy and enhance crop growth under this stress [11,12].

Vesicular arbuscular mycorrhizae are of universal occurrence in roots of 95% of land plants [13,14]. However, during past few decades reports are available about the presence of AM in plant portions other than roots [15,16]. Since than a number of examples have been added to the literature with an emphasis on a much wider occurrence of these associations than reported ever before.

We had planned this study to further add to the information about AM proliferating in roots and non-root portions of some bulbous plants.

Materials and Methods

Sampling of root and non-root portions like scale leaves, dried decaying sheathing leaves on the bulbs of two test bulbous plants viz. *Allium cepa*, *Amaryllis vittata* and *Zaphyranthes citrina*. The sampling was done regularly from February to May. Specific sites (Sites name) were selected for sampling of plants. The root/bulb samples were dug up along with rhizospheric soil. Extreme care was taken to avoid the disturbance of root systems, adhering decayed and semi-decayed scale leaves and epidermis. The samples were brought back to the Lab. in polythene bags. Decaying scale like leaves and fine roots were gently peeled off with the help of forceps, while rest of the bulbous portion along with root system were dipped in a bucket of water for half an hour. The adhering soil was removed with the help of camel hairbrush while washing gently under the tap water. The root system of all samples, scales and other portions were cut up into 1 cm^2 pieces and then fixed in F.A.A. (Formaline: Acetic Acid : Alcohol in 5:5:90 ratio by volume) in properly labeled MacCartney bottles separately. The samples were cleared and stained for analysis of colonization of AM fungi using Phillips and Hayman procedure [15]. The roots were cleared in 10% KOH solution in an autoclave, placed in 0.1N HCl for 2-3 minutes for neutralization and then stained with trypan blue solution (0.05% in lactophenole glycerine). The sample pieces were mounted on the glass slides in a drop of lactic acid and were observed under low power (10x) of the light microscope. Extent of AM infection was recorded with the help of an already calibrated ocular micrometer. This was done by randomly focusing the plant material under the microscope and by measuring the hyphae. The number of vesicles and other structures were also recorded in the same way. Microphotography was done with the help of Minolta X 700 camera with the adapter tube.

Soil adhering to the roots and other portions like bulbs were utilized for screening the associated AM spores. Spore extraction was done by following wet sieving decanting method of Nicolson and Gerdemann [11,12]. Density and diversity of spores was recorded. Spores were identified using synoptic key by Morton [16] and Schenck and Perez [17]. All the data was statistically analyzed by computing Standard

***Corresponding author:** Muhammad Ali, Institute of Agricultural Sciences, University of the Punjab, Lahore-54590, Pakistan
E-mail: muhammadali.mycologist@gmail.com

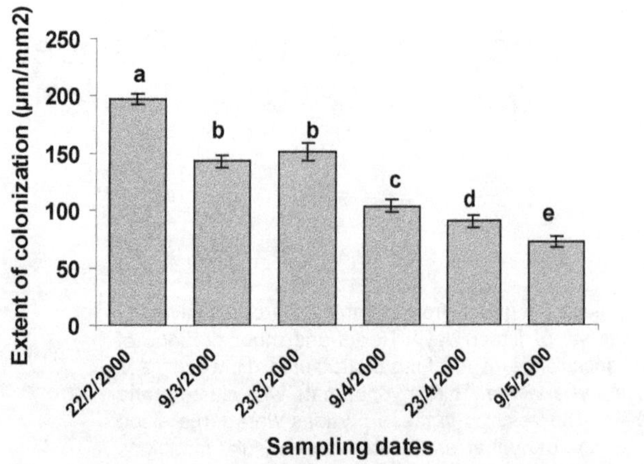

Figure 1: Seasonal changes in extent of AM colonization in root system of *Amayrllis vittata*. Line on data bars show SE of the mean while data bars with different letters show significant difference (p=0.05) as determined by DMRT.

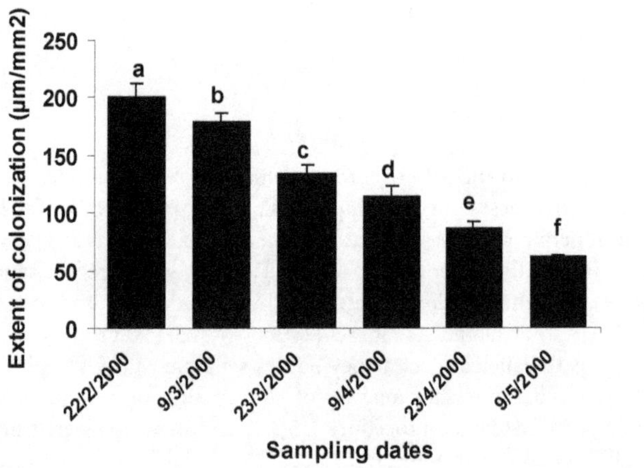

Figure 2: Seasonal changes in status of AM infection in root system of *Zaphyranthes citrina.* Line on data bars show SE of the mean while data bars with different letters show significant difference (p=0.05) as determined by DMRT.

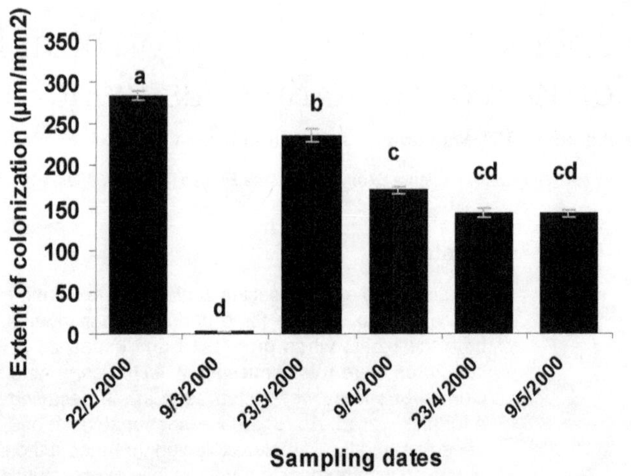

Figure 3: Seasonal changes in status of AM infection in decaying scale leaves of *Zaphyranthes citrina*. Line on data bars show SE of the mean while data bars with different letters show significant difference (p=0.05) as determined by DMRT.

Error, Least Significant Difference; (LSD) and Duncan's New Multiple Range Test (Steel and Torrie [18].

Result and Discussion

AM colonization was completely absent in scales and roots of *Allium cepa*, and roots of *Amaryllis vittata* (members of Amaryllidaceae) may be attributed to existence of immunity to attack by pathogenic fungi. In bulbs of *Allium cepa* resistance to attack by fungus is due to the presence of catecol and protocatechuic acid in the dry, pigmented, outer scale leaves. Members of Amaryllidaceae also produce fungitoxin (Methylene-buty rolactose) in their bulbs, which also diffuses to plant roots (Figures 1-3) and (Tables 1-3).

Sample No	Sampling date	No. of vesicles per mm^2	No. of Intramatrical spores per mm^2	No. of cells filled with DSEF per mm^2
1.	22/2/2000	000.77d ± 0.29	002.44e ± 0.94	008.66e ± 0.66
2.	9/3/2000	002.66b ± 0.577	003.66d ± 1.33	007.33f ± 2.33
3.	23/3/2000	002.66b ± 1.00	005.33c ± 0.19	013.66d ± 1.76
4.	9/4/2000	002.44c ± 0.1	007.99b ± 0.51	015.66c ± 0.66
5.	23/4/2000	002.88a ± 0.39	012.66a ± 3.38	016.00b ± 0.57
6.	9/5/2000	002.44c ± 1.25	012.10a ± 2.27	018.00a ± 4.04
L.S.D		1.36	3.40	3.93

Table 1: Status of various AM structures in decaying scale like leaves of *Amaryllis vittata* at different sampling time.

Sample No	Sampling month	No. of vesicles per mm^2	No. of Intramatrical spores per mm^2	No. of cells filled with DSEF per mm^2
1.	22/2/2000	011.00a ± 0.57	003.66c ± 1.76	--
2.	9/3/2000	010.00b ± 1.52	003.66c ± 0.33	--
3.	23/3/2000	010.33b ± 1.45	003.66c ± 0.33	001.33c ± 0.33
4.	9/4/2000	008.88c ± 3.39	003.89d ± 0.39	006.00b ± 3.00
5.	23/4/2000	006.33d ± 0.03	005.33b ± 0.33	--
6.	9/5/2000	005.33e ± 0.57	006.33a ± 0.36	007.33a ± 0.88
L.S.D		1.77	1.48	2.42

Table 2: Status of various AM structures in root systems of *Zaphyranthes citrina* at different sampling times.

Sample No	Sampling date	No. of vesicles per mm²	No. of Intramatrical spores per mm²	No. of cells filled with DSEF per mm²
1.	22/2/2000	032.55a ± 1.25	004.00e ± 0.57	008.66a ± 2.40
2.	9/3/2000	030.77b ± 0.83	005.33d ± 0.66	004.66c ± 0.33
3.	23/3/2000	026.21c ± 0.29	011.33e ± 1.45	005.33b ± 1.53
4.	9/4/2000	023.88d ± 1.68	022.00d ± 1.52	004.00c ± 1.73
5.	23/4/2000	018.32e ± 1.33	030.33ab ± 1.85	003.66d ± 0.88
6.	9/5/2000	015.55f ± 1.56	032.00a ± 0.57	003.33d ± 1.20
L.S.D		2.35	2.30	2.76

Table 3: Status of various AM structures in scale leaves of *Zaphyranthes citrina* at different sampling times.

References

1. Javaid A, Iqbal SH, Hafeez FY (1994) Effect of different strains of Bradyrhizobium and two types of vesicular arbuscular mycorrhizae (VAM) on biomass and nitrogen fixation in Vigna radiate (L.) Wilzek var. NM 20-21.Sci Inter (Lahore) 6: 265-267.
2. Yao MK, Tweddell RJ, Désilets H (2002) Effect of two vesicular-arbuscular mycorrhizal fungi on the growth of micropropagated potato plantlets and on the extent of disease caused by Rhizoctonia solani. Mycorrhiza 12: 235-242.
3. Kapoor R, Giri B, Mukerji KG (2004) Improved growth and essential oil yield and quality in Foeniculum vulgare mill on mycorrhizal inoculation supplemented with P-fertilizer. Bioresour Technol 93: 307-311.
4. Subramanian KS, Santhanakrishnan P, Balasubramanian P (2006) Responses of field grown tomato plants to arbuscular mycorrhizal fungal colonization under varying intensities of drought stress. Sci Hortic 107: 245-253.
5. Wang FY, Lin XG, Yin R, Wu LH (2006) Effect of arbuscular mycorrhizal inoculation on the growth of Elsholtzia splendens and Zea mays and the activities of phosphatase and urease in a multi-metal-contaminated soil under unsterilized conditions. Appl Soil Ecol 31: 110-119.
6. Al-Karaki GN (2002) Benefit, cost and phosphorus use efficiency of mycorrhizal field grown garlic at different soil phosphorus levels. J Plant Nutrition 25: 1175-1184.
7. Feng G, Zhang FS, Li XL, Tian CY, Tang C, et al. (2002) Improved tolerance of maize plants to salt stress by arbuscular mycorrhiza is related to higher accumulation of soluble sugars in roots. Mycorrhiza 12: 185-190.
8. Chen BD, Zhu YG, Smith FA (2006) Effects of arbuscular mycorrhizal inoculation on uranium and arsenic accumulation by Chinese brake fern (Pteris vittata L.) from a uranium mining-impacted soil. Chemosphere 62: 1464-1473.
9. St. Arnaud M, Hamel C, Caron M, Fortin JA (1994) Endomycorrhize VA et sensibilite des plantes aux maladie. In: Fortin JA, Charest C and Piche Y (eds.) La Symbiose Mycorhizienne, Etat des Connaisances, Orbis, Quebec.
10. Bajwa R, Javaid A, Haneef A (1999) EM and VAM technology in Pakistan N: Response of Cowpea to co-inoculation with EM and VAM under allelopathicstress. Pak. J. Botany. 31: 387-369.
11. Nicolson TH (1967) Vesicular arbuscular mycorrhiza, a universal plant symbionts.
12. Gerdemann JW (1968) Vesicular arbuscular mycorrhizae and plant growth. Annual Review of Phytopathology 6: 39-418.
13. Bagyaraj DJ, Manjunath A, Patal RB (1979) Occurrence of vesicular arbuscular mycorrhiza in some tropical aquatic plants. Trans Brit Mycol Soc 72: 164-167.
14. Nasim G (1990) Vesicular arbuscular mycorrhiza in portions other than roots. In: Bushan L. Jalali, Chand L (eds.) Current trends in mycorrhizal research. Proceedings of the national conference on mycorrhiza. A Haryana Agr Uni Hisar 14-16: 4-6.
15. Philips JM, Gerdeman DS (1970) Improved procedure for clearing and staining parasitic and VAM fungi for the rapid assessment of infection. Irans. Brit. Mycol. Soc., 55: 158-161.
16. Morton JB (1988) Taxonomy of mycorrhizal fungi. Classification, nomenclature and identification, Mycotoxicon 100: 267-3240.
17. Schenck NC, Perez Y (1987) Manual for the identification of VAM fungi. In: VAM, 1453.
18. Steel RDG, Torrie JH (1980) principles and procedure of statistics. McGraw Hill Book Co., Inc. New York USA.

The Role of Cutinase and its Impact on Pathogenicity of *Colletotrichum truncatum*

Adelene SM Auyong[1], Rebecca Ford[2] and Paul WJ Taylor[3]*

[1]*Faculty of Science, The University of Melbourne, Parkville 3010 VIC, Australia*
[2]*School of Environment, Griffith University, Southport 4222 QLD, Australia*
[3]*Faculty of Veterinary and Agricultural Sciences, The University of Melbourne, Parkville 3010 VIC, Australia*

Abstract

The phytopathogenic fungus, *Colletotrichum truncatum* infects and colonises chili fruit through direct penetration of the cuticle. The cutinase gene of *C. truncatum* (*CtCut*1), a cutin degrading enzyme was identified, cloned and shown to be essential in breaching the cuticle of chili fruit. The expression of *CtCut*1 gene was studied through RNA-mediated gene silencing and its impact on fungal pathogenicity was demonstrated. The vector, pAA1 encoding a hairpin RNA of GFP and CtCut1 was constructed and transformed into *C. truncatum* pathotype F8-3B (virulent strain). F8-3B-pAA1 transformants exhibited reduced patterns of infection with one isolate having a 45.8% reduction in cutinase activity (reduction in CtCut1 transcript) in comparison to the wild type. Importantly, CtCut1-deficient strains were unable to infect detached chili and soybean hosts as efficiently as the wild type. There was a delay in the infection period by the transformants. Nevertheless, artificial wounding of the cuticle enabled these F8-3B-pAA1 transformants to infect and colonise host tissues, resulting in typical anthracnose disease lesions. Coupled with microscopy, these data suggested that the defect in pathogenicity was likely due to a failure in penetration of the host cutin. Knowledge of the plant-fungal interactions arose from the development of a fungal transformation system for *C. truncatum* and implementation of RNAi technology. This technology thus provides an alternative genetic tool for studies of gene function, particularly of essential pathogenicity genes.

Keywords: Cutinase, Pathogenicity, *Colletotrichum truncatum*, Anthracnose

Introduction

For successful pathogenicity, fungi employ various infection strategies to breach the host cuticle. While some pathogens use mechanical means through the formation of specialised infection structures [1], others enter by direct penetration of the cuticle through the assistance of extracellular cutinase activity [2,3].

Colletotrichum truncatum, a pathogen of legumes, cereals, grasses and vegetable crops such as *Capsicum* species is an ascomycete that infects its host by direct penetration through the host cuticle [4]. The process of infection/penetration is thought to be triggered by cutin degrading enzymes such as cutinase. Cutinases from many fungi have been isolated and characterised [5-8]. Cutinases have been suggested to participate in carbon acquisition for saprophytic growth [9]. They were also presumed to have a role in fungal penetration [10,11] and in surface adhesion [12].

However, the importance of cutinase in pathogenesis of fungi remains controversial since disruption of the cutinase gene in various fungi resulted in contradictory results. The overexpression of *MfCut1* in *Monilinia fructicola* increased pathogenicity on *Prunus* spp. and targeted gene replacement of *Magnaporthe grisea* and *Pyrenopeziza brassicae* cutinases dramatically reduced pathogenicity on rice, barley and oilseed rape hosts, respectively [11,13,14]. On the contrary, disruption of cutinase from other fungi such as *Nectria haematococca* [15] and *Botrytis cinerea* [16] resulted in conflicting evidence of the role of cutinase in pathogenicity. Pathogenicity of transformants in both fungi was unaltered compared to their wild type and control counterparts. The relevance of cutinase gene in fungal pathogenicity of *C. truncatum* however, has not yet been identified.

Cutinase has been proposed to be the hydrolyzing agent of the plant cuticle and/or elicitor of cutin monomers to trigger signaling pathways within fungal pathogens [17,18]. Despite the identification of cutinase in a range of *Colletotrichum* species such as in *C. gloeoesporioides* [3,6], *C. kahawae* [6], *C. trifolii* [19], *C. lagenarium* [20] and *C. capsici* [5], characterisation of the gene and its function in the *C. truncatum*-chili interaction remains unelucidated.

Analysis of gene function has been traditionally performed by conventional methods such as examining for phenotypic or biochemical changes in response to mutation of the gene of interest [20,21]. The process however, can be laborious and time consuming. Furthermore, the abundant wealth of information provided by the ever increasing sequencing of fungal species genomes requires an efficient method to link the genetic information to its biological function with much more accuracy. This has led to the development of alternative methods such as RNA-mediated gene silencing.

RNA-mediated gene silencing is a powerful tool for functional analysis of genes [22]. Initially discovered in *Caenorhabditis elegans* [23], the strategy has now been applied to a range of organisms across different kingdoms [24]. Termed as Post Transcriptional Gene Silencing (PTGS) in plants, RNA interference (RNAi) in animals and quelling in fungi [25], the phenomenon exploits the endogenous gene regulatory mechanism of eukaryotic cells. Double stranded RNA

***Corresponding author:** Paul WJ Taylor, Faculty of Veterinary and Agricultural Sciences, The University of Melbourne, Parkville 3010 VIC, Australia
E-mail: paulwjt@unimelb.edu.au

(dsRNA) is cleaved by a nuclease (Dicer) of the RNase III family into Smaller Interfering RNAs (siRNAs) of 21 to 28 nucleotides in length, that are then incorporated into a ribonucleoprotein complex forming the RNA-Induced Silencing Complex (RISC). The RISC recognises homologous mRNAs by complementary base pairing and triggers sequence-specific degradation of the homologous mRNA [26].

Studies of gene function through specific inhibition of pathogenicity or plant-fungal interaction gene expression using RNA-mediated gene silencing have been reported in many ascomycete fungi [27] including *Colletotrichum gloeosporioides* [28], *Magnaportha oryzae* and *Colletotrichum lagenarium* [29]. However, such silencing mechanism has not yet been employed in *C. truncatum* to understand pathogenicity.

The aim of this paper was to identify functional cutinase genes in *C. truncatum* and then determine if the cutinase gene, *CtCut1* of *C. truncatum* plays a role in pathogenicity. Utilising a short (515 bp) cDNA fragment of *CtCut1* as the silencing target for gene silencing by RNAi, the efficacy of the pAA1 plasmid harboring the hairpin RNA cassette as a mediator for gene silencing was examined. Coupled with an established *Agrobacterium tumefaciens*-Mediated Transformation (ATMT) system for *C. truncatum* [30], the RNA-mediated gene silencing approach was successfully applied to study the function of *C. truncatum* cutinase gene in response to contact with its hosts.

Materials and Methods

Strains, culture conditions and plant materials

Colletotrichum truncatum pathotype F8-3B [31,32] and its transformant expressing GFP [30] that caused anthracnose disease of chili (*Capsicum* spp.) were grown and maintained on Potato Dextrose Agar (PDA; Difco Laboratories, France) at 24°C with 12 h photoperiod. Medium for the GFP-tagged F8-3B transformant was supplemented with 250 μg/ml of hygromycin B antibiotic. Conidiospores obtained from 14-day-old cultures were adjusted to 10^6 conidiospores/ml and used as the recipient in all transformation experiments. Genomic DNA/RNA extraction was carried out on fungal mycelium grown in 50 ml Czapek-Dox broth (CDB; Difco Laboratories, France) for 7 days under the same temperature and photoperiod described above until a mycelial mat was formed.

Agrobacterium tumefaciens strain AGL1, used as the donor for fungal transformation was grown and maintained in Luria Bertani (LB) broth containing 5 mM glucose (LBG; Sigma Aldrich, USA) at 28°C. One Shot® TOP10 Chemically Competent *Escherichia coli* (Invitrogen, Australia) used as the host for gene manipulation was maintained in LB broth at 37°C.

For pathogenicity assays, *Capsicum annuum* genotype Bangchang plants were established in pots in a glasshouse and grown under controlled conditions of light (18 h photoperiod), temperature (28°C) and humidity (60-70%). One-month-old *Glycine max* genotype Bragg soybean plants were also established in pots and maintained in the glasshouse.

Nucleic acid manipulations

Total RNA was extracted from saprophytic stage (growing in CDB) and pathogenic stage (2-, 6- and 24-hours post-inoculated) mycelium using RNeasy Plant Mini kit (Qiagen, Australia). The RNA extraction was performed following the manufacturer's instructions. Contaminating genomic DNA in all RNA samples were removed by DNase I (Qiagen, Australia) treatment at room temperature for 30 min. The integrity and quantity of the total RNA was confirmed by 1.5% agarose gel electrophoresis and 1 kb Plus DNA Ladder (Invitrogen, Australia). First strand cDNA was performed using the Omniscript RT-PCR Kit (Qiagen, Australia) and the oligo-dT primer (Qiagen, Australia) according to the manufacturer's protocol. Gene transcript was evaluated by reverse transcription polymerase chain reaction (RT-PCR) using Cut1F and Cut1R primers (Table 1), designed based on orthologues of fungal cutinase. The *GAPDH* gene was used to normalise the expression levels of the studied gene. A total of three replications were conducted for the experiment.

Genomic DNA was extracted from fungal mycelium using DNeasy Plant Mini kit (Qiagen, Australia) following the manufacturer's protocol. PCR was conducted to amplify cutinase gene in *C. truncatum* F8-3B using gCUT-F and gCUT-R primers (Table 1). The PCR reaction components consisted of 20 ng genomic DNA, 250 μM dNTPs, 0.5 μM of each primer, 1.5 mM $MgCl_2$, PCR buffer and 0.5 U *Taq* DNA Polymerase (Scientifix, Australia) in a total volume of 25 μl. Conditions for amplification were an initial stage of denaturation at 94°C for 5 min followed by 30 cycles of 94°C for 30 s, 60°C for 30 s and 72°C for 30 s and a final elongation step at 72°C for 5 min. to ensure that any remaining single stranded DNA was fully extended. The PCR reaction was done in PTC-100™ Programmable Thermal Controller (MJ Research, Inc, USA). Both PCR products derived from the cDNA and genomic DNA were cloned and sequenced (AGRF, Australia).

Alignment of the sequence was generated using CLUSTALW (1.81) Multiple Sequence Alignments program [33]. Sequence analysis was carried out using BLAST (Basic Local Alignment Search Tool), accessed through the Internet (http://www.ncbi.nlm.nih.gov/) and the *C. graminicola* Sequencing Project, Broad Institute of Harvard and MIT (http://www.broadinstitute.org/). Six frame translations on the nucleotide sequence were performed using SDSC Biology Workbench 3.2 (http://workbench.sdsc.edu) [34]. A homology search was done at the nucleotide and protein level using BLASTN and

Name	Sequence (5' → 3')[a]	Applications
Cut1F	CTTGTCGCTGCCGCTCCTGT	Reverse transcription PCR
Cut1R	AGAGTCGCTTAGCCTCGTTGAT	Reverse transcription PCR
gCUT-F	AATCCCTTACAACTTTCCTCTGACA	Cutinase gene amplification
gCUT-R	ATCAACCGCCACGATACAGACAA	Cutinase gene amplification
SmaI-CutS-F	TATATA**cccggg**TCATCTCTCTTGCCGTTTC	Cassette construction, Southern analysis
XbaI-CutS-R	TCAGAT**tctaga**GTCCAAGGTTCTGCAGGTTC	Cassette construction, Southern analysis
BamHI-CutAS-F	TAGTAT**ggatcc**GTCCAAGGTTCTGCAGGTTC	Cassette construction
XhoI-CutAS-R	ATTCGA**ctcgag**TCATCTCTCTTGCCGTTTC	Cassette construction
S-GFP-R	TCGCCGATGGGGGTGTTCTGCT	PCR, Southern analysis
5NQ. CUTF	TCTGTTCGGCTACACCAAGAA	Quantitative real-time PCR
5NQ. CUTR	CAAGAAATGCGCCGGCAGGAT	Quantitative real-time PCR
hphF	GCTGCGCCGATGGTTTCTACA	pAA1 analysis
hphR	GCGCGTCTGCTGCTCCAT	pAA1 analysis

[a]Restriction sites are bolded lower case letters

Table 1: List of primers used in the study.

BLASTX [35]. Bioinformatics tools were used to analyse and confirm the cloned gene, which was later designated as *CtCut*1. The protein predition/manipulation tools in the Gene Infinity program (http://www.geneinfinity.org/) were also used to predict the cutinase protein molecular weight and protein isoelectric point.

Vector construction and fungal transformation

Two binary vectors, designated as pJF1_ShhGFP and pAA1 were generated for RNA-mediated gene silencing. For this, the sGFP cassette driven by the *ToxA* promoter from the pJF1 [36] plasmid was removed by digestion at the *Hind*III-*Eco*RI site, forming an open plasmid for cloning of a desired fragment. A 1354-bp DNA fragment containing the regulatory sequences and the sense and antisense of GFP interrupted by *C. truncatum* cutinase (Accession number HQ406775) intron to form hairpin structure was synthesised and cloned into the *Hind*III-*Eco*RI digested site of pJF1, resulting in pJF1_ShhGFP fungal silencing plasmid (DNA2.0, USA). The second fungal silencing vector, pAA1 was constructed by cloning the sense and antisense strand of the cutinase cDNA into *Sma*I-*Xho*I and *Bam*HI-*Xba*I sites of the pJF1_ShhGFP plasmid. The internal cutinase fragment (17 bp to 532 bp) was amplified with primers SmaI-CutS-F and XbaI-CutS-R to generate the sense (540 bp) and with primers BamHI-CutAS-F and XhoI-CutAS-R to generate the antisense (541 bp) (Figures 1a and b). Amplifications were performed in two separate reactions with 20 ng/µl *C. truncatum* cutinase DNA previously cloned into pCR®4-TOPO® vector (Invitrogen, Australia) as template. The amplified antisense fragment was digested

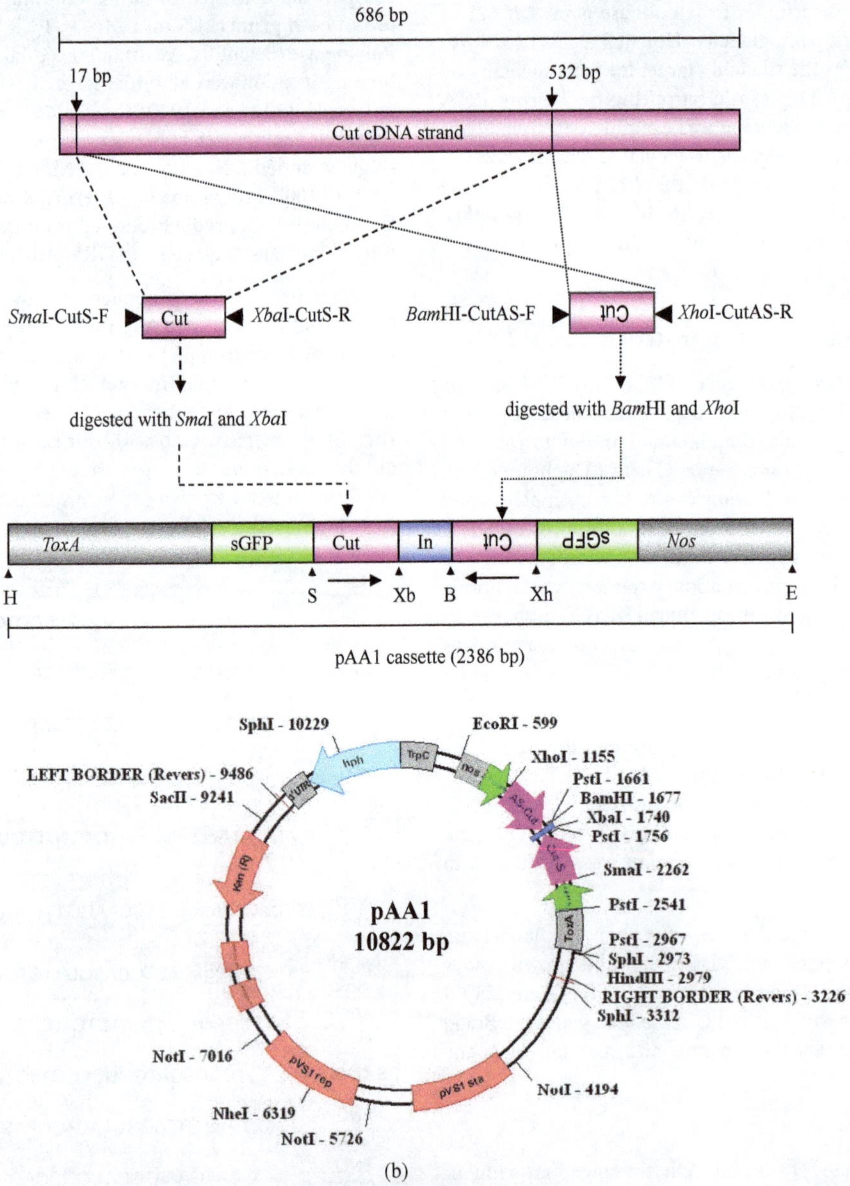

Figure 1: a) Schematic representation map of pAA1 cassette expressing *Cut* hairpin RNA, conferring the cutinase gene in *Colletotrichum truncatum*. The pAA1 recombinant silencing plasmid was derived from pJF1_ShhGFP. pJF1_ShhGFP was a derivative of plasmid pJF1. Arrows indicate the direction of the *Cut* gene fragment. In represents the 57-bp intron from *CtCut*1. *ToxA* represents the *Pyrenophora tritici-repentis ToxA* promoter and *Nos* represents *Nos* terminator. Restriction enzymes site: H, *Hind*III; S, *Sma*I; Xb, *Xba*I; B, *Bam*HI and E, *Eco*RI. The nucleotide sequence of the *cutinase* gene is available from the NCBI database (Accession no. HQ406775). b) Plasmid map of pAA1.

with *Bam*HI and *Xba*I enzymes (New England Biolabs, UK), cloned into *Bam*HI-*Xba*I sites of the pJF1_ShhGFP plasmid and the vector was named pAA1AS. Next, the amplified sense fragment was digested with *Sma*I (New England Biolabs, UK) at 25°C for 16 h and followed by digestion with *Xho*I (New England Biolabs, UK) at 37°C for another 16 h. The fragment was cloned into the *Sma*I-*Xho*I site of the pAA_CutAS plasmid, resulting in the pAA_CutASS plasmid. pAA_CutASS plasmid was later named pAA1.

Fourteen-day-old germinated conidiospores-transformants expressing GFP were transformed with pJF1_ShhGFP and its wild type was transformed with pAA1, respectively by the ATMT method as described [30]. Both pJF1_ShhGFP and pAA1 transformed strains were selected on PDA containing 250 μg/ml hygromycin B antibiotic. Since the transformant-expressing-GFP carried the same selection marker (hygromycin B antibiotic) as the current pJF1_ShhGFP transformant, the double-transformed individuals were unable to be selected on the selective medium. Subsequent analyses were performed only on single-spored putative transformants of pAA1, or also known as F8-3B-pAA1 transformants; with pJF1-ShhGFP transformants used as the control for the experiment.

Microscopy

Mycelium and conidiospores of putative F8-3B-pAA1 transformants containing the inverted repeat cassette within the T-DNA insert were observed using a Leica DMRBE microscope (Leica Microsystems, Germany), for the lack of GFP expression which confirmed the success of the inverted repeat (hairpin) structure formation in the pAA1 plasmid. Similarly, the mycelium and conidiospores derived from infected fruits and leaves in the pathogenicity assays were observed for the lack of GFP expression. Successful transformants were those with no GFP expression. For this, all specimens were placed on a glass slide in a water droplet, covered with a cover slip and observed under the microscope. The images were captured by a Leica DC300F digital camera using Leica IM50 v.4 software. Wild type F8-3B and transformants expressing GFP [30] were used as the controls.

PCR and Southern blot analysis

PCR was used to confirm the presence of T-DNA in ten putative F8-3B-pAA1 (F1, F2.1, F2.2, F4, F5.1, F5.2, F7, F8, F9 and F10) transformants by amplifying an internal 544 bp region of the *hph* gene [37] using hphF and hphR primers (Table 1). Putative transformants of F2.1 and F2.2, and F5.1 and F5.2 were derived from separate single spores of the same transformation event for each isolate respectively. Another PCR was performed with the 207 bp region of the *sgfp* gene using the primers; SmaI-CutS-F and S-GFP-R (Table 1).

Southern blot analysis was performed to confirm the copy number of the endogenous *cutinase* gene in the wild type and the integration of T-DNA containing the hairpin cassette in the transformed fungal genome. Transformed fungal genomic DNA was extracted using a DNA extraction method that utilizes high SDS concentration [38]. A total of 8 μg genomic DNA was digested with 20 U *Hin*dIII (New England Biolabs, UK) at 37°C for 8 h. pAA1 plasmid was prepared using *Hin*dIII and *Eco*RI restriction enzymes. Fragments were separated on 1.0% agarose gel and transferred to a positively charged nylon membrane (Roche Diagnostics Cooperation, Indianapolis, USA). A 540-bp cutinase probe was generated by PCR using the SmaI-CutS-F and XbaI-CutS-R primers (Table 1), labeled with digoxigenin (DIG) according to the DIG High Prime DNA Labelling and Detection Starter Kit II protocol (Roche Diagnostics GmBH, Germany). Pre-hybridization and hybridization of bound fragment(s) was carried out at 42°C for 1.5 h and overnight, respectively. The blot was washed and processed further following the DIG High Prime DNA Labelling and Detection Starter Kit II protocol. The number of bands appeared on the blot indicated the number of copies of the 540 bp cutinase probe that also represented the number of copies of the cutinase genes in the genome. Wild type *C. truncatum* genomic DNA and pAA1 plasmid were used as controls.

Quantitative real-time PCR (qRT-PCR) analysis

The mycelium of 7-day-old F8-3B-pAA1 transformants (F2, F4, F5, F7 and F8) were inoculated onto the surface of 4-week-old young chili leaves of *C. annuum* genotype Bangchang following the procedure by Chen et al. [39]. After 6 h, the inoculated fungal mycelium were collected, freeze-dried and subjected to RNA extraction and cDNA conversion. Wild type F8-3B was used as the control. The transcript levels of *CtCut1* that encoded for the *cutinase* gene of *C. truncatum* in five transformants in response to gene knockdown were quantified. The reactions were performed with ABI qPCR universal master mix (Applied Biosystems, USA) in ABI-Applied Biosystems 7900 HT Thermal Cycler Real Time PCR machine by MCLAB, USA. Triplicate qRT-PCR reactions were performed for all biological replicates. Elongation Factor 1 (EF1) from wild type *C. truncatum* pathotype F8-3B was used as a reference gene for the qRT-PCR experiment because of its stability in diverse developmental stages [40]. Analysis of the qRT-PCR data was performed using the comparative delta Ct (ΔCt) method [41]. Two independent experiments were performed with three technical replicates each. Results from the experiments were statistically analysed using ANOVA.

Pathogenicity bioassays

Fruit of *C. annuum* genotype Bangchang which is susceptible to *C. truncatum* pathotype F8-3B [31] was used in the pathogenicity bioassays. Leaves of *G. max* genotype Bragg, the susceptible leguminous host of *C. truncatum* [42] was alternately used to verify the infection characteristics of the fungus. F8-3B-pAA1 transformants that displayed a single copy gene T-DNA insertion in their genome were randomly selected for use in the pathogenicity assay. Detached fruits and leaves for pathogenicity assays were surface sterilised with 1.5% sodium hypochlorite (NaOCl) for 5 min, followed by several rinses with sterile distilled water before fungal inoculation.

Chili fruit inoculation

Preliminary non-wound chili fruit bioassay was carried out on detached *C. annuum* genotype Bangchang fruit. Agar plugs from 8-day-old cultures consisting of active growing mycelium of transformed strains; F1, F2, F4, F5, F7, F8, F9, F10 and a wild type were excised with a cork-borer and placed onto the surface sterilised fruit. The wild type F8-3B was used as the control and was designated as FW.

Subsequent pathogenicity bioassay using a wild type and two or three F8-3B-pAA1 transformants (F2, F4 and F8) was carried out comparing inoculation method (wound versus non-wound inoculation). Wound inoculation of the detached chili fruit consisted of pricking the surface of the fruit to penetrate the cuticle to a depth of 1 to 2 mm with a sterile needle, then placing agar plugs of actively growing mycelium from 14 to 21-day-old cultures of transformed strains and a wild type over the wound site. Fruits were placed in a plastic box on a tray, containing 900 ml water and sealed to provide humidity and incubated at 28°C in 12 h photoperiod. Three fruit replicates for each treatment was used and the experiment was replicated twice.

Symptoms of the typical anthracnose lesions were evaluated daily and the disease severity on chili fruit caused by F8-3B-pAA1 transformants and wild type F8-3B was scored on a 0-9 rating scale according to the method by Montri et al. [32]. The anthracnose scores were recorded as soon as the first symptom of the disease was observed on the fruit surface between 1 and 9 days after inoculation (DAI) for wound-inoculated fruit and between 1 and 13 DAI for non-wound inoculated fruit. Pathogenicity of the transformed strain was considered retained if the lesion score was similar to the wild type. A consensus rating score was given for each strain based on the predominant score for the three replicates. Ratings were based on comparison to a set of key scale diagrams developed for this disease with each rating representing a range of lesion sizes adjusted for fruit size [32].

Soybean leaf inoculation

Soybean is a very susceptible host to *C. truncatum* and inoculated leaves showed infection within 3 days (unpublished). Hence, soybean leaves of *G. max* genotype Bragg were also used to assess the degree of infection caused by F8-3B-pAA1 transformants (F1, F2, F4, F5, F7, F8, F9 and F10). Surface sterilised detached leaves from 1-month-old soybean plants were placed on moist filter paper disc in Petri dishes. Agar plugs containing actively growing mycelium of F8-3B-pAA1 transformants were inoculated onto the left corner of the abaxial surface of the same detached leaves. Wild type F8-3B, designated as FW was inoculated onto the right corner of the abaxial surface of the detached leaves as the control. Inoculated leaves were incubated at 28°C with 95% RH in 12 h photoperiod. Anthracnose symptoms consisting of sunken and water soaked tissue which developed around the fungal plug on the leaves were evaluated at 5 DAI. Disease severity caused by the transformed strains on the leaves was measured based on the percentage of the diameter lesion produced relative to the wild type. Pathogenicity of the transformants was considered retained if the diameter of the lesion was similar to the wild type. The experiment was conducted with two leaves with each leaf consisting of two replicates of the transformants and a wild type. The experiment was repeated twice and the results obtained were statistically analysed using ANOVA.

Chili leaf inoculation for screening of germ tube and appressoria development

The sterilised abaxial surface of detached young leaves from *C. annuum* genotype Bangchang placed on moist filter paper discs in Petri dishes were inoculated with 10 µl of 2×10^6 conidiospores/ml of F8-3B-pAA1 transformants (F2, F4, F5, F7 and F8). At least two leaves were used per fungal sample and three positions on the abaxial leaf surface (basal end, mid rib and tip end) were inoculated. Wild type F8-3B was used as a control and was designated as FW. The Petri dishes were sealed with parafilm and incubated at 28°C with 12 h photoperiod. Two time points, 6 hours after inoculation (HAI) and 24 HAI were selected to determine the appropriate time for counting germ tube and appressoria structures.

Inoculated tissue samples were excised into 1 cm^2 leaf pieces, and then placed in acetic alcohol solution (absolute ethanol: glacial acetic acid, 1:2) for 24 h to 48 h according to the method by Khan and Hsiang [43] to remove the chlorophyll. The acetic alcohol solution containing the leaf tissues was replaced when the acetic alcohol solution turned dark green. Cleared tissues were removed and rinsed with sterile distilled water, placed on glass slide with the abaxial surface facing up in drops of 0.05% trypan blue (w/v) and examined under the Leica DMRBE microscope (Leica Microsystems, Germany). The images were captured by a Leica DC300F digital camera using Leica IM50 v.4 software.

Each leaf tissue was considered a separate replicate. A random three out of the six replicates were chosen per fungal sample. A total of 100 conidiospores per tissue were observed for conidiospore germination and appressoria formation. Conidiospores were considered germinated when germ tubes were present, regardless of germ tube length. The occurrence of a globular structure either at the conidiospore or at the end of the germ tube with a diameter larger than the germ tube was considered to be an appressorium and these were counted [44]. Results from the experiments were statistically analysed using analysis of variance (ANOVA).

Results

Expression analysis of *C. truncatum* cutinase gene transcript on chili host

Expression analysis of cutinase gene transcript using equal amounts of RNA as the template revealed that the transcript was upregulated during *C. truncatum* infection (Figure 2). Similar RT-PCR results were obtained in all three independent biological replicates.

C. truncatum cutinase 1 (CtCut1)

A 1936-bp cutinase genomic DNA fragment from *C. truncatum* pathotype F8-3B was obtained and was designated as *CtCut1* (Figure 3). The primer binding sites of the sequence were identified. The predicted molecular weight for *CtCut1* DNA was 1196.35 kDa. The nucleotide sequence was found to contain sequence identical to the partial gene obtained previously from the RT-PCR results (data not shown). Sequence analysis of the gene revealed an open reading frame (ORF) of 684 nucleotides and a stop codon (TGA), interrupted by a single 57-bp intron (Figure 3). The predicted ORF encoded a polypeptide of 228 amino acids with a predicted molecular weight of 23.81 kDa and a predicted protein isoelectric point (pI) of 7.82. The nucleotide and predicted amino acid sequences showed significant 99% identity to the *cutinase* of *C. capsici* isolated from papaya, 85% and 84% to *G. cingulata* and *C. gloeosporioides*.

Following the bioinformatics assessment, *CtCut1* DNA was submitted to the NCBI GenBank database and was attributed the accession number, HQ406775.

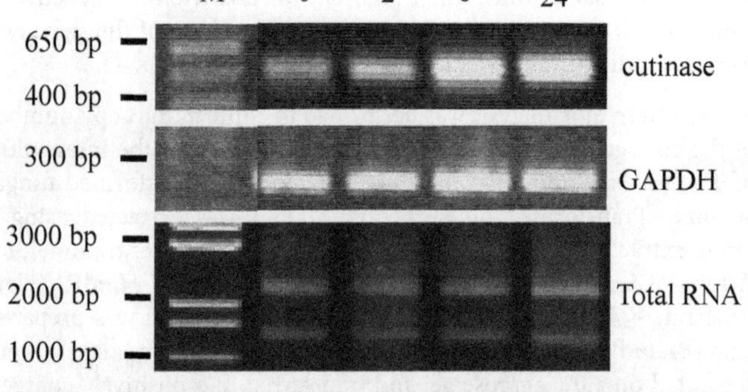

Figure 2: The pattern of *C. truncatum* cutinase gene transcript expressed during the early stages of chili infection. The number on top of the agarose lane denotes the hours after inoculation (HAI) of the fungal mycelium on the host under laboratory conditions. The *GAPDH* gene was used to normalise the expression levels and the total RNA was used to demonstrate the equal loading for cDNA conversion. M, 1 kb Plus DNA marker.

```
1       atcccttacaactttcctctgacatccaccctgagaaagacgtcgcgaaactcgtagtgg  60
61      ttgaccatctctctagatcaccgtcgaatcgagaaagtcataggccgcatttcccctca  120
121     ctccaaagtcatgcttcttccctggcgtcagacggtggaccgaggctggccccttttcct  180
181     cagccaaagagatggaagagtttcttccgaccattcccaatttgatgttctgtcgcgaaa  240
241     agccgatcctcatgtgatctcgcagccttggaacatctaccttgcggactgtgggatcta  300
301     ggtagtcggggatggcttcaagatgcgactggcggaggaatattgtatgtgtttctctgg  360
361     ttgttgagatccggttagtacgcggcttgaatagttcaggagtggacagctagtgagagc  420
421     atcatggaatggattgtaagccggataacgaggcaatcatggccatagacgctccatggc  480
481     atgaagataaatagctcctgaacaacccacgaaaagactcttgatatccagaccatcatc  540
541     agattcaatctctcactcagaccccaaacactctctccttcgctcgaagctcagccagcg  600
601     cgctcactccaaagatcattctcttctcccgcaagtcgtcgcatctgaagaccagtccca  660
                 M  K  F  L  S  I  I  S  L  A  V  S  L  V  A  A  A
661     atacacaaaatgaagttcctcagcatcatctctcttgccgtttcccttgtcgctgccgct  720
        P  V  E  V  G  L  D  T  G  V  A  N  L  E  A  R  Q  S  S  T
721     cctgtcgaagtcggcttggacaccggtgttgccaaccttgaggcccgccagtcctcgacc  780
        R  N  E  L  E  S  G  S  S  S  N  C  P  K  V  I  Y  I  F  A
781     cgcaacgagcttgagtctggcagcagctccaactgccccaaggtcatctacatctttgct  840
        R  A  S  T  E  P  G  N  M                          °
841     cgtgcctctactgagcccggtaacatggtaagtgacagttaagcgtcctcccttccacag  900
                    °        °        G  I  S  A  G  P  I  V  A  D  A  L
901     gatgataactgactcctataacagggcatcagcgcaggccccattgtcgccgacgctctc  960
        E  S  R  Y  G  A  S  Q  V  W  V  Q  G  V  G  G  P  Y  S  A
961     gaaagccgctacggcgcctcacaggtctgggtccagggcgttggcggcccttactctgcc  1020
        D  L  A  S  N  F  I  I  P  E  G  T  S  R  V  A  I  N  E  A
1021    gacctggcctccaacttcatcatacccgagggcacctcccgcgtcgccatcaacgaggct  1080
        K  R  L  F  T  L  A  N  T  K  C  P  N  S  A  V  V  A  G  G
1081    aagcgactcttcacgctcgcaaacaccaagtgccccaactccgccgtcgtcgcaggcgga  1140
        Y  S  Q  G  T  A  V  M  A  S  S  I  S  E  L  S  S  T  I  Q
1141    tacagccagggcacagccgtgatggcgtcctccatctccgagttgagctccacgatccag  1200
        N  Q  I  K  G  V  V  L  F  G  Y  T  K  N  L  Q  N  L  G  R
1201    aaccagatcaagggcgtcgttctgttcggctacaccaagaacctgcagaaccttggacgc  1260
        I  P  N  F  S  T  S  K  T  E  V  Y  C  A  L  A  D  A  V  C
1261    atcccgaacttctcgacatccaagaccgaggtgtactgtgccctcgccgatgctgtgtgc  1320
        Y  G  T  L  F  I  L  P  A  H  F  L  Y  Q  A  D  A  A  T  S
1321    tacggcacgctgtttatcctgccggcgcatttcttgtaccaggccgatgctgctacttct  1380
        A  P  R  F  L  A  A  R  I  G  *
1381    gccccgagattccttgctgctcgcattggctgagctccacgcgagggaaatatttccacg  1440
1441    gacctaggctgaggggagaacgggtctgaacagggctcgaaggacgggcgtccgtggaaa  1500
1501    agatactgagtatctggttgtggctgttgatcggtcagctttggggaggaggtcgaagtg  1560
1561    gatgcggttttcgtttcttatgcttatgctggctttaacactctttttttggtcactct  1620
1621    ttcatgctgtatggtttttgagtattgtagcgagggaagtaaaacgaaatgcaacgtgct  1680
1681    tgtattgctaaggggtatcattgactcattacttctcaagtggtatcatgataatgcctc  1740
1741    acacggcgctctccgacaaaaatctgactgcatctctcaggctctcggtctgccccgcaa  1800
1801    gtacgccgcatgcctctgagaatggtacttgacatcaggggggatgcagcgggttgtacac  1860
1861    tatacgacatgtcttgcattcctcgaatgtaggttttgatgagatgttgatagttgtctg  1920
1921    tatcgtggcggtcgaa  1936
```

Figure 3: Nucleotide sequence of a cloned genomic DNA for *Colletotrichum truncatum* isolate F8-3B *CtCut*1 gene (accession no. HQ406775). The deduced amino acid sequence is shown above the open reading frame. The untranslated sequence gap beginning at nucleotide 868 is a 57-bp intron. The CAAT and TAAATA boxes for transcription as well as the potential polyadenylation site, TAAA are underlined. The three potential AATs are italized. The '*' indicates the stop codon of the ORF (represented by the amino acid sequences). '°' represents the false stop codon in the intron region. Translation was carried out using SIXFRAME (Biology Workbench 3.2, University of California San Diego, 1999). Primer sites used to amplify the gene are shaded in grey. Bolded nucleotide sequence was used as probe in Southern analysis. The cutinase cDNA fragment was also used to generate the hairpin transcriptional unit for RNAi silencing.

Phenotypic analysis of CtCut1 silenced transformants and the integration of pAA1 T-DNA with hairpin structure in *C. truncatum* genome

All *C. truncatum* transformed strains were able to grow on the hygromycin selective medium. Phenotypes of the F8-3B-pAA1 transformants and its wild type (FW) were indistinguishable with respect to colony morphology, growth rate and conidiospore production. The unique feature of the fungus with the formation of concentric rings during sporulation, was exhibited in all individuals. The conidiospores and hyphae of F8-3B-pAA1 transformants and wild type samples did not fluoresce under the GFP filter indicating that the silencing construct have been integrated into the genome and was functioning efficiently.

Single-spore-derived F8-3B-pAA1 transformed strains were subjected to PCR analysis. The hph and sGFP regions of T-DNA insertions were detected through amplification of the expected size fragments. All of the transformants amplified a 544-bp and a 207-bp fragment (Figure 4a). Neither fragment was amplified from the wild type.

Southern blot analysis indicated that 85.7% of the fungal genomes contained the T-DNA insertion (Figure 4b). Restricted genomic DNA of seven F8-3B-pAA1 transformants probed with a DIG-labelled 540 bp cutinase sense fragment confirmed the integration of the hairpin structure into the genome of F8-3B-pAA1 transformants.

*Hind*III was expected to cut the T-DNA sequence once only. All the transformants tested contained one insert of the T-DNA, with the exception of transformant F5, which contained three copies and transformants F4 and F9, which contained two copies of the T-DNA (Figure 4b). A higher molecular endogenous cutinase band was observed in transformant F4, suggesting the integration of T-DNA adjacent to the endogenous *cutinase* gene or the occurrence of partial T-DNA integration into the genome. Transformants F10 however, did not exhibit any additional bands apart from the endogenous cutinase band (Figure 4b). Incidentally, this transformant was able to grow in the selective medium, suggesting that the integration of nicked T-DNA strand containing the hygromycin B resistant fragment into the fungal genome.

The wild type copy of the endogenous *cutinase* gene, *CtCut*1 was visible in all genomes as a hybridising band at ~8 kb (Figure 4b). As *Hind*III and *Eco*RI cut within the T-DNA insert and outside of the exogenous *cutinase* silencing transcriptional unit, a band of 2386 bp which corresponded to the intact pAA1 cassette, was expected and observed (Figure 4b). Undigested and/or partially digested pAA1 plasmid was also seen hybridising at ~10 kb (Figure 4b).

Expression analysis of CtCut1 silenced transformants of *C. truncatum*

Comparative *CtCut*1 transcription of 6 HAI *C. truncatum* mycelium of five F8-3B-pAA1 transformants and one wild type were determined using qRT-PCR. Differential expression of *CtCut*1 that encodes for the *cutinase* gene in *C. truncatum* was detected from these transformants at 6 HAI (Figure 5). In the wild type (FW) mycelium, the amount of *CtCut*1 transcript when normalized to EF1 transcript gave a relative expression value of 2.45 ± 0.11SE compared with 1.33 ± 0.29SE from an average of five transformants. These data indicated that almost one-fold of decrease in the relative expression of *CtCut*1 in F8-3B-pAA1 transformants during infection. The highest significant decrease occurred in transformant F2 followed by F5, F8 and F4 with decrease of *CtCut*1 transcript of 72.0%, 52.7%, 45.1%, and 44.3%, respectively in comparison to the expression in the wild type. In contrast, transformant F7 displayed no significant difference to the wild type although a slight reduction occurred in *CtCut*1 transcript (14.9%). Overall, F8-3B-pAA1 transformants exhibited 54.2% of cutinase activity in comparison to the wild type.

Infection of chili fruits

In the preliminary pathogenicity bioassay with non-wound detached chili fruit, all F8-3B-pAA1 transformants generally showed reduced or no infection compared to the wild type at 13 DAI (data not shown). In a subsequent detailed pathogenicity assay on wounded and non-wounded fruit, the transformed strains were found less aggressive on non-wounded fruit compared to the wild type at both 9 and 13 DAI (Table 2). Very little or no infection occurred in fruit up to 9 DAI for the transformed strains however, by 13 DAI symptom of the disease were displayed at the site of the transformed inoculum (Figure 6), suggesting that there was a delay in the transformants to invade the host tissues. On wounded fruit there was no difference in severity of infection between wild type and transformed strains at 2 or 4 DAI (Table 2).

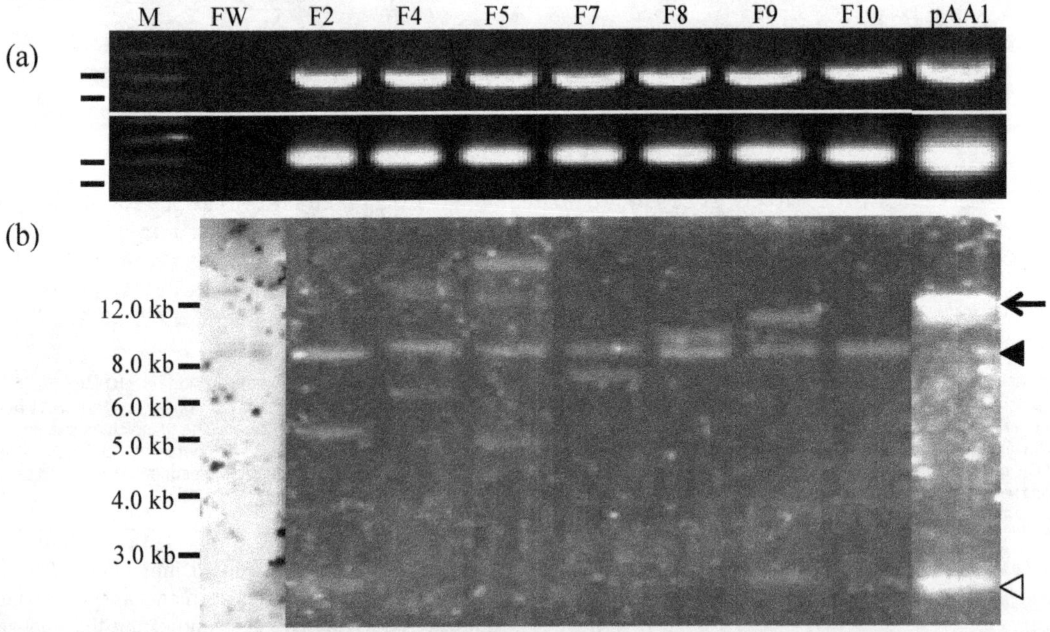

Figure 4: Evidence of pAA1 cassette integration in several independent hygromycin resistant *C. truncatum* pathotype F8-3B fungal genome. (a) PCR analysis of genomic DNA with primers specific for the amplification of an internal 544-bp fragment of the *hph* marker gene (top panel) and the 207 bp *sgfp* gene (lower panel). Wild type, indicated as FW was used as negative control and pAA1 plasmid was used as positive control. M, 1 kb Plus DNA Ladder (bars from top to bottom: 650 bp, 500 bp, 300 bp and 200 bp). (b) Southern analysis of the hygromycin-resistant transformants. Genomic DNA (8 μg) was digested with *Hind*III and probed with DIG-labelled 540 bp cutinase sense fragment. Closed triangle indicates the single copy endogenous *cutinase* gene (hybridised at ~8.0 kb) in the fungal genome. Open triangle represents the exogenous *cutinase* silencing transcriptional unit (2386 bp) within the T-DNA fragment. Non-transformed *C. truncatum* (FW) was used as negative control. *Hind*III+*Eco*RI-digested pAA1 plasmid was used as positive control. Arrow corresponds to the undigested plasmid.

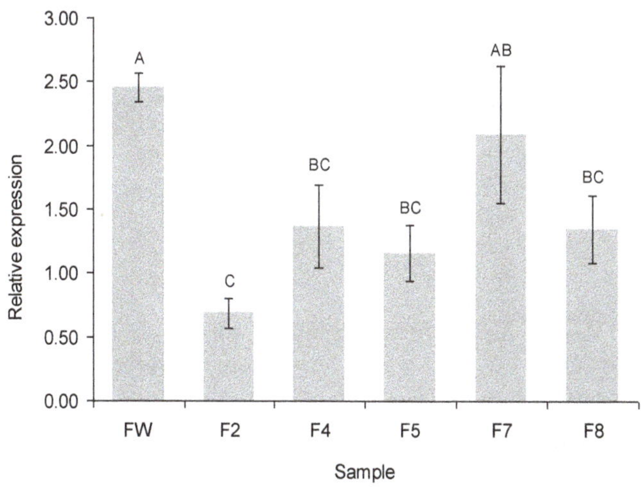

Figure 5: Quantitative gene expression of *CtCut1* transcript in F8-3B wild type (FW) and F8 3B-pAA1 transformants at 6 hours after inoculation (HAI) with *Capsicum annuum* genotype Bangchang. The Y-axis represents the relative expression in fold change of *CtCut1* gene relative to Elongation Factor 1 (EF1) reference gene expression. Bars on graph represent standard error based on mean of two independent experiments. Different letters over the bars indicate significant differences conducted using analysis of variance (ANOVA). LSD (1.034) was calculated at $p<0.05$ (n=2). All data derived from three technical replicates per sample per experiment.

Method	Wounded fruit		Non-wounded fruit	
Isolate / DAI	2	4	9	13
F2	3	5	0	3
F4	nt	nt	1	5
F8	1	5	0	0
FW	3	5	3	7

Note: not tested (nt)

Table 2: Consensus ratings of disease severity based on Montri et al [32] disease scales for pAA1-transformed and wild type *Colletotrichum truncatum* infection of *Capsicum annuum* genotype Bangchang.

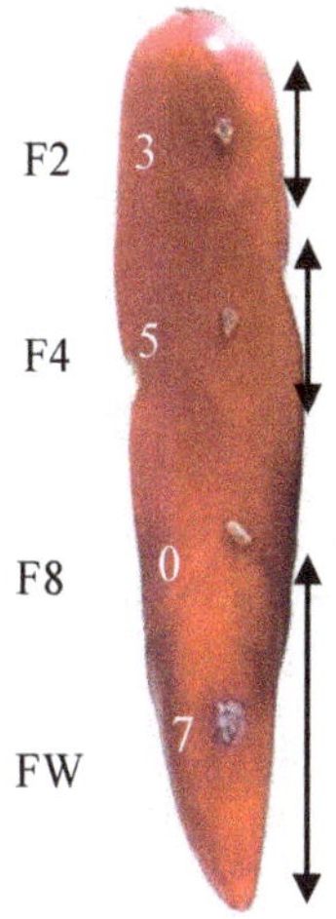

Figure 6: A representational image of the pathogenic development of wild type and transformed *Colletotrichum truncatum* pathotype F8-3B on *Capsicum annuum* genotype Bangchang. Fruit surface of non-wounded chili were inoculated with 14-day-old *C. truncatum* transformants F2, F4 and F8. Wild type (FW) was used as the control. The sample name of the transformants is indicated on the left side of panel. Arrow and the numerical score next to the infected fruit indicate the area of anthracnose sunken lesion.

GFP fluorescence was not detected in all conidiospores isolated from the infected fruits; neither from the wild type nor the transformants (data not shown).

Germ tubes and appressoria development on chili leaves

Germinated conidia formed appressoria by 6 HAI on the leaf of *C. annuum* genotype Bangchang. At 6 HAI, the highest number of appressoria formation occurred in the wild type as opposed to the F8-3B-pAA1 transformants (Figure 7). However, no significant difference in germ tube length was found between the wild type and the transformants, except that the length of germ tubes with appressoria in the transformants were generally longer compared to the wild type (data not shown).

Soybean leaf bioassay

Differences in severity of infection were observed between the F8-3B-pAA1 transformants 5 DAI on soybean leaves compared to the wild type. The transformed strains selected can be categorised into two levels of pathogenicity with high (transformants F2, F5 and F8) and low (transformants F1, F4, F7, F9 and F10) reduction in infection efficiency (Figure 8). These were clearly observed by the different degrees of anthracnose symptom produced in the *inplanta* soybean bioassay (data not shown). Lesion region in soybean leaves demonstrated that transformants with reduced expression of the *cutinase* gene (F2, F5 and F8) displayed a lower degree of pathogenicity compared to the wild type.

Discussion

A full length gene encoding for cutinase, *CtCut1* of *C. truncatum* was successfully isolated and characterised. The amino acid sequence predicted from the nucleotide sequence had a relatively high degree of homology to the cutinases of other fungi in the *Colletotrichum* genus [5,45] as well as of other ascomycetes [46,47]. The similarity of the conserved consensus of *CtCut1* sequence with the other cutinases also suggested that the cloned DNA was cutinase of *C. truncatum*. The GYSQG motif, known to be the signature pattern for cutinase gene [48] was also present in *CtCut1*. The presence of the classical catalytic

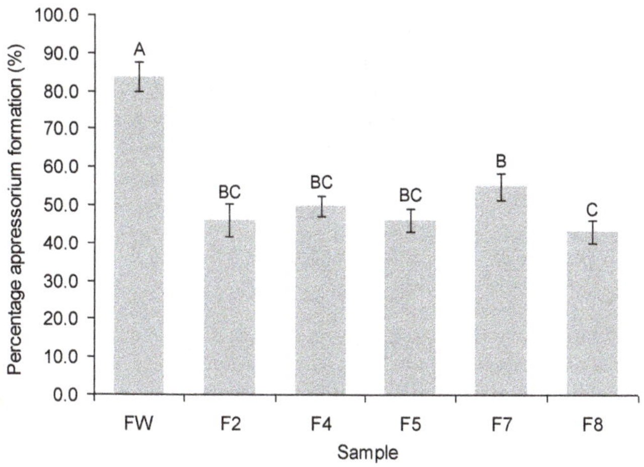

Figure 7: Quantification of appressoria formation in *Colletotrichum truncatum* pathotype F8-3B wild type and transformants on *Capsicum annuum* genotype Bangchang leaf, 6 hours after inoculation (HAI). Different letters over the bars indicate significant differences conducted using analysis of variance (ANOVA). LSD (10.616) was calculated at $p<0.05$ (n=3).

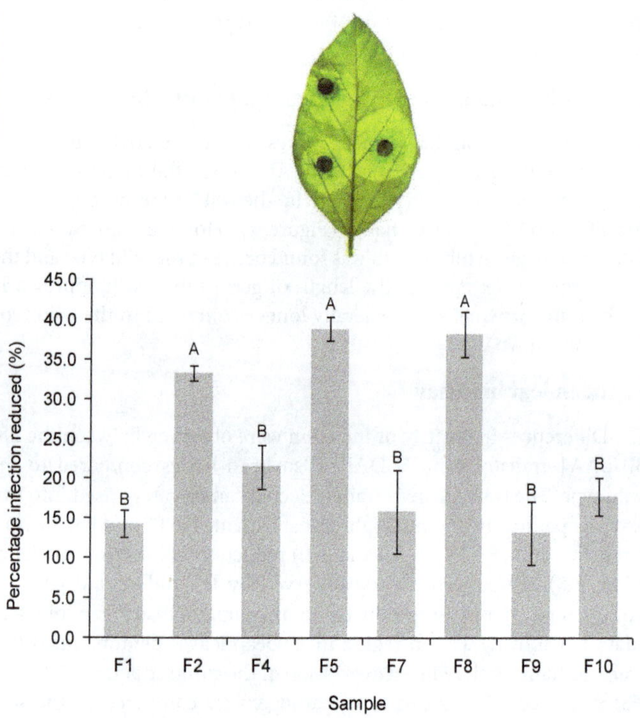

Figure 8: Soybean pathogenicity bioassay for determining the infection efficiency of *Colletotrichum truncatum* pathotype F8-3B-pAA1 transformants lacking cutinase, 5 days after inoculation (DAI). *Cutinase* silenced transformants with different percentage reduction in infection efficiency on soybean leaf (as indicated on the left side of leaf compared with the wild type on the right side of leaf in the picture above the graph). Mean percentage infection reduced values of transformants were normalized against the wild type. Different letters over the bars indicate significant differences conducted using analysis of variance (ANOVA). LSD (8.972) was calculated at $p<0.05$ (n=3).

triad of Asp, Ser and His [49] as well as the essential carboxyl group within the cloned *CtCut*1 also resembled those cutinases from other fungi [5,45,48].

RNA-mediated gene silencing in *C. truncatum* was successfully achieved following the development of a hairpin plasmid, pAA1 using a relatively short fragment of sGFP and CtCut1 as the silencing target. Variable but efficient knock-down of the *cutinase* gene of *C. truncatum* was achieved. The pAA1 plasmid was stably transformed into *C. truncatum* through *Agrobacterium tumefaciens*-mediated transformation with pAA1-dsRNA synthesised *in vitro*. Conidiospores and hyphae of F8-3B-pAA1 transformants were unable to fluoresce under the GFP filter compared to the pJF1 transformants expressing GFP [30]. This was the first indication that the hairpin cassette was incorporated into the fungal genome, and thus had activated the endogenous RNAi machinery within the pathogen.

In addition, the transformed strains were able to survive many rounds of subculture on selective medium, suggesting that the integration was stable. Southern analysis confirmed that the pAA1 cassette had been integrated ectopically into the genome of all transformed strains of *C. truncatum*. Interestingly, the pAA1 hairpin that induced the gene silencing process of CtCut1 did not produce significant off-target effects in *C. truncatum*; albeit the number of T-DNA copies harbouring the pAA1 cassette in the genome. All the transformed strains exhibited no significant changes in their morphology, growth rate and conidiospore production. Therefore, it was likely that the only gene affected by the pAA1 hairpin was the *CtCut*1.

The pAA1 plasmid was versatile as it contained restriction enzyme sites that allowed the creation of hairpin cassette of any gene to be silenced as long as the primers used to amplify the gene of interest were tailored with the compatible restriction enzyme sites. The success of the silencing was predicted to be partially enhanced by the relatively short 57-bp intron sequence in the pAA1 plasmid. Short intron-enhanced gene silencing has been reported to provide stable and high silencing frequencies in *M. oryzae* [29]; which in some cases even led to complete silenced phenotypes [50]. In addition, the intron sequence used to form the hairpin in the pAA1 plasmid derived from the *cutinase* gene of *C. truncatum* was likely to reduce the possibility of organism rejection. Thus, proper regulation of RNA-silencing machinery that enables high levels of gene silencing was predicted to occur. In contrast, Praveen et al. [51] reported that the use of intron sequences from non-related organisms were likely to yield a lower gene silencing efficacy.

Considering the limited detailed genome information of *C. truncatum* and despite the occurrence of partial integration of T-DNA such as observed in transformant F10, the RNAi approach still produced silenced transformants of some degree; which is usually sufficient for practical use [29]. Basically, there were less challenges or effort in the inactivation of the gene such as the need to identify the number of copies or members of the gene in the genome. Nevertheless, the flexibility of RNA-mediated gene silencing to simultaneously silence homologous genes is achievable in *C. truncatum* because the targeted *CtCut*1 gene suppression occurred in a sequence specific and not locus specific manner [29]. This would mean that silencing of *cutinase* gene using this approach was easily performed without the necessity to tailor knockdown per each gene member.

Although further work is needed to identify the presence of cutinase members in *C. truncatum*, Southern analysis revealed that *C. truncatum* had a single copy of the *cutinase* gene in the genome. This may have likely corresponded to the presence of one *cutinase* gene member in the pathogen. Further, there have been reports indicating the divergence of cutinase in other *Colletotrichum* species genomes [52]. In contrast, Skamnioti et al. [53] reported the presence of the multimembered cutinase families in *Magnaportha grisea, Fusarium graminearum, Botrytis cinerea, Aspergilus nidulans* and *Neurospora crassa*. In such event, the RNA-mediated gene silencing of *cutinase* in *C. truncatum* appeared to have simultaneously suppressed the entire gene family (homologous genes) and consequently avoided gene compensation. Baulcombe [54] demonstrated that the co-suppression of gene family occurred during the disruption of *quelling defective* (*qde*-1) gene mutation which resulted in the silencing of a tomato RNA-dependent RNA polymerase homologue in *N. crassa*. Simultaneous silencing of homologues was also observed in the silencing of *OsRac* gene. Three members of the *OsRac* gene family in *Oryza sativa* were suppressed when inverted repeat sequences derived from highly conserved regions of the gene family were used [55]. Not surprisingly, the RNA-mediated gene silencing approach has been adopted in many fungal gene manipulation studies (as reviewed thereafter by Salame et al. [27]).

More importantly, the expression of CtCut1 transcript in the F8-3B-pAA1 transformants was lower than the wild type *C. truncatum*. A wide range of reduced mRNA level of CtCut1 (14.9% to 72%) was observed among the transformed strains. Such variation among transformants using this technique has been observed in other fungal pathogens as well, including *Phanerochaete chrysosporium* [56] and *Botrytis cinerea* [57]. As indicated earlier, the differences could also have been due to the ectopic integration of the T-DNA within the genome of the pathogen. Additionally, the genome complexity of different microorganisms and the environmental changes may perhaps have contributed to the variability [56].

Likewise, the infection and colonisation by F8-3B-pAA1 transformants on detached chili and soybean hosts were inefficient compared to the wild type F8-3B, further suggesting the involvement of the gene during pathogenesis. Transformed strains were unable to infect normally and there was a delay in infection (as indicated by the extremely slow development of anthracnose symptoms on chili fruits). This suggested that the direct penetration through host cuticle during the infection process of *C. truncatum* [4] was largely governed by the presence of cutinase in the fungus.

In contrast, there was no difference in infection and colonisation by F8-3B-pAA1 transformants and wild type on wounded fruit, suggesting that cutinase appeared to no longer have an important role in cell infection after penetration of the cuticle. Hence, there was no disadvantage to the F8-3B-pAA1 transformants having reduced expression of cutinase once they had infected the pericarp tissue. The results were compatible to previously reported work where it was demonstrated that *C. capsici* pathotype F8-3B was able to infect chili fruits if wound inoculation was applied to the fruit [31]. If the hypothesis that cutinase assisted with the penetration of the fungal pathogen, then obviously, the wound inoculation method is unlikely to demonstrate the function of the gene since the host first layer of defence (cuticle and epidermis) has been disrupted.

The reduced level of cutinase expression in the transformants may have interrupted or delayed pathogen recognition of the host surface thus, resulting in the formation of lower number of appressoria. Cutinase may be required to stimulate a host reaction, which in turn is recognised by the pathogen to initiate appressoria formation and penetration. Therefore, the inability of the pathogen to express the effector (cutinase) may have led to poor infection rates. Deising et al. [12] reported that cutinase was involved during the adhesion of *Uromyces viciae* uredospores to the host cuticle. In another report, Woloshuk and Kolattukudy [17] indicated that the expression of cutinase from spores of *F. solani* was triggered upon contact with the plant cuticle. Indeed, Li et al. [58] reported that low levels of cutinase present in the saprophytic phase of *F. pisi* f. sp. *pisi* was essential to induce higher levels of cutinase expression during infection of the fungus to penetrate its host. Since the endogenous cutinase in these F8-3B-pAA1 transformants were suppressed, the regulation of normal cutinase activity was affected. Whether this abnormal/improper regulation of cutinase activity happened due to its incapability to secrete sufficient levels of cutinase at the beginning to induce cutin monomers from the host or from the failure of the pathogen to detect the host cutin monomers is unknown.

The soybean pathogenicity bioassay revealed different levels of soybean leaf infection when inoculated with various selected F8-3B-pAA1 transformants. These observations confirmed that the disturbance of *CtCut*1 gene via RNA-mediated gene silencing (knockdown) altered pathogenicity of the transformants. Indeed, the relative expression of the five chosen F8-3B-pAA1 transformants showed different levels of *CtCut*1 transcription as compared to the control wild type in the qRT-PCR assay. Lower levels of *CtCut*1 transcripts translated to reduced infection. Similar observations were shown in the *Pyrenopeziza brassicae-Brassica napus* system where mutants lacking cutinase transcript expression failed to penetrate the cuticle of *Brassica napus* cotyledon [14]. In another report, Lee et al. [13] showed that the pathogenicity of *Monilinia fructicola* on peach petal was more aggressive when the gene was overexpressed, thus indicating the involvement of the gene during infection.

Taken together, these results demonstrated that pAA1 plasmid was efficient and could be applicable in most genes/*Colletotrichum* species to induce gene silencing. The observation indicated that the silencing event of cutinase had possibly brought an alteration in pathogenicity within the transformants. More durable and novel forms of disease control can be practiced when knowledge of pathogen pathogenicity genes that trigger resistance is achievable/available through this alternative genetic tool approach.

Acknowledgement

This research was supported by the Department of Agriculture and Food Systems. The authors wish to express their gratitude to Dr Annie Wong for providing the soybean plant materials for the pathogenicity assay. Dr Adelene Auyong held scholarships from the University of Melbourne.

References

1. Howard RJ, Ferrari MA, Roach DH, Money NP (1991) Penetration of hard substrates by a fungus employing enormous turgor pressures. Proc Natl Acad Sci U S A 88: 11281-11284.
2. Davies KA, De Lorono I, Foster SJ, Li D, Johnstone K, et al. (2000) Evidence for a role of cutinase in pathogenicity of Pyrenopeziza brassicae on brassicas. Physiological and Molecular Plant Pathology 57: 63-75.
3. Dickman MB, Patil SS (1988) The role of cutinase from Colletotrichum gloeosporioides in the penetration of papaya, in Experimental and Conceptual Plant Pathology, W.M. Hess, R.S. Singh, U.S. Singh, D.J. Weber, Editors. 1988, Gordon and Breach Science Publishers: Montreux, Switzerland 175-182.
4. Ranathunge NP, Mongkolporn O, Ford R, Taylor PWJ (2012) Colletotrichum truncatum Pathosystem on Capsicum spp: infection, colonization and defence mechanisms. Australasian Plant Pathology 41: 463-473.
5. Ettinger WF, Thukral SK, Kolattukudy PE (1987) Structure of cutinase gene, cDNA, and the derived amino acid sequence from phytopathogenic fungi. Biochemistry 26: 7883-7892.
6. Chen Z, Franco CF, Baptista RP, Cabral JM, Coelho AV, et al. (2007) Purification and identification of cutinases from Colletotrichum kahawae and Colletotrichum gloeosporioides. Appl Microbiol Biotechnol 73: 1306-1313.
7. Rubio MB, Cardoza RE, Hermosa R, Gutiérrez S, Monte E (2008) Cloning and characterization of the Thcut1 gene encoding a cutinase of Trichoderma harzianum T34. Curr Genet 54: 301-312.
8. van der Vlugt-Bergmans CJ, Wagemakers CA, van Kan JA (1997) Cloning and expression of the cutinase A gene of Botrytis cinerea. Mol Plant Microbe Interact 10: 21-29.
9. Köller W, Parker DM (1989) Purification and characterization of cutinase from Venturia inaequalis. Phytopathology 79: 278-283.
10. Dickman MB, Kollattukudy PE (1989) Insertion of cutinase gene into a wound pathogen enables it to infect intact host. Nature 343: 446-448.
11. Skamnioti P, Gurr SJ (2007) Magnaporthe grisea cutinase2 mediates appressorium differentiation and host penetration and is required for full virulence. Plant Cell 19: 2674-2689.
12. Deising H, Nicholson RL, Haug M, Howard RJ, Mendgen K (1992) Adhesion Pad Formation and the Involvement of Cutinase and Esterases in the Attachment of Uredospores to the Host Cuticle. Plant Cell 4: 1101-1111.
13. Lee MH, Chiu CM, Roubtsova T, Chou CM, Bostock RM (2010) Overexpression of a redox-regulated cutinase gene, MfCUT1, increases virulence of the brown rot pathogen Monilinia fructicola on Prunus spp. Mol Plant Microbe Interact 23: 176-186.
14. Li D, Ashby AM, Johnstone K (2003) Molecular evidence that the extracellular cutinase Pbc1 is required for pathogenicity of Pyrenopeziza brassicae on oilseed rape. Mol Plant Microbe Interact 16: 545-552.
15. Stahl DJ, Schäfer W (1992) Cutinase is not required for fungal pathogenicity on pea. Plant Cell 4: 621-629.

16. van Kan JA, van't Klooster JW, Wagemakers CA, Dees DC, van der Vlugt-Bergmans CJ (1997) Cutinase A of Botrytis cinerea is expressed, but not essential, during penetration of gerbera and tomato. Mol Plant Microbe Interact 10: 30-38.

17. Woloshuk CP, Kolattukudy PE (1986) Mechanism by which contact with plant cuticle triggers cutinase gene expression in the spores of Fusarium solani f. sp. pisi. Proc Natl Acad Sci U S A 83: 1704-1708.

18. Wang CL, Chin CK, Gianfagna T (2000) Relationship between cutin monomers and tomato resistance to powdery mildew infection. Physiological and Molecular Plant Pathology 57: 55-61.

19. Dickman MB, Ha YS, Yang Z, Adams B, Huang C (2003) A protein kinase from Colletotrichum trifolii is induced by plant cutin and is required for appressorium formation. Mol Plant Microbe Interact 16: 411-421.

20. Bonnen AM, Hammerschmidt R (1989) Cutinolytic enzymes from Colletotrichum lagenarium. Physiological and Molecular Plant Pathology 35: 463-474.

21. Bonnen AM, Hammerschmidt R (1989) Role of cutinolytic enzymes in infection of cucumber by Colletotrichum lagenarium. Physiological and Molecular Plant Pathology 35: 475-481.

22. Ziv C, Yarden O (2010) Gene silencing for functional analysis: assessing RNAi as a tool for manipulation of gene expression. Methods Mol Biol 638: 77-100.

23. Fire A, Xu S, Montgomery MK, Kostas SA, Driver SE, et al. (1998) Potent and specific genetic interference by double-stranded RNA in Caenorhabditis elegans. Nature 391: 806-811.

24. Cogoni C, Macino G (2000) Post-transcriptional gene silencing across kingdoms. Curr Opin Genet Dev 10: 638-643.

25. Fulci V, Macino G (2007) Quelling: post-transcriptional gene silencing guided by small RNAs in Neurospora crassa. Curr Opin Microbiol 10: 199-203.

26. Meister G, Tuschl T (2004) Mechanisms of gene silencing by double-stranded RNA. Nature 431: 343-349.

27. Salame TM, Ziv C, Hadar Y, Yarden O (2011) RNAi as a potential tool for biotechnological applications in fungi. Appl Microbiol Biotechnol 89: 501-512.

28. Shafran H, Miyara I, Eshed R, Prusky D, Sherman A (2008) Development of new tools for studying gene function in fungi based on the Gateway system. Fungal Genet Biol 45: 1147-1154.

29. Nakayashiki H, Hanada S, Nguyen BQ, Kadotani N, Tosa Y, et al. (2005) RNA silencing as a tool for exploring gene function in ascomycete fungi. Fungal Genet Biol 42: 275-283.

30. Auyong ASM, Ford R, Taylor PWJ (2011) Genetic transformation of Colletotrichum truncatum associated with anthracnose disease of chili by random insertional mutagenesis. Journal of Basic Microbiology 52: 372-382.

31. Mongkolporn O, Montri P, Supakaew T, Taylor PWJ (2010) Differential reactions on mature green and ripe chili fruit infected by three Colletotrichum species. Plant Disease 94: 306-310.

32. Montri P, Taylor PWJ, Mongkolporn O (2009) Pathotypes of Colletotrichum capsici, the causal agent of chili anthracnose, in Thailand. Plant Disease 93: 17-20.

33. Thompson JD, Higgins DG, Gibson TJ (1994) CLUSTAL W: improving the sensitivity of progressive multiple sequence alignment through sequence weighting, position-specific gap penalties and weight matrix choice. Nucleic Acids Res 22: 4673-4680.

34. Subramaniam S (1998) The Biology Workbench--a seamless database and analysis environment for the biologist. Proteins 32: 1-2.

35. Altschul SF, Madden TL, Schäffer AA, Zhang J, Zhang Z, et al. (1997) Gapped BLAST and PSI-BLAST: a new generation of protein database search programs. Nucleic Acids Res 25: 3389-3402.

36. Flowers JL, Vaillancourt LJ (2005) Parameters affecting the efficiency of Agrobacterium tumefaciens-mediated transformation of Colletotrichum graminicola. Curr Genet 48: 380-388.

37. Lorang JM, Tuori RP, Martinez JP, Sawyer TL, Redman RS, et al. (2001) Green fluorescent protein is lighting up fungal biology. Appl Environ Microbiol 67: 1987-1994.

38. Brasileiro BTRV, Coimbra MRM, Morais Jr MAd, Oliveira NTd (2004) Genetic variability within Fusarium solani specie as revealed by PCR-fingerprinting based on pcr markers. Brazilian Journal of Microbiology 35: 205-210.

39. Chen X, Shen G, Wang Y, Zheng X, Wang Y (2007) Identification of Phytophthora sojae genes upregulated during the early stage of soybean infection. FEMS Microbiol Lett 269: 280-288.

40. Fang W, Bidochka MJ (2006) Expression of genes involved in germination, conidiogenesis and pathogenesis in Metarhizium anisopliae using quantitative real-time RT-PCR. Mycol Res 110: 1165-1171.

41. Schmittgen TD, Livak KJ (2008) Analyzing real-time PCR data by the comparative C(T) method. Nat Protoc 3: 1101-1108.

42. Backman PA, Williams JC, Crawford MA (1982) Yield losses in soybean from anthracnose caused by Colletotrichum truncatum. Plant Disease 66: 1032-1033.

43. Khan A, Hsiang T (2003) The infection process of Colletotrichum graminicola and relative aggressiveness on four turfgrass species. Can J Microbiol 49: 433-442.

44. Dita MA, Brommonschenkel SH, Matsuoka K, Mizubuti ESG (2007) Histopathological study of the Alternaria solani infection process in potato cultivars with different levels of early blight resistance. Journal of Phytopathology 155: 462-469.

45. Bakar FDA, Cooper D, Zamrod Z, Mahadi NM, Sullivan PA (2001) Cloning and characterisation of the cutinase genomic and cDNA gene from the fungal phytopathogen Glomerella cingulata. Asia-Pacific Journal of Molecular Biology and Biotechnology 9: 119-130.

46. Sweigard JA, Chumley FG, Valent B (1992) Cloning and analysis of CUT1, a cutinase gene from Magnaporthe grisea. Mol Gen Genet 232: 174-182.

47. Yao C, Köller W (1994) Diversity of cutinases from plant pathogenic fungi: cloning and characterization of a cutinase gene from Alternaria brassicicola. Physiological and Molecular Plant Pathology 44: 81-92.

48. Belbahri L, Calmin G, Mauch F, Andersson JO (2008) Evolution of the cutinase gene family: evidence for lateral gene transfer of a candidate Phytophthora virulence factor. Gene 408: 1-8.

49. Martinez C, Nicolas A, van Tilbeurgh H, Egloff MP, Cudrey C, et al. (1994) Cutinase, a lipolytic enzyme with a preformed oxyanion hole. Biochemistry 33: 83-89.

50. Nguyen QB, Kadotani N, Kasahara S, Tosa Y, Mayama S, et al. (2008) Systematic functional analysis of calcium-signalling proteins in the genome of the rice-blast fungus, Magnaporthe oryzae, using a high-throughput RNA-silencing system. Mol Microbiol 68: 1348-1365.

51. Praveen S, Ramesh SV, Mishra AK, Koundal V, Palukaitis P (2010) Silencing potential of viral derived RNAi constructs in Tomato leaf curl virus-AC4 gene suppression in tomato. Transgenic Res 19: 45-55.

52. Liyanage HD, Köller W, McMillan Jr RT, Kistler HC (1993) Variation in cutinase from two populations of Colletotrichum gloeosporioides from citrus. Phytopathology 83: 113-116.

53. Skamnioti P, Furlong RF, Gurr SJ (2008) Evolutionary history of the ancient cutinase family in five filamentous Ascomycetes reveals differential gene duplications and losses and in Magnaporthe grisea shows evidence of sub- and neo-functionalization. New Phytol 180: 711-721.

54. Baulcombe DC (1999) Gene silencing: RNA makes RNA makes no protein. Curr Biol 9: R599-601.

55. Miki D, Itoh R, Shimamoto K (2005) RNA silencing of single and multiple members in a gene family of rice. Plant Physiol 138: 1903-1913.

56. Matityahu A, Hadar Y, Dosoretz CG, Belinky PA (2008) Gene silencing by RNA Interference in the white rot fungus Phanerochaete chrysosporium. Appl Environ Microbiol 74: 5359-5365.

57. Goldoni M, Azzalin G, Macino G, Cogoni C (2004) Efficient gene silencing by expression of double stranded RNA in Neurospora crassa. Fungal Genet Biol 41: 1016-1024.

58. Li D, Sirakova T, Rogers L, Ettinger WF, Kollattukudy PE (2002) Regulation of constitutively expressed and induced cutinase genes by different zinc finger transcription factors in Fusarium solani f. sp. pisi (Nectira haematococca). Journal of Biological Chemistry 277: 7905-7912.

Isolation and identification of Soil Fungi from Wheat Cultivated Area of Uttar Pradesh

Rajendra Kumar Seth*, Shah Alam and Shukla DN

Department of Botany, Bharagawa Agricultural Botany Laboratory, University of Allahabad, Allahabad, U.P., India

Abstract

Experiments were conducted to find out different soil from wheat-cultivated area during 15 April to 10 May, 2013-2014. The soil fungi were isolated following the soil dilution plating technique. The obtained soil fungi from wheat-cultivated area were *Aspergillus* spp., *Penicillum* spp., *Geotrichum* spp., *Gloesporium* spp., *Fusarium* spp., *Mycelia sterilia, Arthrobotrys* spp., *Cladosporium herbarum* in Allahabad district. In Mirzapur district, *Aspergillus* spp., *Penicillum* spp., *Rizoctinia* spp., *Fusarium* spp., *Mucor* spp. were recorded from wheat cultivated area. In Varanasi district, *Aspergillus* spp., *Penicillum* spp., *Rizoctinia* spp., *Fusarium* spp., *Mucor* spp., *Alternaria* spp., *Helminthosporium oryzae,* and *Humicola grisea* were recorded from wheat-cultivated area. *Aspergillus* spp. and *Penicillum* spp. was common fungi presented in three different districts Allahabad, Mirzapur and Varanasi, of Uttar Pradesh.

Keywords: Soil fungi; Wheat; Uttar Pradesh

Introduction

Wheat (*Triticum aestivum* L.) is one of the most important cereals in the world and is part of a staple diet for nearly 35% of the world's population [1]. It is grown in about 102 countries of the world covering about 220.69 million hectares of land which is 32% of the total cultivated land of the world. The area and production increased to 0.83 million hectare and 1.84 million metric tons, respectively in 2000 [2].

Soil fungi play an important role as major decomposers in the soil ecosystem. There are about 75,000 species of soil fungi in the world [3]. Fungi are one of the dominant groups present in soil, which strongly influence ecosystem structure and functioning and thus plays a key role in many ecological services [4]. Therefore, there is a growing interest in assessing soil biodiversity and its biological functioning [5].

The yield was 2.8 t/ha in 2011-2012 cropping year [6] which is very low compared to those in the research farm level (3.5 t/ha to 5.1 t/ha) [2]. Coupled with many other factors, diseases also play an important role in lowering the yield [7,8].

The process of decomposition is governed by the succession of fungi at various stage of decomposition [9-12] nutrient level of soil, crop residue and prevailing environmental conditions [13-17].

The current study was aimed detection of soil fungi from wheat field. The study involved isolation, identification and screening of soil fungi of fungal species prevailed Allahabad, Varanasi, and Mirzapur districts of Uttar Pradesh in India.

Material and Methods

The present studies were carried out at Bhargava Agricultural Botany laboratory, Department of Botany, University of Allahabad, Allahabad. Soil samples were collected from wheat-cultivated areas of selected sites of Allahabad, Varanasi and Mirzapur district during 15 April to 10 May 2013-2014 for detection of soil fungi.

Study area

Three studies area were select, first district Allahabad is situated in Southern Eastern. It lies between the parallels of 24° 47' north latitude and 81° 19' east longitudes, second is Mirzapur district located at 25.15° N and 82.58° E, and third Varanasi is situated at 25.28° N and 82.96° E in Uttar Pradesh, India. Soil taken 15 cm depth and put in small sterilized polythene bags for laboratory analysis.

Isolation of soil fungi

The samples were processed for isolation using the soil dilution plate [18]. The soil fungi were isolated following the soil dilution plating technique of [19]. The moisture content of a certain amount of soil was determined and fresh soil quantities corresponding to 25 gm of oven-dried soil were calculated [20]. Each soil sample was diluted to 1×10^{-4} concentration suspension. Then, 1 ml of the soil suspension (containing 0.0001 g wet soil) was drawn by pipette into a Petri dish (90 mm). A mixture of 25 ml of warm, molten glucose-ammonium nitrate agar (GAN) added with Rose Bengal and streptomycin was poured over the soil suspension and the Petri dish was rotated gently to let the soil suspension mix well with the medium. Five replications were completed for each soil sample (0.0005 gm wet soil). All the Petri dishes were incubated at room temperature (26°C to 28°C) in darkness for 3-5 days or longer.

Identification of the soil fungi

The fungi were identified with the help of literature [21]. The colonies were counted and identified using the soil dilution plate method. The counting and identification procedure was carried out under a stereomicroscope. Then the identified colonies were transferred to Petri dishes containing agar. In the Petri dishes, different types of colonies developed. Identification of the organism was made with the help of the relevant literature [22,23]. For the identification of the isolates, Smith [24] was followed. Identification of the taxa were carried out according to Hasenekoglu [25], Subramanian [26], Ellis [27], Raper and Thom [28], Raper and Fennell [29], Zycha [30], Samson and Pitt [31,32].

***Corresponding author:** Rajendra Kumar Seth, Department of Botany, Bharagawa Agricultural Botany Laboratory, University of Allahabad, Allahabad, 211002, U.P., India
E-mail: aurajendrakumar22@gmail.com

Screening of soil fungi

Screening of soil fungi after each stage the ineffective isolates were excluded from further testing. isolation of microorganisms and primary screening was done according to the method given by Vega et al. [33]. Various soil fungi recorded from different three districts *viz.* Allahabad, Mirzapur and Varanasi.

Results and Discussion

The results obtained of different three district screening of soil fungi wheat cultivated area from the analyses 10 blocks of soil through soil dilution plate methods to determine the screening of soil fungi. Different soil fungi were recorded from wheat cultivated areas *Aspergillus* spp., *Penicillum* spp., *Geotrichum* spp., *Gloesporium* spp., *Fusarium* spp., *Mycelia sterilia*, *Arthrobotrys* spp., *Cladosporium herbarum*. In which *Aspergillus* spp. and *Penicillum* spp. common soil fungi recorded of district Allahabad in Table 1.

S.No.	Isolated Fungi
1	**Aspergillus sp.**
	Aspergillus oryzea (Ahlburg Cohn)
	Aspergillus flavus (Link)
	Aspergillus variecolor (Thom and Church)
	Aspergillus ochraceus (Withelm)
	tspergillus niveus (Blotch)
2	**Penicillum sp.**
	Penicillum variabil (Sopp.)
	Penicillum citrinum (Thom)
	Penicillum notatum (Westling)
	Penicillum steckii (Zaleski)
	Penicillum spp.(Perithecial)
3	*Geotrichum* spp.
4	*Gloesporium* spp.
5	*Fusarium* spp. (Sterile)
6	*Mycelia sterilia* (Four)
7	*Arthrobotrys* spp.
8	*Cladosporium herbarum* (Persoon)

Table 1: Isolation and identification of soil fungi from wheat cultivated area in district Allahabad.

The five *Aspergillus species* were recorded *viz. Aspergillus oryzea, Aspergillus flavus, Aspergillus variecolor, Aspergillus ochraceus, Aspergillus niveus* whereas the five *Penicillum species* were recorded *viz. Penicillum variabil, Penicillum citrinum, Penicillum notatum, Penicillum steckii, Penicillum* sp, in Allahabad district. Saksena, et al. [34] also were recorded soil fungi in Allahabad district. In the experiment detection of soil fungi from wheat, cultivated area consists of 10 blocks in district Mirzapur. The soil fungi were recorded from wheat cultivated area are *Aspergillus* spp., *Penicillum* spp., *Rizoctinia* spp., *Fusarium* spp., *Mucor* spp. *Aspergillus* spp. *Penicillum* spp. and *Fusarium* spp., were common soil fungi recorded (Table 2).

The eight *Aspergillus species* were recorded *viz. Aspergillus niger, Aspergillus flavus, Aspergillus oryzea, Aspergillus luchuensis, Aspergillus terreus, Aspergillus variecolor, Aspergillus awamori, Aspergillus niveus,* whereas the five *Penicillum* species were recorded *viz. Penicillum funiculosum, Penicillum frequentans, Penicillum steckii, Penicillum* spp., *Penicillum variabil,* the two *Rizoctinia* spp. were recorded i.e. *Rizoctinia oryzae, Rizoctinia cohnii,* four *Fusarium* spp. were recorded *of Fusarium* spp., *Fusarium avenaceum, Fusarium oxysporium, Fusarium javanicum*. The two *Mucor* species were recorded i.e. *Mucor fragilis,* and *Mucor jansseni* in district Mirzapur. Saksena, et al. [35] finding these fungi in Mirzapur district.

S. No.	Isolated Fungi
1	***Aspergillus* sp.**
	Aspergillus niger (Tiegh)
	Aspergillus flavus (Link)
	Aspergillus oryzea (Ahlburg Cohn)
	Aspergillus luchuensis (Inui)
	Aspergillus terreus (Thom)
	Aspergillus variecolor (Thom and Church)
	Aspergillus awamori (Nakazawa)
	Aspergillus niveus (Blotch)
2	***Penicillum* sp.**
	Penicillum funiculosum (Thom)
	Penicillum frequentans (Westling)
	Penicillum steckii (Zaleski)
	Penicillum sp.(Perithecial)
	Penicillum variabil (Sopp.)
3	***Rizoctinia* sp.**
	Rizoctinia oryzae (Went and Geerl.)
	Rizoctinia cohnii (Berl. And de Toni)
4	***Fusarium* sp.**
	Fusarium sp. (Sterile)
	F,usarium avenaceum (Fr.)
	Fusarium oxysporium (Schlect. Ex Fr.)
	Fusarium javanicum (Koord)
5	***Mucor* sp.**
	Mucor fragilis (Bain)
	Mucor jansseni (Lendner)

Table 2: Isolation and identification of soil fungi from wheat cultivated area in district Mirzapur.

In district Varanasi, detection of soil fungi from wheat cultivated area consists of 8 blocks. The results were obtained of soil fungi from wheat cultivated area are *Aspergillus* spp., *Penicillum* spp., *Rizoctinia* spp., *Fusarium* spp., *Mucor* spp., *Alternaria* spp., *Helminthosporium oryzae*, and *Humicola grisea*. In which *Aspergillus* spp., *Penicillum* spp. and *Fusarium* spp. were common soil fungi found. In Table 3 the eight *Aspergillus* spp. were recorded *viz. Aspergillus niger, Aspergillus flavus, Aspergillus luchuensis, Aspergillus terreus, Aspergillus variecolor, Aspergillus oryzea, Aspergillus luchuensis, Aspergillus terreus, Aspergillus variecolor, Aspergillus awamori, Aspergillus niveus, Aspergillus sydowi* spp., whereas the five *Penicillum* spp were recorded *viz. Penicillum funiculosum, Penicillum frequentans, Penicillum steckii, Penicillium rubrum Penicillium chrysogenum* spp., one *Rizoctinia* sp was recorded i.e. *Rizoctinia oryzae*, Three *Fusarium* sp. were recorded *viz. Fusarium semitectum, Fusarium oxysporium, Fusarium javanicum,* and one *Mucor* species i.e. *Mucor racemosus*, three *Alternaria* sp. were recorded *Alternaria alternata, Alternaria solani,* and *Alternaria claymydospora* were recorded from wheat cultivated area in Varanasi district.

Conclusion

The two common soil fungi were obtained *Aspergillus* spp. and *Penicillum* spp. in three different districts at Allahabad, Mirzapur and Varanasi, of Uttar Pradesh in India.

Acknowledgements

We are thankful to my sincerely Supervisor Prof. D.N. Shukla Department of Botany, University of Allahabad, Allahabad, India for Providing Laboratory Facilities, and I also thanks to my friend Shah Alam for views and opinions expressed in this article.

S No.	Isolated Fungi
1	***Aspergillus* sp.**
	Aspergillus niger (Tieghem)
	Aspergillus flavus (Link)
	Aspergillus luchuensis (Inui)
	Aspergillus terreus (Thom)
	Aspergillus variecolor (Thom and Church)
	Aspergillus awamori (Nakazawa)
	Aspergillus niveus (Blotch)
	Aspergillus sydowi (Bainier and Sastary)
2	***Penicillum* sp.**
	Penicillum funiculosum (Thom)
	Penicillum frequentans (Westling)
	Penicillum steckii (Zaleski)
	Penicillium rubrum (Stoll)
	Penicillium chrysogenum (Stoll)
3	***Rizoctinia* sp.**
	Rizoctinia oryzae (Went and Geerl.)
4	***Fusarium* sp.**
	Fusarium semitectum (Berkeley and Revenel)
	Fusarium oxysporium (Schlechtendahl)
	Fusarium javanicum (Koord)
5	***Mucor* sp.**
	Mucor racemosus (Fresenius)
6	***Alternaria* sp.**
	Alternaria alternata (Fr.) Keissler
	Alternaria solani (Sorauer)
	Alternaria claymydospora
7	*Helminthosporium oryzae* (Sacc.)
8	*Humicola grisea* (Traaen)
9	*Pythium aphanidermatum* (Edson) Fitzpatrick

Table 3: Isolation and identification of soil fungi from wheat cultivated area in district Varanasi.

References

1. Behl RK, Narula N, Vasudeva M, Sato A, Shinano T, et al. (2006) Harnessing wheat genotype × Azotobacter strain interactions for sustainable wheat production in semi-arid tropics. Tropics 15: 121-133.

2. Hasan MK (2006) Yield gap in wheat production: A perspective of farm specific efficiency in Bangladesh. Ph.D. dissertation, Department of Agricultural Economics, BAU, Mymensingh.

3. Finlay RD (2007) The fungi in soil. In: van Elsas JD, Jansson JK, Trevors JT (Eds.) Modern Soil Microbiology. CRC Press, New York. pp: 107-146.

4. Orgiazzi A, Lumini E, Nilsson RH, Girlanda M, Vizzini A, et al. (2012) Unravelling soil fungal communities from different Mediterranean land-use backgrounds. PLoS One 7: e34847.

5. Barrios E (2007) Soil biota, ecosystem services and land productivity. Ecol Econom 64: 269 285.

6. BBS (2012) Hand book of agricultural statistics January'1994, sector monitoring unit, Ministry of Agriculture, Government of Bangladesh.

7. Saunders D (1990) Report of an on-farm survey: Dinajpur district. Monograph no. 6 Wheat research institute Bangladesh agricultural research institute, Nashipur, Dinajpur.

8. Badaruddin M, Saunders DA, Siddique AB, Hossain MA, Ahmed MU, et al. (1994) Determining yield constraints for wheat production in Bangladesh. In: Saunders, DA, Hettel GP (Eds.) Wheat heat-stressed environments: Irrigated day areas and wheat farming systems CIMMYT, Mexico pp: 265-271.

9. Beare MH, Pohlad BR, Wright DH, Coleman DC (1993) Residue placement and fungicide effects on fungal communities in conventional and no-tillage soils soil science. Soc Am J 57: 392-399.

10. Valenzuela E, Leiva S, Godoy R (2001) Seasonal variation and enzymatic potential of micro fungi associated with the decomposition of Northofagus pumilio leaf litter. Revista Chilena de Historia Natural 74: 737-749.

11. Rai JP, Sinha A, Govil SR (2001) Litter decomposition mycoflora of rice straw. Crop Science 21: 335-340.

12. Santro AV, Rutigliano FA, Berg B, Fioretto A, Puppi G, et al. (2002) Fungal mycelium and decomposition of needle litter in three contrasting coniferous forests. Acta Oecologia 23: 247-259.

13. Nikhra KM (1981) Studies on fungi from Jabalpur soils with special reference to litter decomposition. Ph.D. Thesis, Jabalpur University, India.

14. Coockson WR, Beare MH, Wilson PE (1998) Effect of prior Crop residue Management decomposition. Appl Soil Ecol 7: 179-188.

15. Cruz AG, Gracia SS, Rojas FJC, Ceballos AIO (2002) Foliage decomposition of velvet bean during seasonal drought. Interciencia 27: 625-630.

16. Simoes MP, Madeira M, Gazariani L (2002) Decomposition dynamics and nutrient release of Cistus salvifolius L. and Cistus ladanifer L. leaf litter. J Agri Sci 25: 508-520.

17. Mc Tiernan KB, Couteaure MM, Berg B, Berg MP, de Anta RC, et al. (2003) Changes in chemical composition of Pinus sylvestris needle decomposition along a European coniferous forest climate transect. Soil Biol Biochem 35: 801-812.

18. Waksman SA (1922) A Method for Counting the Number of Fungi in the Soil. J Bacteriol 7: 339-341.

19. Johnson LF, Curl EA, Bond JH, Fribourg HA (1960) Methods for studying soil Mycoflora: Plant disease relationship. Burgess Publishing Co. Minneapolis p: 179.

20. Öner M (1973) A research on the microfungus floras of the Atatürk University Erzurum Çiftlioi, Eoer Da north slope and Trabzon-Hopa coastal lobe. Ankara: Atatürk Üniv. Yaynlar p: 71.

21. Nagamani Kunwar IK, Manoharachary C (2006) Hand book of soil fungi. I.K. International Pvt. Ltd.

22. Thom C, Raper KB (1945) A Manual of Aspergilli and Wilkins Co. Baltimore. Md., USA.

23. Gliman JC (1957) A manual soil fungi. Iowa State College Press, USA.

24. Smith G (1971) An Introduction to industrial mycology. London: Edward Arnold Ltd p: 390.

25. Hasenekoglu I (1991) Toprak mikrofunguslari. Atatürk Ünv. Yay. Vol. 7. No: 689, Erzurum. Soil Microfungi (Turkish).

26. Subramanian CV (1983) Hyphomycetes Taxonomy and Biology. London: Academic Press p: 502.

27. Ellis MB (1971) Dematiaceus hyphomycetes. Kew surrey UK: Commonwealth mycological institute p: 608.

28. Raper KB, Thom C (1949) A manual of the Penicillia. Baltimore: Williams and Wilkins company p: 875.

29. Raper KB, Fennell DI (1965) The genus Aspergillus. Baltimore: Williams and Wilkins Company p: 685.

30. Zycha H, Siepmann R, Linneman G (1969) Mucorales. Lehre: Stratuss and Cramer Gmbh p: 347.

31. Samson RA, Pitt JI (1985) Advances in Penicillium and Aspergillus Systematics. Plenum Press: New York and London p: 483.

32. Samson RA, Pitt JI (2000) Integration of modern taxonomic methods for Penicillium and Aspergillus classification. Amsterdam: Harwood Academic Publishers p: 510.

33. Vega K, Villena GK, Sarmiento VH, Ludeña Y, Vera N, et al. (2012) Production of alkaline cellulase by fungi isolated from an undisturbed rain forest of peru. Biotechnol Res Int 2012: 934325.

34. Saksena RK, Krishna Nand FNI, Sarbhoy AK (1962) Ecology of the soil fungi Uttar Pradesh-I. 29: 207-224.

35. Saksena RK, Krishna Nand FNI, Sarbhoy AK (1966) Ecology of the soil fungi Uttar Pradesh-II. 33: 298-306.

27

Influence of Protoplast Fusion in *Trichoderma* Spp. on Controlling Some Soil Borne Diseases

Hardik N Lakhani[1], Dinesh N Vakharia[1], Abeer H Makhlouf[2], Ragaa A Eissa[3] and Mohamed M Hassan[3,4*]

[1]*Biotechnology Department, Junagadh Agricultural University, India*
[2]*Agricultural Botany Department, Menoufiya University, Egypt*
[3]*Genetics Department, Menoufiya University, Egypt*
[4]*Scientific Research Center, Biotechnology and Genetic Engineering Unit, Taif University, KSA*

Abstract

Protoplasts from *Trichoderma harzianum* NBAII Th 1 and *T. viride* NBAII Tv 23 were isolated using lysing enzymes. Protoplast fusion of *T. harzianum* and *T. viride* was carried out. The fused protoplasts generated and 21 fusant isolates were used to study their antagonistic activity and RAPD-PCR characterization in comparison with their parents. Among the fusant isolates, F7 isolate produced maximum growth inhibition pathogen (one and half-fold increase as compared to the parent strains). All the tested fusants isolates exhibited increased antagonistic activity against *Fusarium oxysporum* than the parent strains except fusant F21. Specific results for fingerprinting were obtained by the seven primers of RAPD. These markers produced different fragment patterns with varied number of bands and yielded a total of 79 distinct bands. Polymorphic bands came to be 16.5% whereas monomorphic bands came to be 83.5%. Moreover, the OPO-13 primer has showed the highest polymorphism 35.3%. While, the OPA-16 primer has shown the lowest polymorphism 9.0%. The Dendrogram based on RAPD marker results grouped the two parent strains and twenty-one fusants into two different clusters with about 83% genetic similarity. The first cluster contained the two parent strains and twenty fusants, and the second cluster contained the fusant F2. Our results in this study suggested that the protoplast fusion technique is useful for developing the superior hybrid strains and enhance antagonistic activity of *Trichoderma* spp. against tested pathogenic fungi.

Keywords: Protoplast fusion; RAPD marker; Biocontrol activity; Pathogenic fungi and *Trichoderma* spp.

Introduction

Trichoderma is one of the well-known fungal used as a biocontrol agent in controlling of plant diseases caused by some fungal pathogens. *T. harzianum* and *T. viride* are filamentous soil fungi that functions as a biocontrol agent for a wide range of economically important aerial and soil borne plant pathogens [1]. The mycoparasitic activity of this organism is attributed to a combination of successful nutrient competition, the production of cell wall-degrading enzymes, and antibiosis [2,3]. Several strains of the genus *Trichoderma* are being tested as alternatives to chemical fungicides [4]. However, full-scale application of *Trichoderma* for biological control of plant pathogens has not been widespread. At a molecular genetic level, attempts to increase the biocontrol ability of *Trichoderma* have been focused on increasing chitinase or proteinase activity either by increasing the number of copies of the appropriate genes or by fusing them with strong promoters (e.g., *pcbh1::ech42*) [3-6]. These strategies have not always resulted in the expected increase in biocontrol activity. Protoplast fusion is a good tool in *Trichoderma* spp. improvement and developing hybrid strains in other filamentous fungi [3,4]. Several reports confirmed that protoplast isolation and regeneration in different fungi. Moreover, strain improvement for high yield of chitinase and cellulase by exploiting protoplast fusion system also reported [7-9]. Protoplast requires specific conditions for isolation and regeneration such as temperature, incubation time and medium, etc. [10,11]. Recent development in molecular techniques have created new possibilities for the selection and genetic improvement of life stocks [5,7,12,13]. RAPD markers are dominant, that is, it is not possible to distinguish between homozygous loci. Heterozygous RAPD loci are observed as different sized DNA segments amplified from the same locus, are detected only rarely. RAPD technique is notoriously laboratory dependent and not reproducible. Template concentration/primer mismatch may result in decreasing the amount of PCR product even total absence of PCR product, and hence the results are difficult to interpret [12,14-16]. The present study aimed to isolation and fusion of protoplast in *Trichoderma harzianum* NBAII Th 1 and *T. viride* NBAII Tv 23 for enhancement their biocontrol activity against some root rot and wilt causing pathogens which cause soil borne diseases. Furthermore, fingerprinting of parent strains and their corresponding fusants using RAPD marker.

Materials and Methods

Trichoderma samples

Two parental strains (*T. harzianum* NBAII Th 1 and *T. viride* NBAII Tv 23) were kindly gift from IARI, New Delhi and from MTCC, Chandigarh those two strains were done and 21 fusants were selected.

Protoplast preparation and fusion

Protoplasts preparation was carried out using fungicide tolerant isolates of *Trichoderma* parental strains (*T. harzianum* NBAII Th 1 and *T. viride* NBAII Tv 23) according to Stasz et al. [17]. Strains of *T. harzianum* (sensitive to carbendazim and resistant to Thiophanate-methyl) and *T. viride* (sensitive to Thiophanate-methyl and resistant to carbendazim) mycelia were subjected to lytic activity with different fungal enzyme such as β-glucuronidase and driselase at different conversation for 2 hrs with intermittent agitation [18]. The enzyme

***Corresponding author:** Mohamed M Hassan, Faculty of Agriculture, Genetics Department, Menoufiya University, Egypt
E-mail: khyate_99@yahoo.com

effectively acted on mycelia mass and produced protoplasts and it released soon after the lytic activity from the mycelia structure were found to be smaller in size and later gradually enlarged to a hyaline spherical structure. Protoplasts were fused according to Hassan [5].

Antagonistic activity of *Trichoderma* strains against some pathogenic fungi

The antagonistic activity of *T. harzianum* NBAII Th 1 and *T. viride* NBAII Tv 23 against *Rhizoctonia solani*, *Sclerotium rolfsii, Fusarium oxysporum* and *Macrophomina phaseolina* was determined by dual culture technique according to Fahmi et al. [19].

DNA extraction

Genomic DNA from *Trichoderma* parents and fusants were isolated as described [20] using CTAB extraction buffer.

RAPD analysis

RAPD analysis was performed using seven different primers. The name of RAPD primers are as the following: OPA-16, OPA-17, OPA-18, OPB-11, OPJ-07, OPK-03, OPO-13. The RAPD amplification reactions were performed as described in [15,21].

Data analysis

RAPD profiles were screened for generating polymorphism among the *Trichoderma* parents and fusants studied. Band position for each isolate and primer combination were scored as either present (1) or absent (0) for phylogenetic analysis using NTSYS-pc (Numerical Taxonomy and Multivariate analysis) system version 2.2 by Exeter software. The dendrogram of the genetic relatedness among *Trichoderma* parents and fusants was produced by means of the unweighted pair group method with arithmetic average (UPGMA) analysis [22].

Results and Discussion

Protoplast isolation

The enzyme effectively acted on mycelia mass and produced protoplasts and it released soon after the lytic activity from the mycelia structure they were smaller in size and later gradually enlarged to a hyaline spherical structure. The enzyme reacted with the cell wall of mycelial mass became swelled up and round hyaline globules by the

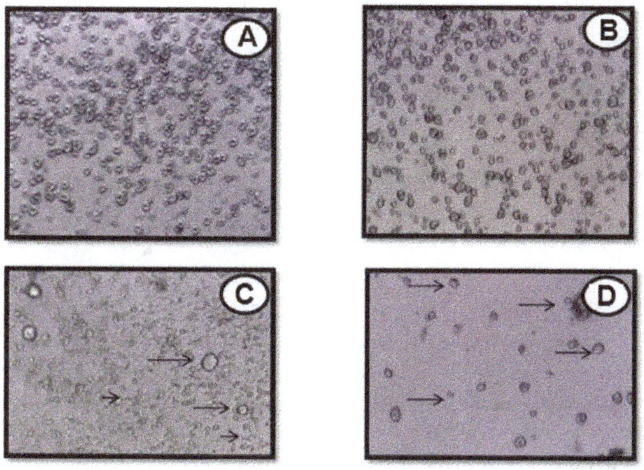

Figure 1: Protoplast isolation A) parental lysis of *T. harzianum* B) parental lysisi of *T. viride* C) complete lysis *T. harzianum* D) complete lysis *T. viride*.

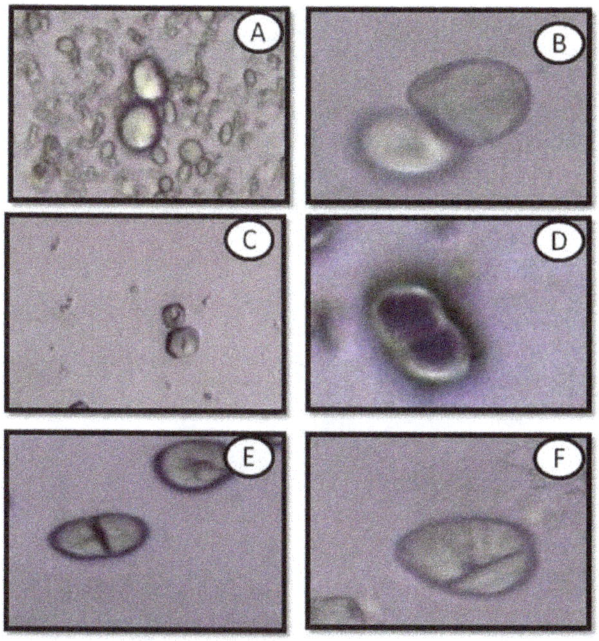

Figure 2: Different stage of protoplast fusion A) cell contact B) attachment to each other C) starting of fusion D) pairing of each other E) cytoplasmic fusion F) fusion completion.

action of these enzymes. The action of lytic enzymes on the mycelial mass depends on the used enzyme, its concentration, the incubation period and age of the mycelial mass [5,8].

The conditions for releasing the protoplasts were similar as reported earlier [5,10]. Swelling and rounding up of cell content were observed at initial stage and the lysis of mycelia started after 90 min (Figures 1A and 1B). Complete lysis of mycelia and release of protoplasts were observed after 3h (Figures 1C and 1D). The protoplasts released initially were smaller in size but later they enlarged to spherical structures.

Interestingly, the protoplasts yield was affected by many conditions as the concentrations of lysing enzymes. At low concentrations, the lysis of fungal mycelium was confined with a minimum release of protoplasts whereas at high concentrations, the mycelium lysed producing a large numbers of protoplasts. Among different concentrations of tested lysing enzymes, KCl as osmotic stabilizer was optimal for the release of protoplasts from different *Trichoderma* spp. However, in previous study highest protoplasts were obtained from *T. harzianum* using Novozym 234 at 10 mg/ml with 0.6 M KCl [18]. Other author used 10 mg/ml of Novozym 234 with 0.6 M sucrose to isolate maximum number of protoplasts from *T. harzianum* and *Trichoderma koningii* [7], and also, obtained maximum protoplasts from *Trichoderma roseum* using Novozym 234 in combination with chitinase and cellulase, each at 5 mg/ml [23].

Protoplast fusion

When adding the PEG solution to the protoplasts, they were attracted to each other and pairs of protoplasts were observed. Later, the plasma membrane at the place of contact dissolved and protoplasmic contents fused together, followed by nuclear fusion in most cases. Finally, the fused protoplasts became single, larger and round or oval shaped structures (Figure 2). Protoplasts fusion in *Trichoderma harzianum* NBAII Th 1 and *T. viride* NBAII Tv 23 has been achieved in the present study using 40% PEG. Similar concentration of PEG was reported as optimum for protoplasts fusion between different *Trichoderma* spp.,

Strains	Antagonistic averaged inhibition percentage %			
	Rhizoctonia solani	*Sclerotium rolfsii*	*Fusarium oxysporum*	*Macrophomina phaseolina*
T. harzanium	70	59	64	84
T. viride	69	63	65	86
Fusant 1	75	68	77	90
Fusant 2	79	76	80	89
Fusant 3	86	70	73	84
Fusant 4	86	76	87	89
Fusant 5	80	69	86	90
Fusant 6	65	69	90	93
Fusant 7	79	67	91	95
Fusant 8	58	72	83	92
Fusant 9	81	76	79	89
Fusant 10	56	78	66	86
Fusant 11	58	76	91	88
Fusant 12	80	76	87	93
Fusant 13	81	70	86	83
Fusant 14	50	56	70	91
Fusant 15	57	73	86	94
Fusant 16	79	78	76	91
Fusant 17	83	69	84	88
Fusant 18	78	66	92	84
Fusant 19	79	79	91	87
Fusant 20	80	72	80	86
Fusant 21	72	57	56	86

Table 1: Antagonistic potential of two parental strains and twenty-one of their fusants against some pathogenic fungi.

from 40% to 60% of PEG is suitable for protoplasts fusion in different fungi [23,24]. However, Pe'er and Chet [18], used 33% PEG for protoplast fusion in *T. harzianum*. Therefore, concentration of PEG is highly critical for effective fusion of protoplasts [5,24].

Antagonistic activity of fusants in dual culture technique

Antagonistic effects of all *Trichoderma* parental and fusant isolates *in vitro* were tested against *Rhizoctonia solani, Fusarium oxysporum, Sclerotium rolfsii* and *Macrophomina phaselona*. The antagonistic capacities of root rot and wilt causing pathogens were tested using dual culture method and the results are presented in Table 1 and Figure 3. The antagonistic activity was determined in the most fusant isolates compared to the parents. Among the fusant isolates, fusant isolate-7 produced maximum growth inhibition. It showed approximately one and half-fold increase over the parent strains. All the tested fusants exhibited increased antagonistic activity against *Fusarium oxysporum* than the parent strains except fusant-21. In all the dual culture plates, the inhibition zone appeared as a curve, with concavity oriented towards pathogens. The curvature of the inhibition area between the inoculation of *Trichoderma* fungi and the inoculation of pathogenic fungi in the same PDA plate depends on the growth rate of the inoculation. If one colony has a faster growth rate than the other, a curve in the inhibition zone will be observed easily. However, if the two inoculations have the same growth rate, a straight line would be observed when mycelia from both fungi come into contact [6,24].

Interestingly, all *Trichoderma* isolates exhibited inhibition of the mycelial growth of all pathogens. This could be due to the production of diffusible components, such as lytic enzymes or water-soluble metabolites. These components, such as chitinase, protease and glucanase, were always secreted by *Trichoderma* at low concentration [5,25]. Therefore, they can act against the pathogenic fungi before mycelial contact, thus increasing the antagonism of *Trichoderma*. From above results, it can be said that the highest growth inhibition of the tested pathogens was detected for fusant F7 followed by fusant F6. It also can be concluded that the tested *Trichoderma* strains reduced the growth of all tested soil borne pathogens. Therefore, can be used for integrated disease management of soil borne plant pathogens. The degree of antagonism varied between and within strains of *Trichoderma* against the soil borne plant pathogens [6,19,24,26].

Molecular characterization of parents and fusants by RAPD

RAPD-PCR has been used to evaluate genetic variation and taxonomic relationships in many fungi such as *Trichoderma* [27]. This technique have been used extensively in detecting genetic variation in different microorganism species because of its simplicity [9,12,15]. Genomic diversity of parents and fusants of *Trichoderma* produced by protoplast fusion was investigated by RAPD analysis. The RAPD results are illustrated in Table 2 and Figure 4. Data showed polymorphic numbers of the genetic bands, which were the electrophoretic products of PCR for parents and fusant strains. RAPD-PCR reactions were performed among two parents (P1 was *T. harzianum* NBAII Th 1, P2 was *T. viride* NBAII Tv 23) and twenty one fusant isolates (F1 to F21) using seven different 10-mer primers, which were preselected for their performance with *Trichoderma* DNA. The seven RAPD primers produced different fragment patterns and numbers. The primers yielded a total of 79 distinct bands (RAPD markers), (16.5%) of which were considered as polymorphic and 83.5% of which were considered as monomorphic.

The total number of bands as shown in Table 2 varied from 17 bands with primer OPO-13 (Figure 4) to 8 bands with primer OPA-18. The total of monomorphic amplicons was 66 and the total of polymorphic amplicons was 79. It can be concluded from our study that RAPD markers are effective in detecting similarity between *Trichoderma* strains and their fusants. In fact, they provide a potential tool for studying the inter-strain genetic similarity and the establishment of genetic relationships. The RAPD-PCR results using primer (OPB-11) has showed the highest polymorphism, 12 bands in these *Trichoderma*

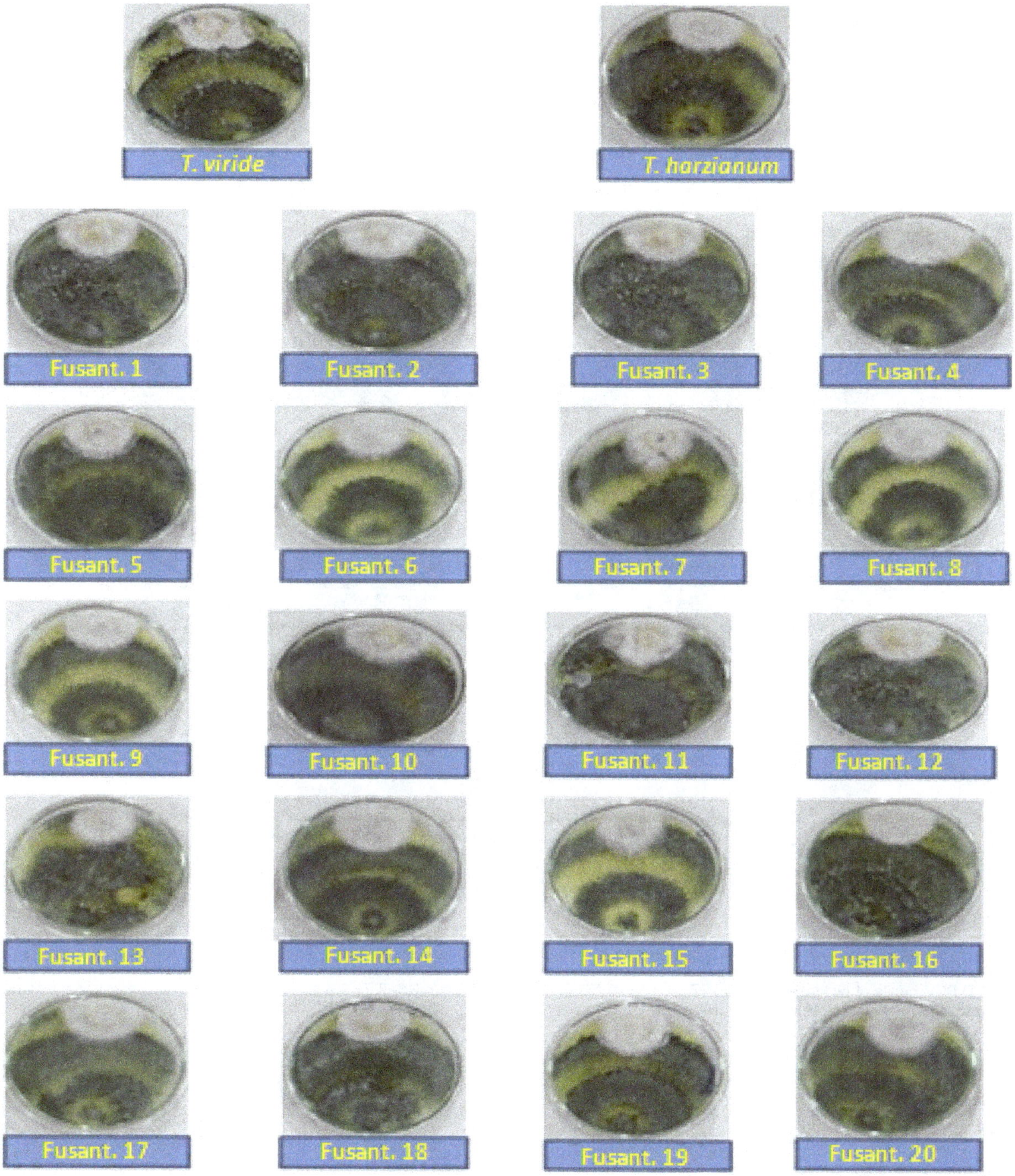

Figure 3: Antagonistic test of two parental strains (*T. harzianum* NBAII Th 1 and *T. viride* NBAII Tv 23) and twenty one of their fusants produced by protoplast fusion against *Fusarium oxysporum*.

Primers	Total Bands	No. of Monomorphic Bands	No. Polymorphic Bands	% Monomorphic bands	% Polymorphic bands
OPA-16	11	10	1	91.0	9.00
OPA-17	9	9	0	100	0.00
OPA-18	8	6	2	75.0	25.0
OPB-11	12	12	0	100	0.00
OPJ-07	9	7	2	77.8	22.2
OPK-03	13	11	2	84.6	15.4
OPO-13	17	11	6	64.7	35.3
Total	79	66	13	83.5	16.5

Table 2: Polymorphic bands of each RAPD primers and percentage of polymorphism in parent strains and their corresponding fusants.

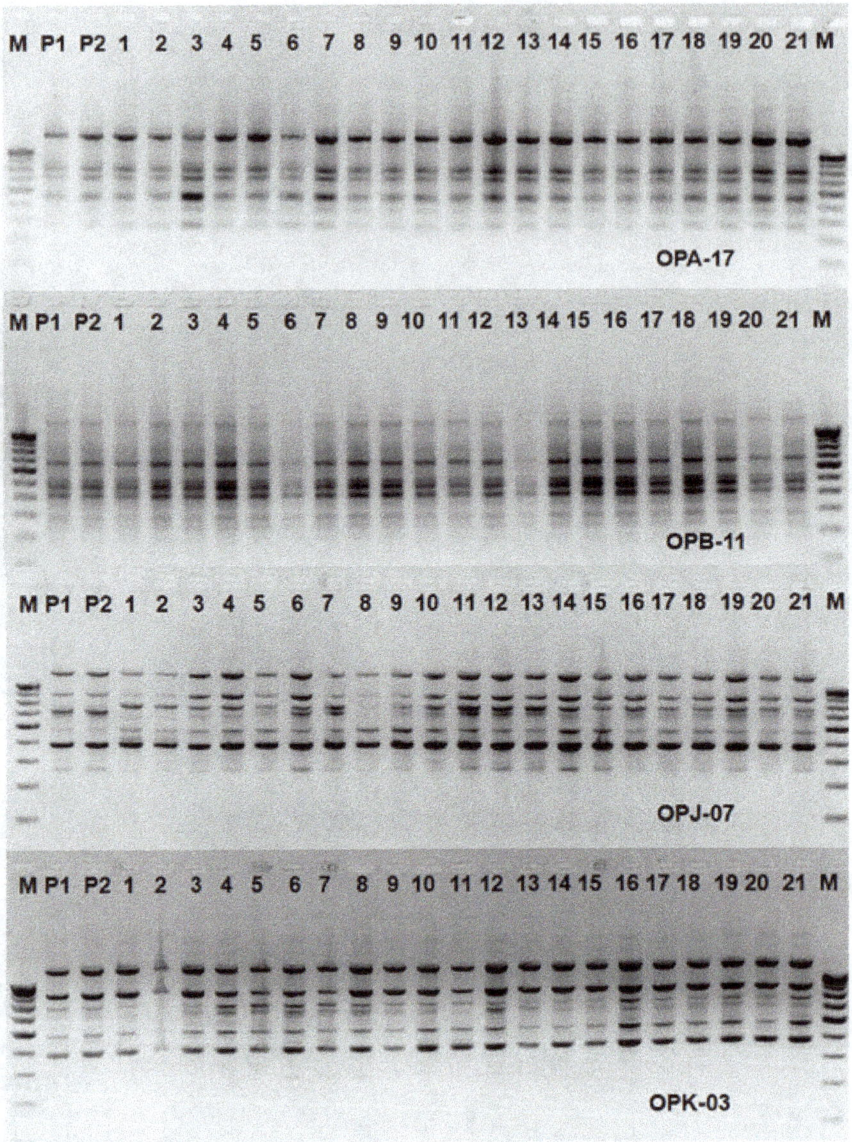

Figure 4: RAPD-PCR profile of two parental strains (*T. harzianum* NBAII Th 1 and *T. viride* NBAII Tv 23) and twenty one of their fusants produced by protoplast generated with seven RAPD primers. M: is 100 bp DNA ladder.

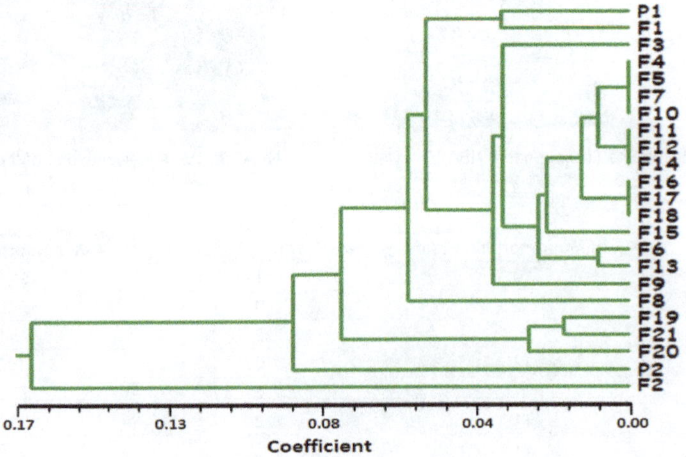

Figure 5: Dendogram of RAPD-PCR profile of two parental strains (*T. harzianum* NBAII Th 1 and *T. viride* NBAII Tv 23) and twenty one of their fusants generated with seven RAPD primers.

strains ranged from 75 bp to 1750 bp. In case of monomorphism, two primers have showed 100 % monomorphism among two parent's strains and twenty one fusants.

Cluster analysis of *Trichoderma* parent strains and their fusants

All fragments from RAPD markers were used for molecular characterization of *Trichoderma* strains by genetic similarities and designing the phylogenetic tree for these *Trichoderma* strains. According to genetic similarity, two parent strains and twenty-one fusants were grouped into two major clusters with about 83% genetic similarity. The two parent strains and twenty fusants were grouped in the first cluster and only fusant-2 was grouped in the second cluster (Figure 5). The overall genetic distance among *Trichoderma* strains was relatively low. The smallest genetic distance (0.008) was estimated between F11 and F12, moreover F20 and F15 relatively showed the highest genetic distance, whereas the genetic distance for parent P1 and fusant F1 was marginally low. It may be suggested that the tested biological and biochemical traits caused by allelic variation of functional genes, whereas RAPD is a genome-wide fingerprinting technique and in most cases target repetitive DNA regions. Sharma et al. [28] also found no correlation between genetic variability assessed by RAPD markers and the ability of *Trichoderma* isolates to antagonize *Sclerotium rolfsii.* However, other author found a relationship between RAPD polymorphisms of *Trichoderma* strains and their antagonism against *Aspergillus niger* [29]. Moreover, the others found a high relationship between some molecular markers of *Trichoderma* strains and their antagonism against many plant pathogens as *Fusarium oxysporium*, *Pythium altimum* and *Sclerotium rolfsii* [5,8,24,30]. Moreover, here and in a previous study, *T. harzianum* was the predominant taxon, *T. harzianum* is the most commonly reported species in the genus, occurring in diverse ecosystems and ecological niches. However, it must be borne in mind that the name "*T. harzianum*" applies to a species complex [31,32]. So we observed that most fusant strains trend to parent 1(*T. harzianum* NBAII Th 1) characters and these haplotype fusant species may be seen to comprise a multiplicity of species when subjected to multilocus phylogenetic analysis.

Conclusion

Our results refer to combined studies, including antagonism test, protoplast fusion and molecular markers, are necessary to select indigenous *Trichoderma* strains that can be used under different environmental conditions. From the two parent strains and twenty one of their fusants produced by protoplast fusion F6 and F7 fusants showed high antagonistic activity against *Rhizoctonia solani, Sclerotium rolfsii, Fusarium oxysporum* and *Macrophomina phaseolina* comparing with parent strains.

References

1. Nosir WS (2016) *Trichoderma harzianum* as a Growth Promoter and Bio-Control Agent against *Fusarium oxysporum* f. sp. Tuberose. Adv Crop Sci Technol 4: 217-224.
2. Brunner K, Zeilinger S, Ciliento R, Woo SL, Lorito M, et al. (2005) Improvement of the Fungal Biocontrol Agent *Trichoderma atroviride* To Enhance both Antagonism and Induction of Plant Systemic Disease Resistance. App Environ Microbiol 71: 3959-3965.
3. Elad Y (2000) *Trichoderma harzianum* T39 preparation for biocontrol of plant diseases control of *Botrytis cinerea, Sclerotinia sclerotiorum* and *Cladosporium fulvum.* Biocont Sci Technol 10: 499-507.
4. Mathivanan N, Srinivasan K, Chelliah S (2000) Biological control of soil-borne diseases of cotton, eggplant, okra and sunflower with *Trichoderma viride.* J Plant Dis Prot 107: 235-244.
5. Hassan MM (2014) Influence of protoplast fusion between two *Trichoderma* spp. on extracellular enzymes production and antagonistic activity. Biotechnol Biotechnological Equip 28: 1014-1023.
6. Parmar HJ, Hassan MM, Bodar NP, Umrania VV, Patel SV, et al. (2015) *In vitro* antagonism between phytopathologic fungi Sclerotium rolfsii and Trichoderma strains. Int J App Scie Biotechnol 3: 16-19.
7. Hassan MM, Eissa RA, Hamza HA (2011) Molecular characterization of intraspecific protoplast fusion in *Trichoderma harzianum*. New York Sci J 4: 48-53.
8. Hassan MM, El-Awady MA, Lakhani HN, El-Tarras AE (2013) Improvement of Biological Control Activity in *Trichoderma* Against Some Grapevine Pathogens Using Protoplast Fusion. Life Science Journal 10: 2275-2283.
9. El-Komy MH, Saleh AA, El-Ranthodi A, Molan YY (2015) Characterization of Novel *Trichoderma asperellum* isolates to select effective biocontrol agents against tomato *Fusarium* wilt. Plant Pathology Journal 31: 50-60.
10. Lalithakumari D, Mathivanan N (2003) Final project report on interspecific hybridization by protoplast fusion in *Trichoderma* spp. for lytic enyzme complex. New Delhi, India: Department of Biotechnology, Government of India.
11. El-Bondkly AM, Talkhan FN (2007) Intrastrain crossing in *Trichoderma harzianum* via protoplast fusion to chitinase productivity and biocontrol activity. Arab J Biotechnol 10: 233-240.
12. Salama SA, Tolba AF (2003) Identification of *Trichoderma reesei* using specific molecular marker and their ability as biocontrol agent. Egy J Genet and Cytol 32: 137-151.
13. Godrat R, Alireza K, Ardeshir NJ, Saeid S (2005) Evaluation of genetic variability in a breeder flock of native chicken based on randomly amplified polymorphic DNA markers. Ir J Biotechnol 3: 231-234.
14. Senthil-Kumar N, Gurusubramanian G (2011) Random amplified polymorphic DNA (RAPD) markers and its applications. Sci Vision 3: 116-124.
15. Hassan MM, Gaber A, El-Hallous El (2014) Molecular and Morphological Characterization of *Trichoderma harzianum* from different Egyptian Soils. Wulfenia Journal 21: 80-96.
16. Castro-Rocha A, Flores-Márgez JP, Aguirre-Ramírez M, Fernández-Pavia S, Osuna-Ávila P, et al. (2014) Traditional and Molecular Studies of the Plant Pathogen *Phytophthora capsici*: A Review. J Plant Pathol and Microbiol 5: 245-252.
17. Stasz TE, Harman GE, Weeden NF (1988) Protoplast preparation and fusion in two biocontrol strains of *Trichoderma harzianum.* Mycol 80: 141-150.
18. Pe'er S, Chet I (1990) *Trichoderma* protoplast fusion; a tool for improving biocontrol agents. Can J Microbiol 36: 6-9.
19. Fahmi AI, Al-Talhi AD, Hassan MM (2012) Protoplast fusion enhances antagonistic activity in *Trichoderma* spp. Nat Sci 10: 100-106.
20. Doyle JJ, Doyle JL (1987) A rapid DNA isolation procedure for small quantities of fresh leaf tissue. Photochem Bull 19: 11-15.
21. Narayan KP, Lata AS, Kotasthane AS (2006) Genetic relatedness among *Trichoderma* isolates inhibiting a pathogenic fungi *Rhizoctonia solani.* Afr J Biotechnol 5: 580-584.
22. Rohlf FJ (2000) NTSYS-PC numerical taxonomy and multivariate analysis system, Version 2.1. Exeter Software, Setauket, New York.
23. Balasubramanian N, Thamil PV, Gomathinayagam S, Lalitha-kumaria D (2012) Fusant *Trichoderma* HF9 with enhanced extracellular chitinase and protein content. App J Biochem Microbiol 48: 409-415.
24. Hassan MM, Ismail IA, Sorour AA (2014) Phylogeny and antagonistic activity of some protoplast fusants in *Trichoderma* and *Hypocrea.* Int J App Sci Biotechnol 2: 146-151.
25. Fahmi AI, Eissa RA, El-Halfawi KA, Hamza HA, Helwa MS (2016) Identification of *Trichoderma* spp. by DNA Barcode and Screening for Cellulolytic Activity. J Microb Biochem Technol 8: 202-209.
26. Selim ME (2015) Effectiveness of *Trichoderma* Biotic Applications in Regulating the Related Defense Genes Affecting Tomato Early Blight Disease. J Plant Pathol Microbiol 6: 311-317.
27. Williams JG, Kubelik AR, Livak KJ, Rafalski JA, Tingey SV (1990) DNA polymorphisms amplified by arbitrary primers are useful as genetic markers. Nucleic Acids Res 18: 6531-6535.

28. Sharma K, Mishira AK, Misra RS (2009) Morphological, biochemical and molecular characterization of *Trichoderma harzianum* isolates for their efficacy as biocontrol agents. J Phytopathol 157: 51-56.

29. Gajera HP, Vakharia DN (2010) Molecular and biochemical characterization of Trichoderma isolates inhibiting a phytopathogenic fungi *Aspergillus niger* Van Tieghem. Physiol Mol Plant Pathol 74: 274-282.

30. Shahid M, Srivastava M, Sharma A, Kumar V, Pandey S, et al. (2013) Morphological, Molecular Identification and SSR Marker Analysis of a Potential Strain of *Trichoderma/Hypocrea* for Production of a Bioformulation. J Plant Pathol Microbiol 4: 204-210.

31. Druzhinina IS, Kubicek CP, Komoń-Zelazowska M, Mulaw TB, Bissett J (2010) The *Trichoderma harzianum* demon: complex speciation history resulting in coexistence of hypothetical biological species, recent agamospecies and numerous relict lineages. BMC Evolu Biol 10: 94.

32. Błaszczyk L, Popiel D, Chełkowski J, Koczyk G, Samuels GJ, et al. (2011) Species diversity of *Trichoderma* in Poland. J App Gene 52: 233-243.

28

Use of Native Local Bio Resources and Cow Urine for the Effective Management of Post-harvest Diseases of Apples in Northwest Himalayan States of India

Manica Tomar[1]* and Harender Raj[2]

[1]*Directorate of extension education, Dr. Y.S.P Univ. of Hort. and forestry, Nauni, Solan H.P, 173230, India*
[2]*Department of Plant Pathology, Dr. Y.S.P Univ. of Hort. and forestry, Nauni, Solan H.P, 173230, India*

Abstract

Fungal pathogens namely *Alternaria alternata, Botrytis cinerea, Glomerella cingulata, Monilinia fructigena, Penicillium expansum* are reported to cause considerable post harvest losses in the state of Himachal Pradesh in India. Bio formulation comprising of six botanicals (seed/leaves) and cow urine were found effective for the management of post-harvest rot in apples. Fruit dip and wrappers impregnated with the bio formulations resulted in 84.7 per cent reduction in the post harvest rot after 75 days of storage at 4°C. Bio formulation treated fruits resulted in better fruit firmness and low TSS (14-16%).

Keywords: Post-harvest rots; Fungi; Cow urine; Plant extracts; Storage; Apples; Himachal Pradesh

Introduction

Apple (*Malus domestica* Borkh.) is an important fruit crop of India grown in about 1.19 lakh hectare area with a production and productivity of 25.85 lakh metric tonnes and 21.8 metric tonnes/hac, respectively (NHB, 2013-14). In India commercial cultivation of apple is confined in the states of Himachal Pradesh, Jammu and Kashmir, Arunachal Pradesh and Uttrakhand. Himachal Pradesh is situated in the heart of western Himalayas. The State is almost wholly mountainous with altitude ranging from 350 meters to 6,975 meters above the mean sea level. The State has a deeply dissected topography and rich temperate flora. About 20% of gross cropped area of Himachal Pradesh is covered with the Horticulture crops, which include the fruits, vegetables, flowers, plantations and spice crops, medicinal, & aromatic plants, roots & tuber crops cover. Apple is the major fruit crop of the state. Post-harvest losses are of major concern in the fruits in India as the total losses are to the tune of 30 percent of the total yield which are valued approximately Rs13,600 crores annually. The Indian horticulture industry is making losses estimated at more than US$32.7bn annually due to poor post-harvest practices and facilities, The Economic Times reported. In apple, post-harvest losses ranges from 10 to 25 percent in Himachal Pradesh and fungal pathogens are dominant cause of these losses. Among the different fungal pathogens, *Alternaria alternata, Botrytis cinerea, Glomerella cingulata, Monilinia fructigena, Penicillium expansum* are the dominant ones causing post-harvest losses. Fruit, due to their low pH, higher moisture content and nutrient composition are very susceptible to attack by pathogenic fungi, which in addition to causing rots may also make them unfit for consumption by producing mycotoxins [1,2]. Eckert and Ratnayake [3] estimated that out of 100,000 species of fungi, less than 10% are plant pathogens and more than 100 species of fungi are responsible for the majority of postharvest diseases. There are different strategies for the management of post-harvest losses due to biotic causes like fungal pathogens. Among these, better post-harvest handling practices and use of pre and post-harvest fungicides are the major management strategies. However, use of chemical fungicides in the management of post-harvest diseases poses a risk of residues in the harvest for the consumers. Use of chemicals in the management of post-harvest diseases is restricted due to their possible carcinogenicity, teratogenicity, high and acute toxicity, long degradation periods, environment pollution and their effects on human beings. On the other hand, bio-chemicals derived from extracts of the plants or other bio-resources have no toxic effects and their use is gaining grounds as alternatives to the prevalent chemical control measures. Plants like *Azadirachta indica, Ocimum sanctum, Eucalyptus spp, Aloe barbadensis, Vitex negundo* etc., contain an array of secondary substances like phenols, flavonoids, quinones, tannins, saponins and sterol which can be exploited for their different anti-fungal properties. The cow urine is capable of treating many curable as well as incurable diseases of human's animals and plants and has been used extensively in ayurvedic preparations since time immemorial as cited in ancient Indian holy texts like *Charaka Samhita, Sushruta Samhita, Vridhabhagabhatt, Atharva Veda, Bhavaprakash, Rajni Ghuntu, Amritasagar*, etc., [4]. Cow urine facilitated rapid and holistic recovery in disease infected combs, promoted the growth of brood, enhanced the efficiency of the worker bees in the colonies, thus revealed that the cow urine can serve as a potential eco-friendly measure for management of European foulbrood (EFB), a serious, bacterial disease of honeybee brood found throughout the world in honeybee colonies and also as an indirect control of mite diseases in colonies [5]. Achliya and coworkers [6] while working on cow urine, found many antimicrobial properties in different fractions of cow urine. Use of antifungal plant products and other bio-resources has not been studied for the management of apple post-harvest diseases. In Himachal Pradesh most of the apples are sent to different parts of the country for sale in cardboard boxes. Some of the annual produce is procured by the cold storage companies like *Adani Fresh* and *Reliance*

***Corresponding author:** Manica Tomar, Directorate of extension education, Dr. Y.S.P Univ. of Hort. and forestry, Nauni, Solan H.P, 173230, India
E-mail: manicatomar22@gmail.com

industries etc. Orchardists of the state face major losses due to post harvest disease during transportation and storage. Therefore, the objective of this study was to determine the effect of extracts of locally available plants and cow urine individually and in combination against major post harvest diseases of apple in India to reduce the losses by eco friendly methods.

Materials and Methods

Periodic surveys were carried out at 15 days interval in different apple growing areas in the state of Himachal Pradesh, India. Incidence of post-harvest loss due to fungal pathogens was also observed in the cold storages. Incidence of losses was recorded in major apple producing villages across the state. Four villages in district Shimla (Rohroo, kotkhai, Sarahan and Gumma), two in district Kinnaur (Nichar and Kalpa) and two in district Kullu (Bajaura and Kullu) were surveyed for recording the incidence of post harvest losses in apple. Terminal fruit markets, locally called "*Mandis*", in each district were surveyed during the season (June-oct) for recording the post-harvest losses in apple due to different post-harvest pathogens. Incidence was calculated by randomly sampling 25 apple boxes during the harvesting season (June to October). In the cold storages, incidence of post-harvest losses was recorded from November to March. Total rotting of apples was calculated by counting the rotted apples per box. The rotted apples from each location were brought to the laboratory and isolations were taken to find the associated pathogens. Each associated fungal pathogens were purified on Potato dextrose agar medium, purified cultures were then stored at 18 ± 1°C for further use.

Twelve plant species and cow urine were evaluated *in vitro* for their anti fungal activities against seven important post-harvest pathogens (*Alternaria alternata, Botrytis cinerea, Glomerella cingulata, Monilinia fructigena, Penicillium expansum, Rhizopus stolonifer* and *Tricothecium roseum*) isolated from the fruits. The fungi were purified on PDA and identified with the help of key of British Mycological society and simplified fungi identification key by Jean William Wood Ward [7]. Water extracts of twelve different locally available plant species (leaves) were prepared for evaluating their efficacy *in vitro*. Freshly harvested leaves of *Bougainvillea glabra, Dedonia viscosa, Eucalyptus globulus, Mentha piperita, Roylea elegans , Ocimum sanctum, Murraya koenigii, Chrysanthemum coronarium, Polyalthia longifolia, Pelargonium graveolens, Lawsonia inermis and seeds of Melia azedarach* were taken and washed twice in sterilized distilled water. Samples of each plant species were grinded in a mixer and grinder by adding equal quantity (w/v) of sterilized distilled water. Extract was filtered twice through a sterilized muslin cloth and Whatman filter paper no. 4, respectively and Refrigerated at 5°C for further use. Each extract was steam sterilized at 15 lbs pressure for five minutes in an autoclave. Plant extracts were tested against the test pathogens *in vitro* by using poisoned food technique [8] to study their inhibitory effect on the mycelial growth. Six plant extracts were selected out of twelve tested based on their efficacy to inhibit the mycelial growth of the fungi. The six selected extracts which gave more than 90 percent inhibitory effect *in vitro* were *Bougainvillea glabra, Dedonia viscosa, Eucalyptus globulus, Mentha piperita, Roylea elegans and Ocimum sanctum.*

These six botanicals were evaluated in two combinations as Field Formulation-I (FF-I) and Field Formulation-II (FF-II) against post-harvest fungal pathogens. Field Formulation-I was prepared by adding equal quantity of the sterilized plant extract of the six effective plants to equal quantity of sterilized distilled water (w/v). In FF-II instead of distilled water, cow urine was added (w/v). *In vitro* evaluation of FF-I, FF-II and Neemazal (neem based commercial formulation) was done at 50 percent concentration while SAAF (Mancozeb12%+Carbendazim 63%) was evaluated at 0.2 percent concentration. The efficacy of these two Field Formulations (FF-1 & FF-2) was compared with commercial neem formulation (Neemazal), edible Wax coating formulation (Carnauba wax) and fungicide (SAAF). These treatments were applied on fruits by giving them a dip treatment and the fruit wrappers were impregnated by the treatments solution.

The efficacy of FF-1, FF-II, Neemazal and SAAF was evaluated as direct skin coating of fruits and also by impregnating the fruit wrappers. Freshly harvested fruits of variety "Golden Delicious" having a TSS of 15 per cent were used for the experiment. The solution of FF-1, FF-II and Neemazal for skin coating was applied at 10 per cent concentration while SAAF at 0.2 percent concentration. Fruits were dipped in the respective solutions for half an hour. These apples were then air dried and packed in the corrugated card board boxes and stored in cold storage at 4°C. Fruit Skin coating by the edible wax was applied by instant dipping of the fruits in the warm (40°C) wax. To prepare the impregnated wrappers, five sheets of blotting paper measuring 75 × 50 cm were placed one over the other on a clean table. On each set of these five papers, 150 ml each of SAAF (500 ppm), FF-1 (10%), FF-2 (10%) and Neemazal (10%) were sprayed with hand sprayer. Uniform soaking of the solution was secured by spreading the solution smoothly over the sheets. The sheets were allowed to air dry in shade. The sheets were cut into 6 equal wraps each measuring 25 cm^2. Similarly, untreated sheets (butter paper and news paper) of the same dimension (25 cm^2) were used for comparison with the impregnated wrappers. The fruits were wrapped singly in each wrap, packed and stored in cold storage (4°C) for 75 days. Each treatment contained thirty fruits with three replications. Observations on per cent rotting of fruits were taken after 30, 45 and 75 days, respectively. Per cent rotting was calculated by the following formula:

$$\text{Rotting (\%)} = \frac{C-T}{T} \times 100$$

Where C is the control treatment where apples were not treated T is the different treatments given to the fruits.

General quality parameters analysis

After 75 days of storage, Total soluble solids (TSS) and firmness of the fruits were recorded. While TSS (%) was recorded with digital refractometre ATAGO Co Ltd (Japan), fruit firmness was recorded with help of Digital force gauge DFIS 50, USA using a piston cylinder of 4mm dia. All the experiments were conducted in completely randomized design and values of $P<0.05$ were considered significantly different.

Microbial analysis

Microbial analysis included the count of fungi, molds and yeasts on the fruit surface by dilution plate method. Potato dextrose agar medium with rose Bengal was used for enumeration of fungi while for bacteria and yeasts Nutrient agar media was used [9]. The incubation temperature for fungi, bacteria and yeast was 25 ± 1°C. Experiments were replicated thrice and results were expressed as $\log_{10}$ CFU per fruit.

Results and Discussion

Twenty one different types of post-harvest fungal pathogens were found associated with rotting of apple fruits in Himachal Pradesh (Table 1) and their average incidence varied between 0.5 to 45.6 percent.

Among the different post-harvest pathogens, Blue mould rot caused by *Penicillium expansum* was found most prevalent with an incidence of 45.6 percent (Table 1). In addition, fungi namely *Monilinia fructigena*,

Common Name of the rot	Causal organism	Percent Prevalence
Alternaria rot	*Alternaria alternata* Keissl.	1.0
Black pox	*Helminthosporium populosum* Berg	0.5
Black rot	*Sphaeropsis malorum* Pk.	0.5
Black mould rot	*Aspergillus niger* van Tieghem	0.5
Blue mould rot	*Penicillium expansum* Thom.	45.6
Bitter rot	Glomerella cingulata (Ston.) Spauld. and Schren	14.0
Brown rot (Apple black)	*Monilinia fructigena* (Aderh. and Ruhl.) Honey	10.8
Cladosporium rot	*Cladosporium herbarum* Lk. ex Fr.	0.5
Core rot	*Penicillium, Alternaria, Fusarium and Trichothecium* spp	2.0
Eye rot	Cylindrocarpon mali (Allesch) Wr.	0.5
Fusarium rot	*Fusarium spp.*	0.5
Gliocladium rot	*Gliocladium viride* Matr.	0.5
Grey mould rot	*Botrytis cinerea* Pers. Ex Fr.	1.5
Pestalotia rot	*Pestalotia hartigii* Tub.	0.5
Phytophtora rot	*Phytophthora cactorum* (Leb. and Cohn) Schroe	0.5
Pink mould rot	*Trichothecium roseum* Link	12.6
Stalk end rot	*Phoma mali* Schulz. and Sacc.	1.0
Stemphylium rot	*Stemphylium congestum* Newton	0.5
Soft rot	*Mucor piriformis* Fisch.	0.5
Sour rot	*Geotrichum candidum* Link	1.0
Whisker's rot	*Rhizopus stolonifer* (Her. Ex Fr.) Lind	5.0

Table 1: Prevalence of Post-harvest rots in Himachal Pradesh.

Penicillium expansum, Glomerella cingulata, Rhizopus stolonifer and *Trichothecium roseum* were also found associated with the rotted apples.

With a mean incidence of 10.02 percent, the cumulative incidence due to various rots varied from 4.3 to12.95 percent as shown in Table 2. Highest disease incidence was recorded in Shimla district, whereas, district Kinnaur had the least incidence (4.35 %). Blanpied and Purnasiri [10] while working on apples also reported maximum incidence of *P. expansum* and *B. cinerea* in storage. Blue mould rot due to *P. expansum* was found to be the main cause of post-harvest rot both in terminal markets and in cold storage with a mean per cent incidence of 2.86 percent.

Brown rot (*Monilinia fructigena*) and bitter rot (*Glomerella cingulata)* were found to be the next important post harvest rotting diseases in the state with an incidence of 2.24 and 1.42 percent, respectively. Whisker's rot (*R. stolonifer*) and pink mould rot (*Trichothecium roseum*) were found to be more prevalent in the warmer locations of the state with an incidence of 1.03 and 0.36 percent, respectively While *M. fructigena* was of common occurrence in terminal fruit markets and cold storage units of Shimla district (>6000 a.m.s.l) indicating the affinity of this pathogen for cooler temperatures (<18°C) prevailing in these areas during September to October months. Its incidence was highest in Khashdhar (14.2%) followed by Gumma (11.50%) in Shimla district. Ivic et al. [11] while working on dynamics and intensity of apple disease development during storage found that *M. fructigena, P. expansum* and *R. stolonifer* were the major fungi causing maximum losses to apple during storage.

The efficacy of both the field formulations (FF-1and FF-2) was compared with fungicide SAAF and Neemazal in the form of skin coating and impregnated wrappers for the management of post-harvest rotting of apple due to the fungal pathogens (Table 3).

Impregnation of wrappers with these bio-formulations (FF-1 FF-2, and Neemazal) was found effective in reducing the post-harvest rotting. Perusal of data revealed that treatment of wrappers with SAAF proved to be most effective with 81.2 percent reduction in the post-harvest rot of apple after 75 days of storage at 4°C.

However, among bio-formulations, wrappers impregnated with

District/location	Altitude (ft) a.m.s.l	G. cingulata	M. fructigena	P. expansum	R. stolonifer	T. roseum	Others
Shimla							
Khasdhar*	8000	2.00	14.20	0.00	0.00	0.00	2.00
Dhambari*	6800	1.50	7.30	3.00	1.00	0.00	1.50
Sandasu*	6500	2.10	5.00	1.50	0.50	0.50	1.50
Rohru+	6000	3.50	6.20	0.50	0.00	0.00	2.50
Kotkhai+	5200	2.70	4.10	3.20	1.50	0.50	3.50
Gumma**	4750	0.00	11.50	3.60	0.00	0.00	1.00
Sarahan+	7750	4.20	2.00	1.50	0.00	0.00	3.30
Deothi*	7000	3.50	1.90	3.50	0.50	1.30	2.70
Thanedhar**	7500	0.00	3.00	1.00	0.00	0.00	0.00
Navbahar**	7000	0.00	3.00	0.50	0.00	0.00	0.00
Shimla+	7000	1.90	2.10	4.50	0.50	1.00	3.10
Mean		**1.94**	**5.48**	**2.07**	**0.36**	**0.30**	**1.92**
Kinnaur							
Nichar*	7200	0.00	3.00	0.50	0.00	0.50	2.70
Kalpa+	9000	0.00	0.00	1.00	0.00	0.00	1.00
Mean		**0.00**	**1.50**	**0.75**	**0.00**	**0.25**	**1.85**
Kullu							
Bajaura+	5165	3.90	1.70	2.60	0.50	0.50	4.00
Kullu*	5250	1.60	0.90	3.50	1.50	0.50	3.50
Mean		2.75	1.30	3.05	1.00	0.50	3.75
Overall Mean		1.42	2.24	2.86	1.03	0.36	2.11

Table 2: Incidence of post-harvest diseases of apple at various locations in Himachal Pradesh.

Wrappers impregnation treatment	Per cent rotting at different durations (days)				
	30	45	60	75	Mean
SAAF	0.00 (0.71)	0.00 (0.71)	0.00 (0.71)	5.0 (2.23)	1.25 (1.09)
Neemazal	0.00 (0.71)	6.66 (2.58)	8.33 (2.88)	11.07 (3.32)	6.51 (2.37)
Field Formulation 1	0.00 (0.71)	3.33 (1.82)	5.0 (2.23)	8.33 (2.88)	4.16 (1.45)
Field Formulation 2	0.00 (0.71)	0.00 (0.71)	3.33 (1.82)	6.66 (2.58)	2.49 (1.45)
Butter paper (without any treatment)	2.73 (1.65)	6.66 (2.58)	12.73 (3.56)	18.33 (4.28)	10.11 (3.01)
Newspaper (without any treatment)	5.0 (2.23)	8.33 (2.88)	15.0 (3.87)	21.07 (4.59)	12.35 (3.39)
Control	6.07 (2.46)	12.73 (3.56)	20.0 (4.47)	26.6 (5.16)	16.35 (3.91)

Figures in the parentheses are square root transformed values.
C.D. (0.05)
Treatment (T) 1.28
Duration (D) 1.16
Interaction (T X D) 1.48

Table 3: Effect of different impregnated wrappers on fruit rotting in storage at 4°C.

FF-2 were found most effective with 75.1 percent reduction in the fruit rotting and both the treatments FF-1 and FF-2 were found statistically at par. Untreated wrappers (butter paper and news paper) resulted in 43.9 to 54.4 per cent more rotting in comparison to apples wrapped in FF-2 impregnated wrappers. SAAF fungicide gave 92.35 percent control of rots as compared to control. Field formulation FF-2 containing cow urine was significantly better over Neemajal. Achliya and co-workers [6] while working on Antimicrobial activity of different fractions of Cow Urine reported the inhibition of several fungi and bacteria.

As percent studies cow urine has proved to be an effective pest controller and larvicide when used alone and also in combination with different plant preparations by enhancing the efficacy of different herbal preparations [12-14].

Bio-formulations were found more effective when used as skin coating as compared to their impregnation in the wrappers (Table 4).

All the bio-formulations were found effective in the management of post-harvest rotting of apple in storage. Skin coating with FF-2 was found effective with 84.7 percent reduction in the post-harvest rotting in the storage after 75 days at 4°C. However, the fruits dipped in fungicidal solution Saaf gave maximum protection against post-harvest rots. Skin coating with FF-2 resulted in more reduction in the post-harvest rotting in comparison to impregnation of wrappers with FF-2. Skin coating with Neemazal was also found effective with 71.1 percent reduction in the post-harvest losses in the storage. The fruits coated with edible carnauba wax gave 75.02 percent reduction in disease as compared to control. Bio formulations were found to be equally effective against the rots.

The mechanism of disease suppression by plant products and biocontrol agents have suggested that the active principles present in them may either act on pathogen directly or induce systemic resistance in host plants resulting in reduction of disease development [15]. The antifungal compounds present in this leaf extract may have prominent effect in inhibiting the mycelial growth of the pathogen [16].

Plants have been shown to produce pectinase and proteinase inhibitors which restrict the microbial development [17-20]. Albersheim and Anderson [17] showed that proteins from plant cell walls inhibited polygalacturonases (PG) secreted by plant pathogens. Fielding [18] found similar inhibitors in extracts of peach and plum fruit. The level of the PG-inhibitor activity was correlated negatively with the rate of fungal rot development in apple fruits.

To develop an eco-friendly strategy for the management of post-harvest diseases of apple, most effective treatments were combined to test their efficacy against post harvest rots in storage and also to check the quality parameters like shelf life, Fruit Firmness and Total Soluble Substances (TSS). Perusal of data in Table 5 reveals that fungicidal solution of SAAF was most effective in controlling storage rots as compared to other treatments. However, Field Formulation-2 when given as fruit dip was able to reduce drastically with 80 percent disease control after 75 days of fruit storage at 4°C (Table 5). FF-2 containing cow urine was significantly helpful in managing storage rots when applied on fruits as well as on wrappers. Edible fruits wax was found to be significantly less effective in managing storage rots as compared to other treatments.

Quality parameters of the fruits like TSS and fruit firmness were found to be better in the treated fruits. Wax coating of fruits with carnauba wax showed best Fruit firmness (15.0 lbs per square inch) after 75 days of storage. Fruit firmness of 14.0 lbs per square inch after 75 days of storage was recorded in fruits which were dipped in with FF-2 as compared to 11.0 lbs per square inch in control.

Similarly, fruit wrapped in FF-2 impregnated wrappers were also effective with fruit firmness of 14.5 lbs per square inch. Fruits dipped in these bio-formulations were having a good shining and less wrinkles on the fruit skin in comparison to the other treatments.

Total soluble solids were the lowest (13.5) in wax coated fruits. In the fruits dipped in Field Formulation 2, level of TSS increased from 13.0 to 14.5 after 75 days of storage, indicating a slower ripening of the fruits. Fruits dipped in fungicidal solution SAAF indicated maximum TSS of 16.5 percent. These results indicated that a coating of field formulation on fruits not only prevented the losses from storage rots but also helped in maintaining the firmness and quality of fruits.

Firmness is an important quality parameter for fresh fruit, which decreases during storage as a result of cell wall degradation and loss of turgor. Rojas-Graü et al. [19-21] found that alginate edible coatings enriched with vanillin (up to 6%) and oregano (1%) applied to fresh-cut apples were effective in improving firmness. However, lemongrass containing coatings and oregano at 5% induced severe texture softening. They attributed it to the lower pH of those edible coating

Treatment	Storage rot(%) due to post-harvest diseases		
	30 days	45 days	75 days
Fruits coated with wax	2.66 (1.82)*	5.33 (2.49)	6.66 (2.74)
Fruits dipped in Neemazal (0.1%)	1.33 (1.41)	5.33 (2.49)	8.0 (3.0)
Fruits dipped in *Saaf* (0.2%)	0.00 (1.0)	1.33 (1.41)	2.66 (1.82)
Fruits dipped in Field Formulation-1 (10.0%)	1.33 (1.41)	4.0 (2.23)	6.66 (2.74)
Fruits dipped in Field Formulation-2 (10.0%)	1.33 (1.41)	2.66 (1.82)	4.0 (2.07)
Control	9.33 (3.20)	17.33 (4.27)	26.67 (5.23)

Figures in the parentheses are square root transformed values.
C.D. (0.05)
Treatment (T) 0.48
Duration (D) 0.23
Interaction (T X D) 0.58

Table 4: Effect of different dip treatments on fruit rot in storage.

Treatments	Storage rot (%) due to post-harvest diseases			Fruit firmness (lbs) at zero day	Fruit firmness (lbs) after 75 days of storage	TSS (%) at zero day	TSS (%) after 75 days of storage
	30days	45days	75days				
Fruits dipped in Field Formulation-2 (10.0%)	0.00 (1.0)*	2.85 (1.58)	5.71 (2.16)	16.0	14.0	13.0	14.5
Wrappers impregnated with Field Formulation-2 (10.0%)	2.85 (1.58)	5.71 (2.16)	8.56 (2.74)	16.0	14.5	13.0	15.0
Fruits coated with wax	2.85 (1.58)	5.70 (2.16)	11.42 (3.32)	16.0	15.0	13.0	13.5
Fruits dipped in Saaf (0.2%)	0.00 (1.00)	2.85 (1.58)	4.54 (1.39)	16.0	13.0	13.0	16.5
Control	8.56 (2.74)	20.0 (4.52)	28.57 (5.36)	16.0	11.0	13.0	20.0

*Figures in the parentheses are square root transformation values.
CD (0.05)
Treatment 0.93
Days 0.72
Interaction (Treatment × Days) 1.14

Table 5: Effect of best post- harvest fruit treatments on fruit rot due to diseases and shelf life in storage.

Treatments	No. of fungal and bacterial microorganisms (Log_{10} $CFUg^{-1}$) on the fruit surface per cm^2 after different durations (days) in storage					
	0		45		75	
	Fungal	Bacteria	Fungal	Bacteria	Fungal	Bacteria
Fruits dipped in Field Formulation-2 (10.0%)	6.8 ± 0.0	7.9 ± 0.0	0.0 ± 0.0	0.0 ± 0.0	0.8 ± 0.3	0.7 ± 0.3
Wrappers impregnated with Field Formulation-2 (10.0%)	6.8 ± 0.0	7.9 ± 0.0	7.4 ± 0.0	8.0 ± 0.0	7.5 ± 0.0	8.1 ± 0.0
Fruits coated with wax	6.8 ± 0.0	7.9 ± 0.0	7.9 ± 0.3	8.9 ± 0.4	8.3 ± 0.0	9.2 ± 0.0
Fruits dipped in Saaf (0.2%)	6.8 ± 0.0	7.9 ± 0.0	0.0 ± 0.0	0.0 ± 0.0	1.0 ± 0.0	0.8 ± 0.4
Control	6.8 ± 0.0	7.9 ± 0.0	7.9 ± 0.1	9.0 ± 0.1	9.1 ± 0.0	11.0 ± 0.0

*Figures in the parentheses are logarithmic transformation values.
C.D. (0.05)
Treatment 0.290
Days 0.225
Interaction (T × D) 0.355

Table 6: Effect of different post- harvest fruit treatments on fungal and bacterial surface micro-flora.

solutions. Also, lemongrass has as main compound citral, confirming our results. Guerreiro et al. [22] also reported a better effect of Eugenol than Cit on *Arbutus unedo* fresh fruit storage when using alginate based edible coatings.

According to Duan et al. [23] for blueberries, Soluble Solid Content (%) was not significantly affected by cold storage or coating (sodium alginate and chitosan) treatments. These results differ from those reported by Gol et al. [24] and Velickova et al. [25] who showed a decrease in the total soluble solids content in strawberries, at the end of storage, and attributed it to respiration, when using other edible coatings.

In our case the fruits before storage had a TSS of 13.0 percent which increased significantly in untreated control as compared to those dipped in cow urine based plant extract formulation. Cow's urine is not a toxic effluent as 95% of its content being water, 2.5% urea and the remaining 2.5%, a mixture of minerals, salts, hormones and enzymes [26]. The biochemical estimation of cow urine has shown that it contains sodium, nitrogen, sulphur, Vitamin A, B, C, D, E, minerals like manganese, iron, silicon, chlorine, magnesium, calcium salts, phosphate, lactose, carbolic acid, enzymes, creatinine and hormones [27]. Fractions of cow urine obtained by solvent extraction possess antimicrobial activity due to presence of aforesaid components those are solely responsible for the action [28].

The fungal and bacterial count increased with time in some of the treated fruits in storage (Table 6). In control this increase was constant up to 75 d storage. However, when FF-2 was applied on fruits, no fungal and bacterial development on fruits was recorded after 45 d of storage but 0.8 ± 0.3 and 0.7 ± 0.3.

Discussion

Chemical treatment of plant diseases especially in edible commodities has been drawing concerns due to their residual properties and human health issues. This has led to the introduction of eco friendly holistic approaches for the management of these diseases. The mechanism of disease suppression by plant products and bio control agents have suggested that the active principles present in them may either act on pathogen directly or induce systemic resistance in host plants resulting in reduction of disease development [15]. Plant extracts have been considered as an alternate and efficient way of plant disease management. In the present studies, twelve plant extracts were studied *in vitro* against the major post harvest pathogens of apple out of which only best six were further used. The present study also incorporated the use of cow urine for plant disease management. In India Cow is considered as pious animal and cow urine is known for its therapeutic properties. The cow urine is not only used against human ailments as therapeutic agent but also has several other uses as in agriculture and sericulture sectors [5]. The biochemical estimation of cow urine has shown that it contains sodium, nitrogen, sulphur, Vitamin A, B, C, D, E, minerals, manganese, iron, silicon, chlorine, magnesium, citric, succinic, calcium salts, phosphate, lactose, carbolic acid, enzymes, creatinine and hormones [27]. The present studies showed that the formulations with cow urine inhibited the fungal growth more efficiently as compared to the formulations containing water. Wrappers impregnated with FF-2 containing cow urine were found most effective with 75.1 percent reduction in the fruit rotting. Though the chemically treated fruits were better than cow urine based field formulation but statistically they were at par with each other. However, the results of water-based formulation (FF-1) were statistically different from both the chemically treated fruits as well as cow urine based formulation treated ones, there by clearly indicating that cow urine based formulation had more inhibitory effect on the fungi which was quite similar to the inhibition rendered through chemicals. Fruit dip in Saaf, Neemajal, edible wax, FF-1 and FF-2 gave maximum control of post harvest diseases through Saaf followed by cow urine based FF-2. Achliya and co-workers [6] while working on Antimicrobial activity of different fractions of Cow Urine reported the inhibition of several fungi and bacteria. As per recent studies cow urine has proved to be an effective pest controller and larvicide when used alone and also in combination with different plant preparations by enhancing the efficacy of different herbal preparations [12-14].

Shelf life of apples is an important attribute in storage. The apples in India are stored in cold storage chains and then supplied in the markets at an appropriate time for earning profits. During marketing of storage apples, TSS and fruit firmness decide the fruit price. In the present study it was observed that though the chemically treated fruits had comparatively lesser storage rots but its firmness had decreased over a period of 75 days in storage and the TSS content had also risen from 13.0 to 16.5 percent, indicating the ripening process of the fruits. However, The TSS and fruit firmness was 14.5 percent and 14.0 lbs respectively, in FF-2 treated apples which were comparatively better than other treatments used. In agrarian country like India, where majority of rural population have cows as their additional source of income, Cow urine based formulations would definitely prove to be a potential medicine, which in turn would reduce the pressure on the existing use of chemicals.

Conclusion

The present study tried to integrate the holistic approaches for the important storage rots of apples in the North West Himalayan states of India. Use of cow urine for management of plant diseases has been explored in this research. There is need for further research in this regard, as cow urine can be easily available, cheap, unharmful potential substitute for chemicals especially in the edible products. Protection of stored apples from post harvest diseases by chemicals lead to residual toxicity in human body. Therefore, plant extracts and cow urine can be utilised as potential source of inexpensive and efficient source of crop protection from post-harvest diseases.

References

1. Phillips DJ (1984) In: Moline HE (Ed.), Post-harvest pathology of fruits and vegetables, California, pp. 50-54.
2. Moss M (2002) Mycotoxin review: Aspergillus and Penicillium. Mycologist 16:116-119.
3. Eckert JW, Ratnayake M (1983) In: Lieberman M, (Ed.), Post-harvest physiology and crop preservation, Plenum Press, New York.
4. Pathak ML, Kumar A (2003) Cow praising and importance of Panchyagavya as medicine. Sachitra Ayurveda 5: 56-59.
5. Mohanty I, Manas Ranjan S, Jena D, Pallai S (2014) Diversified uses of cow urine. International journal of pharmacy and Pharmaceutical sciences 6: 20-22.
6. Achliya GS, Meghre VS, Wadodkar SG, Dorle AK (2004) Antimicrobial activity of different fractions of Cow Urine. Indian J Nat Prod 20: 14-16.
7. Williams-Woodward J (2001) Simplified Fungi Identification Key. Special Bulletin 37, The University of Georgia.
8. Nene VL, Thaplyal PN (1987) Fungicides in Plant Disease Control. Oxford & IBH Publ. Co. Pvt. Limited, New Delhi, India. p. 507.
9. Downes FP, Ito K (2001) Compendium of Methods for the Microbiological Examination of Foods, 4thedn, American Public Health Association, Washington, D.C
10. Blanpied GD, Purnasiri A (1968) Penicillium and Botrytis rot of McIntosh apples handled in water. Plant Disease Reporter 52: 865-867.
11. Ivic D, Cvjetkovic B, Milicevic T (2006) Dynamics and intensity of apple disease development during its storage. Agriculture 12: 36-41.
12. Chawla PC, Risorine R (2010) A Novel CSIR Drug Curtails TB Treatment, CSIR News. 3: 60-152.
13. Mandavgane SA, Rambhal AK, Mude NK (2005) Development of cow urine based disinfectant. Nat Prod Rad 4: 410-415.
14. Ahirwar RM, Gupta MP, Banerjee S (2010) Field efficacy of natural and indigenous products on sucking pests of Sesame. Indian J Nat Prod Resources 1: 221-226.
15. Paul PK, Sharma PD (2002) *Azadirachta indica* leaf extract induces resistance in barley against leaf stripe disease. Physiology and Molecular Plant Pathology 61: 3-13.
16. Kagale S, Marimuthu T, Thayumanavan B, Nandakumar R, Samiyappan R (2004) Antimicrobial activity and induction of systemic resistance in rice by leaf extract of Datura metel against *Rhizoctonia solani*. Physiology and Molecular Plant Pathology 65: 91-100.
17. Albersheim P, Anderson AJ (1971) Proteins from plant cell walls inhibit polygalacturonases secreted by plant pathogens. Proc Natl Acad Sci U S A 68: 1815-1819.
18. Fielding AH (1981) Natural inhibitors of fungal polygalacturonase in infected fruit tissue. J Gen Microbiol 123: 377-381.
19. Brown AE, Adikaram NKB (1983) Role of pectinase and protease inhibitors in fungal rot development in tomato fruits. Phytopath Z 106: 225-239.
20. Cervone F, Hahn MG, De Lorenzo G, Darvill A, Albersheim P (1989) Host-Pathogen Interactions : XXXIII. A Plant Protein Converts a Fungal Pathogenesis Factor into an Elicitor of Plant Defense Responses. Plant Physiol 90: 542-548.
21. Rojas-Graü MA, Raybaudi-Massilia RM, Soliva-Fortuny RC, Avena-Bustillos RJ, McHugh TH, et al. (2007) Apple puree-alginate edible coating as carrier of antimicrobial agents to prolong shelf-life of fresh-cut apples. Postharvest Biol Technol 45: 254-264.
22. Guerreiro AC, Gago CML, Faleiro ML, Miguel MGC, Antunes MDC (2015) The effect of alginate-based edible coatings enriched with essential oils constituents on *Arbutus unedo* L. fresh fruit storage. Postharvest Biol Technol 100: 226-233.
23. Duan J, Wu R, Strik BC, Zhao Y (2011) Effect of edible coatings on the quality of fresh blueberries (Duke and Elliott) under commercial storage conditions. Postharvest Biol Technol 59: 71-79
24. Gol NB, Patel PR, Rao TVR (2013) Improvement of quality and shelf-life of strawberries with edible coatings enriched with chitosan. Postharvest Biol Technol 85: 185-195.
25. Velickova E, Winkelhausen E, Kuzmanova S, Alves VD, Moldão-Martins M (2013) Impact of chitosan-beeswax edible coatings on the quality of fresh strawberries (*Fragaria ananassa* cv *Camarosa*) under commercial storage conditions. LWT-Food Sci Technol 52: 80-92.
26. Bhadauria H (2002) Cow Urine- A Magical Therapy. Vishwa Ayurveda Parishad. Int J Cow Sci 1: 32-36.
27. Jain NK, Gupta VB, Garg R, Silawat N (2010) Efficacy of cow urine therapy on various cancer patients in Mandsaur District, India - A survey. Int J Green Pharm 4: 29-35.
28. Jarald E, Edwin S, Tiwari V, Garg R, Toppo E (2008) Antioxidant and antimicrobial activities of cow urine. Global J pharmacol 2: 20-22.

Management of Banana (Musa Paradisiaca 1 L) Fruit Rot Diseases using Fungicides

Kedar Nath[1]*, Solanky KU[1] and Madhu Bala[2]

[1]*Department of Plant Pathology, N.M. College of Agriculture, Navsari Agricultural University, Navsari-396450, India*
[2]*Department of Plant Breeding and Genetics, College of Agriculture, Junagadh Agricultural University, Junagadh-362001, India*

Abstract

Banana suffers from several diseases at all the stages of its life. Finger rot and fruit rot caused by the fungus *Lasiodiplodia theobromae* (Pat.) Griffth and Maubl are the most important diseases in field as well as post-harvest of banana fruits. In this study, the antifungal activity of total seven fungicides was tested under *in-vitro* condition against *L. theobromae* and under *in-vivo* condition. The results of present study showed that six fungicides at all tested concentrations were a significantly check the fungal growth. At lowest tested concentration (250 ppm) carbendazim and propiconazole were completely inhibited fungal growth. Copper oxychloride at all tested three concentration were stimulated the mycelia growth of *L.theobromae*. Results of field experiment showed that carbendazim @0.5 gL^{-1} 17 and propiconazole @1mlL^{-1} 18 were completely reduced the percent disease index (PDI) and gave cent percent reduction of the finger rot disease followed by SAAF (97.36%). One hand containing ten fruits were selected from each treated bunch brought to laboratory, kept for ripening under natural Condition up to eating ripening stage the results showed that propiconazole @1mlL^{-1} 21 was highly reduction of PDI (1.50%) and gave highest reduction of fruit rot disease (98.20%) followed by carbendazim (4.005) with increased shelf life of banana fruits. Fruits were dipped in fungicides solution for 2 minutes and kept for ripening results showed that minimum PDI was observed in propicanazole and SAAF (1.00%) treated fruits with 25 maximum reduction of fruit rot disease (98.76%) followed by carbendazim (2.50%) with 96.79 percent reduced fruit rot disease.

Keywords: Banana; Fruit rot; Finger rot; Pathogen; Fungicides; *L. theobromae*

Introduction

Banana (*Musa paradisiaca L.),* fruit is one of the most important commercial fruit and vegetable crops grown all over the world in the tropical and subtropical areas. It is the second largest fruit crop, belongs to family *Musaceae* in order *Scitamineae.* It can be grown round the year and it is widely adopted in India. Apart from this it is considered as potential 'Dollar Earning crop'. Major banana producing countries are India, China, Philippines, Brazil, Ecuador, Indonesia, Costa Rica, Mexico, Thailand and Colombia. It is cultivated on an area of 4.81 Mha with an average production of 100.9 MT. in world, India produced 25.6% of total banana production of the world during 2012-13 (FAO) [1]. It shared 32.6 percent of total national fruit production during 2012-13 [2]. It ranks third in terms of area and first in production with a second in productivity of 34.2 mt/ha [2]. Gujarat share 17.1% of total national banana production with highest productivity (62.3t/ha.). Bananas are highly perishable commodities with post-harvest losses estimated to 25-30% [3]. Banana fruit suffers from many serious diseases such as fruit rot, crown rot, finger rot, cigar-end rot and pitting disease. The current postharvest problems of bananas are mainly concerned with storage and marketing. It is necessary to identify the most prevalent pathogen causing above said diseases and ultimately to reduce the yield loss as well as post-harvest loss of the banana fruit. The aim of this study was to evaluate the common and easily available fungicides in the markets for determine minimum inhibitory concentration (MIC) values of different level to find out the most suitable for field and post-harvest applications to reduced yield and post-harvest losses due to fruit rot of banana.

Materials and Methods

Isolation, identification and proving pathogenicity 50 of banana fruit rot pathogens

Diseased samples of banana fruit (cv. Grand naine) with pulp rot, crown rot, tip end rot, red spots, peeling injury/bruising and finger rot were collected from fruit markets, fruit stalls and domestic store-rooms of South Gujarat (viz., Navsari, Gandevi, Surat, Bardoli and Vyara) and field of soil and water management project, department of plant pathology and fruit research station Gandevi and brought to laboratory, examined visually and microscopically. The symptoms were observed were up to complete rotting of banana fruits. Repeated isolations were carried out from the crown portion, rotted pulp, reddish spot on pericarp and dried tip end rot, after washing thoroughly with tap water. The infected tissue were cut into small bits, surface sterilized with 2% sodium hypochlorite solution for 30 second followed by three subsequent washings with sterilized distilled water and then transferred aseptically on potato dextrose agar (PDA) medium in petriplates. The petriplates were incubated at room temperature for development of fungus growth. The plates were observed daily, the initial growth observed was picked up aseptically and it was transferred to PDA slants. The pure culture thus obtained was further purified by aerial mycelia tip technique. The pure cultures of isolated fungi were stored in slants in the refrigerator at 4^{0}C and used for further investigations.

Identification of isolates

The pure culture isolates obtained from the diseased banana fruits were used for the purpose of identification. Each isolate was subjected

***Corresponding author:** Kedar Nath, Department of Plant Pathology, N. M. College of Agriculture, Navsari Agricultural University, Navsari-645039, India
E-mail: drkdkushwaha@nau.in

to colony and microscopic examinations during which their structural features were observed. After purification, each fungus was allowed to sporulate. The sporulating cultures were identified on the basis of morphological characters of somatic and reproductive structures including spores/conidia. This was followed by a slide mount of each isolate under the lacto phenol cotton blue stain. The characteristics observed were matched with those available in manuals of Barnett and Hunters [4]. They were then identified accordingly.

Pathogenicity test

To prove the Koch's postulate, mature and semi ripen healthy banana fruits (cv. *Grandnaine*) were collected from field as well as from fruit market of Navsari and brought to the laboratory. The fruits were then surface sterilized by 2% sodium hypo chlorite solution for 2 minute followed by three washings with sterilized water and air dried then separately inoculated with each of the isolated fungus by Pin- Pricking method. Five fruits were separately inoculated with each of the isolated fungus. The inoculated as well as inoculated fruits were placed in sterilized, loosely tied polythene bags. A piece of sterilized wet absorbent cotton was placed inside each bag and the bag was kept at room temperature (24-28^{0}C) in an incubation room for symptoms development, inoculated fruits were observed regularly. Reisolation of pathogenic fungi from the diseased fruits was done. Morphological as well as cultural characters of reisolated fungi were compared with those of previously isolated from diseased banana fruits.

Management of Banana fruit rot diseases

The antifungal activity of seven different systemic, non-systemic and combination product (systemic+non systemic) fungicides at three different concentrations were evaluated against *L. theobromae* under in vitro and *in vivo* condition.

In vitro evaluation of fungicides against *L. theobromae*

The aim of this objective was to determine minimum inhibitory concentration (MIC) values of different fungicides to find out the most suitable fungicide with minimum concentration for field as well as post-harvest application to reduce yield and post-harvest losses due to *L.theobromae*. The antifungal activity of seven different fungicides evaluated, in these, three non-systemic (viz., mancozeb, copper oxychloride, chlorothalonil) @1500, 2000, 2500 ppm, three systemic (viz., carbendazim, propiconazole, hexaconazole) @ 250, 500, 1000 ppm and one combination (carbendazim 12 %+mancozeb 63 %, SAAF 75 WP) @1500, 2000, 2500 ppm were carried out against most frequently isolated fungus pathogen viz., *L.theobhomae in vitro* by poisoned food technique method described by Nene and Thapliya, [5]. The measured quantities of fungicides were incorporated in the melted sterilized PDA medium aseptically to obtain desired concentration (minimum inhibitory concentrations, MICs; 250 to 2500 ppm) of different fungicides at the time of pouring into borosil glass petri plates ($^{\varnothing}$90 mm). The 60 ml medium with fungicide was shaken well to give uniform dispersal of fungicides. Than the 20 ml medium with fungicides were poured in each of the Petri plates. After solidification, 5 mm discs of 7 days old culture of *L.theobromae* was placed in the center of test plates and arranged in completely randomized design with three repetitions. The plates were

Incubated at 25 ± 2°C temperature. The plates without fungicides served as control. After 48 and 72 hr of incubation, diameter of fungal growth was measured in each case, by averaging two diameter of fungal colony at right angle to one another and the percent inhibition was calculated by using the formula given by Bliss, [6].

Field evaluation

The field experiments were carried out at a commercial banana field "Soil and Water Management Project", situated at Navsari Agricultural University, Navsari, district Navsari of the Gujarat. Mancozeb, carbendazim, propicanazole, mancozeb+carbendazim (SAAF) that gave the best inhibition of *L.theobromae in vitro* were further evaluated in the field conditions. Efficacy of the fungicides namely mancozeb @ 3.33 gL^{-1}, carbendazim @ 0.5 gL^{-1}, propicanazole @ 1mL^{-1} and mancozeb+carbendazim (SAAF) @ 2gL^{-1} were dissolved in water to get a final concentration of 2500, 250, 250 and 250 ppm respectively which were used for field and post-harvest treatment. Experiment was laid out in randomized block design (RBD) with five treatments in four replications. Variety Grand naine was used for experimentation. The plants were thoroughly sprayed two times, first at bunch emergence and second spray before 15 days of harvest. Before the first spray the plants were tagged and all dead leaves were removed by cutting. Fertilizer application was done as per the recommended dose @ 200:90:200 NPK g/ plants. Irrigation was given as and when required with tube well water. Isolation was done from the bunches of treated and control plants before each spray to determine the infection on each plant. The effect of fungicides on banana bunch was also evaluated by assessing the percent disease index before and after treatment. Percent disease index was evaluated with the help of a model given by Rose [7]. Percent disease index (PDI) was recorded as procedure followed. Percent disease index was evaluated before the harvesting and 10 days after 2nd spray of each fungicide. Disease severity was recorded on the basis of percent fruit area infected under following assessment key (Figures 1 and 2).

Percent disease index (PDI) was calculated using to the formula [7].

$$\text{Percent disease index} = \frac{\text{Sum of all numerical ratings}}{\text{Total number of fruit examined} \times \text{maximum rating}} \times 100$$

One hand containing ten fruits were selected from each treated bunch brought to laboratory, kept for ripening under natural condition at room temperature up to eating ripening stage. Untreated fruits from each untreated bunch served as control. Each treatment was repeated four times, containing forty fruits. All treatments of the uninoculated fruits as well as untreated fruits were packed in sterilized polythene bags and stored at 25-28°C and 90-95% RH. After 12 days, PDI was calculated at eating ripening stage for all treatments as mentioned above. Efficacy (E) of each chemical treatment was calculated as under

$$E = \frac{\text{PDI of control fruits} - \text{PDI of treated fruits}}{\text{PDI of control fruits}} \times 100$$

Mancozeb @ 3.33 g/lit., carbendazim @ 0.5 g/lit., propicanazole @ 1ml/litre and mancozeb+carbendazim (SAAF) @ 2 g/lit. Were dissolved in water to get a final concentration of 2500, 250, 250 and 1500 ppm respectively which were used for the post-harvest dips treatments. Banana (c. v. Grand naine) fruits were harvested at uniform maturity stage and were treated by dipping for 2 minutes in the respective fungicides solutions. Before treatments, fruits were surface sterilized with 2% sodium hypochlorite solution for 2 minutes then 3 times rinsed with sterilized water. A randomized complete block design with four replicates considering one hand as one replication having 10^{-1}2 healthy fruits. The fruit samples were subjected to the above treatments and placed in tray for natural ripening at ambient temperature (25-

Figure 1: Assessment key for banana finger rot disease.

Figure 2: Assessment key for banana fruit rot (ripe) disease.

28°C) up to full ripening stage. Percent disease index (PDI) was worked out as above mentioned procedure.

Results

Isolation, identification and proving pathogenicity of banana fruit rot pathogens. Different fungi were successfully isolated from different banana fruit rots included, crown portion, rotted pulp, reddish spot on pericarp and dried tip end rot. The mixed infection of *L. theobromae* Pat., *F. moniliformae* Shield, *Fusarium* sp., *A. niger* VanTiegh., *Acremonium* sp. And *Curvularia* sp. at different stages of field, market and storage was observed of all the isolated fungi *L. theobromae* pat and *F. moniliformae* Sheld. Were predominantly infected banana fruits with *L. theobromae* being the most virulent, exhibiting dried tip end rot and pulp rot Cultural and morphological characters of isolated fungus were studied on PDA medium. On the basis of cultural and morphological characters, the isolates were identified as *L. theobromae* Pat., *F. moniliformae* Sheld, *Fusarium* sp., *A. niger*, *Acremonium* sp. and *Curvularia* sp. (Figure 3) with the help of illustrated genera of imperfect fungi [4]. Among these isolates, *L. theobromae* Pat and *F.monniliformae* were frequently isolated and well responsible for finger rot, crown rot and fruit rot disease in field as well as storage condition. However, for detail identification the purified cultures were confirmed at Agharkar Research Institute, Pune (No.3/426-2008). Immature healthy banana fingers were inoculated with seven days old culture of *L. theobromae* and left for symptom development. Results observed that Symptoms usually begin at the lower-end of the finger at the inoculated site. The decay spreaded uniformly causing black brownish discolaration of the peel and softening of the pulp. The affected area of the peel becomes wrinkled and encrusted with pycnidia. The pulp was reduced to a soft, rotten mass and a dark-grey mold grew on the peel surface under high humidity. The rate of disease development increased with maturity and spreaded to adjacent fingers. Infected clusters tend to ripe prematurely and fully mature fruits are most susceptible to infection. The infection occurred through tissues at the flower-end of the fingers and wounds were created by the insects. Thus, results of our study corroborate with previous workers reported by Goos et al. [8]. They found association of *Botryodiplodia theobromae* (Pat.) with finger rot disease and reported that infection occurs through tissues at the flower-end of the fingers and wounds. Slabaugh [9] found that finger rot disease caused by *Botryodiplodia theobromae* has been reported from Central and South America, the Caribbean Islands, India, Taiwan, and the Philippines. Symptoms produced at the lower-end of the finger or at the wound site resulted in to black- brownish discolaration of the peel and a softening of the pulp and peel becomes wrinkled. The *L. theobromae* was inoculated on healthy banana (var. Grand naine) fruits, produced brownish- black discolaration on the peel and a softening of the pulp on unripe banana fruits. On the ripe fruits water–soaked brownish discoloration observed in the infected area. Under moist conditions white to light grey cottony mycelium covered the infected tissues. Pycnidiospore formation takes place rather late (Figure 4). Water soaked brown discolaration that appeared later turned to dark brown and pulp got rotted slowly, rotting was fast in high humid condition due to *F. moniliformae* (Figure 5). *A. niger* produced light brown discoloration on peel and rotted pulp when fruits were over riped. *Curvularia* sp. produced minute pinkish spots when fruits fully ripened, or at over ripen stage, it extended in size, gone up to pericarp, but not infected pulp. *Acremonium* sp. produced brownish internal rotting at blossom end portion of fruits. The pathogenicity test showed that *Lasiodiplodia theobromae* as the most virulent pathogen, and in culture, it was in the most abundant. This suggests that *L.*

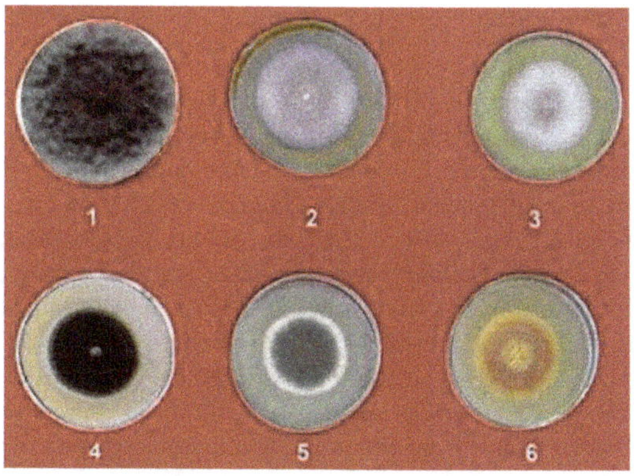

1. Lasiodiplodia theobromae
2. Fusarium moniliformae
3. Fusarium sp. 4. Aspergillus niger 5. Acremonium sp. 6. Curvularia sp.

Figure 3: Fungi associated with banana fruit rot.

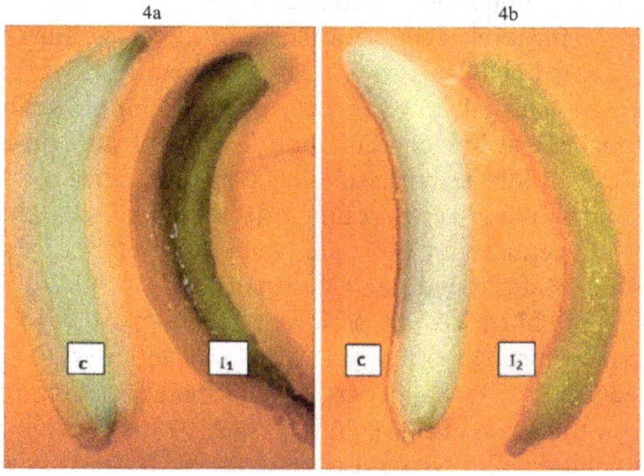

Figure 4: Symptoms of banana pulp rot.**I1**-Banana pulp rotted due to *L. theobromae* (Pat.) Griffth and Maubl. **I2**- Banana pulp rotted due to *F. moniliformae* Sheld **C**- healthy banana pulp.

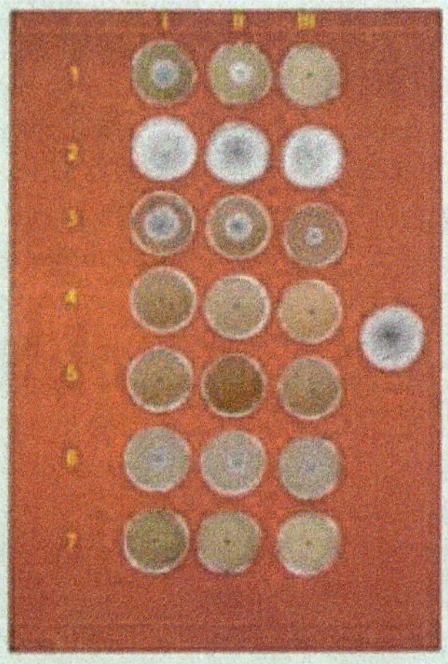

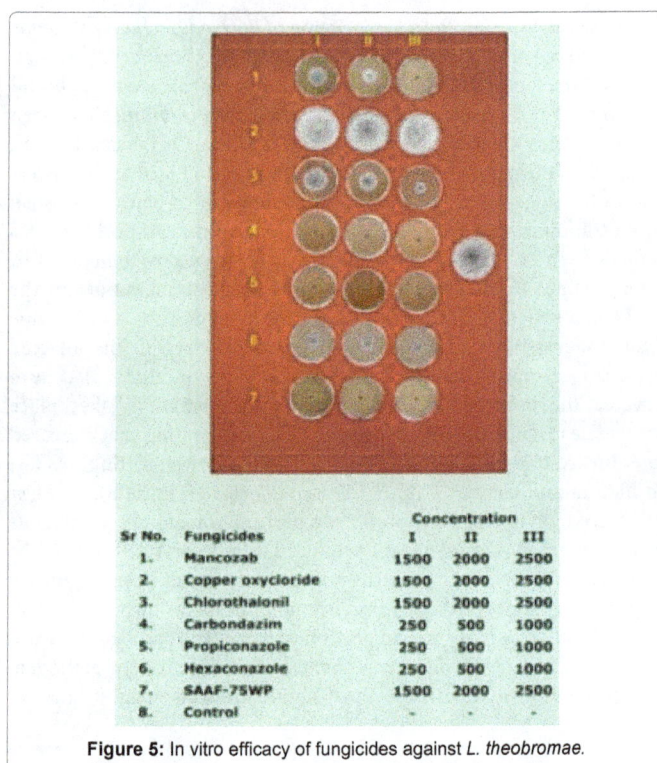

Sr No.	Fungicides	Concentration I	II	III
1.	Mancozab	1500	2000	2500
2.	Copper oxycloride	1500	2000	2500
3.	Chlorothalonil	1500	2000	2500
4.	Carbondazim	250	500	1000
5.	Propiconazole	250	500	1000
6.	Hexaconazole	250	500	1000
7.	SAAF-75WP	1500	2000	2500
8.	Control	-	-	-

Figure 5: In vitro efficacy of fungicides against *L. theobromae*.

theobromae Pat. Could be the leading cause of postharvest fruit rot in banana especially in South Gujarat from where all the samples were obtained. Crown rot disease was the major post-harvest disease due to fungal pathogens viz ; *L. heobromae, F. moniliformae* and *A. niger* the frequency of occurrence of these pathogens was 72% on ripe banana fruits of Grand Naine variety. The next fruit rot disease was due to *L. theobromae* and *F. moniliformae* with 10% fréquency occurrence. *L. theobromae* was the most frequently (43%) isolated in both the diseases followed by *F. moniliformae* (21.5%) from both Navsari and Surat market places. This is suggested that *L. theobromae* could be the leading cause of finger rot as well as post-harvest fruit rot in banana especially here in South Gujarat from where all the samples were obtained.

Management of banana fruit rot diseases

***In vitro* evaluation of fungicides:** Efficacy of seven fungicides at three different concentrations was evaluated for their comparative efficacy against mycelial growth of *L. theobramae* through "Poisoned Food Technique"under *in vitro* condition. All the evaluated fungicides exhibited varying level of efficacy against *L. theobramae*. The perusal of results presented in Table 1 revealed that all the fungicides tried were inhibitory to the fungal growth except coper oxychloride. Among these system fungicide at low concentration (250 ppm) carbendazim and propiconazole, while hexaconazole at 500 ppm inhibited cent percent fungus growth. In non-systemic fungicide, mancozeb and chlothalonil at highest concentration (2500 ppm) completely inhibited the fungal growth. In case of carbendazim 12% +mancozeb 63% (SAAF, 75 WP) completely inhibited growth at lowest (1500 ppm) concentration after 48 hrs of incubation. Copper oxychloride at lowest (1500 ppm) stimulated (12.40%) the fungus growth and stimulation of growth was increased with increased concentration. After 72 hrs of incubation the fungus growth was observed that systemic fungicides, carbendazim and propiconazile at lowest concentration completely inhibited the fungus growth, while hexaconazole at 500 ppm can't completely inhibited the fungus growth that's why the effect of hexaconazole was reduced after 48 hrs of incubation. In non-systemic fungicide, mancozeb at higher concentration (2500 ppm) completely inhibited the growth of fungus followed by chlorothalonil (87.59%), while in case of copper oxychloride the fungus over grew the petri plates i.e., growth was stimulated. SAAF retained their fungitoxicity up to 72 hrs petri plats incubated gave cent percent inhibition. These were four fungicides namely; carbendazin and propiconazole at 250 ppm concentration and mancozeb at 2500 ppm and SAAF at 1500 ppm were found superior in cent percent growth inhibition and proved statistically superior over the rest of fungicides tested. But in case of all tests concentrations for copper oxychloride the pathogen over grew the petri plates i.e., growth was stimulated.

Field evaluation: Out of evaluated fungicides, mancozeb @ 3.33 gL^{-1}, carbendazim @ 0.5 gL^{-1}251 , propiconazole @ 1mlL^{-1}, and SAAF @ 2gL^{-1} dissolved in tap water to make final concentration at 2500 ppm, 250, 250 and 1500 ppm respectively were selected as field and post-harvest treatments on the basis of their inhibitory effects under *in vitro* condition. The result presented in Table 2 revealed that PDI of finger rot was in these treatments ranged between 0.0 to 0.22 as compared to control (2.23%). PDI in banana finger rot was completely checked by carbendazim @ 0.5 gL^{-1} and propicinazole @ 1mlL^{-1}, followed by SAAF (0.08%) @ 2 gL^{-1} amd mancozeb (0.22.) @ 3.33 gL^{-1}. The overall finger rot disease reduction ranged from 91.48 to 100.00% in treated bunches as compared to control. Cent percent reduction of finger rot disease was in propiconazole and carbendazim treated bunches followed by SAAF and mancozeb by 97.36% and 91.48% respectively. All the tested fungicides were significantly reduced finger rot disease under field condition up to harvesting of banana bunches. Propiconazole treated bunches takes 3-5 days more for fruit maturity and increased 5-8% bunch weight as compaired to other tested fungicides as well as untreated bunch.

Storage condition: One hand containing ten fruits were selected from each treated bunch brought to laboratory, kept for ripening under natural condition up to eating ripening stage the results showed that PDI of fruit rot was in these treatments ranged between 1.50 to 11.50 % (Table 3). A minimum PDI of 1.50% was recorded in carbendazim @ 0.5 gL^{-1} followed by 4.0 % inpropiconazole @1mlL^{-1}treated banana bunches. While, 5.50 % and 11.50% PDI was observed in SAAF@ 2 gL^{-1} and mancozeb 3.33 gL^{-1} treated bunches as compared to control (83.00%). In term of percent disease reduction, the overall disease reduction was observed ranged from 86.12% to 98.20%. Maximum reduction of fruit rot disease was observed in propiconazole (98.20%) followed by carbendazim (95.16%) and SAAF (93.45%) treated bunches. While mancozeb was found moderate reduction of banana fruit rot disease as compared to control. Present study also indicated that propiconazole treated fruits take 2-3 day more time to eating ripe stage as compared to other treatments and pulp rotting was rare when stored up to 16 days after harvesting. The present results suggested that propiconazole and carbendazim were the best fungicide to control the field as well as storage disease and also increased shelf life of banana fruits upto16 days after harvest. The post-harvest application result presented in Table 4 all tested fungicides showed the greatest activity with significantly PDI reduction. Minimum PDI was observed in propiconazole and SAAF (1.0%) treated fruits and maximum PDI (4.0%) was observed in mancozeb treated fruits as compared to controlled (76.50%) up to eating ripened stage and16 days of storage. Maximum fruit rot disease reduction was observed in propiconazole and SAAF (98.76%) treated fruits followed by carbendazim (96.79%). Mancozeb and carbendazim produced brownish discoloration on fruit skin after 4-6 days of storage, but there was no effect on pulp. Overall,

Sr.No.	Fungicides	Conc. (ppm)	Growth after 48hrs of incubation		Growth after 72hrs of incubation	
			Growth (mm) *	Growth inhibition (%)	Growth (mm)	Growth inhibition (%)
1.	Mancozeb (Dithane M-45 75%WP)	1500	16.00* (4.06) **	60.49 (10.99) ***	32.33 (5.73) **	64.07 (8.03)
		2000	6.67 (2.68)	83.54 (12.00)	20.67 (4.60)	77.04 (8.80)
		2500	0.00 (0.71)	100.00 (12.67)	0.00 (0.71)	100.00 (10.02)
2.	Copper oxychloride (Blue copper 50 WP)	1500	45.50 (6.78)	-12.40 (6.94)	90.00 (9.51)	0.00 (0.71)
		2000	51.83 (7.23)	-28.00 (5.69)	90.00 (9.51)	0.00 (0.71)
		2500	61.67 (7.88)	-52.30 (2.84)	90.00 (9.51)	0.00 (0.71)
3.	Chlorothalonil (Kavach 75 WP)	1500	19.33 (4.45)	52.27 (10.62)	41.17 (6.45)	54.26 (7.40)
		2000	9.33 (3.13)	76.97 (11.72)	25.67 (5.09)	71.48 (8.48)
		2500	0.00 (0.71)	100.00 (12.67)	11.17 (3.41)	87.59 (9.39)
4.	Carbendazim (Bavistin 50 WP)	250	0.00 (0.71)	100.00 (12.67)	0.00 (0.71)	100.00 (10.02)
		500	0.00 (0.71)	100.00 (12.67)	0.00 (0.71)	100.00 (10.02)
		1000	0.00 (0.71)	100.00 (12.67)	0.00 (0.71)	100.00 (10.02)
5.	Propiconazole (Tilt 25 % EC)	250	0.00 (0.71)	100.00 (12.67)	0.00 (0.71)	100.00 (10.02)
		500	0.00 (0.71)	100.00 (12.67)	0.00 (0.71)	100.00 (10.02)
		1000	0.00 (0.71)	100.00 (12.67)	0.00 (0.71)	100.00 (10.02)
6.	Hexaconazole (Contaf 5 % EC)	250	7.83 (2.88)	80.65 (11.88)	18.67 (4.38)	79.26 (8.93)
		500	0.00 (0.71)	100.00 (12.67)	12.33 (3.58)	86.30 (9.31)
		1000	0.00 (0.71)	100.00 (12.67)	9.33 (3.13)	89.63 (9.49)
7.	Carbendazim 12 % + Mancozeb 63 % (SAAF 75 WP)	1500	0.00 (0.71)	100.00 (12.67)	0.00 (0.71)	100.00 (10.02)
		2000	0.00 (0.71)	100.00 (12.67)	0.00 (0.71)	100.00 (10.02)
		2500	0.00 (0.71)	100.00 (12.67)	0.00 (0.71)	100.00 (10.02)
8.	Control	-	40.50 (6.40)		90.00 (9.51)	
	S.Em.±	0.036		0.08	0.079	0.05
	C.D. at 5%	0.10		0.22	.22	0.15
	C.V. %	2.49		1.20	3.68	1.17

Table 1: Efficacy of fungicides on growth of *L. theobromae* under *in vitro* condition. * Average of three repetition. ** Figures are $\sqrt{X}$ + 0.5 transformed values *** Figures are $\sqrt{X}$ + 60 transformed values.

Sr. No.	Fungicides (ppm)	Per cent disease index (PDI)**	Disease reduction (%)
1.	Mancozeb (Dithane M-45 75%WP) (2500)	0.22** (4.86)*	91.48** (73.55)*
2.	Carbendazim (Bavistin 50 WP) (250)	0.00 (4.05)	100.00 (90.00)
3.	Propiconazole (Tilt 25 % EC) (250)	0.00 (4.05)	100.00 (90.00)
4.	Carbendazim 12 % + Mancozeb 63 % (SAAF 75 WP) (1500)	0.08 (4.36)	97.36 (81.59)
5.	Control	2.23 (9.51)	0.00 (4.05)
S.Em.±		0.042	0.12
C.D. at 5 %		.013	0.37
C.V.%		9.09	3.01

* Figures in the parentheses are angular transformed (X+0.5) values.
** Average of four replication.

Table 2: Effect of fungicides on banana fruit (finger) rot disease development under field Conditions.

the carbendazim @ 0.5 gL^{-1} and propicinazole @ 1mlL^{-1} were found to control the finger rot as well as post-harvest diseases of banana when plants were spayed two times, first just after bunch emergence and second spray before 15days of harvest.

Discussion

Finger rot and fruit rot caused by the fungus *Lasiodiplodia theobromae* (Pat.) Griffth and Maubl. are the most important diseases in field as well as post-harvest of banana fruits in south Gujarat condition. *In vitro* results showed that carbendazim, propiconazole, hexaconazole, carbendazim 12% +mancozeb 63% (SAAF, 75 WP) at lowest tested concentration and mancozeb at highest tested concentration (2500 ppm) were completely inhibited the fungal growth and proved statistically superior over the rest of fungicides tested. Copper oxychloride was found to stimulate the growth of *L.theobromae*. The present results are in agreement with the finding of several studies (Sabalpara, [10]; Thakore, [11]; Godara, [12]) showed that bavistin (0.025%) and dithane M-45 (0.05%) were effective against B. *theobromae* under *in vitro* condition. Ahmad et al. revealed that carbendazim (0.1%) and mancozeb (0.25%) were highly fungitoxic to *L. theobromae* in both solid and liquid media. Banik et. al. [13] observed the complete inhibition of mycelial growth of *B. theobromae* causing mango fruit rot by carbendazim (400 ppm), followed by captan (450 ppm), thiophanate methyl (450 ppm), ziram (600 ppm) and chlorothalonil (650 ppm). Yadav and Majumdar [14] reported effectiveness of carbendazim and mancozeb against *L. theobromae* (Guava isolate). Copper oxychloride at lowest (1500 ppm) stimulated (12.40%) the *L. theobromae* growth and stimulation of growth was increased with increased concentration. Muhammad et. al. [15] also reported that carbendazim and thiophanate methyl when used @ 1 ppm a.i. or more significant inhibition of mycelial growth of *L. theobromae*. Whereas, copxykil, cuprocaffaro and thiovit failed to inhibit the mycelial growth of *L. theobromae*. The results of field experiment showed that the spraying of carbendazim and propiconazole were completely reduced PDI of finger rot disease of banana over the rest fungicide treatment as compared to control (2.23%). Maximum percent reduction of finger rot disease was found in carbendazim 0.5 gL^{-1} and propiconazole @ 1 mlL^{-1} spraying bunches followed by SAAF (97.36%) and mancozeb as compared to control. All the tested

Sr. No	Fungicides (ppm)	Per cent disease index (PDI)	Disease reduction (%)
1	Mancozeb (Dithane M-45 75%WP) (2500)	11.50* (20.27)**	86.12* (68.55)**
2	Carbendazim (Bavistin 50 WP) (250)	4.00 (12.25)	95.16 (77.98)
3	Propiconazole (Tilt 25 % EC) (250)	1.50 (8.13)	98.20 (83.45)
4	Carbendazim 12 % + Mancozeb 63 % (SAAF 75 WP) (1500)	5.50 (14.18)	93.45 (75.76)
5	Control	83.00 (67.62)	0.00 (4.05)
	S.Em.±	1.52	1.39
	C.D. at 5 %	4.57	4.29
	C.V.%	13.05	9.68

Table 3: Effect of pre-harvest application of fungicides on banana fruit rot disease development under storage conditions.

Sr. No.	Fungicides (ppm)	Per cent disease index (PDI)	Disease reduction (%)
1	Mancozeb (Dithane M-45 75%WP) (2500)	4.00* (11.53)**	94.70* (77.34)**
2	Carbendazim (Bavistin 50 WP) (250)	2.50 (9.97)	96.79 (80.52)
3	Propiconazole (Tilt 25 % EC) (250)	1.00 (7.03)	98.76 (85.06)
4	Carbendazim 12 % + Mancozeb 63 % (SAAF 75 WP) (1500)	1.00 (7.03)	98.76 (85.06)
5	Control	76.50 (61.34)	0.00 (4.05)
S.Em.±		1.83	1.03
C.D. at 5 %		5.51	3.11
C.V.%		20.22	9.95

Table 4: Effect of post-harvest application of fungicides on banana fruit rot disease development under storage conditions * Figures in the parentheses are angular transformed (X+0.5) values. ** Average of four replication Fruit assessment was done at the "eating" stage.

fungicides were significantly controlled finger rot disease as compared to control. Similar results were reported from the control of mango die-back caused by *L.theobromae* by spraying with carbendazim @ 0.1% at fortnight interval [16]. Fruits harvested from treated plant, which were kept for natural ripening at room temperature. Minimum PDI was observed in propicinazole (1.50%) followed by carbendazim (4.00%) over the rest fungicides as compared to control (83.00%). Maximum per cent reduction (98.76%) of fruit rot was observed in propiconazole treated fruits followed by SAAF (98.67%) and carbandzim (95.64%) under storage condition up to eating ripe stage. No thytotoxic effect found in any treatment. In post-harvest treatment the results showed that minimum PDI (1.00%) was observed in propiconazole and SAAF treated fruits fillowed by carbendazim (2.50%). Maximum percent disease reduction (98.20%) was observed in propiconazole followed by SAAF and carbendazim treated fruits. Mancozeb and carbendazim showed phytotoxic effect producing brownish discoloration after 4-6 days of storage, but there was no effect on pulp. From these post-harvest experiment suggested that propiconazole was best fungicide to reducing post-harvest loss, while carbendazim and macozeb were other best fungicide if its applied as 10-15 days before the harvest. The present results are more or less in agreement with the results obtained by Khanna and Chandra [17] who reported benomyl and aretan were highly toxic as they completely checked the banana (Var. Harichal) fruits rot pathogen viz., *F. moniliformae* and *F. roseum* as pre and post inoculation treatment up to 8 days. Ved Ram and Dharam vir [18] got complete control of banana fruit rot decay caused by *A. flavus* and *A. fumigatus* by treating the fruits with thiophanate methyl, benlate, thiobendazole, bavistin, propionic acid and sodium metabisulphite at 2000 ppm up to 8 days of storage. Treatments of banana frui ts at three di fferent concent rat ion of thiophanate-methyl and benomyl which inhibi ted the di fferent rot in bananan frui ts range from 89.6 to 100.0% also reported by Latchmeah and Santchurn [19]. The results of field and post-harvest treatment suggested that most common and easily available fungicide i.e., propiconazole @1mlL^{-1} was completely reduced the finger rot disease in field and fruit rot disease in post-harvest disease is considered to be very important in the present day situation because both disease caused by *L. theobromae*. Similarly to the *in vitro* efficacy, the outcome of the field and post-harvest evaluation was highly encouraging propiconazole significantly reduced the post-harvest disease.

Acknowledgments

The expert technical contributions of Dr. K.U. Solanky and Dr. Mahesh Kumar Mahatma are gratefully acknowledged. We acknowledged to Department of plant pathology, Navsari Agricultural University, Navsari, for providing technical assistance. Soil and Water Management Research Station, N.A.U, Navsari for providing experimental field. Department of statistic for statistical analyses.

References

1. FAO (2013) Climatic Database.
2. NHB (2013) Horticulture data base.
3. Kachhwaha M, Chile A, Khare, Mehta A, Mehta P (1991) A new fruit rot disease of Banana. Indian Phytopathology 43: 211.
4. Barnett HL, Hunters BB (1985) Illustrated Genera of Imperfect Fungi (4thedn). Macmillan Incorporation 201.
5. Nene YL, Thapliya AS (1993) Fungicides in plant disease control, Oxford and IBH Publishing Co. Pvt. Ltd. New Delhi pp: 525-540.
6. Bliss CA (1934) The method of probits analysis. Science 79: 38-39.
7. Rose DH (1974) Diseases of apple fruits in the market. Bull US Dep. Agric 1253: 24.
8. Goos RD, Cox EA, Stotsky G (1961) *Botryodiplodia theobromae* and its association with Musa species. Mycologia 53: 262-277.
9. Slabaugh WR (1994) In: ploetz RC, Zentmyer GA, Nishijima WT, Rohrbach KG, Ohr HD (eds.) A *Botryodiplodia* finger rot. Pp: 387 5-6. Compendium of Tropical Fruit Diseases. APS Press. Paul St, MN.pp: 111.
10. Sabalpara AN (1983) Investigations regarding twig blight and die-back disease of mango caused by *Botryodiplodia theobromae* Pat M. Sc. (Agri) thesis, G.A.U, Sardar Krishinagar (Unpublished).
11. Thakore RA (1983) Studies on post-harvest disease of sapota (*Achras sapota L.*) occurring in South Gujarat and their control. M. Sc. (Agri.) thesis, G.A.U, Sardar Krishinagar (Unpublished).
12. Godara SL (1994) Studies on post-harvest 363 diseases of orange fruits. Ph.D. Thesis (Unpublished) Submitted to Rajasthan Agricultural University Campus, Udaipur.
13. Banik AK, Kaiser SIKM, Dhua RS (1998) Evaluation of some systemic and non-systemic fungicides against *Botryodiplodia theobromae*, the cause of dieback disease of mango. J Soil & Crops 8: 199-222.
14. Yadav RK, Majumdar VL (2004) Efficacy of plant extracts, biological agents and fungicides against Lasiodiplodia theobrome incited die back of guava (*Psidium guajava*). J Mycol Pl Pathol 34: 415-417.
15. Muhammad AK, Lodhi AM, Saleem S (2005) Chemical control of *Lasiodiplodia theobromae* the causal agent of mango decline in Sindh. Pak J Bot 37: 1023-1030.
16. Rawal RD (1998) In: Arora RK, Ramanatlia Rao V (Eds.) Management of fungal diseases in tropical fruits. Tropical fruits in Asia: Diversity, Maintenance, Conservation and Use. Proceedings of the IPGRI-ICAR-UTFANET Regional training course on the conservation and use of germplasm of tropical fruits in Asia held at Indian Institute of Horticultural Research Bangalore, India.

17. Khanna KK, Chandra S (1976) Control of Banana fruit rot caused by *Fusarium moniliforme* and *Fusarium roseum*. Indian Phyto Path 29: 20-21.

18. Ved Ram, Dharamvir (1984) Post-harvest chemical treatment for prevention of *Aspergilus flavus* and *A. fungigatus* on banana fruits. Pesticides 18: 65-66.

19. Latchmeah RS, Santchurn D (1991) Chemical control of common postharvest disease of banana cv. Naine in Mauritius. Revue Agricole et Sucriere del Ile Mauric 70: 8-11.

Integrated Management of Garlic White Rot (*Sclerotium cepivorum* Berk) Using Some Fungicides and Antifungal *Trichoderma* Species

Chemeda Dilbo[1], Melaku Alemu[2], Alemu Lencho[3] and Tariku Hunduma[1*]

[1]*Ethiopian Institute of Agricultural Research, Ambo Plant Protection Research Center, P.O. Box 37, Ambo, Ethiopia*
[2]*Ethiopian Institute of Agricultural Research, National Agricultural Biotechnology Research Laboratory, P.O. Box 31, Holetta, Ethiopia*
[3]*College of Agriculture and Veterinary Sciences, Department of Plant science, Ambo University, P.O. Box 19, Ambo, Ethiopia*

Abstract

White rot (*Sclerotium cepivorum* Berk), is one of the most destructive soil borne pathogens that pose significant threat to production of garlic and other *Allium* species in Ethiopia and all over the world. Since most of the conventional control methods are not effective, the development of eco-friendly and cost effective integrated management method is critically required. A study was then conducted with completely randomized design and three replications that consist of all possible combinations of 31 treatments. The study was conducted during 2013/14 under greenhouse condition with the objective of evaluating the effect of two fungicides, Apron Star 42 WS and Tebuconazole, and in combination with four *Trichoderma* species namely *T. hamatum, T. harzianum, T. oblongisporum* and *T. viride*. The results of this study revealed that the efficacy of both fungicides, when tested alone, against *S. cepivorum* was lower than those treated with *Trichoderma spp.* alone and the fungicide combined treatments. Among all treatments, T16 (Apron Star 42 WS fungicide combined with *T. hamatum* and *T. viride*) has provided the best antagonistic activity against *S. cepivorum* with no disease incidence, followed by *T. viride* (T8) alone and Tebuconazole combined with *T. hamtum* (T21) (both 11.1% incidence). This was well correlated to the level of foliar, stem base and bulb rots symptoms as well as to plant growth and biomass of garlic plant parts. The results suggested that integration of fungicides and *Trichoderma* species is better than applying them alone, which could be attributed to the synergistic and additive growth promoting effects of combined treatments besides controlling the disease. This integrated approach appears to be the first report in Ethiopia, which has never been tested before.

Keywords: Garlic; White rot (*Sclerotium cepivorum*); *Trichoderma spp*; Fungicides

Introduction

Garlic (*Allium sativum* L.) is a monocotyledonous plant and belongs to the family *Alliaceae*. It is the second most widely cultivated vegetable next to onion and widely produced for its medicinal and nutritional properties and has been recognized in almost all the cultures for its culinary properties. Garlic is an excellent source of several minerals and vitamins that are essential for health and has medicinal role for centuries such as antibacterial, antifungal, antiviral, antitumor and antiseptic properties [1]. In Ethiopia, the total area under garlic production in 2011/12 reached 13,278.55 ha and the production is estimated to be over 123,961.46 tons annually [2]. Production of garlic is done on sandy soil with higher organic matter content, pH 6-7 at altitude of 1800-2500 m.a.s.l, rainfall 600-700 mm and temperature of 15-24°C [3,4]. Economic significance of garlic in Ethiopia is fairly considerable and contributes to the national economy as export commodity [5] and important for small holder farmers [6]. It was reported that heavy damage to garlic due to fungal diseases, in later years, has become very important in major production areas of garlic [7-11].

Of the fungal diseases, white rot (*Sclerotium cepivorum* Berk) is the most destructive disease of garlic, and other *Allium* species throughout the world. It attacks leaves, roots, and bulbs of *Allium spp.* and can survive in the soil for nearly 20 years. Sclerotia are the only reproductive structures of *S. cepivorum* has no perfect stage has not yet been described and no asexual spores are produced. The sclerotia are stimulated to germinate only by *Allium*-specific root exudates (alkyl-cysteine sulphoxides) which are broken down by soil microorganisms to form thiols and sulphide compounds and then stimulate *S. cepivorum* sclerotia to germinate, indicating that the host range is limited to *Allium* species [10]. White rot causes important economic losses in garlic production worldwide and can cause losses from 1 to 100% [11]. In Ethiopia, the yield loss has been found to range between 20.7% and 53.4 % [12]. Once it is established permanently renders a field unusable for a garlic production. In spite of its importance garlic productivity in many parts of the world, is low due to the lack of improved variety and, traditional production system besides diseases and pest problems. The use of low quality seeds, imbalanced fertilizers, inappropriate agronomic practices, uneven irrigations and marketing facilities are the main constraints [9,10].

Management of diseases caused by soil borne pathogens like *S. cepivorum*, is very difficult and need a multi-pronged management strategy [13]. The earliest methods used to control garlic and onion white rot were cultural and physical practices of field hygiene and sanitation and crop rotation were used for primary inoculums reduction. These have been viewed as impractical for *Allium* white rot control due to long persistence nature of the sclerotia for more than 20 years. Soil flooding, soil solarisation and sterilization, biological control agents,

***Corresponding author:** Tariku Hunduma, Ethiopian Institute of Agricultural Research, Ambo Plant Protection Research Center, P.O. Box 37, Ambo, Ethiopia
E-mail: tarikuh2012@gmail.com, tarikuh2002@yahoo.com

sclerotia germination stimulants (diallyl disulfides, DADS), composted onion waste, host resistant were also found moderately effective at varying degrees [14-16]. It has been found that systemic as well as non-systemic fungicides significantly reduced garlic white rot disease development and resulted in improved garlic yield. Several effective fungicides have been recommended against this pathogen. Among these, Tebuconazole was also effective in reducing the incidence and in increasing the yield when applied as a clove treatment [17,18].

Recent efforts have focused on developing economically safe, long lasting and effective bio-control methods for the management of plant diseases. Use of biocontrol agents has been shown to be eco-friendly and effective against many plant pathogens. Among the fungal antagonists, *Trichoderma* is considered as the most important because it controls various soil borne and seed diseases caused by a wide range of fungal pathogen [19,20]. *Trichoderma* grows rapidly when inoculated in the soil as it is naturally resistant to many toxic compounds including herbicides, fungicides and insecticides such as DDT and phenolic compounds. The resistance to toxic compounds may be due to the presence of ABC transport systems in *Trichoderma* strain. The biocontrol mechanisms exercised by *Trichoderma* could be attributed to mycoparasitism, competition for nutrients, release of toxic metabolites and extra cellular hydrolytic enzymes [21].

In Ethiopia, research effort on host resistant against garlic white rot is very limited. It was reported that systemic as well as non-systemic fungicides significantly reduced incidence of white rot, its progress rate and severity that also resulted in improved garlic yield [7,9]. Study revealed that some of the *Trichoderma* species are endowed with great potential in controlling the garlic white rot [22].

The most effective control systems to date have involved the integration of a number of systems for managing garlic white rot [13,23]. The combined use of biocontrol agents and chemical pesticides has attracted much attention as a way to obtain synergistic or additive effects in the control of soil-borne pathogens. Seed treatment with *Trichoderma* along with compatible fungicide is common practice among the farmers for economic and effective management of seed and soil-borne plant diseases. Combination of *Trichoderma* with reduced levels of fungicide promotes the degree of disease suppression without risk on non- target organisms similar to that achieved with full dose of fungicide application [24-26]. *Trichoderma harzianum* C52 was found to be compatible with some fungicides and determined to be effective biocontrol agent of the onion white rot pathogen [27]. It was found that *T. viride* combined with either Tebuconazole or onion compost resulted in enhanced white rot control (>90%) and was better than any treatment alone [23,28].

However, attempt has not been made in Ethiopia to determine the effect of integrating various control measures with *Trichoderma* species for the management of white rot in garlic. Hence, the present study was undertaken on the management of garlic white rot with the integration of four selected *Trichoderma spp.* (*T. hamatum, T. harzianum, T. oblongisporum* and *T. viride.*) and two recommended fungicides (Apron Star 42 WS and Tebuconazole) under pot culture condition. In this paper the results of this integrated management of garlic white rot under pot culture condition is described.

Materials and Methods

Experimental design

The experiment was conducted in a Completely Randomized Design (CRD) with three replications and 31 treatments consisting of all possible combinations with the objective to achieve integrated management of garlic white rot using four *Trichoderma spp* of PPRC isolates and two recommended fungicides [Apron Star 42 WS and Tebuconazole (Folicur 250 EC] under greenhouse condition. The *Sclerotium cepivorum* sclerotia propagules were maintained and undertaken in pot experiment (Seedling bioassay), as described earlier by [23,29]. Inoculated local garlic clove with *S. cepivorum* and un-inoculated alone were used as positive and absolute control, respectively.

Culturing of *Sclerotium cepivorum*

Culture specimens of *S. cepivorum* preserved in the Mycology Section of Ambo Plant Protection Research Centre (APPRC) were used for this study. Stock culture was inoculated onto sterile potato dextrose agar (PDA) plates and incubated at 25°C for 2 days and then examined for the growth of the fungus. After incubation, the appearance of colonies on the medium was observed which proved the viability of preserved isolates of the *S. cepivorum*. The well-grown mycelium was selected for further study.

Mass production of *Sclerotium cepivorum*

The sclerotia of *S. cepivorum* isolate were first produced on PDA in 9-cm diameter petri dishes by incubating at 20°C for 5 days. Since the pathogen doesn't have functional spores, a small, round, seed-like structure known as sclerotia was initially produced. The refreshed *S. cepivorum* sclerotia were further inoculated on whole wheat grains [30]. Fifty grams of the inoculated whole wheat grains were added to each of twenty, 250 ml conical flasks, the content of the flasks were treated with 45 millilitres of 0.0025 % (w/v) Chloramphenicol and the flasks were left overnight at room temperature. The treated flasks of wheat (50 g each) were autoclaved at 121°C and 15 psi for 30 min, and this was repeated for three consecutive days. After cooling to room temperature, each flask was inoculated with four, 5 mm disks of *S. cepivorum* taken from the actively grown edge of a 5 day old culture grown on PDA. The flasks were incubated at 20°C in the dark for 6 to 8 weeks and shaken at weekly intervals to ensure an even distribution of mycelium. During the first three weeks of incubation, 0.5 ml of sterile distilled water (SDW) was added if the flasks appeared dry, to encourage mycelia growth .

Harvesting *Sclerotium cepivorum* sclerotia

The sclerotia of *S.cepivorum* were harvested from the wheat grains using progressive wet sieving through 850 μm, 500 μm and 250 μm sieves [31,32]. Only healthy sclerotia was retained on the 500 μm sieve which was air dried on sterilized Whatman No. 1 filter paper for 24 h before they were used or conditioned. The sclerotia used after this stage was termed "fresh". Before using for the greenhouse study, both the fresh and conditioned sclerotia viability were resolved by taking a sample of 100 sclerotia and surface sterilized in 0.25% sodium hypochlorite (NaOCl) for 1min. Subsequently, it was washed in five changes of sterile distilled water (SDW), then spreaded over Whatman No. 1 filter paper to absorb excess liquid. Then, it was placed onto PDA in petri-dishes. The petri-dishes were sealed with polythene wrap and then incubated at 20°C in the dark and the sclerotial viability/ germination was examined for 10 days. The number of germinated sclerotia was recorded to reach >96%. Once the viability of the sclerotia germination percentage and competence were decided, 100 g of sclerotia/kg of sterilized moist soil was incorporated into the *in vivo* experiment. This is based on the fact that 0.01-0.1 g sclerotia/g of soil resulted in infection of less than or equal to 85-100% and 100% incidence of disease in onion and garlic plants, respectively. This is similar to the finding that only one sclerotia per kilogram of soil can provoke a 50%, and 10-20 sclerotia per kilogram can result in infection of essentially all plants (as the disease severity depends on sclerotia levels in the soil at the time of planting [33].

Mass production of *Trichoderma spp.*

The *Trichoderma spp.* used in this study were obtained from the culture specimen collections of APPRC, that, previously isolated from soils characterized in Ethiopia and preserved in culture collection [34]. These *Trichoderma spp.* were found to be effective in controlling faba bean fungal disease, *Fusarium solani* [35]. Furthermore, out of seven *Trichoderma* species tested under *in vitro* and *in vivo* antifungal activities against white rot of garlic, four of them registered high percentage inhibition zone ranging from 51.7 to 59.3%. [26] Therefore these four potent species were selected for the present study viz., *T. hamantum, T. harzianum, T. oblongisporum and T. viride.*

Mass multiplications of *Trichoderma spp.* were carried out according to standard procedures [36,37]. Thus, spore suspensions of *Trichoderma spp.* were prepared by adding 20 ml sterile distilled water to a three-week-old petri-dish cultures and scraping gently with a sterile spatula. The harvested spore suspension of *Trichoderma spp.* were inoculated into a sterilized one litre jar containing wheat bran, sand and water medium or sorghum grain and incubated for three days at 20°C.

In vivo efficacy test

The experiment was conducted under greenhouse condition using the local cultivar of garlic. The appropriate soil composition were made proportionally with the composition of sand, compost and sandy clay loam soil mixed at (1:1:2 ratios) and then sterilized. Each pot (21 cm top diameter and 9 cm height) were filled with 3 kg of mixed soil. The pots were arranged and placed in saucers so that all watering were from below, then after, the cloves of garlic were first surface sterilized using 70% ethanol for five mins and rinsed three times with SDW. Then cloves were dressed with recommended fungicides (Apron Star 42 WS (3gm of Apron Star 42 WS powder with 10 ml of water) and Tebuconazole (2.1 ml of Tebuconazole with 15 ml water) [4] by partial and/or with combinations of both fungicides and then soaked for one hour.

The treated cloves were planted at 3 cm depth into the moist soil thoroughly incorporated with 100 g sclerotia propagules/kg of soil in the pot (5 cloves/pot were planted and two of them were thinned after germination) immediately under greenhouse condition at 12-15°C minimum and 26-30°C maximum temperature. Each *Trichoderma spp.* spore suspension were prepared by diluting with SDW at the rate of 10 g *Trichoderma spp.* mass produced/2 litre of water were mixed. Subsequently, 300 ml adjusted spore suspension of *Trichoderma spp* were drenched on the planted soil of each pot after seven days and continued within three days intervals [38]. The emerging garlic plants were assessed for symptoms of white rot every week up to 18 weeks. The treatments were arranged as (i) four *Trichoderma spp.* each alone or (ii) two fungicides each alone or (iii) fungicides combined with one or more *Trichoderma spp.* These were evaluated for their potential to control garlic white rot on garlic under greenhouse condition. Thus, the effect of partial and combined treatments for the control of *Sclerotium* cepivorum was examined and the result was compared with un-inoculated treatment. White rot disease incidence and severity was recorded in each pot.

The following treatments were applied for the experiment:

1. -(ve) absolute control
2. +(ve) control (inoculated with *S. cepivorum)*
3. Apron Star 42 WS+*S. cepivorum*
4. Tebuconazole+*S. cepivorum*
5. *T. hamatum+S. cepivorum*
6. *T. harzianum+S. cepivorum*
7. *T. oblongisporum+S. cepivorum*
8. *T. viride+S. cepivorum*
9. *S.cepivorum+T. hamatum+T. harzianum+T. oblongisporum+ T. viride*
10. Apron Star 42 WS+*S. cepivorum+T. hamatum*
11. Apron Star 42 WS+*S. cepivorum+T. harzianum*
12. Apron Star 42 WS+*S. cepivorum+ T. oblongisporum*
13. Apron Star 42 WS+*S. cepivorum+ T. viride*
14. Apron Star 42 WS+*S. cepivorum+T. hamatum* and *T. harzianum* combination
15. Apron Star 42 WS+*S. cepivorum+ T. hamatum* and *T. oblongisporum* combination
16. Apron Star 42 WS+*S. cepivorum+ T. hamatum* and *T. viride combination*
17. Apron Star 42 WS+*S. cepivorum+ T. harzianum* and *T. oblongisporum* combination
18. Apron Star 42 WS+*S. cepivorum+ T. harzianum* and *T. viride combination*
19. Apron Star 42 WS+*S. cepivorum+ T. oblongisporum* and *T.viride combination*
20. Apron Star 42 WS+*S. cepivorum+ T. hamatum+ T. harzianum+ T. oblongisporum+ T. viride*
21. Tebuconazole+*S. cepivorum+ T. hamatum*
22. Tebucunazole+*S. cepivorum+ T. harzianum*
23. Tebucunazole+*S. cepivorum+ T. oblongisporum*
24. Tebucunazole+*S. cepivorum+ T. viride*
25. Tebucunazole+*S. cepivorum+ T. hamatum* and *T. harzianum* combination
26. Tebucunazole+*S. cepivorum+ T. hamatum* and *T. oblongisporum* combination
27. Tebucunazole+*S. cepivorum+ T. hamatum* and *T.viride combination*
28. Tebucunazole+*S. cepivorum+ T. harzianum* and *T. oblongisporum* combination
29. Tebucunazole+*S. cepivorum+ T. harzianum* and *T. viride combination*
30. Tebucunazole+*S. cepivorum+ T. oblongisporum* and *T. viride combination*
31. Tebucunazole+*S. cepivorum+ T. hamatum+ T. harzianum+ T. oblongisporum+ T. viride*

Data analysis

Data on initial and final plant stand count at emergence, disease incidence and severity were collected every week from the experiment. All garlic bulbs were hand-harvested from each pot. Average of plant height, shoot length, root length and bulb biomass were recorded at soggy/moist phase and also after drying the samples in air for 7 days. Furthermore, 5 bulbs were randomly collected from which bulb diameter were measured, weight of cloves per bulb/plant were determined and number of cloves per bulb/plant were counted as described by [39]. Severity was assessed using a scale from 0 to 5 [40] and Disease Severity Index (DSI) was calculated.

Plants were uprooted separately from pots of each replication and determined for *mycelium expansion*; bulb and root rots were undertaken. A disease severity index based on symptoms observed and a disease severity formula was used to rate garlic treatments for their resistance to *S. cepivorum*. The Analysis Of Variance (ANOVA) of the data was separately subjected to SAS version 9.0 for further analysis and also the treatment mean were further separated by Duncan's Multiple Range Test (DMRT) at 5% significance level.

Results and Discussion

Effect of treatments on the foliar, stem base and bulb rot symptoms

The treated garlic plants showed slightly yellowing and wilting of delicate leaves and thin stems appeared after germination and also very few elongated roots developed on bulbs. White rot incidence was evaluated every week from the first appearance of the disease. Infected plants were examined as a small patch of plants or single plant more chlorotic than surrounding plants 45 days after artificial inoculation. The symptoms appeared as chlorosis as of lower leaves beginning at the tips, followed by a necrosis and collapse of the affected leaves of the aerial parts of the seedlings (Figure 1).

Observation of bulbs infections were carried out after harvest 127 days after artificial inoculation. Thereafter, the development of mycelia mat around stem base and sclerotial emerged on the bulbs of different treatments were seen (Figure 2). The seedlings exhibited characteristic garlic white rot symptoms including the blueing foliage, leaf tip dieback and a patchy distribution of diseased seedlings within each pot as shown in (Figure 3).

The disease symptom was initiated early in the trial just three weeks after planting. Some seedlings were discoloured, collapsed and lying on the soil surface of the pot. White rot symptoms appeared at about 45 days after planting the inoculated garlic bulbs and it's foliar and stem base symptoms incidence assessment was recorded weekly. Initially, disease symptoms were observed in all treatments, while the disease increased slowly in treatments and then become conspicuous just at the three to five leaves stage. The observations of assessments made at three stages (i). Foliar symptoms (every week until 98 days), (ii) Stem base symptoms (84-98 days), and (iii) Bulb rots symptoms (126 days) are presented in Table 1. When there is no stunting, no leaves colour change, no chlorosis, no wilting and collapsing and no stem base rotting, the bulb is designated as "Healthy".

The diseased tissues were diagnosed and the pathogen was re-isolated as an evidence of the disease development in the trial. Over 50% of the total white rot infections were recorded in the first seven weeks. The number of diseased seedlings in all treatments increased slowly for the duration of the trial and after 12 weeks. Among 279 seedlings, 126 were infected with garlic white rot. These seedlings showed characteristic garlic white rot symptoms including the blueing foliage, leaf tip dieback and a patchy distribution of diseased seedlings within each pot. After 16 weeks, more than 80% of the seedlings were diseased in the pathogen control; a significantly greater ($p>0.05$) amount of disease than on both fungicides treatment applications (Table 2). In the positive control (T2), the whole plants were extremely affected within less than 6 weeks and the seedlings were completely died.

Effect of treatments on the growth and biomass of garlic plants

The effect of different treatments on the Shoot length (cm), Plant height (cm), Fresh biomass wt. (g), Root length (cm), Fresh bulb biomass (g), Bulb diameter (cm), plant dry biomass (g), Number of cloves/bulb, Wt. of cloves/bulb revealed that weight of cloves/bulb

Figure 1: Symptoms of garlic white rot appeared early on the seedlings.

(2a) 2b)

Figure 2: Mycelia mat around stem base and emergence of *Sclerotia* on the bulbs. (2a) top view of the mycelia mat, (2b) closer view of stem base and bulbs.

Figure 3: Blueing foliage, leaf tip dieback and a patchy distribution of diseased seedlings.

Assessment time after inoculation	Score	Description	Treatments
1. Foliar symptoms (every week until 98 days)	0	Healthy leaves (no disease symptom)	T8, T16 and T21
	1	one to two leaves infected	T2, T3, T4, T5, T6, T7, T8, T9, T10, T11, T12, T13, T14, T15 and T21
	2	Two to three leaves infected	T1,T2, T3, T4, T5, T6, T7, T9, T10, T11, T12, T13, T14, T15,T17, T18, T19, T20, T22, T23...T31.
	3	Three to four leaves infected	T1,T2, T3, T4, T5, T6, T7, T9, T10, T11, T12, T13, T14, T15,T17, T18, T19, T20, T22, T23, T31.
	4	Four to five leaves infected	T2, T3, T4, T5, T6, T7, T9, T10, T11, T12, T13, T14, T15, T17, T18, T19, T20, T22, T23...T31
	5	Five leaves infected	T1,T2, T3, T4, T5, T6, T7, T9, T10, T11, T12, T13, T17, T18, T19, T20, T22, T23, T31.
2. Stem base symptoms (84-98 days)	0	Stem base free from mycelium and sclerotia	T8, T16 and T21
	1	Mycelium and sclerotia absent on stem base	T2, T3, T4, T5, T6, T7, T9, T10, T11, T12, T13, T14, T15, T17, T18, T19, T20, T22, T23...T31.
	2	Mycelium present, sclerotia absent on stem base	T2, T3, T4, T5, T6, T7, T9, T10, T11, T12,T13, T14, T15, T17, T18, T19, T20, T22, T23....T31
	3	Mycelium and sclerotia present on stem base	T2, T3, T4, T5, T6, T7, T9, T10, T18, T19, T20, T22,. T31
	4	Mycelium absent, sclerotia present on stem base	T2, T3, T4, T5, T6, T7, T9, T10, T18, T19, T20, T22, T23.
	5	Only sclerotia present on stem base	T2, T3, T4, T5, and T6.
3. Bulb rot symptoms (126 days)	0	Healthy stem base and bulbs	T8, T16 and T21
	1	Mycelium present, sclerotia absent on bulbs	T1,T9, T10, T11, T12, T13, T14
	2	Mycelium absent, sclerotia present on bulbs	T2, T3, T4, T5, T19, T20, T22... T30.
	3	Mycelium and sclerotia present on bulbs	T2, T3, T4, T5, T6, T7, T9, T10, T11, T12, T13, T14, T18, T19, T20, T22, T23...T31
	4	Only mycelium presents on all bulbs	T1,T2, T3, T4, T5, T6, T7, T9, T10, T11, T12, T13, T14, T15, T17, T18, T19, T20, T22, T23, T31
	5	Only sclerotia present on all bulbs	T2, T3, T4, T5, T6, T20, T23, T24 T25, T26, T27, T28, T29 and T31

Table 1: Assessment of foliar, stem base and bulb rot symptoms observed in different treatments.

Treatments	Disease severity score/ Replication			Total plants infected	Sum of disease severity score	Mean of disease severity	Disease Incidence (%)
	Rep. I	Rep. II	Rep. III				
T1	5.000	4.160	5.000	8.5	14.160	1.6	94.4
T2	5.000	5.000	5.000	9	15.000	1.7	100
T3	3.333	5.000	5.000	8	13.333	1.5	88.9
T4	3.333	5.000	5.000	8	13.333	1.5	88.9
T5	1.666	3.333	3.333	5	8.332	0.9	55.6
T6	1.666	1.666	3.333	4	6.665	0.7	44.4
T7	0.000	1.666	3.333	3	4.999	0.6	33.3
T8	0.000	0.000	1.666	1	1.666	0.2	11.1
T9	1.666	0.000	3.333	3	4.999	0.6	33.3
T10	1.666	1.666	3.333	4	6.665	0.7	44.4
T11	0.000	3.333	3.333	4	6.666	0.7	44.4
T12	3.333	3.333	0.000	4	6.666	0.7	44.4
T13	1.666	0.000	5.000	4	6.666	0.7	44.4
T14	0.000	1.666	3.333	3	4.999	0.6	33.3
T15	1.666	1.666	1.666	3	4.998	0.6	33.3
T16	0.000	0.000	0.000	0	0.000	0.0	0 .0
T17	1.666	3.333	3.333	5	8.332	0.9	55.6
T18	1.666	5.000	5.000	7	11,666	1.3	77.8
T19	0.000	5.000	3.333	5	8.333	0.9	55.6
T20	0.000	1.666	5.000	4	6.666	0.7	44.4
T21	1.666	0.000	0.000	1	1.666	0.2	11.1
T22	1.666	5.000	3.333	6	9.999	1.1	66.7
T23	5.000	3.333	1.666	6	9.999	1.1	66.7
T24	0.000	5.000	5.000	6	10.000	1.1	66.7
T25	1.666	5.000	3.333	6	9.999	1.1	66.7
T26	1.666	0.000	5.000	4	6.666	0.7	44.4
T27	1.666	3.333	3.333	5	8.332	0.9	55.6
T28	3.333	5.000	0.000	5	8.333	0.9	55.6
T29	1.666	5.000	3.333	6	9.999	1.1	66.7
T30	3.333	5.000	3.333	7	11.666	1.3	77.8
T31	5.000	1.666	1.666	5	8.332	0.9	55.6

Table 2: Disease incidence (%) and severity of garlic seedlings bioassay recorded in each treatment.

in the uninoculated control T1 is 1.12 whereas T16 (Apron star 42 WS+*S.cepivorum*+ *T. hamatum & T. viride* combination) has yielded 8.44 g, suggesting that, it may have synergistic and additive growth promoting effect on garlic in addition to controlling the white rot disease (Table 3). The same beneficial growth promoting effect can also be clearly deduced in treatment T8 (*T. viride*+*S. cepivorum*) and T21 (Tebuconazole+*S.cepivorum*+ *T. hamatum*). It is to be noted that growth promotion effect is one of the mechanism of *Trichoderma spp* exerted for control of phytopathgenic diseases [41-43].

Effect of treatments on the disease incidence and severity on garlic plants

Significant differences on disease incidence was observed at all assessment times among the treatments (p>0.05) (Table 2). The highest incidence and severity was recorded on negative and positive control (T1 and T2) (94.4% and 100%, respectively) and with both fungicides (T3, T4) (88.9%) and Apron Star 42 WS combined with *T. harzianum* and *T. viride* (T18) (77.8). Tebuconazole combined with *T. oblongisporum* and *T. viride* (T30) has (77.8%) disease incidence under similar conditions. The highest disease incidence and severity observed in uninoculated (-ve absolute control) treatment (T1) was assumed to occure from mycelia remained in the cloves after surface sterilization. Apron Star 42 WS treated with both *T. hamatum* and *T. viride* (T16) has provided efficient and highly significant disease control as compared with uninoculated check. Whereas, Tebuconazole combined with *T. hamatum* (T21) and *T. viride* alone (T8) were the next treatments that showed lower percentage (11.1%) of disease incidence as compared to all other treatments except "Apron Star 42 Ws combined with *T. hamatum* and *T. viride*". Therefore, these two treatments are relatively the promising bioagent for antagonising garlic white rot next to Apron Star 42 WS combined with *T. hamatum* and *T.viride* (T16).

When the Tebuconazole combined with *T. harzianum, T. oblongisporum* and *T. viride* alone (T22 and T23) and Tebuconazole integrated with *T. harzianum* and *T. viride* combination (T29) exhibited disease incidence of 66.6%. On other hand, Apron Star 42 WS combined with *T. harzianum* and *T. viride,* (T18) and Tebuconazole integrated with *T. hamatum* and *T. harzianum* (T25) and *T. oblongisporum* and *T. viride* (T30) disease incidence of (77.8%) were recorded, while both fungicides alone (T3, T4) provided 88.9%. However, garlic cloves treated by *T. harzianum* alone (T6) and the mixed-up of Apron Star

Treatments	Shoot length (cm)	Plant height (cm)	Fresh biomass wt. (g)	Root length (cm)	Fresh bulb biomass (g)	Bulb diameter (cm)	Dry biomass wt. of plants (g)	Number of cloves bulb^{-1}	Wt. of cloves/bulb/ plant (g)
1	4.57fg	11.57cd	10.43c	6.50c	12.50d	0.70klm	6.93c	0.80lm	1.12kl
2	0.00g	0.00h	0.00n	0.00h	0.00e	0.00m	0.00h	0.00m	0.00 l
3	9.90abcdef	5.90efg	5.57defghij	2.67defg	0.87e	2.57lm	0.38h	1.78jkl	1.21ijk
4	7.10bcdef	6.93defg	0.87mn	0.60gh	0.10e	0.53lm	0.47h	1.22kl	1.12jk
5	6.67bcdef	9.63cdef	3.73ijklm	1.97defgh	0.50e	2.57bcd	0.35h	1.67jkl	1.50fghijk
6	9.83abcdef	10.27cdef	3.10jklmn	1.93defgh	0.33e	2.87b	0.93h	2.34hijk	1.38ghijk
7	11.53abc	7.23defg	7.77cdef	3.30de	1.13e	2.80bc	1.29h	3.78defg	2.07cdefghij
8	13.23a	34.67ab	38.23ab	13.57b	26.07b	5.50a	27.72b	8.89c	7.69a
9	7.43abcdef	7.70defg	5.10fghijk	1.93defgh	0.77e	2.53bcd	2.21defgh	3.67defg	2.39bcdefgh
10	11.33abcd	10.20cdef	5.37efghij k	2.13defg	0.80e	2.80bc	3.71defg	4.22def	2.59bcdef
11	8.90abcdef	11.10cde	7.40cdefg	2.00defgh	1.57e	2.23bcdef	3.85def	4.55d	2.69bcde
12	6.50bcdef	11.77cd	8.80 cd	3.27de	1.57e	2.40bcde	4.27cde	4.11defg	2.96bcd
13	5.50defg	8.63cdefg	7.10defgh	2.20defg	1.00e	2.67bc	4.54cd	3.67defg	2.52bcdef
14	5.97bcdefg	11.93cd	8.63cde	2.83def	1.77e	2.65bcde	4.08de	4.22def	3.42b
15	7.77abcdef	13.43 C	6.73defghi	3.30de	1.57e	2.43bcde	4.19cde	4.33de	3.17bc
16	11.83ab	38.27a	40.87a	14.57b	33.93a	5.57a	34.40a	11.78a	8.44a
17	8.10abcdef	10.37cdef	2.97jklmn	1.83defgh	0.97e	1.47ghij	2.07defgh	4.00 defg	2.50bcdef
18	5.67cdefg	5.97efg	2.37jklmn	1.00fgh	0.23e	1.40hijk	1.17fgh	4.33de	3.15bc
19	9.70abcdef	8.00defg	5.23fghijk	2.60defg	0.87e	1.57fghij	1.74efgh	4.34de	2.42bcdefg
20	10.67abcde	8.63cdefg	8.67cd	2.70def	1.07e	2.17fbcdefg	4.34cde	4.34efghi	2.98bcd
21	11.07abcd	32.83b	37.43b	20.70a	18.50c	5.10a	29.80b	10.11b	7.46a
22	4.67efg	5.77fg	1.63lmn	1.17fgh	0.27e	1.13ijkl	0.70h	3.44defgh	1.90defghijk
23	7.97abcdef	8.47cdefg	2.60jklmn	1.23efgh	0.37e	1.07jkl	1.24fgh	3.33fghi	2.40bcdefgh
24	7.13abcdef	4.30h	2.40jklmn	1.10fgh	0.30e	0.57lm	1.13fgh	3.44defgh	2.00defghij
25	9.07abcdef	9.40cdefg	2.23klmn	1.07fgh	0.70e	0.53lm	0.88h	3.00ghi	2.29defghi
26	10.87abcd	9.30cdefg	4.33hijkl	3.33d	0.57e	1.50fghij	2.37defgh	3.22efghi	1.68efghijk
27	9.83abcdef	7.33defg	3.13 jklmn	1.53defgh	0.37e	1.77fghij	0.95h	3.22efghi	1.92defghij
28	7.90abcdef	5.57fg	3.83ijklmn	1.47defgh	0.50e	2.13bcdefgh	2.08defgh	2.44hij	1.62efghijk
29	5.67cdefg	6.85defg	2.57jklmn	1.23efgh	0.33e	2.00cdefgh	1.21fgh	1.67jkl	1.49fghijk
30	5.40defg	5.40fg	2.10lkmn	0.87fgh	0.30e	1.77efghij	0.98h	2.22ijkl	1.31hijk
31	10.23abcdef	7.83defg	4.10hijkl	1.33defgh	0.77e	1.83efghi	2.11defgh	3.78defg	2.41bcdefgh
LSD (5%)	6.0146	2.001	3.2858	2.084	3.0388	2.003	2.7703	1.1034	2.002
CV (%)	45.30	30.19	25.42	37.34	52.17	21.72	34.58	25.71	16.53

Means in every column with the same letters are not significantly different at 5% significance level.
LSD: Least Significant Difference; CV: Coefficient of Variation.

Table 3: Effects of different treatments on the growth and biomass of garlic plant parts.

42 WS (T11) with individual four *Trichoderma spp.* (T10, T11, T12, T13) and also T20 integrated with all four bioagents as one and Tebuconazole combined with *T. hamatum* and *T. oblongisporum* (T26) all showed medium disease control (44.4%). In the garlic plants treated with Tebuconazole combined mutually with all four *Trichoderma spp.* (T31) had the disease incidence of (55.6%).

The results revealed that there was no significant difference in the percentage of diseased seedlings (88.9%-77.8%) between the garlic applied with both fungicides partly and with combination of *Trichoderma spp.* in disease control measures. In the remaining treatment, disease prévalence percentage range were categorized in ascending order as : 0% (T16); 11.1% (T8 and T21); 33.3% (T9, T14 and T15); 44.4% (T10, T11, T12, T13, and T26); 55.6% (T6, T17, T20, T27, T28 and T31); 66.7% (T5, T7, T19, T22, T23, T25 and T29); 77.8% (T18, T24 and T30) ; 88.9% (T3 andT4) ; 94.4% - 100% (T1 and T2) (Figure 4)

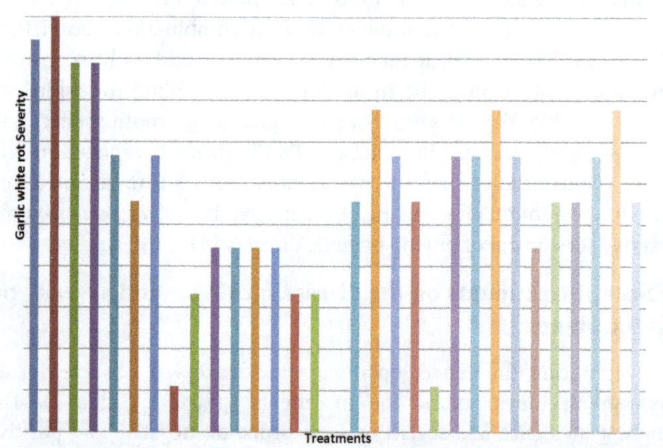

Figure 4: Effect of different treatments on disease severity of *S. cepivorum*.

The combination of fungicide with one or two species of *Trichoderma* bioagent has provided similar effective control of *Allium* white rot (AWR) caused by *S.cepivorum* pathogen under greenhouse condition. *T. harzianum* has earlier been reported to reduce *S. cepivorum* infection from 84% to 29% in greenhouse trials [44]. Similarly, *T. koningii* was reported to reduce onion white rot disease by 60% when incorporated in a millet formulation and added to soil at seed planting [45] and *T. koningii*, the same isolate as used in biological control study, provided 79% disease control of onion white rot, when incorporated into the soil in a sand: bran mix [32]. Similar to previous research results, this study has also showed the best significant disease control when combining Apron Star WS fungicide with two *Trichoderma species* (*T. hamatum* and *T.viride*) (T16); Tebuconazole fungicide along with *T. hamatum* integration (T21) and *T. viride* alone (T8) gave a better result in greenhouse trial (Figure 5).

Figure 5: Effect of different treatments T8, T3 and T1 on disease severity of *S. cepivorum*. (5a)=T. *viride* (T8), (5b)=Tebuconazole (T3) and (5c)=Un-inoculated (T1).

It was observed that cloves/seedlings treated with Tebuconazole showed a suppressive effect on the developments of the whole plant and roots formation and even delayed the germination by a week. In all treatments receiving Tebuconazole, the plant stands was weak, very thin and fragile. The roots were very few in numbers and very thin, shrivelled, elongated and sheath paled off easily and bulbs were tiny.

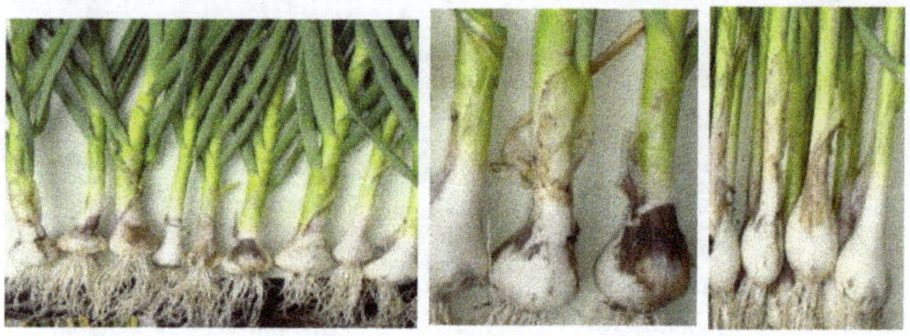
Figure 6: Effect of treatment T16 on disease severity of *S. cepivorum*.

The beneficial effect of the fungicide in combination with *Trichoderma spp.* at planting was effective to suppress the sclerotial germination. These results indicate that, combining of fungicide with the bio-agent(s) provides a good prospect for garlic growers as *Trichoderma spp.* are safe for animals, human beings and environment. The results indicate a break from susceptible cropping, and integration of fungicide with biocontrol agents was the strategy providing greater reduction in suppressing the viability of *S. cepivorum* inocula in the soil resulting in minimum incidence of white rot (T16) (Figure 6). The magnitude of economic benefit and synergistic value of *Trichoderma spp.* after combination remains to be determined, as does the ecological one, when combining these practices. It can be seen that, the degree of disease control achieved by treatment with Tebuconazole alone was lower than that of all the treatments combined with Apron Star 42 WS fungicide, and mixed-up of one or more mixture of four *Trichoderma spp.* each other when compared to un-inoculated and T16 which has absolutely controlled the pathogen.

It was reported that treatment of garlic cloves with Tebuconazole and base spray provided significant reduction in the rate of disease progress and the final of plant mortality by *S. cepivorum* [17]. Eighty five percent disease incidence reduction was reported in Tebuconazole treated plots compared with untreated plots in onion [46]. Other researchers also reported that combinations of Tebuconazole and a biocontrol agent enhanced the control of onion white rot [23]. Even though an indication of antagonism has been obtained from this greenhouse study and previous literature based on similar isolates [32], the level of control did vary with the same isolates when similar methodology was used. In general, an important factor in biocontrol agent effectiveness is the rate at which the propagules/mycelium dilution amounts proliferate when applied to the potting mix. To predict and successfully use biological control agents for soil borne disease control, it is critical that their biology and ecology be more completely understood. It was beyond the scope this study to determine the individual components of the three types of potting mixed-up in relation to microbial carrying capacity as an indicative difference to antagonize the white rot pathogen activities. Thus, integration of fungicides and biological control agents may enable the number of fungicide sprays to be reduced, while providing control of garlic white rot.

Conclusion

The results obtained in this study revealed that the two *Trichoderma spp.* (*T. hamatum* and *T. viride*) in combinations with two fungicides can have substantial antagonistic activity against garlic white rot pathogen. This could be attributed to their synergistic and additive growth effects that yielded better biomass besides controlling the disease. The findings also suggest that *T. hamatum* and *T. viride* are playing an important role in controlling garlic white rot pathogen better than the two fungicides alone. This is highly advantageous in light of the fact that the use of *Trichoderma*-based products is not only safe for the farmers and consumers but is also environmentally friendly. Tebuconazole has been frequently reported as effective fungicide against this aggressive pathogen worldwide, including in Ethiopia, whereas Apron Star 42 WS is reported here in Ethiopia for the second time, while it is not common elsewhere. Since the compatibility of *Trichoderma spp* with these two fungicides is now proved in this study, the method can be tested for control of other diseases.

References

1. Deresse D (2010) Antibacterial effect of garlic (Allium sativum) on Staphyloccus aureus: An in vitro study. Asian Journal of Medical Sciences 10: 62-65.
2. FDRE/CSA (The Federal Democratic Republic of Ethiopia, Central Statistical Agency). (2011/12) Area and production of Major crops Statistical Bulletin volume I.
3. EIAR-DZARC (Ethiopian Institute of Agricultural research-Debre Zeit Agricultural Research Centre) (1995) Garlic Production System. Ethiopia Agricultural Research Organization, Debre Zeit Research Center.
4. Getachew T, Eshetu D, Tebkew D (2011) Management system of onion and garlic production and productivity technique by using fungicides (in Amharic) Debre Zeit Agricultural Research Centre (2ndedn.) pp. 29-34.
5. Fekadu M, Dandena G (2006) Review of the status of vegetable crops production and marketing in Ethiopia. Uganda Journal of Agricultural Sciences 12: 26 -30.
6. FAO (Food and Agriculture Organization) (2006) Food Security in Ethiopia, Agriculture and Consumer Protection Department.
7. Tamire Z, Chemeda F, Parshotum PK, Sakhuja PK, Ahmed S (2007) Association of white rot (Sclerotium cepivorum) of garlic with environmental factors and cultural practices in the North Shewa highlands of Ethiopia. Journal of Crop Protection 26: 1566-1577.
8. Mengistu H (1994) Research on vegetables disease in Ethiopia. In: Herath E, Dessalegne L (eds.) Proceedings of the Second National Horticultural Workshop of Ethiopia, 1-3 December 1994. IAR, Addis Ababa, Ethiopia, pp. 209-215.
9. Zeray S, Mohammed Y (2013) Searching and evaluating of cost effective management options of garlic white rot (Sclerotium cepivorum Berk) in Tigray, Northern Ethiopia. J. Plant Pathol. Microb. 4: 189.
10. Amin M, Tadele S, Thangavel S (2014) White rot (Sclerotium cepivorum-Berk)- an aggressive pest of onion and galic in Ethiopia: An overview. Journal of Agricultural Biotechnology and Sustainable Development 6: 6-15.
11. Pinto CMF, Mafia LA, Casali VWD, Berger RD, Cardoso AA (2000). Production components and yield losses of garlic cultivars planted at different times in a field naturally infested with Sclerotium cepivorum, Int. J. Pest Manage. 46: 67-72.
12. Tamire Z, Chemeda F, Sakhuja PK (2007) Management of white rot (Sclerocium cepivorum) of garlic using fungicides in Ethiopia. Journal of Crop Protection 26: 856-866.
13. Ulacio-Osorio D, Zavaleta-Mejía E, Martínez-Garza A, Pedroza-Sandoval A (2006) Strategies for management of Sclerotium cepivorum Berk in garlic. Journal of Plant Pathology 88: 253-261.
14. Tyson JL, Fullerton RA, Elliott GS, Reynolds PJ (2000) Use of diallyl disulphide for the commercial Control of Sclerotium cepivorum. New Zealand Plant Protection 53: 393-397.
15. Coventry E, Noble R, Mead A, Whipps JM (2002) Control of white rot (Sclerotium cepivorum) with coopted garlic waste. Journal of Soil Biochemistry 34: 1037-1045.
16. Prados-Ligero AM, Bascon-Fermandez J, Calvet-Pinos C, Corpas- Hervias C, Ruiz AL, et al. (2002) Effect of different soil and clove treatments in the control of white rot of onion. Journal of Applied Biology 140: 247-253.
17. Melero-Vara JM, Prados-Ligero AM, Basallote-Ureba MJ (2000) Comparison of physical, chemical and biological methods of controlling onion white rot (Sclerotium cepivorum Berk). European Journal of Plant Pathology 106: 581-588.
18. Duff AA, Jackson KJ, Donnell WEO (2001). Tebuconazole (Folicur) potential in the control of white rot (Sclerotium cepivorum) in onion subtropical. Second International Symposium on edible Alliaceae, Acta Horticulture, Queensland, Australia, 555: 247-250.
19. Lo CT, Nelson EB, Harman GE (1996) Biological control of turf grass diseases with a rhizosphere competent strain of Trichoderma harzianum. Plant Disease. 80: 736-741.
20. Mathivanan N, Srinivasan K, Chellaiah S (2000). Field evaluation of Trichoderma viride pers. Ex. SF. Gray and Pseudomonas fluorescens Migula against foliar diseases of groundnut and sunflower. J Bul Cont 14: 31-34.
21. Harman GE, Howell CR, Viterbo A, Chet I, Lorito M (2004) Trichoderma species--opportunistic, avirulent plant symbionts. Nat Rev Microbiol 2: 43-56.
22. Arega F (2012) Evaluation of Trichoderma species for the control of garlic white rot (S. cepivorum Berk). MSc. Thesis. Ambo University. Ambo, Ethiopia.
23. Clarkson JP, Scruby A, Mead A, Wright C, Smith B (2006) Integrated control of Allium white rot with Trichoderma viride, tebuconazole and composted onion waste. Journal of Plant Pathology 5: 375-386.
24. Gowdar, SB, Ramesh-Babu HN, Nargund VB, Krishnappa M (2006) Compatibility of fungicides with Trichoderma harzianum. Agric Sci Digest 26: 279 - 281.
25. Saxena D, Tewari AK, Rai D (2014) The in vitro effect of some commonly used fungicides, insecticides and herbicides for their compatibility with Trichoderma harzianum pbt23. World Applied Sciences Journal 31: 444-448.
26. Thoudam R, Dutta BK (2014) Compatibility of Trichoderma atroviride with fungicides against blackrot disease of tea: an in vitro study. Journal of International Academic Research for Multidisciplinary 2: 25-33.
27. Mclean KL, Hunt J, Stewart A (2001) Compatibility of the biocontrol agent Trichoderma harzianum C52 with selected Fungicides. New Zealand Plant Protection 54: 84-88.
28. Noble R, Coventry E (2004) Suppression of soil-borne plant pathogens using composts: a review. Biocontrol Science and Technology 15: 3-20.
29. Hunger SA, Mclean KL, Eady CC, Stewart A (2002) Seedling infection assay for resistance to Sclerotium cepivorum in Allium species. New Zealand Plant Protection.
30. Alexander BJR, Stewart A (1994) Survival of sclerotia of Sclerotinia ans Sclerotium spp .in New Zealand horticultural soil. Soil Biology & Biochemistry 26: 1323-1329.
31. Backhouse D, Stewart A (1989) Interactions between Bacillus species and sclerotia of Sclerotium cepivorum. Soil Biology and Biochemistry. 21: 173-176.
32. Kay SJ, Stewart A (1994) Evaluation of fungal antagonists for control of onion white rot in soil box trials. Plant Pathology 43: 371-377.
33. Crowe FJ, Hall DH, Greathead AS, Baghott KG (1980) Inoculum density of Sclerotium cepivorum and incidence of white rot of onion and garlic. Journal of Phytopathology 70: 64-69.
34. Temesgen B, Kubicek CP, Druzhinina IS (2010) The rhizosphere of Coffea Arabica in its native highland forests of Ethiopia provides a niche for a distinguished diversity of Trichoderma. Diversity 2: 527-549.
35. Belay H (2010) In-vivo Evaluation of Indigenous Trichoderma species for Bio-control of Faba bean Root rot Caused by Fusarium solani Pest Management. Journal of Ethiopia 14: 62-67.
36. Rini CR, Sulochana KK (2007) Substrate evaluation for multiplication of Trichoderma spp. Journal of Tropical Agriculture 45: 58-60.
37. Yadav LS (2012) Antagonistic activity of Trichoderma sp. and evaluation of various agro wastes for mass production. Indian Journal of Plant Sciences 1: 109-112.

38. Maketon M, Apisitsantikul J, Siriraweekul C (2008) Greenhouse evaluation of Bacillus subtilis AP-01 and Trichoderma harzianum AP-001 in controlling tobacco diseases. Braz J Microbiol 39: 296-300.

39. Abouziena HF, El-Saeid HM (2013) Developmental changes in growth, yield and volatile oil of some chinese garlic lines in comparison with the local cultivar "Balady". Pak J Biol Sci 16: 1138-1144.

40. Rengwalska MM, Simon PW (1986) Laboratory evaluation of Pink root rot and Fusarium basal rot resistance in Garlic. Plant Disease 70: 670-672.

41. Benítez T, Rincón AM, Limón MC, Codón AC (2004) Biocontrol mechanisms of Trichoderma strains. Int Microbiol 7: 249-260.

42. Sharma R, Joshi A, Dhaker RC (2012) A brief review on mechanism of Trichoderma fungus Use as biological control agents. International Journal of Innovations in Bio-Sciences 2: 200-210.

43. Gajera H, Domadiya R, Patel S, Kapopara M, Golakiya B (2013) Molecular mechanism of Trichoderma as bio-control agents against phytopathogen system - a review Current Research in Microbiology and Biotechnology 1: 133-142.

44. Abd-El-Moity TH, Shatla TH (1981) Biological control of white rot disease of onion (Sclerocium cepivorum) by Trichoderma harzianum. Phytopathologische Zeitschrift 100: 29-35.

45. Wong JAL, Schupp P, Archer C, Lacey M, Williams W (1995) Sclerotial fungi research in Tasmania. In: Porter IJ (ed.) Proceedings of the Australasian Sclerotial fungi Workshop 10th Biennial Australasian Plant Pathology Society Conference. 27 August 1995, Lincoln University, New Zealand.

46. Fullerton RA, Stewart A, Slade, EA (1995) Use of dethylation inhibiting fungicide (DMIs) for the control of onion white rot (Sclerocium cepivorum Berk.). New Zealand Journal of Crop and Horticultural Science 23: 121-125.

Role of Total Soluble Sugar, Phenols and Defense Related Enzymes in Relation to Banana Fruit Rot by *Lasiodiplodia theobromae* [(Path.) Griff. and Maubl.] During Ripening

Kedar Nath[1,2*], Solanky KU[1], Mahatma MK[3,4], Madhubala[2] and Rakesh M Swami[3]

[1]Department of Plant Pathology, N. M. College of Agriculture, Navsari Agricultural University, Navsari-645039, India
[2]Regional Rice Research Station, Navsari Agricultural University, Vyara 39 4650, India
[3]Department of Plant Molecular Biology and Biotechnology, Navsari Agricultural University, Navsari 39 450, India
[4]Directorate of Groundnut Research, Indian Council of Agricultural Research, P. B. No.5, Junagadh 362 001, India

Abstract

During storage conditions banana fruits get infected by several fungal diseases like finger rot, fruit rot, crown rot, cigar–end rot and pitting disease etc. Among these diseases fruit rot caused by *Lasiodiplodia theobromae* [(Path.)Griff. and Maubl.] is most serious post harvest disease under South Gujarat condition and it causes changes in biochemical contents of banana pulp and peel during ripening. Sugar, Phenols Phenylalanine ammonia lyase (PAL), Polyphenol oxidase (PPO) and Peroxidase (POX) are said to play important role in plant disease resistance. Total sugar, phenolic content, phenylalanine-ammonia lyase, polyphenol oxidase and peroxidase activities were determined in infected and uninfected banana fruits during ripening at 0, 48, and 72 h after incubation. The results showed that total soluble sugar content was increased with the ripening stages but it was decreased in infected fruits as compared to uninfected fruit. Reduction in PAL activity and enhancement in PPO and POX activity may be correlated with reduction of phenol content during ripening stage but it was still increased in infected banana fruits.

Keywords: Banana fruit rot; *Lasiodiplodia theobromae;* Sugar; Phenol; PAL; Polyphenol oxidase; Peroxidase

Abbreviations: PAL: Phenylalanine Ammonia Lyase; PPO: Polyphenol Oxidase; POX: Peroxidase; APMC: Agricultural Produce Market Committee; GMFU: Green Mature Fruit Uninfected; GMFI: Green Mature Fruit Infected; SRFU: Semi-ripe Fruit Uninfected; SRFI: Semi-Ripe Fruit Infected; RFU: Ripe Fruit Uninfected; RFI: Ripe Fruit Infected

Introduction

Banana (*Musa paradisiaca L.*), fruit is one of the most important commercial fruit grow all over the world in the tropical and subtropical areas. It can be grown round the year and it is widely cultivated in India. Apart from these it is considered as potential 'Dollar Earning crop'. It is known since the dawn of ancient history as one of the delicious fruit in the world. The ripening is then done artificially in the various ways, such as exposing the bunches to the sun, placing them over a hearth, wrapping them with green leaves and piling them in a heap, storing them in a closed godowns or smoking them or treated with ethylene. The fruit takes 48 to72 h to ripening. Banana fruit suffers from many serious diseases such as fruit rot, crown rot, finger rot, cigar–end rot and pitting disease. The most important disease problem of bananas is finger rot as well as fruit rot caused by *Lasiodiplodia theobromae* in field and storage condition in South Gujarat. Finger rot diseases may involve several fungal species, but the most commonly associated organism is *Lasiodiplodia theobromae* [1-3]. The genus *Lasiodiplodia*, *L. theobromae,* is geographically widespread but is most common in the tropics and subtropics regions which are associated approximately with 500 hosts [4]. Biochemical changes observed in carbohydrates are predominant chemical transformations during ripening of climacteric fruit with a decrease in starch and an increase in sugar content during the ripening of most fruits [5]. Banana fruits at the unripe stage have a relatively higher starch content which is almost completely hydrolysed during ripening to simple sugars, glucose, fructose and sucrose [6]. Glucose and fructose are predominant sugars in post-climacteric banana fruits [7]. Transformation of starch to glucose in bananas fruit during ripening [8]. Total sugar content is increased and total phenol decreased in ripe non-spongy fruits. However, in spongy-white pulp caused by *Staphylococcus* bacteria, a decrease in reducing, non-reducing sugars and total phenol [9]. However, as the spongy tissue affected pulp turned brownish black due to increase Phenolic content over the non-spongy ripe fruits. Polyphenol oxidase has been found in most higher plants, and is responsible for enzymatic browning of raw fruits and vegetables [10]. This reaction is important in food preservation and processing, and generally considered to be an undesirable reaction because of the unpleasant appearance and concomitant development of an off flavour. Peroxidase activity has been very low in unripe non-spongy mango fruits however it is increased by several fold in non-spongy fruits upon ripening. The development of black color indicated changes in phenolics content either by POX or polyphenol oxidase (PPO) enzymes [9]. In the present studies to determined the total sugar and total phenol content and defense related enzymes viz., Phenylalanine ammonia lyase (PAL), Polyphenol oxidase (PPO) and Peroxidase (POX) activity during banana fruit ripening and in relation to banana fruit rot by *L. theobromae* in during storage.

Materials and Methods

Biochemical studies were determined to post harvest changes in

***Corresponding author:** Kedar Nath, Department of Plant Pathology, N. M. College of Agriculture, Navsari Agricultural University, Navsari-645039, India
E-mail: drkdkushwaha@nau.in

uninfected and infected banana fruit with fungus in storage. The isolate of the entophytes fungus *Lasiodiplodia theobromae* (Griff. and Maubl.) was identified on cultural and morphological characters were studied on PDA medium. The isolate was identified as *L. theobromae* with the help of illustrated genera of imperfect fungi [11]. The identification of isolate has been confirmed by mycologist, Agharkar Research Institute. Pune (No.3/426-2008), India. Isolate was cultured on potato dextrose agar (PDA) medium and spore suspensions were prepared by washing the surface of 15 days cultures on petriplates with sterile distilled water. The spore suspension was adjusted in sterile water at 10^6 /ml by using haemocytomere and used as inoculum. The banana fruits (cv. Grand naine) at three different stage *i.e.*, mature green fruits, semi-ripe fruits and ripe fruits were obtained from stock holder at APMC, Navsari. The fruits were surface disinfected with 2% sodium hypochlorite for 2 min., rinse with two times with sterile distilled water than dried under ceiling fan for an hour and then fruits were infected with spore suspension @ 0.2 ml in each fruit by injection at tip end region. Five fruits in each replication at each stages are infected and maintained three replication in same. Infected and uninfected fruits were kept separately into plastic tray (15 × 25 cm size) then covered with plastic sheet and incubated at room temperature. Biochemical analysis of pulp and pericarp from infected and uninfected banana fruits were separately determined at 0, 48 and 72 h after incubation.

Samples were collected from healthy and rotted fruits with pulp and pericarp at three different ripening stage. The samples were either used immediately or kept frozen in liquid nitrogen and stored at -20°C until further use for analysis of biochemical constituents *viz.*, total sugar, phenol content and defense related enzymes such as Phenylalanine ammonia lyase (PAL), Polyphenol oxidase (PPO) and peroxidase assay in relation to disease development.

Total soluble sugar content

Homogenized pulp with pericarp (100 mg) were extracted with 5 ml of 80% ethanol and centrifuged at 3000 rpm for 10 minutes. Extraction was repeated 3 times with 80% ethanol and supernatants were collected into 25 ml volumetric flasks. Final volume of the extract was made to 25 ml with 80% ethanol. The extract (0.3 ml) was pipetted from each treatment into separate test tubes, and then tubes were placed in to boiling water bath for 3 minutes to evaporate the ethanol. One ml of MillQ water and 4 ml of 0.2% anthrone reagent (200 mg in 100 ml H_2SO_4) were added in each test tube and placed in ice cold water. Reagent blank was prepared by adding 1 ml of distilled water and 4 ml of anthrone reagent. The intensity of colour was read at 600 nm on spectrophotometer. A standard curve was prepared using 10 mg glucose per 100 ml distilled water [12].

Total soluble sugar (mg/g) = Sample O.D × Standard O.D. × Dilution factor

Extraction and estimation of total phenolic compounds

One gram of fruit tissue (pulp with pericarp) homogenized in a pre-chilled mortar and pestle in 10 ml of 80 per cent methanol and the extracts left for 24 h at room temperature before centrifuge at 15,000 rpm for 10 minutes [4]. One ml methanolic extract was added to 5 ml of distilled water and 250 μL of Foline-ciocalteu reagent and the solution was kept at 25°C for 3 min. Then 1 ml of a saturated solution of Na_2Co_3 (20 per cent solution of Na_2Co_3) and 1 ml of distilled water were added and the mixture were incubated for 1 h at 25°C. The absorption of the developed blue color was measured using spectrophotometer at 725 nm of single wavelength. The total phenol content was calculated by comparison with a standard curve obtained from using pyrocatachol ranging between 0-25 μg concentration.

The amount of phenols present in the sample was calculated as below formula.

Phenol (mg/g) = Sample O.D × Standard O.D. × Dilution factor

Defense related enzymes

Phenylalanine ammonia lyase (PAL): Three hundred milligram of fruit tissue homogenized in a pre-chilled mortar and pestle in 3 ml of extraction buffer containing 50 mM borare-HCL buffer (pH 8.5) and 0.04 per cent β-mercaptoethanol. The homogenate was centrifuge at 10,000 rpm for 15 min. at 4°C. The clear supernatant was used as the enzyme source for Phenylalanine ammonia lyase assay [13,14].

The reaction mixture containing 3.0 ml of 0.1 M sodium borate buffer (pH 8.8), 0.5 ml of 0.1 M phenylalanine (dissolved in 0.1 M sodium borate buffer, pH 8.8). The reaction was initiated by the addition of 0.1 ml enzyme extract. The tubes were incubated at 37°C for 2 h. The blank was also set with substrate (containing 3.0 ml of 0.1 M sodium borate buffer, pH 8.8 and 0.5 ml of 0.1 M phenylalanine). The O.D. read at 290 nm after 2 h. The enzyme activity was expressed as U/h/g protein.

Polyphenol oxidase (PPO): Three hundred milligram of fruit tissue were grounded with a pre-chilled mortar and pestle in 3 ml of 0.1 M sodium phosphate buffer, pH 6.0. The homogenate was centrifuge at 10,000 rpm for 15 min. at 4°C. The clear supernatant was used for PPO activity assay. The reaction mixture contained 2.9 ml of catechol (0.01 M catechol in 10 mM phosphate buffer, pH 6.0) and reaction was initiated by the addition of 0.1 ml enzyme extract. The changes in the colour due to the oxidized catechol was read at 490 nm for one minute at an interval of 15 seconds. The enzyme activity was expressed as change in O.D./min/g protein [15].

Peroxidase (POX): The banana fruit tissue (1 g) was homogenized in a pre-chilled mortar and pestle in 2 ml of 0.1 M phosphate buffer, pH 7.0 at 4°C. The homogenates were centrifuged at 12,000 rpm at 4°C for 15 min. The supernatant was used as enzyme source for POX activity assay. The reaction mixture consisted 2 ml of 50 mM phosphate buffer (pH7.0) and 0.1 mM EDTA, 10 mM guaiacol 450 μl, 10 Mm H_2O_2 450 μl and 100 μl enzyme extract. POX activity was determined in the homogenates by measuring the changes in absorbance at 420 nm due to the formation of tetraguaiacol (ε=26.6 mM^{-1} cm^{-1}) in a reaction mixture were recorded at 15 seconds intervals for 1 minute. The enzyme activity was expressed as changes in the absorbance of the reaction mixture min^{-1} g^{-1} on fresh weight basis [16].

Results and Discussion

In this study, the changes in the levels of total sugar and phenol as well as the activities of Phenylalanine ammonia lyase (PAL), polyphenol oxidase (PPO) and peroxidase (POX) enzymes during banana fruit ripening and fruit rot disease development were determined.

Total soluble sugar

In presented result, showed that the total soluble sugar varied in uninfected and infected banana fruits with *L. theobromae* at different ripening stages. The total soluble sugar content was increased with ripening in infected as well as uninfected banana fruits, but it was lower in semi ripe and ripe infected fruits than uninfected fruits after 48 and 72 h of incubation. Total soluble sugar content was higher (85.84 mg/g fresh weight) in ripe fruits while lowest (2.01 mg/g fresh weight) in green mature fruits at initial time. Total sugar content was increased in the beginning of the fruit ripening with increased of storage period.

It was higher (211.57 mg/g fresh weight) in ripe uninfected fruits and still further decreased (203.99 mg/g fresh weight) in infected fruits and lowest in uninfected green mature fruits after 72 h of storage (Figure 1a). Total soluble sugar content was increase in green mature infected fruits (9.07 fold) followed by green mature uninfected fruits (4.62 fold). However in semi ripe and ripe it was increased with 3.58 and 2.46 fold respectively after 72 h of storage (Table 1). These results are corroborated with earlier worker found transformation of starch to glucose in bananas during ripening [6,8]. However, higher starch content in unripe banana fruits completely hydrolysed during ripening to simple sugars, glucose, fructose and sucrose. Catalase activity is decrease in healthy as well as in the fruits pulp infected by *Aspergillus flavus* and increase invertase activity which helped in accumulation of soluble sugars making pulp more soft [17]. Accumulation of sugar neat the infection court which increase the pectolytic and cellulolytic enzymes activity which are essential for pathogenesis and play key role in disintegration of tissue which caused pulp rot development Similarly total sugar content was in results indicated that decreasing of total soluble sugar content in infected fruits related with disease development because it utilized by the pathogen as food for their

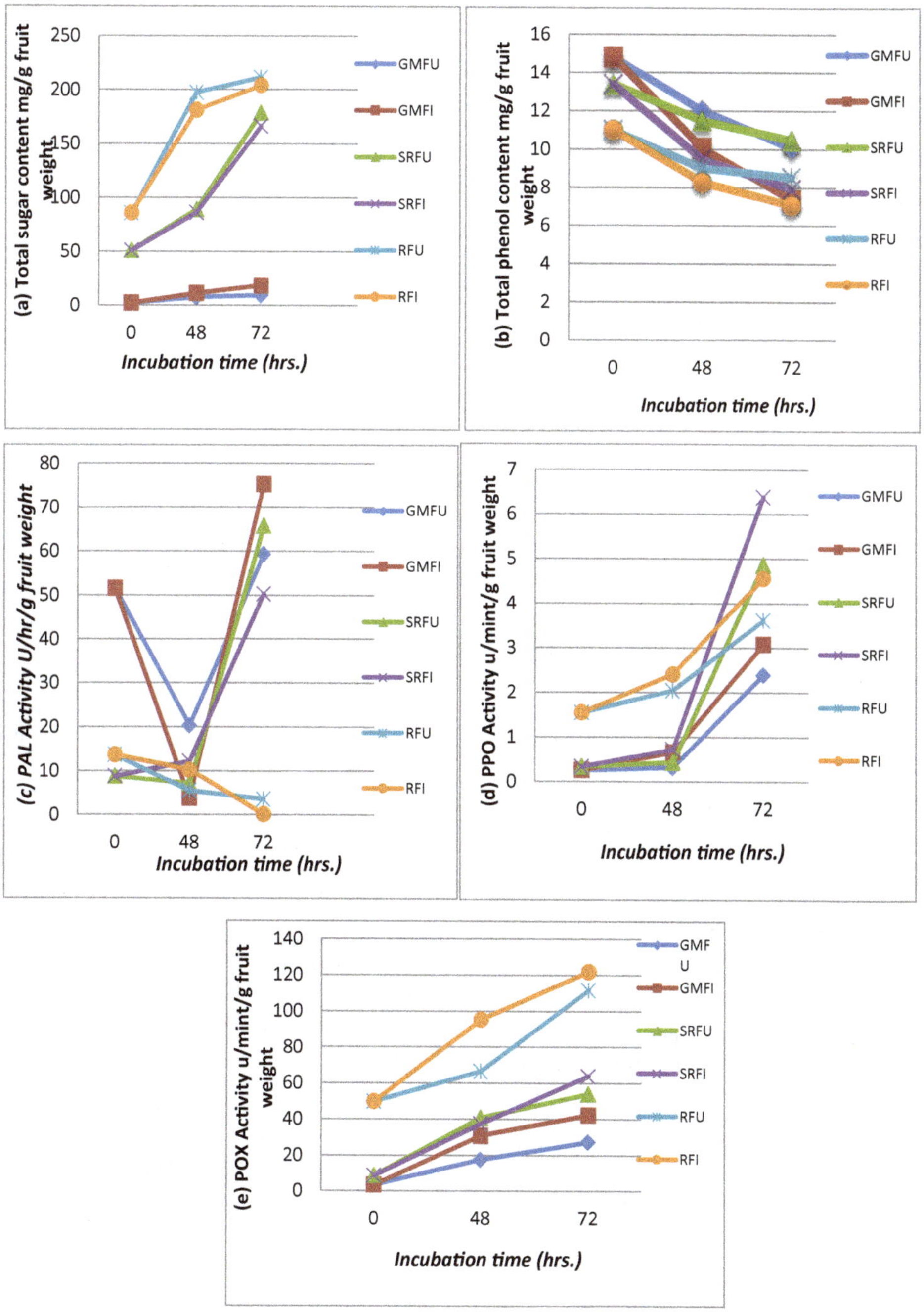

Figure 1: Biochemical changes in infected and uninfected banana fruit (Vart. Grand Naine), (a) Total sugar content mg/g fruit weight, (b) Total phenol content mg/g fruit weight, (c) PAL Activity U/hr./g fruit weight, (d) PPO Activity U/mint./g fruit weight, (e) POX Activity U/mint/g fruit weight.

(a) Total phenol content mg/g fruit weight

Sr. No.	Sample	0 hrs.	48 h after		72 h after			
		Uninfected	Uninfected	Infected	Uninfected	Infected	Fold *	Fold**
1	Green mature fruits	2.016 ± 0.79	7.514 ± 0.79	11.176 ± 0.59	9.327 ± 0.34	18.28 ± 0.23	+4.62	+9.07
2	Semi ripe fruits	50.819 ± 2.90	89.139 ± 0.79	86.14 ± 1.32	178.54 ± 2.01	166.14 ± 5.40	+3.58	+3.27
3	Ripe fruits	85.841 ± 1.06	197.32 ± 2.87	181.05 ± 0.99	211.57 ± 56.42	203.99 ± 5.21	2.46	+2.38

(b) Total sugar content mg/g fruit weight

Sr. No	Sample	0 hrs.	48 h after		72 h after			
		Uninfected	Uninfected	Infected	Uninfected	Infected	Fold	Fold
1	Green mature fruits	14.802 ± 0.26	12.028 ± 0.72	10.05 ± 0.53	10.065 ± 1.08	7.358 ± 1.19	-1.47	-2.01
2	Semi ripe fruits	13.375 ± 1.43	11.483 ± 1.07	9.476 ± 0.16	10.456 ± 0.11	7.911 ± 0.07	-1.28	-1.69
3	Ripe fruits	10.984 ± 0.39	9.008 ± 0.65	8.266 ± 0.59	8.522 ± 0.91	7.0547 ± 0.73	-1.29	-1.56

(c) PAL Activity U/hr/g fruit weight

Sr. No	Sample	0 hrs.	48 h after		72 h after			
		Uninfected	Uninfected	Infected	Uninfected	Infected	Fold	Fold
1	Green mature fruits	51.551 ± 3.44	20.331 ± 2.70	3.777 ± 0.77	59.328 ± 4.91	75.326 ± 2.0	+1.15	+1.61
2	Semi ripe fruits	8.777 ± 0.34	7.1103 ± 1.0	12.221 ± 1.29	65.738 ± 6.37	50.328 ± 3.28	+4.81	+3.68
3	Ripe fruits	13.666 ± 1.39	5.444 ± 0.27	10.221 ± 1.05	3.555 ± 3.01	6.11 ± 1.17	+3.27	+1.04

(d) PPO Activity u/mint/g fruit weight

Sr. No	Sample	0 hrs.	48 h after		72 h after			
		Uninfected	Uninfected	Infected	Uninfected	Infected	Fold	Fold
1	Green mature fruits	0.253 ± 0.13	0.319 ± 0.05	0.660 ± 0.23	2.398 ± 1.03	3.08 ± 0.28	+9.48	+12.17
2	Semi ripe fruits	0.330 ± 0.09	0.429 ± 0.10	0.726 ± 0.10	4.862 ± 1.35	6.380 ± 0.46	+14.73	+19.33
3	Ripe fruits	1.551 ± 0.31	2.046 ± 0.41	2.409 ± 0.50	3.619 ± 1.02	4.565 ± 1.57	+2.33	+2.94

(e) POX Activity u/mint/g fruit weight

Sr. No	Sample	0 hrs.	48 h after		72 h after			
		Uninfected	Uninfected	Infected	Uninfected	Infected	Fold	Fold
1	Green mature fruits	3.4176 ± 0.23	17.468 ± 1.73	30.759 ± 1.87	27.341 ± 1.34	42.151 ± 4.84	+8.0	+12.33
2	Semi ripe fruits	8.734 ± 0.74	40.632 ± 2.73	37.214 ± 2.40	53.923 ± 3.95	63.796 ± 3.98	+6.17	+7.30
3	Ripe fruits	49.746 ± 3.41	66.454 ± 4.21	95.314 ± 5.40	111.64 ± 6.98	121.90 ± 4.12	+2.24	+2.45

Table 1: Activity of total soluble sugar, phenols and defense related enzymes in relation to banana fruit rot by *Lasiodiplodia theabromae* [(Path.) Griff. and Maubl.] and ripening stages. Mean of three replication, *Change over time in uninfected, **Change over time in uninfected on infected.

growth and development. [5,7,18,19]. Total sugars also play a major role in disease resistance since sugars are the precursors for the synthesis of phenolics and phytoalexins which suppress the pectolytic and cellulolytic enzymes that are essential for pathogenesis.

Total phenols

Phenolic compounds are the most important group implicated in both constitutive and induced resistance and a distinct correlation between the degree of plant resistance and phenolics present in plant tissue has been demonstrated.

The phenol content was decreased with increasing storage period in uninfected and further decreased in infected in all three stages. Total phenol content was higher in green mature fruits (14.80 mg/g fruit weight) followed by semi- ripe (13.38 mg/g fruit weight) and ripe fruits (10.98 mg/g fruit weight) at initial period. Lowest total phenol content was observed in ripened infected fruits(7.05 mg/g fruit weight) followed by semi-ripe and green mature infected fruits with 7.91 and 7.38 mg/g fruit weight as compared to uninfected fruits after 72 h of incubation (Figure 1b). Highest phenol content was decreased in mature green fruits ranging from 14.80 to 7.35 mg/g fruit weight during ripening and it was further still decreased in infected same fruit. However, phenol content was least decreased in ripe fruits more or less similar in uninfected and infected. The total phenol contents change over the period it was more decreased in infected mature green fruits with -2.01 fold but in uninfected same fruit it increased with -1.47 fold after 72 h of incubation. In semi ripe fruits -1.69 fold changes over infected fruits was observed and further -1.28 fold change increased over uninfected fruits. However, total phenol content was decreased as the fruits ripened, and in rotted fruits decreased further. Our results are consistence with earlier workers, found that Phenolic content was significantly decreased in sapota fruit infected by *Aspergillus niger* [20]. This suggested that during ripening in healthy fruits phenol synthesis is decreased. During mango fruits ripening, the initial total phenolics content of unripe non-spongy fruits decreased by about 30 per cent [9]. From these results suggested that decrease phenol content during ripening will lead to fruit rot disease development.

PAL activity

The results demonstrated that enzyme activity was higher in green mature infected fruits (75.33 Units h^{-1} g^{-1} protein) than uninfected (59.33 Units h^{-1} g^{-1} protein) which was decreased in ripe infected fruits(9.11 Units h^{-1} g^{-1} protein) and uninfected (3.56 Units h^{-1} g^{-1} protein) after 72 h of storage (Figure 1c). These results suggests that constitutive level of PAL was higher in green mature fruits then decreased with the ripening stages. It was first decrease then increased in infected green mature fruits. The PAL activity was increased after 24 and 72 h storage in green mature and semi-ripe fruits irrespective of infection, while decreased in ripe fruits as compared to initial period. Moreover, PAL activity was 4.81 fold higher in uninfected semi ripe fruits as compared to infected semi ripe fruits after 72 h of incubation (Table 2). Our results collaborated with recently workers reported activities of the defense related enzymes phenylalanine ammonia Lyase (PAL) was enhanced in benzothiadiazole (BTH) and in Methyl jasmonate (MeJA) treated banana fruits irrespective of inoculated with *Colletotrichum musae* or

Sr. No.	Biochemical activity	Activity folds change over 72 h of incubation					
		Green mature fruits		Semi ripe fruits		Ripe fruits	
		uninfected	infected	uninfected	infected	uninfected	infected
1	Total phenol content mg/g fruit weight	-1.47	-2.01	-1.28	-1.69	-1.29	-1.56
2	Total sugar content mg/g fruit weight	+4.62	+9.07	+3.58	+3.27	+2.46	+2.38
3	PAL activity U/hr./g fruit weight	+1.15	+1.61	+4.81	+3.68	+3.27	+1.04
4	PPO activity U/mint./g fruit weight	+9.48	+12.17	+14.73	+19.33	+2.33	+2.94
5	POX activity U/mint/g fruit weight	+8.0	+12.33	+6.17	+7.30	+2.24	+2.45

Table 2: Activity of total soluble sugar, phenol and defense related enzymes in banana fruits uninfected and infected by *L. theobromae* at different ripening stages.

not [21]. The results suggest that post harvest decay in banana fruits can be controlled by BTH and MeJA. PAL is induced by stress such as wounding and fungal attack [22,23]. From this results suggested that enhanced the PAL activities was correlated with pathogenesis.

PPO activity

PPO activity was found lower in green mature infected and uninfected fruits at all storage period of determination although, the enzyme activity of uninfected (2.4 Units min^{-1} g^{-1} protein) was more or less similar to infected (3.08 Units min^{-1} g^{-1} protein) after 72 h of storage (Figure 1d). At initial period of incubation the PPO activity was (1.55 Units min^{-1} g^{-1} protein) higher in ripe fruits as compared to semi-ripe fruits (0.33 Units min^{-1} g^{-1} protein) andt it was increased with the incubation period in both infected and uninfected fruits. However, in uninfected semi ripe fruits PPO activity was increased about 14.73 fold after 72 h incubation, while in infected semi-ripe fruits it was higher (19.33 fold). PPO activity was increased with the ripening stages and it was further increased in infected fruits of all stages but highest activity found in infected semi ripe fruit then in uninfected fruits. Our results suggests that PPO activity was increased beginning the fruit ripening and further increased with the increasing storage time but increment was more pronounced in the inoculated fruits. These results are consistant with earlier workers they also observed increased polyphenol oxidase in banana fruit infected by *Botryodiplodia theobromae* [24]. Polyphenol oxidase and peroxidase activity gradually decreased during the development of the fruits followed then by an increase during the ripening period. The level of sugar gradually increased during fruit development and ripening which affects the taste of medlar fruits [21,25]. These observations suggest that the increase in PPO and POD activities as well as in sugar and protein contents has an important role in reducing the astringent taste of the fruits. Our results showed that PPO activity was induced during pathogenesis of *L. theobromae*.

POX activity

POX activity was found higher in ripe fruits (49.75-121.9 μmol guaiacol min^{-1} g^{-1} protein) than in semi ripe fruits (8.73-63.8 μmol guaiacol min^{-1} g^{-1} protein) up to72 h of storage (Figure 1e). The POX activity was enhanced after infected with *L. theobromae* in all ripening stages. The degree of increase in infected green mature, semi ripe and ripe was 12.33, 7.30 and 2.45 fold higher respectively than uninfected fruits (Table 1). Our results showed that POX activity was enhanced during host-pathogen interaction. These results are corroborated with earlier findings [26] indicated the green and red pepper fruit with *Colletotrichum gloeosporiodes*, the peroxidase genes were strongly activated in the green fruit by anthracnose disease but not in red fruit resistant to anthracnose disease. These results suggest that the peroxidase genes may be inducible during the pathogenesis of the anthrachose disease rather than in the disease resistance response. Similar results were observed in banana fruit infected by *Botryodiplodia theobromae* [24]. Change in peroxidase levels in mango fruits at various stages of ripening. The development of black color in ripened fruits indicated increased activity of POX or PPO enzymes [9,21]. In present study reduction in PAL activity and enhancement in PPO and POX activity in ripe fruits may be correlated with reduction of phenol content in ripe fruits. Higher activity of these enzymes in inoculated fruits as compared to uninoculated fruits reflects the invasion of pathogen.

Polyphenol oxidase and Peroxidase activity gradually decreased during the development of the fruits followed then by an increase during the ripening period [25]. These observations suggest that the increase in PPO and POD activities as well as in sugar and protein contents has an important role in reducing the astringent taste of the medlar fruits. Changes in peroxidase levels at various stages of ripening and spongy mango fruits. In unripe non-spongy fruits, a very low POX activity was observed. POX activity increased by several fold in non-spongy fruits upon ripening, however, in spongy fruits there was a further 1.5-fold increase in activity over that in ripe nonspongy fruits. The development of black color indicated changes in phenolics content either by POX or PPO enzymes [9].

Summary and Conclusion

Total sugar was higher in uninfected ripe fruits and decreased in fruits infected with *L. theobromae* and lowest in uninfected green mature fruits after 72 h of storage. A nine-fold increase in total sugar content was observed in green mature inoculated fruits followed by green mature uninfected and semi ripe fruits. Total phenol content was higher in green mature fruits than decline in semi ripe and ripe fruits at initial time. The phenol content was decreased with increasing storage period in uninfected and further decreased infected in all three stages.

Phenylalanine ammonia lyase (PAL) activity was higher in green mature infected fruits than uninfected which was decreased in ripe infected and uninfected fruits after 72 h of storage. PPO activity of uninfected fruits was more or less similar to infected fruits after 72 h of storage. However, in uninfected semi ripe fruits PPO activity was increased about 14.72 fold after 72 h of incubation, while in infected semi-ripe fruits it was higher (19.33 fold) as compared to mature green and ripe fruits. POX activity was found higher in ripe fruits than in semi ripe fruits up to72 h of incubation. The POX activity enhanced after inoculation with *L. theobromae* in all ripening stages. The degree of increase in infected green mature, semi ripe and ripe was 12.33, 7.30 and 2.45 fold higher respectively than uninfected fruits.

During ripening stages the level of total phenol gradually decreased in both infected and uninfected fruits, but it was more decreased in infected green mature fruits. The level of total sugar gradually increased during fruit ripening. PAL activity decreased in ripening stages but it increased in infected fruits. PPO and POD activity gradually increased in the ripening stage but degree of increase was higher in infected fruits. Reduction in PAL activity and enhancement in PPO and POX activity may be correlated with reduction of phenol content in ripening stages but it still increased in infected fruits. PAL, PPO and POX activity may be induced during pathogenesis of *L. theobromae* and stress during ripening.

Acknowledgments

The expert technical contributions of Dr. K.U. Solanky and Dr. Mahesh Kumar Mahatma are gratefully acknowledged. We acknowledged to Prof. and Head, Department of Plant Biotechnology, ASPEE College of Horticulture and Forestry, and Department of plant pathology, Navsari Agricultural University, Navsari, for providing technical assistance. Specially acknowledged to Department of statistic for statistical analyses.

References

1. Goos RD, Cox EA, Stotsky G (1961) *Botryodiplodia theobromae* and its association with *Musa* species. Mycologia 53: 262-277.
2. Wade NL, Kavanagh EE, Sepiah M (1993) Effects of modified atmosphere storage on banana postharvest diseases and the control of bunch main-stalk rot. Postharvest Biology and Technology 3: 143-154.
3. Ploetz R (1998) Banana disease in the subtropics: a review of their importance, distribution and management. Acta Horti 490: 263-276.
4. Punithalingam E (1980) Plant diseases attributed to *Botryodiplodia theobromae* Pat. J Cramer Vaduz 42-43.
5. Biale JB (1964) Growth, Maturation, and Senescence in Fruits: Recent knowledge on growth regulation and on biological oxidations has been applied to studies with fruits. Science 146: 880-888.
6. Biale JB (1960) The post-harvest biochemistry of tropical and subtropical fruits. Adv Food Res 10: 293.
7. Lizana LA (1976) Quantitative evaluation of sugars in banana fruit ripening at normal to elevated temperatures. Acta Hortic 57: 163.
8. Terra NN, Garcica E, Lajolo FM (1983) Starch sugar transformation during banana ripening. J Food Sci 48: 1097.
9. Janave MT (2008) Biochemical changes induced due to *Staphylococcal* infection in spongy alphonso mango (*Mangifera indica* L.) fruits. J Crop Sci Biotech 10: 167-174.
10. Mathew AG, Parpia HAB (1971) Food browning as a polyphenol reaction. Adv Food Res 19: 75-145.
11. Barnett HL, Hunter BB (1972) Illustrated genera of imperfect fungi. (3rdedn) Mixneapolis, Burgess Pub, p. 188.
12. Franscistt W, David FB, Robert MD (1971) The estimation of the total soluble carbohydrate in cauliflower tissue. Experiment in plant phyisiology, Van, Nostrand. Reinhold Camp, New York, p. 16.
13. Swain T, Hills WE (1959) The phenolic constituents of *Prunus domestica* I. the quantitative analysis of phenolic constituents. J Sci Food Agri 10: 63-68.
14. Mahadevan A, Sridhar R (1996) In: Methods in physiological plant pathology. Kalyani Publisher.
15. Malik CP, Sing BM (1980) In: *Plant Enzymology and Histo-Enzymology.* Kalyani Publishers New Delhi, p. 286.
16. Hammerschmidt R, Nuckles EM, Kuc J (1982) Association of enhanced peroxidase activity with induced systemic resistance of cucumber to *Colletotrichum lagenarium.* Physiol Plant Pathol 20: 73-82.
17. Singh R, Saxena VG (1991) Catalase and invertase activity in Banana fruit infected by *Aspergillus niger.* Indian J mycol Pl Pathol 21: 211-212.
18. John M, Robinson M, Karkari SK (1981) Starch and sugar transformation during the ripening of plantains and bananas. J Sci Food Agric 32: 1021.
19. Kadioglu A, Yavru I (1998) Changes in the chemical content and polyphenol oxidase activity during development and ripening of cherry laurel. Phyton (Horn, Austria) 37: 241-251.
20. Wagh PM, UN Bhale (2012) Changes in phenolic contents of sapota pulp (*Achras sapota*) due to different isolates of *Aspergillus niger.* Bioscience Discovery 3: 263-265.
21. Zhu S, Ma B (2007) Benzothiazole or methyl jasmonate induced resistance to *colletotrichum musae* in harvested banana fruit is related to elevated desense enzyme activities. Hort Sci & Bio-tech 82: 500-506.
22. Morello JM, Romero MP, Ramo T, MJ Motilva (2005) Evaluation of L-Phenylalanine ammonia-lyase activity and phenolic profi le in olive drupe (*Olea europaea* L.) from fruit setting period to harvesting time. Plant Science 168: 65-72.
23. Pereyra L, Roura SI, CE Del Valle (2005) Phenylalanine ammonia lyase activity in minimally processed Romaine lettuce. Lebensmittel Wissenschaft und Technologie 38: 67-72.
24. Chakraborty N, Nandi B (1978) Enzyme activity in banana fruits rot by *Botrydiplodia theobromae* pat. Acta Agrobotanica 31: 41-46.
25. Aydin N, Kadioglu A (2001) Changes in the chemical composition, polyphenol oxidase and peroxidase activities during development and ripening of Medlar fruits (*Mespilus germanica L.*). Bulg. J. Plant Physiol 27: 85-92.
26. Lee SC, Lee YK, Kim KD, BK Hwang (2000) In situ hybridization study of pathogen and organ-dependent expression of a novel thionin gene in pepper (*Capsicum annum*). Physiol Plantarum 110: 384-392.

Interactions between Four *Fusarium* Species in Potato Tubers and Consequences for Fungal Development and Susceptibility Assessment of Five Potato Cultivars under Different Storage Temperature

Boutheina Mejdoub-Trabelsi[1,2]*, Hayfa Jabnoun-Khiareddine[2] and Mejda Daami-Remadi[2]

[1]*Institut Supérieur Agronomique de Chott-Mariem, Université de Sousse, 4042, Chott-Mariem, Tunisia*
[2]*UR13AGR09, Production Horticole Intégrée au Centre Est Tunisien, Centre Régional des Recherches en Horticulture et Agriculture Biologique de Chott-Mariem, Université de Sousse, 4042, Chott-Mariem, Tunisia*

Abstract

Fusarium dry rot of potato caused by *F. sambucinum*, *F. oxysporum*, *F. solani* and *F. graminearum* is particularly prominent in Tunisia resulting in partial or complete tuber decay during storage. This fungal complex can occur within the same potato tuber. Cultivar's reaction to different mixtures depending on the temperature of storage used can give additional information on their relative aggressiveness. Thus, interactions between these *Fusarium* species was investigated using single and mixed infection onto five local potato cultivars (Spunta, Oceania, Nicola, Mondial and Atlas) under two temperatures (20 and 30°C). Data indicated that the lesion diameter and the penetration of dry rot, noted 21 days post-inoculation, varied significantly depending on cultivars, inoculation treatments and temperatures of storage tested and their interactions. The combination of *F. sambucinum* and *F. solani* (C2-1) was found to be the most aggressive inoculation treatment. This treatment was followed by the association of *F. sambucinum* with *F. oxysporum* (C2-4) and the combination of *F. sambucinum* with *F. solani* and *F. graminearum* (C3-4). However, the four *Fusarium* species, when considered individually, exhibited globally reduced aggressiveness as compared to the tested complexes suggesting occurrence of synergistic interaction. Overall, all mixed inoculums including *F. sambucinum* showed increased aggressiveness levels on the majority of cultivar x temperatures combinations. Potato cultivars exhibited differential response to the different Fusarium mixtures tested depending on the temperature of storage used. None of the cultivars tested was completely tolerant to all inoculation treatments and only cvs. Spunta and Oceania exhibited lesser susceptibility to four mixtures. The isolation frequency of *Fusarium* species was also variable depending on single or mixed inoculum used for tuber infection and according to cultivar × temperature combination considered. This relative predominance may reflect their competitive potential in mixture and their relative involvement in dry rot development and severity.

Keywords: Aggressiveness; Competition; *Fusarium* dry rot; Mixed infections; Synergy; Susceptibility

Introduction

Dry rot caused by *Fusarium* species is an important soil borne potato disease known worldwide by its serious post-harvest tuber rotting and seed piece decay after planting. This disease can reduce crop establishment by affecting the developing potato sprouts and causing crop losses estimated to up to 25% while more than 60% of tubers can be infected during storage [1]. Post-harvest tuber losses can be as high as 28% in China and 88% were mainly attributed to dry rot disease [2]. A recent study showed that in Great Britain crop losses attributed to this disease have been estimated to about 95% [3].

Fusarium dry rot (FDR) initial symptoms appear on tuber at wound sites as shallow small brown lesions after approximately one month of storage. The lesions enlarge in all directions and the periderm eventually sinks and may wrinkle in concentric rings as the underlying dead tissue desiccates [4]. Cavities underneath the rotted area are usually lined with *Fusarium* mycelia and spores of various colors with abundant white or carmine-colored mycelium. Fully rotted tubers become shriveled and mummified [5].

FDR is particularly prominent in Tunisia resulting in partial or almost complete loss of stored potatoes occurring especially under traditional storage conditions [6]. The past decay has been one of rapid change in the potato industry with the widespread use of refrigerated storage [3]. Although these losses can be greatly reduced by storage at low temperatures (1-5°C), dry rot is still considered a serious threat and causes important crop losses because open field traditional storage were the most practiced by Tunisian farmers.

Thirteen *Fusarium* species were reported to be involved in potato dry rot disease worldwide [7]. In Tunisia, *F. sambucinum*, *F. solani*, *F. oxysporum*, and *F. graminearum* were the most predominant. They were mostly isolated in combinations of two or more species [5,8-9]. Disease incidence appears to be affected by soil type, potato cultivars, *Fusarium* species and local climate conditions such as temperature. In fact, *Fusarium* species exhibited variable degrees of aggressiveness depending on storage temperatures. They have both thermal picks of aggressiveness one at lower (10-15°C) and one at higher (30-35°C) temperatures depending on the fungal complex involved in disease development [10]. Due to their large adaptive potential to temperature variation, *Fusarium* species are characterized by their co-occurrence

***Corresponding author:** Boutheina Mejdoub-Trabelsi, Institute Supérieur Agronomique de Chott-Mariem, Université de Sousse, 4042, Chott-Mariem, Tunisia, E-mail: boutheinam2002@yahoo.fr

within the same potato tuber and most probably by the variation of their relative dominance in refrigerated and non-refrigerated stores.

Microbial ecologists have shown that different types of interactions can occur among species, depending on biotic and abiotic factors [11]. Positive correlations can result either from direct synergy between species on the same infection site, or from indirect associations facilitated by micro-climatic conditions conducive to several species. Many reports have shown the importance of interspecific interactions between *Fusarium* species on many host plants such as pea [12], maize [13] and especially wheat [14] but few data were available on potato.

The most effective and environmentally approach to control FDR is the utilization of resistant potato cultivars but only cursory germplasm evaluation for resistance to this disease has been reported in literature [15, 16]. In Tunisia, several reports have focused on the predominant species involved in FDR disease, the response of the local potato cultivars to the four species and the effect of temperature variation on their relative aggressiveness These screenings revealed the absence of potato cultivar combining resistance to the whole fungal disease complex and only some cultivars exhibited reduced susceptibility to at least two *Fusarium* species [9,17]. For a sustainable behavior of potato tubers towards the four *Fusarium* species involved in disease development, the type of the interaction between *F. sambucinum, F. solani, F. graminearum*, and *F. oxysporum* f. sp. *tuberosi* should be more elucidated on tubers based on standardized inoculation methods. In addition, environmental conditions, especially temperature, required for infection and subsequent tissue colonization may depend upon *Fusarium* species used for tuber inoculation. Thus, this abiotic factor i.e., temperature should be considered during qualifications of interactions between *Fusarium* species and breeding for multiple resistances.

In Tunisia, interactions between *Fusarium* species and some potato pathogens have been observed under natural conditions. In fact, *F. graminearum, Rhizoctonia solani, Colletotrichum coccodes, Pythium* spp. and *Verticillium* spp. were frequently isolated from potato plants exhibiting early dying syndrome [18-20]. Daami-Remadi et al. [21] also reported synergistic interactions occurring between *V. dahliae, F. oxysporum* f. sp. *tuberosi* and *Meloidogyne javanica* and leading to reduced plant growth and increase in vascular wilt severity, galling index, egg masses number and female fecundity.

Nevertheless, regarding FDR pathosystem, the relationships between *Fusarium* species has not been explored in detail. Therefore, this pathogen-cultivar pathological investigation may lead to a better understanding of how interactions between the four *Fusarium* species may affect disease incidence and severity. By understanding the interspecific interaction within the FDR complex, the obtained information will be useful for reliable disease predictions and management. Thus, the current study was set up to mimic natural infections and the overall objective is to (i) clarify the type of interaction between the four *Fusarium* species and assess its impact on disease development and severity based on single and mixed inoculations and (ii) determine their relative aggressiveness under two different temperatures and depending on the plant material used (i.e., potato cultivars).

Materials and Methods

Fungal inoculum

Fusarium spp. isolates were obtained from dry rotted tubers and from partially or totally wilted plants of different potato cultivars. These isolates belonged to a *Fusarium* complex composed of *F. solani, F. graminearum, F. sambucinum*, and *F. oxysporum* f. sp. *tuberosi. Fusarium* species were identified based on macromorphological, micromorphological, pathological characteristics and molecular tools based on sequencing of ITS region [22-24]. These *Fusarium* isolates were cultured on Potato Dextrose Agar (PDA) medium supplemented with 300 mg/l of streptomycin sulphate (Pharmadrug Production Gmbh, Hamburg, Germany). Their virulence was maintained by bimonthly inoculation of freshly wounded and apparently healthy tubers and re-isolation on PDA plates. For their preservation up to 12 months, monoconidial cultures were maintained at -20°C in a 20% glycerol solution.

Inoculum was composed of a single *Fusarium* species or a mixture of two, three or four species selected beforehand based on their aggressiveness (Mejdoub-Trabelsi, unpublished data). Spore suspensions were prepared by culturing isolates in Potato Dextrose Broth (PDB). After incubation for 7 days at 25°C under continuous shaking at 150 rpm, the cultures were filtered through two layers of cheesecloth to remove mycelium and then through two layers of Whatman No. 1 filter paper. The final conidial concentration in the filtrates was adjusted to 10^7 conidia /ml using a Malassez cytometer. An equal volume (100 µl) of inoculum was used for tuber inoculation.

Potato cultivars

Potato (*Solanum tuberosum* L.) tubers without physical injuries or visible infection were chosen for the bioassay. They belonged to five cultivars namely Atlas, Mondial, Nicola, Oceania, and Spunta. These cultivars were subscribed in the list A of the Tunisian varietal assortment and were kindly provided by the Technical Center of Potato and Artichoke of Tunisia. They were stored for two months in darkness at 6°C and brought to room temperature three hours before use. Prior to inoculation, tubers were superficially disinfected with a 10% sodium hypochlorite solution (Aiglol Production, Zaouiet Sousse, Tunisia) during 5 min then rinsed with sterile distilled water and air dried until use.

Inoculation and incubation

Before inoculation, tubers were wounded by removing a tuber plug (6 mm in diameter and depth) with a sterile cork-borer. These wounds were challenged with 100 µl of a single or mixed conidial suspension. For mixed inoculations, equal volumes of each *Fusarium* species suspension were mixed to obtain three types of mixtures (binary, ternary or quaternary) ready for tuber infection. Inoculation treatments were composed of combinations of two (six fungal treatments I,. C2-1, C2-2, C2-3, C2-4, C2-5, and C2-6), three (four treatments C3-1, C3-2, C3-3, and C3-4) or four *Fusarium* species (Treatment C4). Individual inocula consisted of four single treatments composed of each *Fusarium* species (*Fusarium solani* C1-1, *F. sambucinum* C1-2, *F. oxysporum* C1-3, and *F. graminearum* C1-4).

The inoculated tubers (two replicates of five tubers per elementary treatment) were placed in plastic bags to maintain a high humidity and then incubated for three weeks at two different temperatures (20 or 30°C). It should be mentioned that at 30°C, the difference in aggressiveness of *Fusarium* species is more evident whereas at temperatures inferior or equal to 20°C only two groups can be distinguished Daami-Remadi et al. [10]. Furthermore, Mejdoub-Trabelsi et al. [9] demonstrated that when dry rot was assessed depending on temperature effect, *F. sambucinum* and *F. graminearum* developed the most severe symptoms at 15, 20 and 25°C whilst at 30°C, *F. solani* caused the severest rot.

After the incubation period, tubers were cut along the longitudinal axis across the inoculation sites. For disease assessment, symptomatic lesions were measured both externally and internally.

For the external evaluation, the two perpendicular diameters of the lesion were recorded and the mean diameter was calculated for each site of inoculation by using the following formula: Mean lesion diameter (mm)=(d1+d2) /2 where d1 and d2 two perpendicular diameters.

Dry rot severity was also estimated internally through the extent of the induced decay i.e., maximal width (w) and depth (d) of the rotted tissue. The pathogen penetration into tubers was calculated based on Lapwood et al. [25] formula as follow:Penetration (mm)=(w/2 + (d-6))/2

Potato cultivars were then ranked for their susceptibility to different *Fusarium* treatments based on the following below scale:

✓ Tolerant: Lesion diameter ≤ 11 mm;

✓ Moderately susceptible: 11 mm<Lesion diameter<15 mm;

✓ Highly susceptible: Lesion diameter ≥ 15 mm.

Inoculation treatments were ranked for their relative aggressiveness based on the following scale:

- Less aggressive (LA): Lesion diameter ≤ 11 mm
- Moderately aggressive (MA): 11 mm<Lesion diameter<14 mm
- Highly aggressive (HA): Lesion diameter ≥ 14 mml

Statistical analyses

A factorial analysis of variance (ANOVA) was performed to determine the significance of the main factors and their interactions using a completely randomized factorial design with three factors i.e., potato cultivars, fungal treatments (tubers inoculated with single or mixed inoculum and the non-inoculated tubers) and temperatures of storage. Mean separations were performed by Duncan's multiple range test (at $P < 0.05$).

Results

Inoculated tubers of all cultivars exhibited typical external dry rot lesions expressed as brown to black flecks on the tuber surface (Figure 1). Wrinkled concentric rings appeared around the inoculation site. When cut open, infected tubers also showed light to dark brown or black rot and internal cavities containing molds within rotted tissues whereas non-inoculated control tubers remained symptomless. The intensity of these symptoms varied depending on cultivars used, inoculation treatments and temperatures of incubation tested.

Effect of single or mixed inoculation with *Fusarium* species on dry rot lesion diameter

The mean diameter of dry rot lesion, noted 21 days post-inoculation, varied significantly (at $P \leq 0.05$) depending on cultivars, inoculation treatments and temperatures of storage tested and their interactions. Therefore, differences between single and mixed inoculation treatments were performed for each cultivar and at each temperature (Figure 2). Overall, disease score as estimated by lesion mean diameter shows that cultivar response to *Fusarium* species inoculated singly or in combination varied upon the temperature used for tuber storage. In fact, for cv. Spunta stored at 20°C, the highest lesion diameter (> 15 mm) recorded 21 days-post-inoculation was noted on tubers challenged with C1-2, C2-4, and C3-3 inoculation treatments. However, when stored at 30°C, the severest dry rot lesions were observed on tubers inoculated with C4 and C3-1 treatments (composed of four and three *Fusarium* species, respectively). For cv. Oceania, C1-2, C2-6, C2-1, and C2-4 were found to be the most aggressive inoculation treatments at 20°C whereas at 30°C, C2-1, C2-4, C4, and C2-6 induced the highest lesion diameter. Inoculated to tubers belonging to cv. Mondial and incubated at 20°C, the mixed inoculation treatment C3-3 caused the most severe dry rot lesions (diameter of about 17 mm). However, when stored at 30°C, 12 out of the 16 inoculation treatments tested including single and mixed inoculum (namely C1-4, C3-4, C2-4, C2-3, C2-1, C4, C2-5, C1-1, C1-2, C2-2, C3-1, and C3-2) induced a significantly similar dry rot severity where the lesion diameter, noted after 21 days post-inoculation, ranged from 12.5 to 15 mm. As shown in Figure 2, cv. Atlas tubers stored at 20°C and challenged with the different inoculation treatments exhibited dry rot lesions varying in intensity (diameter 13-14.5 mm) where the most severe ones were incited by some binary (C2-1, C2-3, and C2-6) or ternary (C3-1 and C3-4) *Fusarium* mixtures. However, at 30°C, three single infections (C1-2, C1-3, and C1-4), four binary (C2-1, C2-3, C2-4, and C2-6), four ternary (C3-1, C3-2, C3-3, and C3-4) and one quaternary inoculation treatments induced significantly similar dry rot lesions (diameter ranging between 10.5 and 13.5 mm). Tested on cv. Nicola tubers and incubated at 20°C, the inoculation treatments C2-1, C1-2, C2-4, C3-1, and C4 led to the development of the most severe lesions (diameter 15-17 mm). However, when stored at 30°C, the binary mixtures C2-1, C2-2 and C2-3 were found to be the most aggressive (diameter ranging from 15 to 17 mm). It should be highlighted that the treatment C2-3, ranked among the least aggressive treatments on cv. Nicola when incubated at 20°C, moved to the most aggressive group when incubated at 30°C.

Data shown in Figure 2 also revealed that the response of the potato cultivars to the inoculation treatments tested varied depending on temperature of incubation. Moreover, the ranking of single and combined treatments, based on their relative aggressiveness, varied significantly depending on the plant material used. By examining pooled data for all cultivar × temperature combinations tested, it can be noted that some inoculation treatments were ranked within the most aggressive group at many cases. In fact, at 20°C, treatments C1-2,

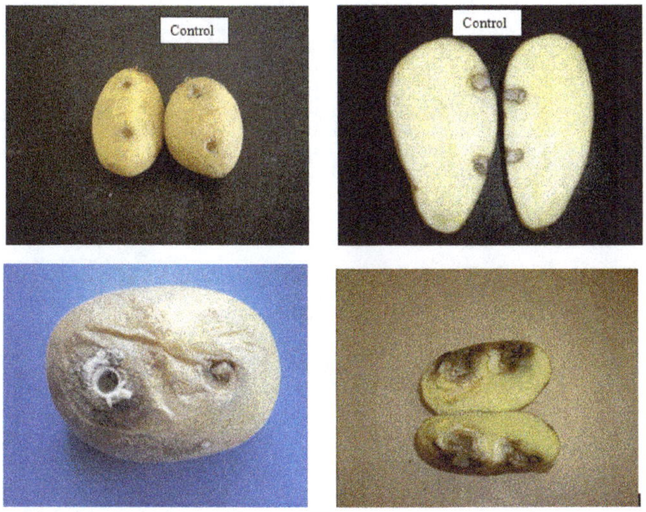

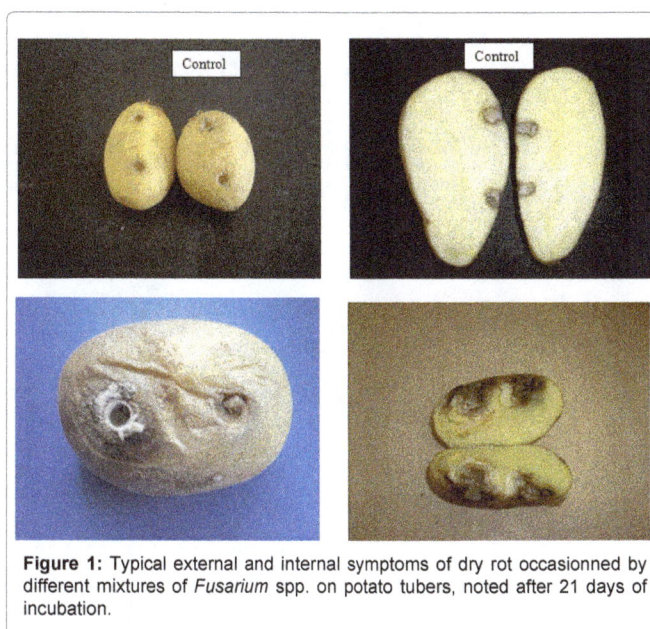

Figure 1: Typical external and internal symptoms of dry rot occasionned by different mixtures of *Fusarium* spp. on potato tubers, noted after 21 days of incubation.

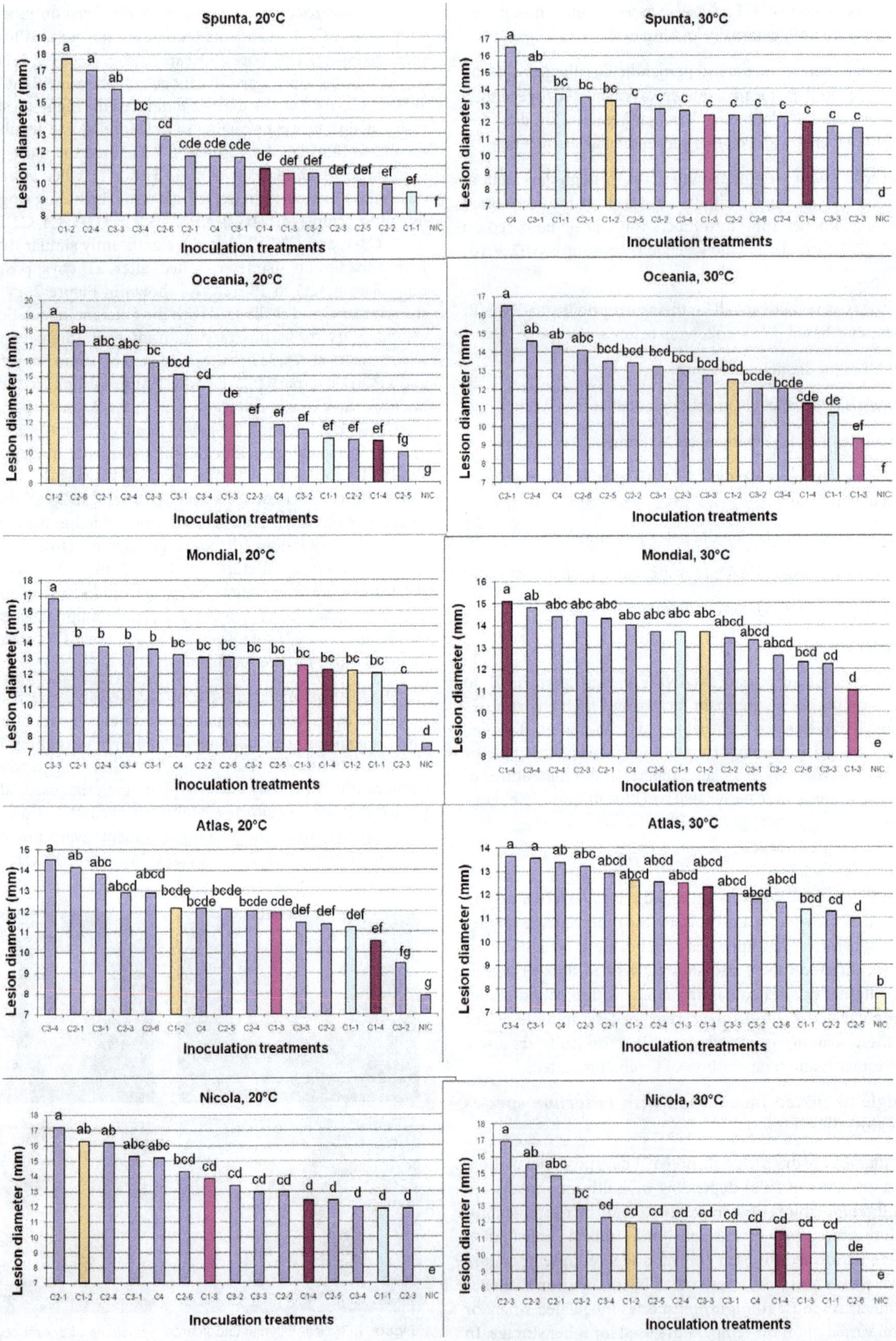

Figure 2: Lesion diameter of dry rot incited by single or combined inoculation treatments with four *Fusarium* species tested on five potato cultivars and noted 21 days post-incubation at 20 and 30°C. Bars affected by the same letter are not significantly different according to Duncan's multiple range tests at $P \leq 0.05$.

C2-1 and C2-4 exhibited the highest aggressiveness on 3 out of the 5 cultivars tested (cvs. Spunta, Oceania and Nicola for C1-2 and C2-4 and cvs. Oceania, Nicola, Atlas for C2-1). The combined treatments C2-6, C3-1, and C3-3 caused the severest dry rot lesions on 2 cultivars (cvs. Nicola and Oceania for C2-6 and C3-1 and cvs. Spunta, Mondial for C3-3) whereas C2-3, C3-4 and C4 induced the highest dry rot severity on only one cultivar (Atlas for C2-3 and C3-4 and Nicola for C4). However, at 30°C, C2-1 and C4 were ranked within the group of most aggressive treatments on 4 cultivars out the 5 tested (cvs. Spunta, Oceania, Mondial and Nicola for C2-1 and cvs. Spunta, Oceania, Mondial and Atlas for C4). Inoculation treatments C2-3, C2-4, and C3-1 were classified within this group on 3 cultivars (cvs. Mondial, Atlas and Nicola for C2-3, cvs. Mondial, Atlas and Oceania for C2-4 and cvs. Spunta, Atlas, Oceania for C3-1) whereas C1-2, C2-2, C2-6 and C3-4 single and mixed inoculums exhibited their highest aggressiveness on 2 cultivars (cvs. Mondial and Spunta for C1-2, cvs. Oceania and Nicola for C2-2, cvs. Oceania and Mondial for C2-6, and cvs. Atlas and Mondial for C3-4). C1-4 and C3-3 treatments showed their highest aggressiveness on only one cultivar (Mondial for C1-4 and Oceania for C3-3). Moreover, some single or mixed inoculation treatments (such as C1-1, C2-5 and C3-2) led to reduced dry rot severity for the majority of cultivar × temperatures combinations tested whereas other single treatments exhibited comparable aggressiveness as combined ones under given cultivar and temperature conditions as detailed above.

Effect of single or mixed inoculation with *Fusarium* species on penetration

The level of tuber decay estimated based on mean pathogen penetration depended on cultivars, fungal treatments used for inoculation and temperatures of incubation as a significant interaction was recorded between the three fixed factors. The ANOVA showed the effect of single factors and their interactions on mean penetration.

Results shown in Figure 3 indicated the mean penetration incited by the different inoculation treatments, recorded after 21 days of incubation, depending on potato cultivars used and temperatures of storage. For cv. Spunta tubers incubated at 20°C, the most aggressive treatments were C1-2, C2-4 and C2-6 where the mean penetration ranged between 10 and 12 mm. However, when stored at 30°C, the ranking of the inoculation treatments, based on the severity of the tuber internal decay, has changed. In fact, the most aggressive treatments at 30°C were found to be C4, C1-1, C3-1 and C2-1. Thus, some single treatments such as C1-1 exhibited significantly similar aggressiveness on cv. Spunta tubers as some binary (C2-1), ternary (C3-1) and quaternary (C4) treatments. For cv. Oceania stored at 20°C, the highest mean penetration (8-10 mm) was incited by C2-4, C1-2, C2-1, C3-3, and C3-4 inoculation treatments whereas at 30°C, the most severe internal tuber decay was noted on tubers challenged with C2-1, C4, C3-1, C2-5, C2-4, C2-2, and C2-6 mixed inoculums. It should be highlighted that some binary treatments (C2-1 and C2-4) were ranked within the most aggressive inoculum at both temperatures tested while C2-5 showing the least penetration at 20°C moved to the most aggressive treatment when incubated at 30°C. For tubers belonging to cv. Mondial and stored at 20°C, the severest internal tuber decay was incited by C3-3, C2-4, C1-2, C4, C1-4 and C2-1 inoculation treatments. However, at 30°C, 10 out of the 15 treatments tested exhibited significantly comparable mean penetration. These most aggressive treatments are composed of single (C1-1, C1-2, and C1-4) inoculum and some binary (C2-1, C2-2, C2-3, C2-4, and C2-5), ternary (C3-4) and quaternary (C4) mixed infections. Data shown in Figure 3 also indicated that, at 20°C, the most severe internal decay was noted on cv. Atlas challenged with C3-4 and C2-3 mixed inoculum while at 30°C, the significantly highest penetration was incited by C2-1, C3-4, C3-1, C3-2, C4, C2-4, and C1-2 treatments. It should be mentioned that the combined inoculum C3-4 was ranked among the most aggressive treatments on this cultivar and at both temperatures tested whereas C2-3 treatment showed reduced aggressiveness, as compared to 20°C, when incubated at 30°C. It is also important to note that the ternary combination C3-2, classified among the most aggressive treatments at 30°C, has induced surprisingly the least internal tuber decay when incubated at 20°C. Furthermore, cv. Nicola data presented in Figure 3 also indicated that some single inoculums (C1-2), and some binary (C2-1 and C2-4), ternary (C3-1) and quaternary (C4) mixed infections led to the development of the most severe internal tuber decay as compared to the other inoculation treatments tested. However, at 30°C, the highest penetration was noted on tubers challenged with C2-1, C2-2 and C2-3 binary inoculum. It should be highlighted that C2-1 was ranked within the most aggressive group of treatments on this cultivar and under both temperatures of storage tested. However, some treatments, as is the case of C1-2 showed reduced aggressiveness on cv. Nicola when incubated at 30°C as compared to 20°C. Furthermore, C2-2 and C2-3 treatments classified within the most aggressive ones at 30°C showed reduced mean penetration (< 5 mm) when incubated at 20°C.

Data shown in Figure 3 also indicated variable reaction of the potato cultivars to the different inoculums tested varied depending on temperature of incubation. Furthermore, the ranking of single and combined *Fusarium* mixtures, based on their relative aggressiveness, depended significantly upon the plant material used. Pooled data for all cultivar × temperature combinations tested revealed that some tested treatments were classified among the most aggressive group at many cases. In fact, at 20°C and based on mean penetration records, treatments C1-2 and C2-4 exhibited the severest internal tuber decay on 4 (Spunta, Oceania, Nicola and Mondial) out of the 5 cultivars tested. The combined treatment C2-1 led to the highest penetration on 3 cultivars (Nicola, Oceania and Spunta); C3-3, C3-4 and C4 incited the most severe internal dry rot symptoms on 2 cultivars (cvs. Oceania and Mondial for C3-3, cvs. Oceania and Atlas for C3-4, and cvs. Oceania and Spunta for C4) whereas C1-4, C2-3, C2-6, and C3-1 caused the most important tuber decay on only one cultivar. However, at 30°C, C2-1 belonged to the most aggressive group on all the 5 cultivars tested. The treatment C4 was classified within this group on 4 cultivars (Spunta, Oceania, Mondial and Atlas) whereas C2-4 and C3-1 exhibited the severest penetration on 3 cultivars (cvs. Oceania, Atlas and Mondial for C2-4 and cvs. Oceania, Atlas and Spunta for C3-1). C1-1, C1-2, C2-2, C2-3, C2-5, and C3-4 showed their highest aggressiveness on 2 cultivars (Mondial and Atlas for C1-2 and C3-4, Mondial and Spunta for C1-1, Mondial and Nicola for C2-3, Mondial and Oceania for C2-5, Nicola and Oceania for C2-2) whilst C1-4 and C2-6 were able to induce severe tuber decay only on one cultivar (Mondial for C1-4 and Oceania for C2-6). It should be mentioned that some single or mixed inoculation treatments (such as C1-3, C2-3 and C2-6) led to reduced tuber internal decay for the majority of cultivar × temperatures combinations tested whereas some single treatments (such as C1-1 an C1-2) exhibited comparable aggressiveness as some combined inoculums (as is the case of C2-1 and C2-4) under given cultivar and temperature conditions.

Consequently, all inoculation treatments were ranked into aggressiveness classes based on their disease severity estimated via the lesion diameter noted on the five cultivars at both temperatures tested. It was revealed from the results (Table 1) that the mixture C2-1 composed of *F. sambucinum* and *F. solani* was found to be the most

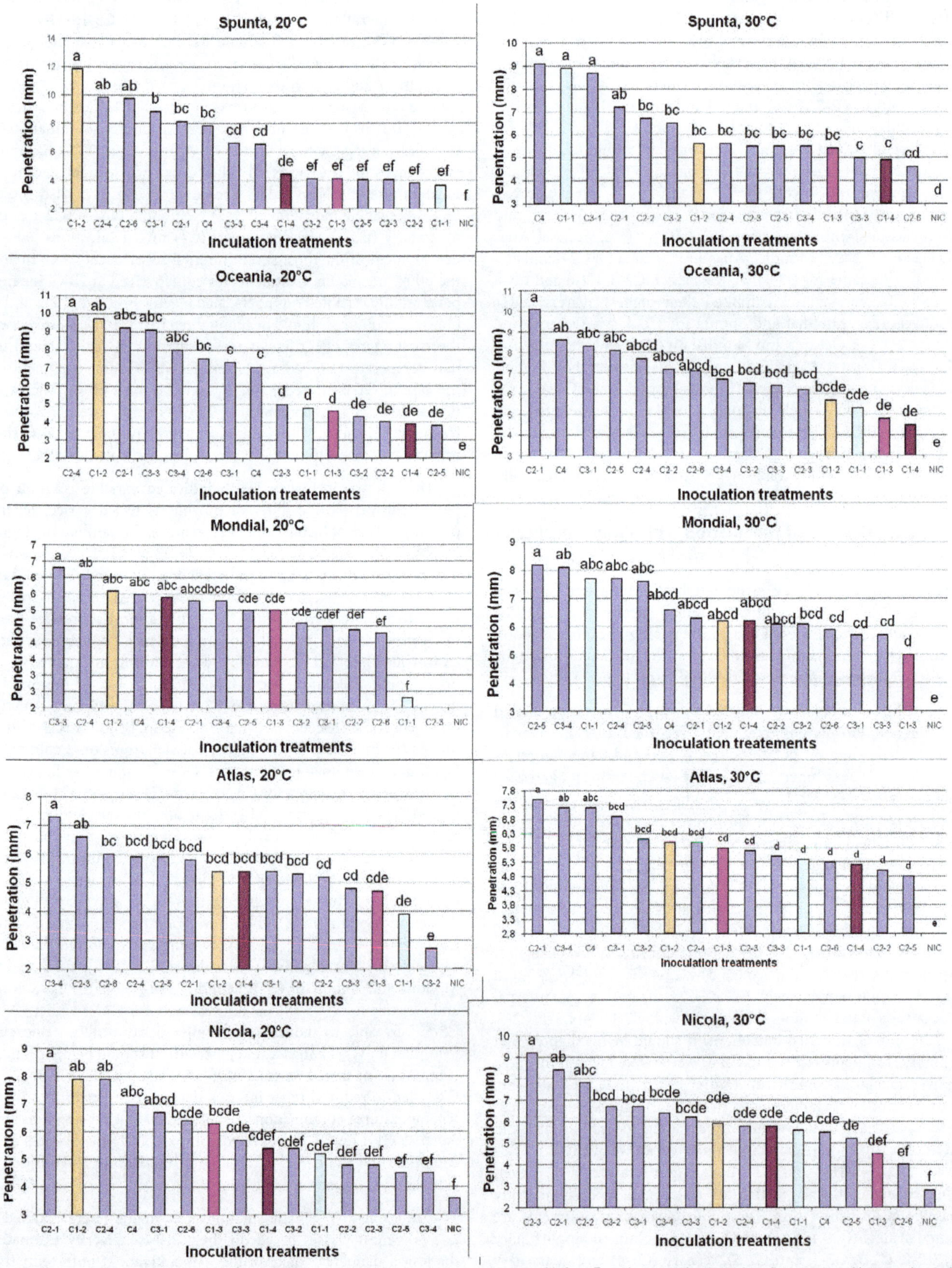

Figure 3: Mean penetration of dry rot incited by single or combined inoculation treatments with four *Fusarium* species tested on five potato cultivars and noted 21 days post-incubation at 20 and 30°C. Bars affected by the same letter are not significantly different according to Duncan's multiple range test at $P \leq 0.05$.

aggressive treatment and behaved as highly aggressive on 5 cultivar × temperature combinations out the 10 tested and as moderately aggressive on the 5 other remaining ones (Figure 4). This treatment was followed by two other mixed inoculation treatments namely C2-4 and C3-4. The association of *F. sambucinum* with *F. oxysporum* (i.e., C2-4) and the combination of *F. sambucinum* with *F. solani* and *F. graminearum* (i.e., C3-4) were highly aggressive on 4 out the 10 cultivar × temperature combinations tested and moderately aggressive on the

Cultivar/Temperatures	**Spunta**		**Oceania**		**Mondial**		**Atlas**		**Nicola**		Ranking
	20°C	30°C	20°C	30°C	20°C	30°C	20°C	30°C	20°C	30°C	Order
C1-1: *F. sol*	LA	MA	LA	LA	MA	MA	MA	MA	MA	LA	6
C1-2: *F. sam*	HA	MA	HA	MA	LA	MA	LA	MA	HA	MA	3
C1-3: *F. gra*	LA	MA	MA	LA	LA	MA	M	MA	MA	LA	6
C1-4: *F. oxy*	LA	MA	LA	M	MA	HA	LA	MA	MA	MA	5
C2-1: *F. sam+F. sol*	M	MA	HA	HA	MA	HA	HA	MA	MA	HA	1
C2-2: *F. sol+F. oxy*	LA	MA	LA	MA	MA	MA	MA	MA	MA	HA	5
C2-3: *F. sol+F. gra*	LA	MA	MA	MA	LA	HA	MA	MA	MA	HA	4
C2-4: *F. sam+F. oxy*	HA	MA	HA	MA	MA	HA	MA	MA	HA	MA	2
C2-5: *F. oxy+F. gra*	LA	MA	LA	MA	MA	MA	MA	LA	MA	MA	6
C2-6: *F. sam+F. gra*	M	MA	HA	MA	MA	MA	MA	MA	MA	LA	5
C3-1: *F. sam+F. sol+F. oxy*	M	HA	HA	MA	MA	MA	MA	MA	MA	MA	4
C3-2: *F. sol+F. oxy+F. gra*	LA	MA	LA	MA	MA	MA	LA	MA	MA	MA	6
C3-3: *F. sam+F. oxy+F. gra*	HA	MA	HA	MA	HA	MA	MA	MA	MA	MA	3
C3-4: *F. sam+F. sol+F. gra*	HA	MA	HA	MA	MA	HA	HA	MA	MA	MA	2
C4: *F. sam+F. sol+F. oxy+F. gra*	MA	HA	MA	MA	MA	HA	MA	MA	HA	MA	3

Table 1: Ranking of aggressiveness of single or combined inoculation treatments tested on potato tubers belonging to five cultivars and incubated at two temperatures based on lesion diameter of Fusarium dry rot. HA: Highly aggressive: Lesion diameter ≥ 14 mm; MA:Moderately aggressive: 11 mm<Lesion diameter<14 mm; LA: Less aggressive: Lesion diameter ≤ 11 mm, *F. gra: F. graminearum; F. oxy: F. oxysporum; F. sam: F. sambucinum; F. sol: F. solani.*

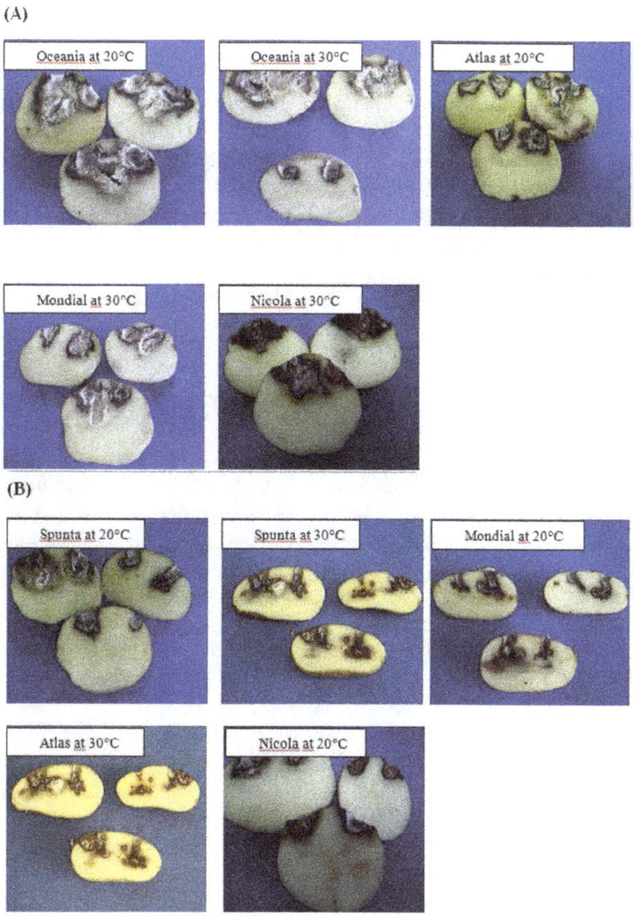

Figure 4: Dry rot occasioned by the most aggressive mixture of *F. sambucinum* and *F. solani* which behaved as highly aggressive (A) to moderately aggressive (B) on 5 different cultivar x temperature combinations.

rest of combinations. *F. sambucinum* combined with *F. oxysporum* and *F. graminearum* (i.e., C3-3) or associated with all *Fusarium* species (C4) the obtained fungal complex behaved as highly aggressive on three cultivars by temperature combinations and as moderately aggressive on the remaining ones. Also, *F. sambucinum* mixed with *F. oxysporum* and *F. solani* (I,. C3-1) was ranked as highly aggressive on two combinations and as moderately aggressive on the rest. Table 1 also revealed that the four *Fusarium* species, when considered individually, were classified as less, moderately or highly aggressive pathogens depending on the cultivar × temperature combination tested but exhibited globally reduced aggressiveness as compared to the tested binary, ternary or quaternary complexes. It should be highlighted that all mixed inoculums including *F. sambucinum* showed increased aggressiveness levels on the majority of cultivar × temperatures combinations.

Cultivar responses to inoculation treatments tested

Table 2 shows that the differential response of the potato cultivars to the different *Fusarium* mixtures tested depending on the temperature of storage used. In fact, cv. Spunta behaved as highly susceptible and moderately susceptible on 5 and 18 out of 30 inoculation treatment x temperature combinations tested, respectively, while it has been ranked as tolerant on the remaining ones. This cultivar was found to be highly susceptible to C1-2, C2-4, C2-6 and C3-3 treatments at 20°C and to C3-1 and C4 when stored at 30°C. It seemed to be tolerant C1-1, C1-3, C1-4, C2-2, C2-3, C2-5 and C3-2 infections when stored at 20°C. Cv. Oceania has been classified as highly susceptible and moderately susceptible on 6 and 17 out of 30 inoculation treatment x temperature combinations tested, respectively, and tolerant to the rest of combinations. This cultivar was ranked as highly susceptible to C2-1 at both temperatures tested and to C1-2, C2-4, C2-6 and C3-3 inoculation treatments when stored at 20°C. Cv. Nicola behaved as highly susceptible to C1-2 and C2-4 when stored at 20°C and to C2-1 and C2-3 when maintained at 30°C whereas it has been ranked as tolerant to C1-1, C1-3 and C2-6 infections when incubated at 30°C. Cv. Mondial was found to be highly susceptible to C1-4 and C3-3 based treatments at 30 and 20°C, respectively and moderately susceptible at all the remaining inoculation treatment x temperature combinations tested. However, cv. Atlas tolerated C1-4 and C3-2 infections at 20°C and C2-5 treatment at 30°C and behaved as moderately susceptible to all the other treatments regardless of the temperature used.

Isolation frequency of *Fusarium* species depending on inoculation treatments tested

As shown in Table 3, the isolation frequency of *Fusarium* species was variable depending on single or mixed inoculum used for tuber infection and according to cultivar × temperature combination considered. In fact, for tubers challenged with single inoculum, the isolation frequency of the different *Fusarium* species ranged between 62.5 and 100% depending on cultivars and temperatures of storage tested. It should be mentioned that some tuber tissues pieces plated on PDA are occasionally colonized by some other fungal species such as *Aspergillus* spp., *Trichoderma* spp. and *Penicilllium* spp. that inhibit *Fusarium* development and its isolation at 100%.

Table 3 also revealed that for all binary mixtures tested, the relative isolation frequency of each *Fusarium* species varied from 12.5 to 75% depending on cultivars and temperatures of incubation used. It should be mentioned that *F. sambucinum* and *F. solani* have been isolated at 25 to 75% from all binary mixtures including them whereas this isolation frequency ranged from 12.5 to 50% for *F. graminearum* and from 12.5 to 62.5% for *F. oxysporum*.

Concerning the ternary mixtures, the isolation frequency of the different pathogens included in the *Fusarium* complex tested varied depending on inoculation treatments, cultivars and temperatures used. In fact, data shown in Table 3 indicated that the isolation frequency varied from 25 to 75% for *F. solani*, from 12.5 to 50% for *F. sambucinum*, from 12.5 to 25% for *F. graminearum* and *F. oxysporum* and this for all cultivar × temperature combinations and all mixed inoculum comprising them.

For the quaternary mixture, Table 3 also showed that all *Fusarium* species were successfully re-isolated with a predominance of two major species (*F. sambucinum* and *F. solani* or *F. sambucinum* and *F. oxysporum*). The isolation frequency of each *Fusarium* species varied from 12.5 to 50% depending on cultivars and temperatures tested. Indeed, the isolation frequency varied from 12.5 to 50% for *F. solani*, from 25 to 50% for *F. sambucinum*, from 12.5 to 25% for *F. oxysporum* whereas *F. graminearum* was scarcely recovered (12.5%).

It can be also concluded that *F. solani*, *F. sambucinum* and at a lesser degree *F. oxysporum* were the most frequently re-isolated agents from the majority of the ternary and quaternary inoculation mixtures tested whereas *F. graminearum* isolation exceeded rarely 12.5%. This may reflect their competitive potential in mixture and consequently, their relative involvement in disease development and dry rot signs scored.

Discussion

The evaluation of local potato cultivars for resistance to FDR at different temperatures was previously investigated in Tunisia [9,10]. However, the present work is the first detailed study, in the world, which reports a susceptibility ranking of local potato cultivars to *Fusarium* species using single and mixed inoculation treatments and different temperatures. Since *Fusarium* species frequently co-exist in the same plant or tuber, comparative susceptibility of potato cultivars to FDR assessed based on artificial inoculation with different mixtures of four species seems compulsory. Our findings highlighted the competitive potential in mixture of *F. sambucinum, F. solani* and *F. oxysporum.* Moreover, our data are in agreement with previous data demonstrating

Treatments /	C1-1		C1-2		C1-3		C1-4		C2-1		C2-2		C2-3		C2-4		C2-5		C2-6		C3-1		C3-2		C3-3		C3-4		C4	
Temperatures	20°C	30°C	20°C	30°C	20°C	30°C	20°C	30°C	20°C	30°C	20°C	30°C	20°C	30°C	20°C	30°C	20°C	30°C	20°C	30°C	20°C	30°C	20°C	30°C	20°C	30°C	20°C	30°C	20°C	30°C
Spunta	T	MS	HS	MS	T	MS	T	MS	MS	MS	T	MS	T	MS	HS	MS	T	MS	MS	MS	MS	HS	T	MS	HS	MS	MS	MS	MS	HS
Oceania	T	T	HS	MS	MS	T	T	MS	HS	HS	T	MS	MS	MS	HS	MS	T	MS	HS	MS	MS	MS	T	MS	HS	MS	MS	MS	MS	MS
Mondial	MS	MS	MS	MS	MS	MS	MS	HS	MS	MS	MS	MS	MS	MS	MS	MS	MS	MS	MS	MS	MS	MS	MS	MS	HS	MS	MS	MS	MS	MS
Atlas	MS	MS	MS	MS	MS	MS	T	MS	MS	MS	MS	MS	MS	MS	MS	MS	MS	T	MS	MS	MS	MS	T	MS	MS	MS	MS	MS	MS	MS
Nicola	MS	T	HS	MS	MS	T	MS	MS	MS	HS	MS	MS	MS	HS	HS	MS	MS	MS	MS	T	MS	MS	MS	MS	MS	MS	MS	MS	MS	MS

Table 2: Differential response of five potato cultivars to single and combined inoculation treatments based on lesion diameter of FDR noted at two temperatures of storage HS:Highly susceptible: Lesion diameter ≥ 15mm; MS: Moderately susceptible: 11< mm Lesion diameter < 15mm; T: Tolerant: Lesion diameter ≤ 11 mm *F. sol: F. solani; F. sam: F. sambucinum:F. gra: F. graminearum; F. oxy: F. oxysporum.* C1-1: *F. sol; C1-2: F. sam; C1-3: F. gra; C1-4: F. oxy; C2-1: F. sam+F. sol; C2-2: F. sol+F. oxy; C2-3: F. sol+F. gra; C2-4: F. sam+F. oxy* C2-5: *F. oxy+F. gra*; C2-6: *F. sam+F. gra*; C3-1: *F. sam+F. sol+F. oxy*; C3-2: *F. sol+F. oxy+F. gra*; C3-3: *F. sam+F. oxy+F. gra*; C3-4: *F. sam+F. sol+F. gra*; C4: *F. sam+F. sol+F. oxy+F. gra.*

		Spunta				Oceania				Mondial				Atlas				Nicola			
		F.sol	*F. sam*	*F. gra*	*F. oxy*	*F.sol*	*F. sam*	*F. gra*	*F. oxy*	*F.sol*	*F. sam*	*F. gra*	*F. oxy*	*F.sol*	*F. sam*	*F. gra*	*F. oxy*	*F.sol*	*F. sam*	*F. gra*	*F. oxy*
C1-1: *F. sol*	20°C	87.5				87.5				87.5				87.5				87.5			
	30°C	100				100				100				87.5				100			
C1-2: *F. sam*	20°C		100				100				87,5				87,5				100		
	30°C		100				87.5				87.5				87.5				100		
C1-3: *F. oxy*	20°C				62.5				75				75				62.5				75
	30°C				62.5				62.5				62.5				62.5				62.5
C1-4: *F. gra*	20°C			75				75				87.5				87.5				100	
	30°C			87.5				75				87.5				87.5				100	
C2-1: *F. sam+F. sol*	20°C	50	50			50	50			25	75			25	75			25	75		
	30°C	75	25			75	25			50	50			50	50			75	25		
C2-2: *F. oxy+F. sol*	20°C	37.5			25	25			25	37.5			37.5	37.5			62.5	25			25
	30°C	50			25	37.5			25	37.5			37.5	50			25	50			25
C2-3: *F. sol+F. gra*	20°C	37.5		12.5		75				37.5		37.5		50		50		75		12.5	
	30°C	50		12.5		75		12.5		37.5		25		50		25		75		12.5	
C2-4: *F. sam+F. oxy*	20°C		62.5		37.5		50		50		62.5		25		50		25		62.5		37.5
	30°C		50		37.5		37.5		50		50		37.5		37.5		37.5		50		37.5
C2-5: *F. gra+F. oxy*	20°C			37.5	25			12.5	37.5			25	37.5			37.5	12.5			37.5	25
	30°C			25	25			12.5	25			12.5	25			25	25			25	25
C2-6: *F. sam+F. gra*	20°C		62.5	25			50	25			50	25			50	25			37.5	25	
	30°C		50	12.5			50	12.5			37.5	12.5			50	25			37.5	25	
C3-1: *F. sam+F. sol+F. oxy*	20°C	37.5	25		12.5	25	37.5		12.5	25	25		25	25	25		12.5	25	25		12.5
	30°C	37.5	37.5		12.5	37.5	25		12.5	37.5	25		25	37.5	37.5		12.5	37.5	25		12.5
C3-2: *F. sol+F. gra+F. oxy*	20°C	37.5		12.5	25	75		0	0	25		12.5	25	50		12.5	12.5	37.5		12.5	25
	30°C	37.5		12.5	25	50		12.5	25	37,5		12.5	12.5	37.5		25	25	37.5		12.5	12,5
C3-3: *F. sam+F. gra+F. oxy*	20°C		50	12.5	25		37.5	25	25		50	12.5	25		37.5	0	12.5		37,5	12.5	25
	30°C		12.5	12.5	25		25	12.5	25		25	12.5	25		37.5	25	12.5		25	12.5	25
C3-4: *F. sam+F. sol+F. gra*	20°C	37.5	37.5	12.5		50	37.5	12.5		50	37.5	12.5		50	25	12.5		37.5	37.5	0	
	30°C	25	25	12.5		37.5	25	12.5		50	25	12.5		37.5	25	25		37.5	25	12.5	
C4: *F. sam+F. sol+F. oxy+F. gra*	20°C	12.5	50	12.5	25	0	50	12.5	25	25	50	0	12.5	37.5	37.5	0	25	12.5	37.5	12.5	25
	30°C	37.5	25	0	25	37.5	25	0	12.5	50	25	12.5	25	50	25	12.5	12.5	37.5	37.5	12.5	12.5

Table 3: Percentage of re-isolates of each component from individual, binary, ternary and quaternary mixtures of *F. sambucinum (F. sam)* or *F. graminearum (F. gra)* or *F. solani (F. sol)* or *F. oxysporum (F. oxy)* re-isolates from five cultivars at two temperatures. The percentage refer to eight fragments with symptoms placed on PDA plates, 2 replicates for each treatment.

that *F sambucinum* was implicated in the increased incidence of potato dry rot in USA [6], Great Britain [3]; Tunisia [8-10], Turkey [26] and Iran [27].

In this paper, inoculation with mixtures of species exhibited variation in *Fusarium* aggressiveness between the five cultivars used which significantly differed in terms of their resistance to dry rot development induced by single or combined species. Novelty of our findings lies in the fact that cultivars have a stronger resistance to *Fusarium* using single inoculum for tuber inoculation. Indeed based on cultivar differences to various *Fusarium* species, we would expect that resistance to *Fusarium* mixture was less than that observed following inoculation with each species alone, and this was the case. Mixed infections with *Fusarium* spp. resulted in a qualitative change of some cultivars from resistant to susceptible. This was in accordance with Bukhart et al. [28] demonstrating that although there is genetic variation for resistance to FDR, additive genetic variance is lacking or minimal, therefore genetic gains will be minimal or nonexistent with mixed inoculations. The only exception is *F. sambucinum* which even alone it decreases the resistance of tested cultivars. This supports the findings of Mejoub-Trabelsi et al. [9] indicating the higher aggressiveness of this pathogen. Our study suggests that double or multiple resistances did not exist in the tested cultivars. This rich activity makes the ranking of cultivar's susceptibility more difficult and instable.

This study demonstrated that in many cases, some combined inocula were more aggressive than single ones suggesting occurrence of synergistic interaction between *Fusarium* species. This interspecific relation, expressed by an additional effect on the severity parameters, can explain that FDR severity has been positively affected by the combinations of *Fusarium* spp. Overall, mixtures of *Fusarium* showed higher aggressiveness than single inoculation treatments. The most aggressive mixture containing *F sambucinum* was not surprising. For instance, *F. sambucinum* was always the most aggressive comparing with the other species when tested alone. In this study, its predominance with *F. solani* and at a lesser degree *F. oxysporum* can be explained by its ability to consume nutrients before the other species (competitive exploitation), at least at these temperatures. Accordingly, Waalwijk et al. [29] suggested that *F. culmorum* has only competitive advantage in mixtures with *F. graminearum* in years characterized by cooler temperatures.

Current findings suggest that *F. solani* is the first to benefit from colonization by *F. sambucinum*, followed by *F. oxysporum* f. sp. *tuberosi*. Xu et al. [30] qualified this phenomenon by synergy, which occurs when host infection by one species leads to the host becoming more vulnerable to infection by another species. Our investigation makes clear that many mixed inoculum increased FDR severity. However, synergism cannot occur in all combinations of *Fusarium* species because some binary and ternary mixtures as is the case of C2-2, C2-3, C2-5, and C3-2 infections led to least disease severity records. Therefore, the additive effect does not characterize the interactions between *Fusarium* species in all the cases. Indeed, with the increase of the number of concomitant species, the synergistic interactions decreased while the antagonistic interaction increased [14]. This inconsistency might be due to variability in competitiveness [31] and toxin-producing capability [32] between and within pathogen species. In the same way, Xu et al. [14] demonstrated that, when wheat inoculated with single or combinations of *Fusarium* species under various combinations of temperature and duration of wetness, the interspecific competition led to large reductions in fungal biomass as compared to single isolate inoculations whereas mycotoxin production increased dramatically in the co-inoculations by as much as 1000 times.

The current study revealed that, among the 11 combined inoculum tested, some mixtures containing *F sambucinum* did not exhibit any increase in the aggressiveness level . Thus, the most likely reason for the observed positive association among species is indirect interaction facilitated by general weather conditions conducive for disease development. Indeed, synergistic interaction occurred with some *Fusarium* spp. more than others. Between *F sambucinum* and *F graminearum*, there were evidence of competition. This may reflect the fact that *F. graminearum* and *F. sambucinum* are more closely related than *F. sambucinum* and *F. solani* and mainly in terms of temperature and moisture requirements for infection. In previous studies realized in Tunisia, Daami-Remadi et al. [17] showed that at less than 25°C, *F. sambucinum* and *F. graminearum* were the most aggressive, while *F. solani* was the most aggressive at temperature equal or superior than 30°C. The competitive advantage of *F. sambucinum* may be related to a better growth and spore germination rates over a wide range of temperature and potato cultivars. Indeed, the occurrence of *Fusarium* species is known to be largely influenced by the climatic conditions such as temperature and relative humidity [33]. The survey results strongly indicate that *F. sambucinum* make a tuber more suitable for *F. solani, F. oxysporum* f. sp. *tuberosi* and with less importance for *F. graminearum*. Similarly, it was reported by Picot et al. [34] that *F. verticilloides* had competitive advantage over *F. graminearum*, when inoculated simultaneously onto maize ears. If we focused in the fact that *F. sambucinum* and *F. graminearum* were always the most aggressive compared to *F. solani* and *F. oxysporum* f. sp. *tuberosi*, as demonstrated in previous work of Mejdoub- Trabelsi et al. [9], the interference competition could be a true explanation.

Since cultivar's resistance to FDR has been positively correlated with single inoculation, the mixture of species, make selection of cultivars with combined resistance extremely difficult. In the accordance with previous findings, several other authors have reported difficulty in combining high resistance to two *Fusarium* species in cultivars [35]. To determine the contribution of each pathogen to the visual disease severity at each mixture, the re-isolation frequency showed that *F. sambucinum, F. solani* and at lesser degree *F. oxysporum* were the most frequently re-isolated. *F. sambucinum* dominated over its mixture component except when added to *F. solani*. In fact, the percentage of re-isolation of the last pathogen was higher (75%) than *F. sambucinum* mainly at 30°C from tubers belonging to cvs. Spunta, Nicola and Oceania. Growing conditions could have favored the isolation of *F. solani*. Accordingly, Goktepe et al. [36] demonstrated that nine isolates of *F. solani* responded similarly to temperature variables with an optimal growth at 30°C. The large interaction effect of cultivar and *Fusarium* species on the degree of rotting indicated that cultivar resistance to one *Fusarium* species does not confer resistance to all *Fusarium* species. None of cultivars tested in this study was completely tolerant to all inoculation treatments tested. Only cvs. Spunta and Oceania showed lesser susceptibility to at the most four mixtures (C2-2, C2-3, C2-5 and C3-2 for cv. Spunta and C2-2, C2-5 and C3-2 for cv. Oceania). Since none of these mixtures contained *F. sambucinum*, our results confirmed the higher aggressiveness of this species. This was in one part, in accordance with what was previously demonstrated in the assessment of local cultivars using single species [9]; none of the cultivars tested was completely resistant to all *Fusarium* species and only some of them showed lesser susceptibility to at the most one species. This is the case of cvs. Spunta, and Nicola, the most cultivated in Tunisia, which tolerated at least one species of *Fusarium*: *F. oxysporum* for the first cultivar and *F. solani* for the second. While cv. Mondial, which tolerated one species in a precedent investigation, presented in this study moderately susceptibility to all mixtures tested suggesting that this cultivar cannot be used for *Fusarium* management.

Several factors may be involved in the observed positive interactions. This interspecific relation may be outcome of complex interactions with the other factors such as varietal susceptibility, inoculum concentrations or particularly climatic conditions. In the same way, Von Der Ohe and Miedaner [32] found that aggressiveness among species and mixtures was significantly different. These differences depended mainly on the year and not on the level of host resistance. Consequently, our results emphasize the high influence of temperature on the mixture performance; otherwise, the ranking of mixtures may change with storage conditions. Thus, we can suggest that the profile of FDR disease complex, depending on cultivars tested may also be significantly affected by temperatures. Indeed, Theron and Holz [37] signaled that the different cultivars did not react uniformly to *Fusarium* spp. at different temperatures and they suggested that when evaluating cultivars for disease resistance or for effectiveness of some disease control measures, tests should be performed at standardized temperature. In the same way, Saremi and Burgess [38] reported the effect of temperature on the community structure of *Fusarium*. They found that communities of *Fusarium* species were significantly different at different temperatures

It can be concluded from the current study that disease development and severity is expected to be more augmented when more than one *Fusarium* species were present. What remains unclear is whether this is the only factor involved for detection of resistance to FDR or whether there are others. It was thought that the mycotoxins produced by two species may be a competitive factor and that the production of secondary fungal metabolites is ecologically significant and confers increased fitness to the producing organism. Since several surveys suggested that *F. solani*, *F. graminearum* and *F. sambucinum* produced one or more trichothecenes, such as deoxynivalenol in North-Central United States [39], further studies are necessary in order to fully understand the diffusion of trichothecenes into potato tubers, to compare difference in trichotechenes' accumulation between resistant and susceptible cultivars and to determine the influence of temperature on the accumulation of these mycotoxins in potato tubers.

Acknowledgments

This work was done as part of research work carried out in the research unit UR13AGR09 titled Integrated Horticultural Production in the Tunisian Centre *East* and funded by the Ministry of Higher Education and Scientific Research of Tunisia.

References

1. Wharton P, Kirk W, Berry D, Tumbalam P (2007) Seed treatment application timing options for control of Fusarium decay and sprout rot of cut seed pieces. Am J Potato Res 84: 237-244.
2. Du M, Ren X, Sun Q, Wang Y, Zhang R (2012) Characterization of *Fusarium* spp. causing potato dry rot in China and susceptibility evaluation of chinese potato germplasm to the pathogen. Potato Research 55: 175-184.
3. Peters JC, Lees AK, Cullen DW, Sullivan L, Stroud GP, et al. (2008) Characterization of *Fusarium* spp. responsible for causing dry rot of potato in Great Britain. Plant Pathol 57: 262-271.
4. Stevenson WR, Loria R, Franc GD, Weingartner DP (2001) Compendium of Potato Diseases, 2nd ed. The American Phytopathological Society, St. Paul.
5. Daami-Remadi M, El Mahjoub M (1996) Potato *Fusarium* species in Tunisia - II: Behaviour of potato cultivars to some Fusarium varieties. Annales de l'TNRAT 69: 113-130.
6. Secor GA, Salas B (2001) Fusarium dry rot and Fusarium wilt. In: 2nd Compendium of potato diseases. American Phytopathological Society Press St. Paul MN.
7. Cullen DW, Toth IK, Pitkin Y, Boonham N, Walsh K, et al. (2005) Use of quantitative molecular diagnostic assays to investigate *fusarium* dry rot in potato stocks and soil. Phytopathology 95: 1462-1471.
8. Chérif M, Raboudi A, Souissi S, Hajlaoui M (2000) Séléction de Trichoderma antagonistes vis-à-vis de l'agent de la pourriture des tubercules de pomme de terre Fusarium roseum var. sambucinum. Revue de l'INAT 15: 115-130.
9. Mejdoub-Trabelsi B, Jabnoun-Khiareddine H, Daami-Remadi M (2012) Effects of *Fusarium* species and temperature of storage on the susceptibility ranking of potato cultivars to tuber dry rot. Pest Technol 6: 41-46.
10. Daami-Remadi M, Jabnoun-Khiareddine H, Ayed F, El Mahjoub M (2006b) Effect of temperatures on aggressivity of Tunisian *Fusarium* species causing potato (*Solanum tuberosum L.*) tuber dry rot. J Agron 5: 350-355.
11. Kennedy P (2010) Ectomycorrhizal fungi and interspecific competition: species interactions, community structure, coexistence mechanisms, and future research directions. New Phytol 187: 895-910.
12. Ondrej M, Talov Rd, Trojan R (2008) Evaluation of virulence of *Fusarium solani* isolates on pea. Plant Protect Sci 44: 9-18.
13. Dorn B, Forrer HR, Schuürch S, Vogelgsang S (2009) *Fusarium* species complex on maize in Switherland: occurrence, prevalence impact and mycotoxins in commercial hybrids under natural infection. Eur J Plant Pathol. 125: 51-56.
14. Xu X, Nicholson P, Ritieni A (2007) Effects of fungal interactions among *Fusarium* head blight pathogens on disease development and mycotoxin accumulation. Int J Food Microbiol 119: 67-71.
15. Esfahani MN (2005) Susceptibility assessment of potato cultivars to *Fusarium dry rot* species. Potato Res 43: 215-226.
16. Valluru, R, Christ BJ, Haynes KG, Vinyard BT (2006) Inheritance and stability of resistance to *Fusarium* tuber rot in tetraploid potatoes. Am J Potato Res 83: 335-341.
17. Daami-Remadi M, Ayed F, Jabnoun-Khiareddine H, El Mahjoub M (2006a) Comparative susceptibility of some local potato cultivars to four *Fusarium* species causing tuber dry rot in Tunisia. J Plant Sc. 1: 306-314.
18. Ayed F, Daami-Remadi M, Jabnoun-Khiareddine H, El Mahjoub M (2006) Effect of potato cultivars on incidence of *Fusarium oxysporum* f. sp. *tuberosi* and its transmission on progeny tubers. J Agron 5: 400-430.
19. Jabnoun-Khiareddine H, Daami-Remadi M, Hibar K, Ayed F, El Mahjoub M (2006) Pathogenicity of Tunisian isolates of three Verticillium species on tomato and eggplant. Plant Pathol J 5: 199-207.
20. Daami-Remadi M, Zammouri S, El Mahjoub M (2008) Effect of the level of seed tuber infection by Rhizoctonia solani at planting on potato growth and disease severity. Afr J Plant Sci Biotech 2: 34-38.
21. Daami-Remadi M, Sayes S, Horrigue-Raouami N, Hlaoua-Ben Hassine W (2009) Effects of Verticillium dahliae Kleb., *Fusarium oxysporum* Schlecht. f. sp. *tuberosi* Synder, Hansen and *Meloidogyne javanica* (Treub.) Chitwood inoculated individually or in combination on potato growth, wilt severity and nematode development. Afr J Microbiol Res 3: 595-604.
22. Tivoli B, Torres H, French ER (1988) Inventory, frequency and aggressivity of different *Fusarium* species or varieties occurring on potato or in its environment in different agroecological zones of Peru. Potato Res 31: 681-690.
23. Mishra PK, Fox RT, Culham A (2003) Development of a PCR-based assay for rapid and reliable identification of pathogenic Fusaria. FEMS Microbiol Lett 218: 329-332.
24. Leslie JF, Summerell BA (2006) The Fusarium laboratory manual. UK Blackwel Publishing Ltd.
25. Lapwood DH, Read PJ, Spokes J (1984) Methods for assessing the susceptibility of potato tubers of different cultivars to rotting by *Erwinia carotovora* subsp. *atroseptica* and *carotovora*. Plant Pathology 33: 13-20.
26. Gachango E, Kirk W, Wharton PS, Schafer, R (2012) Evaluation and comparison of biocontrol and conventional fungicides for control of postharvest potato tuber diseases. Biological Control 63: 115-120.
27. Mogadem BS, Hosseinzadeh AA (2013) Study of *Fusarium* species causing dry rot of potatoes in Ardabil Province. Intl J Agron Plant Prod 4: 1226-1233.
28. Bukhart CR, Barbara JC, Kathleen GH (2007) Non-additive genetic variance governs resistance to *Fusarium* dry rot in a diploid hybrid potato population. Am J Potato Res 84: 199-204.
29. Waalwijk C, Kastelelin P, de Vries I, Kerenyi Z, Van der lee T, et al. (2003) Major changes in *Fusarium* spp. in wheat in the Netherlands. Eur J Plant Pathol 109: 743-754.
30. Xu XM, Parry DW, Nicholson P, Thomsett MA, Simpson DR (2008) Within-field variability of Fusarium head blight pathogens and their associated mycotoxins. Eur J Plant Pathol 120: 21-34.
31. Miedaner T, Cumagun CJR, Chakraborty S (2008) Population genetics of three important head blight pathogens *Fusarium graminearum, Fusarium pseudograminearum* and F. *culmorum*. J Phytopathol 156: 129-139.
32. Von Der Ohe C, Miedaner T (2011) Competitive aggressiveness in binary mixtures of *Fusarium graminearum* and *F. culmorum* isolates inoculated on spring wheat with highly effective resistance QTL. J Phytopathol 159: 401-410.
33. Xu XM, Monger M, Ritieni A, Nicholson P (2007b) Effects of temperature and duration of wetness during initial infection periods on disease development, fungal biomass and mycotoxins concentrations on wheat inoculated with single or combinations of *Fusarium* species. Plant Pathol 56: 934-956.
34. Picot A, Hourcade-Marcolla D, Barreau C, Pinson-Gadais L, Caron D, et al. (2012) Interactions between *Fusarium verticilloides* and *Fusarium graminearum*

in maize ears and consequences for fungal development and mycotoxin accumulation. Plant Pathol 61: 40-151.

35. Huaman Z, Tivoli B, Lindo L (1989) Screening for resistance to *Fusarium* dry rot in progenies of cultivars of S. *tuberosum* sp. andigena with resistance to *Erwinia chrysanthemi*. Am Pot J 66: 357-364.

36. Goktepe F, Seijo T, Deng Z, Harbaugh BK, Peres NA (2007) Toward breeding for resistance to *Fusarium* tuber rot in Caladium: Inoculation technique and sources of resistance. Hortscience 42: 1135-1139.

37. Theron DJ, Holz G (1989) *Fusarium* species associated with dry and stem-end rot of potatoes in South Africa. Phytophylactica 21: 175-181.

38. Saremi H, Burgess LW (2000) Effect of soil temperature on distribution and population dynamics of *Fusarium* species. Journal of Agriculture Science and Technology 2: 119-125.

39. Delgado JA, Schwarz PB, Gillespie J, Rivera-Varas VV, Secor GA (2010) Trichothecene mycotoxins associated with potato dry rot caused by *Fusarium graminearum*. Phytopathology 100: 290-296.

Variation in Chitosan and Salicylic Acid Efficacy Towards Soil-borne and Air-borne Fungi and their Suppressive Effect of Tomato Wilt Severity

Hayfa Jabnoun-Khiareddine[1]*, Riad SR El-Mohamedy[2], Farid Abdel-Kareem[2], Rania Aydi Ben Abdallah[1,3], Mouna Gueddes-Chahed[1] and Mejda Daami-Remadi[1]

[1]UR13AGR09-Integrated Horticultural Production in The Tunisian Centre-East, Regional Center of Research on Horticulture and Organic Agriculture, University of Sousse, 4042, Chott-Mariem, Tunisia

[2]Plant Pathology Department, National Research Center, Dokki, Giza, Egypt

[3]National Agronomic Institute of Tunisia, 1082 Tunis Mahrajène, University of Carthage, Tunisia

Abstract

Two resistance inducers (RIs), chitosan and salicylic acid (SA), were assessed *in vitro* for their antifungal activity against ten tomato phytopathogenic fungi i.e. *Fusarium oxysporum* f. sp. *lycopersici*, *F. oxysporum* f. sp. *radicis-lycopersici*, *F. solani*, *Verticillium dahliae*, *Rhizoctonia solani*, *Colletotrichum coccodes*, *Pythium aphanidermatum*, *Sclerotinia sclerotiorum*, *Botrytis cinerea*, and *Alternaria solani*. The impact of these RIs, applied as soil drench, on Verticillium wilt, Fusarium wilt, and Fusarium Crown and Root Rot severity and on growth parameters of tomato cv. Rio Grande plants were also investigated. Chitosan (0.5-4 mg/ml) and SA (1-25 mM) inhibited mycelial growth of all pathogens in Potato Dextrose Agar (PDA) medium in a concentration-dependent manner, with the greatest inhibition achieved using the highest chitosan and SA concentrations. Inter specific variations in sensitivity to chitosan and SA were detected. *P. aphanidermatum* and *S. Sclerotiorum* were the most sensitive to both RIs. Single treatments with chitosan (4 mg/ml) and SA (10 mM) resulted in varied degree of protection against wilt diseases. Chitosan-and SA-based treatments resulted in 42.1-73.68, 60.86-78.26 and 45-50% reductions in wilt severity, as compared to VD-, FOL- and FORL-inoculated and untreated controls, respectively. All growth parameters noted were enhanced using RIs compared to pathogen-inoculated controls. In fact, SA-based treatment had significantly increased plant height, root and aerial part fresh weights by 17.94, 52.17 and 33.33%, by 23.01, 55.40 and 29.72%, and by 17.72, 50 and 46.84%, respectively, while compared to VD-, FOL- and FORL-inoculated and untreated plants. Chitosan-treated plants showed increment in their height, root and aerial part fresh weights by 13.81, 62.16 and 38.97%, respectively, compared to FORL-inoculated and untreated control. Results from this investigation showed that SA and chitosan may be used as potential inducers of systemic acquired resistance for successfully controlling fungal tomato diseases in Tunisia.

Keywords: Chitosan; Salicylic acid; Antifungal activity; Tomato pathogens; Wilt severity

Introduction

Tomato (*Solanum lycopersicum* L., formerly, *Lycopersicon esculentum* Mill.), is one of the most important vegetable crop worldwide after potato, regarding areas under cultivation and ranked second after eggplant in terms of production [1,2]. In Tunisia, tomato is regarded as one of the most important crops in terms of both value and cropped areas. In fact, average areas of 29 000 ha/year are devoted to the growing of open field and protected tomatoes covering about 17% of areas cultivated with vegetables, with an average annual production of about 1.2 million tons representing about 39% of national production of vegetable crops [3]. Tomato is grown both on small- and large-scales commercial crop. However, diseases are ones of the main problems of tomato cropping in Tunisia and all over the world leading to considerable production decrease [4-6].

In Tunisia, several fungal diseases are known to affect tomato during all stages of plant development resulting in severe damage in roots and/or crown, stems, leaves and fruits. *Fusarium oxysporum* f. sp. *lycopersici* (FOL), *F. oxysporum* f. sp. *radicis-lycopersici* (FORL) and *Verticillium dahliae* (VD) are highly destructive soil-borne pathogens causing wilt and root rot diseases in both greenhouse and field-grown tomatoes in warm producing areas. In fact, yield losses up to 50 and 90% due to Verticillium wilt and Fusarium Crown and Root Rot disease, respectively, have been reported on severely infected tomato cultivars [5,7]. Furthermore, *F. solani, R. solani, C. coccodes, P. aphanidermatum,* and *S. sclerotiorum*, causing root rots and stem decay, are among the most well-known soil-borne pathogenic fungi which are reportedly responsible for severe growth reduction and yield losses in Tunisia. In addition, grey mold caused by *B. cinerea* and early blight incited by *A. solani* are among the most important diseases of tomato aerial parts leading to lower quantity and quality of fruit yields [8,9].

Many strategies have been developed for controlling tomato fungal diseases over years such as cultural practices, chemical treatments, use of resistant cultivars, grafting, soil solarisation, biological control, etc., but serious losses still occur largely because the effectiveness of these approaches is variable and often short lived [5,7]. In fact, the exceptionally long survival of resting structures of soilborne pathogens (chlamydospores, oospores, sclerotia, microsclerotia, etc) along with

***Corresponding author:** Hayfa Jabnoun-Khiareddine, UR13AGR09-Integrated Horticultural Production in the Tunisian Centre-East, Regional Center of Research on Horticulture and Organic Agriculture, University of Sousse, 4042, Chott-Mariem, Tunisia, E-mail: jkhayfa@yahoo.fr

the continuous cropping of susceptible tomato cultivars and the emergence of new races and pathotypes of pathogens make difficult the control of these diseases and lead to an intense use of agrochemicals. In fact, tomatoes are one of the highest pesticide-sprayed vegetables worldwide. Hence, the excessive use of synthetic fungicides has resulted in an increased risk of fungicide resistance, enhanced pathogen resurgence and development of resistance/cross-resistance, toxicological implications to human and animal health, and increased environment pollution [10,11].

Therefore, recent efforts have focused on developing environmentally safe, long lasting and effective alternative methods for the management of these tomato diseases, such as the use of resistance inducers (RIs). In fact, in addition to basal resistance, plants are capable of developing an induced resistance that is a physiological state of acquired defensive capacity elicited by specific environmental stimuli, by which plants' innate defences, are potentiated against subsequent biotic challenge [10-12]. Chemical inducers of plant resistance possess quite different modes of action as compared to synthetic biocides as they have no direct toxicity to pathogens, plants and animals; no negative effects on plant growth, development and yield; broad spectrum of defence; low loading amount; long lasting protection; low economical cost for farmers and good profit for producers [13-15]. A large array of natural and chemical RIs have been reported to induce resistance in many plants including tomato such as benzothiadiazole, beta-aminobutyric acid (BABA), 2,6-dichloroisonicotinic acid, salicylic acid (SA), organic and inorganic salts, chitosan and chitin, etc [13,14,16-19].

Among natural elicitor compounds, chitosan offers a great potential as a biodegradable substance that has both anti-microbial and eliciting activities [20]. Its fungicidal potential has been reported against various species of fungi and oomycetes involved in many pre- and post-harvest diseases of horticultural commodities [21,22].

Among the most commonly tested chemical elicitors, salicylic acid has been shown to play an important role in expression of both local resistance, controlled by major genes, and systemic induced resistance developed after an initial pathogen attack [23]. In fact, SA has been used successfully to control several plant diseases such as Fusarium wilt and crown and root rot of tomato [24-26], tomato root rots [27], root rot/wilt of sesame [28], Verticillium wilt of eggplant [29], Fusarium wilt of chickpea and asparagus [30,31].

Therefore, the objectives of the present work were to (i) evaluate the *in vitro* antifungal activity of chitosan and SA against ten tomato pathogens causing wilts, root rots, stem decay and fruit rots, (ii) to assess their suppressive effects of Verticillium and Fusarium wilts and (iii) to elucidate their impacts on tomato growth, under growth chamber conditions.

Materials and Methods

Plant material

Tomato seeds (cv. Rio Grande) were superficially disinfected by immersion in absolute ethanol for 2 min, followed by extensive rinsing in sterile distilled water. Seeds were sown in alveolus plates filled with previously sterilised peat. Seedlings were grown in a growth chamber at 24-26°C with 12-h photoperiod and 70% humidity. They were watered daily and fertilized once a week with a standard nutrient solution according to [32]. Experiments were performed using 30 days-old tomato plants.

Fungal species

The fungal pathogens used in this study were: FOL, FORL, *F. solani*, VD, *C. coccodes*, *R. solani*, *P. aphanidermatum*, *S. sclerotiorum*, *B. cinerea*, and *A. solani*. These pathogens were isolated from roots, crowns, stems, leaves and fruits of diseased tomato plants and were held in the fungal culture collection of the laboratory of Phytopathology in the Regional Center of Horticulture and Organic Agriculture of Chott Mariem- Tunisia. Cultures of each fungus were maintained on Potato Dextrose Agar (PDA; Difco, Detroit, MI, USA) and were stored in PDA slants at 5°C for further use.

Liquid cultures used for substrate inoculation were prepared on Potato Dextrose Broth (PDB) and incubated at 25°C under continuous shaking at 150 rpm during 4 to 5 days. Concentration of the conidial suspension used was adjusted to 10^7 conidia/ml using a Malassez haemocytometer.

Resistance inducers tested

The RIs tested in this study are listed in Table 1.

To prepare chitosan (Sigma–Aldrich, St. Louis, MO, USA) stock solutions (10 mg/ml), 2 g of high molecular weight chitosan (viscosity=800-2000 cps and >75% deacetylation) were dissolved in 100 ml of distilled water with 2 ml of acetic acid (stirred for 24 h), and the volume was taken up to 200 ml with distilled water. The pH was adjusted to 5.6 by the addition of sodium hydroxide 1.0 N [33]. Chitosan solution was autoclaved for 15 min. The corresponding aliquots were taken to obtain different chitosan concentrations (0.5, 1.0, 1.5, 2.0, 2.5, 3.0 and 4 mg/ml).

SA was tested at concentrations of 1, 5, 10 and 25 mM. The corresponding aliquots were taken from a 1 M stock solution in sterile distilled water.

In vitro antifungal activity of the resistance inducers against tomato pathogens

The inhibitory effect of chitosan and salicylic acid on the mycelial growth of tomato pathogens was evaluated on PDA medium supplemented with streptomycin sulfate (300 mg/l).

The desired quantities of the tested inducers were added to autoclaved and molten PDA medium to achieve the targeted concentrations tested. For each compound, a 10 ml aliquot of amended PDA medium was aseptically poured into a Petri plate (9 cm in diameter), with an unamended PDA dish used as a control. Three agar plugs (6 mm in diameter) cut from 7-day-old fungal cultures were plated at 2 cm from the edge of the Petri plate and equidistantly spaced from each other by 3 cm. The plates were sealed with parafilm and incubated in the dark at 25°C for varying times depending on growth rate of each organism and time necessary for reaching the edge of the plate. This was as briefly as 1 day for fast-growing organisms like *P. aphanidermatum* and as long as 7 days for slower growing ones like VD. Colony diameters were measured at two perpendicular points and the mean was determined. The mycelial growth inhibition percentage was calculated according to Tiru et al. formula [34] as follows: I %=[(C2-C1) / C2] × 100 with C2:

Compound	Chemical formula	Molecular weight	Company
Salicylic acid	$C_7H_6O_3$	138.12 (g/mol)	Merck (Darmstadt, Germany)
Chitosan	$C_{12}H_{24}N_2O_9$	Mw=30.7 kDa	Sigma–Aldrich (St. Louis, MO, USA)

Table 1: Resistance inducers tested in this study.

Mean diameter of the control colony and C1: Mean pathogen colony diameter in the presence of the tested compound. The experiment was repeated twice, with three replicates per treatment.

Statistical analyses were performed, for each RI, following a completely randomised factorial design where the pathogens and the inducer concentrations were the two fixed factors. Six replicates were used for each individual treatment and means were separated using Fisher's protected LSD or Duncan's Multiple Range tests (at $P \leq 0.05$). The whole experiment was repeated twice but only the data of one essay is presented in the present study. Statistical analyses were performed using SPSS software version 16.

Effect of the resistance inducers on Verticillium and Fusarium wilts severity under growth chamber conditions

Chitosan (4 g/l) and salicylic acid (10 mM) were tested *in vivo*, as soil drench, for controlling Verticillium wilt, Fusarium wilt and Fusarium crown and root rot.

Healthy 30 days-old tomato seedlings (cv. Rio Grande) were carefully removed from alveolus plates and re-potted in peat contained in 17 cm diameter-pot. Two days after transplanting, plants were watered by 100 ml of each inducer aqueous solution, as close as possible to the root system. This quantity was sufficient to cover tomato roots without excess. Plants were watered only three days post-treatment, in order to allow a maximum root wetting in the RIs aqueous solution. Five days post-treatments, plants were challenged by 100 ml of each fungal conidial suspension close to the root system.

For each individual treatment, five plants were used and the experiment was repeated twice. Five uninoculated and untreated plants and five inoculated and untreated others were used as controls.

All tomato plants were maintained in a growth chamber at 15-30°C during 60 days and regularly watered and fertilized with a standard nutrient solution according to [32].

Assessment of disease severity was performed 60 days post inoculation (DPI) of tomato plants challenged with pathogens (VD, FOL and FORL) using a disease index recorded on each plant according to wilt and leaf yellowing intensity and a mean value was calculated and considered as disease severity score. For evaluation of wilt symptom, each plant was observed and rated as follows: 0=no symptoms, 1=1-25% plant wilting and yellowing; 2=26-50% wilting and yellowing; 3=51-75% wilting and yellowing; 4=76-99% and 5=dead plant. Furthermore, Plant height and root and aerial part fresh weights were also noted for all tomato plants.

For all parameters measured (disease severity, plant height, root and aerial part fresh weight), statistical analyses were performed for each pathogen used (VD, FOL and FORL) following a completely randomised design where treatments (RIs, inoculated and untreated control and the uninoculated and untreated control) represented the only fixed factor. Five replicates were used per individual treatment and means were separated using Duncan's Multiple Range test (at $P \leq 0.05$). Statistical analyses were performed using SPSS software version 16.

Results

In vitro evaluation of the antifungal activity of resistance inducers against tomato pathogens

The inhibitory effect of two RIs, chitosan and salicylic acid, tested at different concentrations on the mycelial growth of various tomato pathogenic fungi was assessed *in vitro*.

For each inducer, ANOVA analysis revealed a significant (at $P \leq 0.01$) variation in the average fungal colony diameter depending on the tomato pathogens tested and RI concentrations used.

Data shown in Table 2 indicated that chitosan inhibited mycelial growth of all pathogens in solid PDA medium, in a concentration-dependent manner. It is clear that, linear growth of tested fungi decreased significantly with the increase of chitosan concentrations. In fact, as compared to the untreated control, chitosan had inhibited fungal mycelial growth (combined data of all fungi tested) at variable rates depending on concentrations uses where the inhibition ranged between 47.71 and 100% using chitosan at 4 mg/ml compared to 0-24.19%, 2.15-36.97%, 10.92-79.21%, 11.27-83.87%, 21.03-100%, and 36.27-100% achieved using this RI at 0.5, 1, 1.5, 2, 2.5 and 3 mg/ml, respectively.

	Chitosan concentration used (g/l)								
Fungi tested	0	0.5	1	1.5	2	2.5	3	4	Average per fungus tested[a*]
FOL	3.97	2.98	2.50	2.20	1.93	1.85	1.78	1.40	2.33 e
FORL	4.05	3.75	3.22	2.07	1.92	1.65	1.55	1.52	2.46 d
F. solani	3.63	3.55	3.22	2.55	2.45	2.42	2.02	1.90	2.72 c
VD	3.32	3.13	2.62	2.00	1.83	1.60	1.42	1.20	2.14 f
C. coccodes	3.57	3.47	3.00	2.98	2.97	2.82	2.02	1.50	2.79 c
R. solani	4.73	4.48	4.38	4.22	4.20	3.62	3.02	2.15	3.85 a
S. sclerotiorum	4.65	4.65	4.55	0.97	0.75	0.35	0	0	1.99 f
P. aphanidermatum	4.48	4.42	3.50	2.30	1.30	0	0	0	2.00 f
B. cinerea	4.72	3.83	3.70	3.18	2.87	2.23	2.10	1.05	2.96 b
A. solani	4.03	3.68	3.15	1.68	1.30	1.25	1.12	0.98	2.15 f
Average per Chitosan concentration[b*]	4.12 a	3.80 b	3.38 c	2.42 d	2.15 e	1.78 f	1.50 g	1.17 h	

LSD (Fungal pathogens x Chitosan concentrations) = 0.388 cm at $P \leq 0.05$.
[a] Mean mycelial growth per fungal pathogen for all chitosan concentrations combined.
[b] Mean mycelial growth per chitosan concentration tested for all fungal pathogens combined.
[*] For fungal pathogens and chitosan concentrations tested, values (means) followed by the same letter are not significantly different according to Duncan's Multiplue Range test at $P \leq 0.05$.
Pathogens were cultured on PDA medium and incubated at 25°C for 1 d (*P. aphanidermatum*), 2 d (*R. solani* and *S. sclerotiorum*), 3 d (*B. cinerea*, FOL, FORL and *F. solani*), 4 d (*A. solani* and *C. coccodes*) and 7 d (VD).

Table 2: Effect of different chitosan concentrations on the *in vitro* mycelial growth of tomato fungal pathogens cultured on PDA medium at 25°C.

Tomato pathogens showed clear differences when grown in chitosan-amended PDA at different concentrations. In fact, increasing chitosan concentration from 0.5 to 4 mg/ml resulted in greater inhibition of mycelial growth of FOL, FORL and *F. solani* from 24.79 to 64.71%, from 7.41 to 62.55 %, and from 2.29 to 47.71%, respectively. For the two airborne pathogens, *B. cinerea* and *A. solani*, these inhibition rates ranged between 18.73-77.74% and 8.68-75.62 %, respectively. Decreases in mycelial growth of *R. solani*, *C. coccodes* and VD varied from 5.28-54.58%, 2.80-57.94% and 5.53-63.82%, respectively, when treated with chitosan at 0.5 to 4 mg/ml, compared to 0-100% and 1.49-100% noted respectively for *S. sclerotiorum* and *P. aphanidermatum*. Thus, *P. aphanidermatum* and *S. sclerotiorum* were the most sensitive fungal species to chitosan as their growth was totally suppressed at concentrations of 2.5 mg/ml and 3 mg/ml or greater, respectively while when tested at 0.5 mg/ml, chitosan did not affect *S. sclerotiorum* growth but only a slight inhibitory activity was observed at 1 mg/ml. In contrast, *F. solani* was the least sensitive to chitosan where 47.71% decrease in pathogen growth was achieved with the highest chitosan concentration tested (4 mg/ml).

Applied at 2 mg/ml, chitosan reduced by more than 50% the mycelial growth of five fungi out of the ten tested while at 4 mg/ml, inhibition rates ranged between 50 and 65% for *R. solani*, *C. coccodes*, FORL, VD, and FOL and between 75 and 100% for *A. solani*, *B. cinerea*, *P. aphanidermatum* and *S. sclerotiorum*.

Results shown in Table 3 indicated that SA inhibited mycelial growth of all pathogens grown in amended-PDA medium, in a concentration-dependent manner. In fact, as compared to the untreated control, this RI decreased pathogens' growth by 2.41-31.2% at 1 mM compared to 9.69-100% and 35.45-100 %, recorded at 5 and 10 mM, respectively. However, when tested at 25 mM, SA had totally suppressed *in vitro* growth of all tested pathogens.

Moreover, when SA concentration increased from 1 to 10 mM, inhibition of radial growth of FOL, FORL and *F. solani* augmented from 19.03 to 35.45%, from 23 to 53.31% and from 31.27 to 49.48 %, respectively, compared to 2.14-100%, 6.20-100% and 7.11-100% noted on *R. solani*, *C. coccodes* and VD cultures, respectively. When treated with these increasing SA concentrations, mycelial growth of *B. cinerea* and *A. solani* was also reduced by 1.79-48.93% and 20.93-63.57%, respectively, compared to inhibitions of 15-100% and 21.21-100% noted for *S. sclerotiorum* and *P. aphanidermatum*, respectively. In fact, these two last pathogens were the most sensitive fungal species to SA as their radial growth was completely suppressed at 5 mM and greater followed by *R. solani*, *C. coccodes* and VD for which total inhibition was achieved using SA at 10 mM and more.

Applied at 10 mM, SA had suppressed totally the growth of five out of the ten fungi tested (*S. sclerotiorum*, *P. aphanidermatum*, *R. solani*, *C. coccodes* and VD) and reduced by 35.45- 63.57% the remaining pathogens (FOL, FORL and *F. solani*, *B. cinerea*, and *A. solani*).

SA used at 1 mM led to great variation in terms of mycelial growth inhibition ability depending on fungi tested. In fact, the greatest inhibitory effect was recorded against *F. solani* (31.27%) followed by FORL (23%), *P. aphanidermatum* (21.21%), *A. solani* (20. 93%), FOL (19.03%), *S. sclerotiorum* (15.77%), VD (7.11%), *C. coccodes* (6.20%), *R. solani* (2.14%), and *B. cinerea* (1.79%).

Effect of the resistance inducers on wilt severity and plant growth

The *in vivo* effect of chitosan and SA applied as soil drench, 5 days before single inoculation with three soil-borne fungi (VD, FOL and FORL) was evaluated on wilt severity and plant growth in comparison to untreated and inoculated or non-controls.

Wilt severity

All pathogen-inoculated tomato plants showed 60 DPI typical wilt symptoms while uninoculated and untreated plants were symptomless. However, for each pathogen, wilt severity varied significantly (at $P \leq 0.05$) depending on treatments tested (Table 4). In fact, Verticillium wilt severity noted on SA- and chitosan-treated plants was significantly reduced by 42.1 and 73.68%, respectively, compared to VD-inoculated and untreated control ones. Application of chitosan, as soil drench, provided 31.57% greater protection against Verticillium wilt severity than SA-based treatment.

Varied degree of protection of tomato plants against Fusarium wilt was recorded with RIs tested. In fact, SA- and chitosan-based treatments led to 78.26 and 60.86% lower disease severity, as compared to FOL-inoculated and untreated control. Exogenously supplied SA

	Salicylic acid concentration used (mM)					
Fungi tested	0	1	5	10	25	Average per fungus tested[a*]
FOL	4.47	3.62	2.95	2.88	0	2.78 ab
FORL	4.78	3.68	3.13	2.23	0	2.77 ab
F. solani	4.85	3.33	3.05	2.45	0	2.74 ab
VD	3.28	3.05	2.68	0	0	1.80 d
C. coccodes	4.30	4.03	3.88	0	0	2.44 bc
R. solani	4.82	4.72	3.63	0	0	2.63 abc
S. sclerotiorum	4.97	4.18	0	0	0	1.83 d
P. aphanidermatum	4.40	3.47	0	0	0	1.57 d
B. cinerea	4.67	4.58	3.03	2.38	0	2.93 a
A. solani	4.30	3.40	2.40	1.57	0	2.33 c
Average per salicylic acid concentration[b*]	**4.48 a**	**3.81 b**	**2.48 c**	**1.15 d**	**0.00 e**	

LSD (Fungal pathogens x SA concentrations) = 0.631 cm at $P \leq 0.05$.
[a] Mean mycelial growth per fungal pathogen for all salicylic acid concentrations combined.
[b] Mean mycelial growth per salicylic acid concentration for all fungal pathogens combined.
*For fungal pathogens and salicylic acid concentrations tested, values (means) followed by the same letter are not significantly different according to Duncan's Multiple Range test at $P \leq 0.05$.
Pathogens were cultured on PDA medium and incubated at 25°C for 1 d (*P. aphanidermatum*), 2 d (*R. solani* and *S. sclerotiorum*), 3 d (*B. cinerea*, FOL, FORL and *F. solani*), 4 d (*A. solani* and *C. coccodes*) and 7 d (VD).

Table 3: Effect of different salicylic acid concentrations on the *in vitro* mycelial growth of tomato fungal pathogens cultured on PDA medium at 25°C.

Treatment/Pathogen	Disease severity		
	VD	FOL	FORL
Salicylic acid	2.2 b	1 c	2 b
Chitosan	1 c	1.8 b	2.2 b
Inoculated control	3.8 a	4.6 a	4 a
Uninoculated control	0 d	0 d	0 c

*For each pathogen tested, values (means) followed by the same letter are not significantly different according to Duncan's Multiple Range test at $P \leq 0.05$.

Table 4: Effect of chitosan- and salicylic acid-based treatments on disease severity noted on tomato cv. Rio Grande plants inoculated with *Verticillium dahliae* (VD), *Fusarium oxysporum* f. sp. *lycopersici* (FOL) and *F. oxysporum* f. sp. *radicis-lycopersici* (FORL) noted 60 days post-inoculation under growth chamber conditions. Chitosan (4 mg/ml) and SA (10 mM) were applied as soil drench five days before inoculation. Wilt severity was assessed based on 0-5 scale (where 0 = no symptoms and 5 = dead plant)

improved protection against Fusarium wilt by 17.39% compared to chitosan based-treatment.

SA and chitosan *applied* through soil drench had statistically equivalent capacity to reduce Fusarium Crown and Root Rot severity by 50 and 45%, respectively, compared to FORL-inoculated and untreated control.

Plant height

Plant height noted 60 DPI on tomato plants varied significantly (at $P \leq 0.05$) depending on treatments tested. In fact, as given in Table 5, SA treatment had significantly increased plant height by 17.94% compared to VD-inoculated plants and was statistically comparable to the untreated and uninoculated control. Chitosan-treated plants showed significantly comparable height to that noted on both untreated controls (inoculated or non) and also on SA-treated plants.

SA-treated plants showed a slight increase of about 5.12%, even statistically insignificant, of their height in comparison to the untreated and uninoculated control.

SA-treated plants showed significant increase by 23.01 and 20.18% in their height compared to FOL-inoculated and non-controls, respectively. However, chitosan-treated plants showed statistically comparable height than that noted on both untreated controls and on SA-treated plants. In fact, slight height increases of about 18.4 and 15.4%, even statistically insignificant, were noted on chitosan-treated plants in comparison to uninoculated or non-controls.

SA- and chitosan-treated plants showed significantly improved height of about 17.72 and 13.81%, respectively, compared to FORL-inoculated and untreated control and which were significantly similar to the uninoculated control. Plants inoculated with FORL showed significantly decreased height of about 9.95% compared to uninoculated control.

Root fresh weight

Root fresh weight noted 60 DPI on tomato plants varied significantly (at $P \leq 0.05$) depending on treatments tested. Indeed, data given in Table 6 showed that SA treatment had significantly increased root fresh weight by 52.17% compared to VD-inoculated plants and induced a slight increase of root weight of about 23.91%, even statistically insignificant, when compared to the untreated and uninoculated control.

Chitosan-treated plants showed significantly comparable root fresh weight to that noted on both untreated controls (VD-inoculated or non) and also on SA-treated plants. An increase of about 42.10% in root fresh weight, even statistically insignificant, was noted on chitosan-treated plants in comparison to the untreated and uninoculated control.

Treatment/Pathogen	Plant height (cm)		
	VD	FOL	FORL
Salicylic acid	46.8 a	53 a	44 a
Chitosan	42.2 ab	50 ab	42.4 a
Inoculated control	38.4 b	40.8 b	36.2 c
Uninoculated control	44.4 ab	42.3 b	40.2 ab

*For each pathogen tested, values (means) followed by the same letter are not significantly different according to Duncan's Multiple Range test at $P \leq 0.05$.

Table 5: Effect of chitosan- and salicylic acid-based treatments on height of tomato cv. Rio Grande plants inoculated with *Verticillium dahliae* (VD), *Fusarium oxysporum f. sp. lycopersici* (FOL) and *F. oxysporum f. sp. radicis-lycopersici* (FORL) noted 60 days post-inoculation under growth chamber conditions. Chitosan (4 mg/ml) and SA (10 mM) were applied as soil drench five days before inoculation.

Treatment/Pathogen	Root fresh weight (g/plant)		
	VD	FOL	FORL
Salicylic acid	18.4 a	14.8 a	11.2 ab
Chitosan	15.2 ab	11.2 ab	14.8 a
Inoculated control	8.8 b	6.6 b	5.6 c
Uninoculated control	14 ab	11.4 ab	8.8 bc

*For each pathogen tested, values (means) followed by the same letter are not significantly different according to Duncan's Multiple Range test at $P \leq 0.05$.

Table 6: Effect of chitosan- and salicylic acid-based treatments on root fresh weight of tomato cv. Rio Grande plants inoculated with *Verticillium dahliae* (VD), *Fusarium oxysporum* f. sp. *lycopersici* (FOL) and *F. oxysporum* f. sp. *radicis-lycopersici* (FORL) noted 60 days post-inoculation under growth chamber conditions. Chitosan (4 mg/ml) and SA (10 mM) were applied as soil drench five days before inoculation.

Treatment/Pathogen	Aerial part fresh weight (g)		
	VD	FOL	FORL
Salicylic acid	73.8 a	81.4 a	76 a
Chitosan	82 a	68.8 b	66.2 a
Inoculated control	49.2 b	57.2 b	40.4 b
Uninoculated control	72.2 a	66.2 b	60.2 ab

*For each pathogen tested, values (means) followed by the same letter are not significantly different according to Duncan's Multiple Range test at $P \leq 0.05$.

Table 7: Effect chitosan- and salicylic acid-based treatments on aerial part fresh weight of tomato cv. Rio Grande plants inoculated with *Verticillium dahliae* (VD), *Fusarium oxysporum* f. sp. *lycopersici* (FOL) and *F. oxysporum* f. sp. *radicis-lycopersici* (FORL) noted 60 days post-inoculation under growth chamber conditions Chitosan (4 mg/ml) and SA (10 mM) were applied as soil drench five days before inoculation.

In comparison to FOL-inoculated plants, SA treatment had significantly improved root fresh weight by 55.40% and was statistically comparable to the untreated and uninoculated control. Chitosan-treated plants showed significantly comparable root fresh weight to that noted on both untreated controls (FOL-inoculated or non) and also on SA-treated plants.

An increase of about 41.07% in root fresh weight, even statistically insignificant, was noted on chitosan-treated plants in comparison to the untreated and uninoculated control ones.

Chitosan-treated plants showed significantly enhanced root fresh weight by about 62.16 and 40.54%, compared to FORL-inoculated and non-controls, respectively. Root fresh weight noted on SA-treated plants was improved by 50%, which was significantly comparable to that of Chitosan-treated plants and that of the uninoculated control.

Aerial part fresh weight

Aerial part fresh weight, noted 60 DPI on tomato plants cv. Rio Grande, varied significantly (at $P \leq 0.05$) depending on treatments tested. In fact, aerial part fresh weights recorded on SA- and chitosan-treated plants were statistically comparable and were increased by 33.33 and 40% compared to VD-inoculated plants and were also statistically similar to that of the uninoculated plants.

Exogenously applied SA-based treatment resulted in 29.72% increase in plant's aerial part fresh weight in comparison to FOL-inoculated and untreated control plants. Chitosan-treated plants showed aerial part fresh weight statistically comparable to that recorded on both FOL-inoculated or non-control plants; a slight increase of about 16.86% was noted compared to uninoculated control.

Aerial part fresh weight recorded on SA- and chitosan-treated plants was significantly increased by 46.84 and 38.97% compared to FORL-inoculated ones and was also statistically similar to uninoculated plants. These treatments had also improved, even statistically insignificant, the aerial part fresh weight by 20.78 and 9% compared to the uninoculated control plants.

Discussion

In Tunisia, little attention has been paid to the antifungal activity of chitosan towards tomato pathogens. Indeed, the present study demonstrates that this natural compound applied at concentrations varying from 0.5 to 4 mg/ml is effective in inhibiting the radial growth of several fungi, i.e., *F. oxysporum* f. sp. *lycopersici*, *F. oxysporum* f. sp. *radicis-lycopersici*, *F. solani*, VD, *C. coccodes*, *R. solani*, *P. aphanidermatum*, *S. sclerotiorum*, *B. cinerea* and *A. solani*.

An increasing number of studies have been focused on the effect of chitosan on several phytopathogenic fungal species and its ability to reduce their *in vitro* growth [35-38]. In this regard, El Ghaouth, et al. [39] showed that chitosan used at 3.0 mg/ml reduced the radial growth of *B. cinerea*, *Rhizopus stolonifer*, *A. alternata* and *C. gloeosporioides*. Recently, Al-Najada and Gherbawy [40] found that four different concentrations of chitosan (20, 30, 50, and 100 mg/l) led to highly significant decrease in the average radial growth of 22 species of spoilage fungi.

In the present study, we demonstrated that the percentage of fungal growth inhibition recorded in chitosan-amended PDA medium was dependent on concentrations used, and that the highest mycelial growth decrease was achieved using chitosan at 4 g/l. This result is on line with those reported in various studies where a dose-response relationship was generally observed for each tested fungus expressed as average fungal growth decrease with the increase of chitosan concentrations. In fact, El Ghaouth, et al. [39] demonstrated that chitosan reduced markedly the radial growth of *B. cinerea*, *Rhizopus stolonifer*, *A. alternata* and *C. gloeosporioides* with a greater effect at higher concentrations. Wade and Lamondia [41] also noted a linear decrease of *R. solani* growth with the gradual increase from 0.5 to 6.0 mg/ml of chitosan concentration. Bell, et al. [42] mentioned that all isolates of *F. oxysporum* f. sp. *apii* were progressively and uniformly inhibited by concentrations of chitosan up to 3 mg/ml. Trotel-Aziz, et al. [43] showed that the level of mycelial growth inhibition of *B. cinerea* is highly correlated with chitosan concentration. Liu, et al. [44] also found that inhibitory effects of chitosan against *B. cinerea* and *Penicillium expansum* varied significantly depending on chitosan concentrations. Badawy and Rabea [45] found that the radial growth of *A. alternata*, *B. cinerea*, *C. gloeosporioides*, and *R. stolonifer* decreased with the increase of chitosan concentration (750-6000 mg/l).

In the current study, interspecific variations in sensitivity to chitosan was apparent. In fact, applied at 2.5 and 3 mg/ml, chitosan-based treatment led to complete inhibition of *P. aphanidermatum* and *S. sclerotiorum* growth, respectively, while at the highest concentration tested (4 g/l), *F. solani* and *R. solani* were inhibited by 47.71 and 54.58%, respectively. Furthermore, using the same chitosan concentrations, significant differences were noted between *Fusarium* species. For instance, at 0.5 mg/ml chitosan, *F. solani* and FOL were inhibited by 2.29 and 24.79%, respectively. Furthermore, at 2.5 mg/ml chitosan, *F. solani* and FORL were inhibited by 33.49 and 59.26%, respectively. Our results are in agreement with several studies reporting variation in sensibility to chitosan between fungal species [39,46-48]. Palma-Guerrero, et al. [49] mentioned varied tolerance to chitosan between fungi and found that *F. oxysporum* f. sp. *radicis-lycopersici*, *P. ultimum* and *R. solani* were the most sensitive to chitosan. Xu, et al. [50] also found that VD was the most tolerant to chitosan among nine plant-pathogenic fungi exhibiting varied responses to chitosan-based treatments. Similar effect was reported by El Ghaouth, et al. [39] who showed that *R. stolonifer* was the least sensitive to chitosan compared to *B. cinerea*, *A. alternata* and *C. gloeosporioides* at the concentration of 3 mg/ml. El-Ghaouth, et al. [46] reported that chitosan used at 400 mg/l totally inhibited *P. aphanidermatum* mycelial growth but total suppression of *F. oxysporum*, *R. stolonifer*, *P. digitatum*, and *C. gloeosporioides* was achieved using at 3% [47,48].

Variation in chitosan inhibitory effects was reported also between closely related microorganisms. In fact, when testing sensitivity of many fungal species to chitosan, Al-Najada and Gherbawy [40] found that 50 mg/l of chitosan completely inhibited growth of *A. consortialis*, *Neofusicoccum parvum* and many other fungi, while 100 mg/l of chitosan totally inhibited growth of *A. tenuissima*, *Cladosporium cladosporioides*, *F. solani*, *Macrophomina phaseolina*, *R. solani*, and others.

In the present study, 4 mg/ml chitosan was insufficient to completely inhibit *B. cinerea* mycelial growth which is not the case in many other studies. In fact, El-Ghaouth, et al. [39] explained that differences in results could originate from differences in methods used for incorporation of chitosan into growth medium. Guerra-Sánchez, et al. [51] also mentioned that the culture medium may influence on the inhibitory effects of the chitosan as some of them allow measuring the released compounds after chitosan addition without interference.

Moreover, Song, et al. [52] found that toxicity induced by chitosan was dependent on concentration, molecular weight, degree of acetylation, solvent, pH and viscosity.

Many explanations have been postulated for the mode of action of chitosan against fungi [21,39,53,54]. In fact, the polycationic nature of this compound is considered a key to its antifungal properties and the length of the polymer chain enhances its antifungal activity (Hirano and Nagao, 1989). Chitosan also induces marked morphological changes and structural alterations of the fungal cells [21,39,54].

In the current study, SA inhibited hyphal growth of the phytopathogenic fungi tested in a dose dependant manner. These results are consistent with many known reports. Indeed, similar effect was reported in [31] study where a significant negative correlation was detected between SA concentration and *F. oxysporum* f. s. *ciceri* Rs1 mycelial growth in a Petri-plate assay. These authors found that using SA at 100 µg/ml, no inhibition was observed but at 2000 µg/ml, pathogen mycelial growth was completely stopped. In the same sense, Abdel-Monaim, et al. [55] showed that SA tested at 50, 100 and 200 ppm had significantly inhibited radial growth, mycelial dry weight and spore formation of *F. oxysporum* f. sp. *lycopersici* but at different degrees depending on concentrations tested.

Wu, et al. [56] found that this RI decreased hyphal growth and biomass of *F. oxysporum* f. sp. *niveum* where the dry weight of mycelia, noted using the highest concentration (800 mg/l), showed 52% decrease in liquid culture. The obtained result was also in agreement with [57] where SA used at 270 mg/l exhibited fungitoxicity toward *Monilinia fructicola* and significantly inhibited it's *in vitro* growth.

The present finding was in accordance with those of Amborabé, et al. [58] reporting that SA inhibitory activity toward *Eutypa lata*, noted in solid and liquid culture media, also varied in a concentration-dependent manner with the threshold concentration being fixed at 13.8 mg/l. Amborabé, et al. [58] also reported that the antifungal efficiency of SA was higher when the experimental pH was brought to more acidic values (pH 4).

In the present study, SA used at 1 mM caused only 19 and 23% decrease in FOL and FORL mycelial growth, respectively. This is in line with [26,59] results where the three SA concentrations tested (100 μM, 200 μM and 300 μM) had significantly inhibited FOL and FORL growth compared to the control. Similar effects were reported for other pathogens such as *R. stolonifer*, *F. oxysporum*, *R. solani*, *S. rolfsii*, *M. phaseolinae*, *Pythium* sp., and *Phytophthora* using SA at a minimum concentration of 2.5 mM [60,61]. Also, da Rocha Neto, et al. [62] found that SA was able to completely suppress germination of *P. expansum* conidia at 2.5 mM and that this compound caused leakage of the pathogen's proteins to the medium, measurable lipid damage, and intracellular disorganization.

To our knowledge, this is the first report on the *in vitro* antifungal activity of chitosan and salicylic acid against major fungal pathogens infecting tomato in Tunisia.

In the present study, SA and chitosan were tested as soil drench 5 days before tomato plants were challenged with three soil-borne pathogens. In fact, different application methods have been used for the chemical induction of systemic resistance in plants. For example, Thulke and Conrath [63] showed that SA applied either as a seed treatment, spray or soil drench induced systemic resistance in cacao against *Phytophthora palmivora*. Furthermore, for tomato, the period between application of the inducer and the effective activation of defence responses were observed within a short period, 3–7 days post-inoculation [64,65]. In the same sense, Guzmán-Téllez, et al. [66] reported that SA applications on tomato should be performed within a minimum interval of eight days in order to maintain SA concentration related with the increase in plant tolerance to environmental stress.

Interesting results from our *in vitro* essays were also confirmed by the *in vivo* experiments where single treatments with chitosan and SA resulted in varied degree of protection against Verticillium wilt, Fusarium wilt and Fusarium Crown and Root Rot. In fact, *chitosan-based* treatment resulted in 73.68, 60.86, and 45% reductions in wilt severity, as respectively compared to VD-, FOL- and FORL-inoculated and untreated controls. The protective effect of chitosan against severe tomato wilt pathogens has been reported in numerous investigations on a range of crops where the antifungal activity of chitosan has been proven both *in vitro* and *in vivo* [21]. [67] also reported that chitosan used to control plant pathogens has been extensively explored with more or less success depending on pathosystems, used derivatives, concentrations, degree of deacetylation, viscosity, and applied formulation (*i.e.*, soil amendment, foliar application, chitosan alone or in association with other treatments). Chitosan has also been extensively utilized as a seed treatment and as soil amendment to control *F. oxysporum* in many host plants [68]. In fact, applied at an optimal concentration, this biomaterial is able to induce a delay in disease development leading to reduced plant wilting [69]. For example, chitosan was shown to protect tomato plants from Fusarium crown and root rot [69] when used as seed treatment and from Fusarium wilt when applied as foliar spray [64]. In soilless tomato, Benhamou and Lafontaine [70] reported that root rot caused by *F. oxysporum* f. sp. *radicis-lycopersici* was suppressed using chitosan-based amendments.

Studies of Cretoiu, et al. [71] demonstrated that amendment of chitin, which is a precursor of chitosan, improved soil suppressiveness to VD. Recently, Amini [72] also found that chitosan affected VD growth *in vitro* and reduced disease severity and significantly increased weight of potato tubers under greenhouse conditions. In fact, chitosan is often used for plant disease control as a powerful elicitor. When applied to plant tissues, chitosan often agglutinates around penetration sites and forms a physical barrier preventing pathogen from spreading and invading other healthy tissues and it is able to bind various materials and initiate fast wound healing process. Chitosan and derivatives are known to act as potent inducers, enhancing a battery of plant responses both locally around the infection sites and systemically to alert healthy parts of the plant. These include early signaling events as well as the accumulation of defence-related metabolites and proteins such as phytoalexins and PR-proteins [44,45,67,73].

The results of the current study are also in concordance with others studies showing that chitosan has a double effect: it acts as antimicrobial abiotic agent and it also activates several plant defense mechanisms during host-pathogen interactions. These combined effects led to reduced disease severity due to callose deposition, lignification, synthesis of abscisic acid, phytoalexins, and pathogenesis-related proteins as reported in previous studies [21,74].

The present study highlights the protective effect of SA against the economically important fungal diseases. In fact, exogenous application of SA (10 mM) as soil drench had reduced by 42.1, 78.26 and 50% wilt severity, as respectively compared to VD-, FOL- and FORL-inoculated and untreated controls. Similar results were reported in Mandal, et al. [59] study where exogenous application of SA at 200 mM, through root feeding and foliar spray, led to reduced vascular browning and leaf yellowing caused by FOL on tomato plants. In fact, SA-treated plants already challenged with FOL exhibited increased levels of peroxidases and phenylalanine ammonia-lyase activities and also endogenous accumulation of free SA, showing that the root system might assimilate and distribute SA throughout the plant and ultimately activate systemic disease resistance [59,75]. In this regard, Abdel-Monaim [76] reported that SA protected tomato plants from Fusarium wilt disease and induced resistance with increased concentrations. Ojha and Chatterjee [77] also indicated that soil application of SA following inoculation with FOL resulted in maximum peroxidase and polyphenol oxidase activity in tomato leaves on the 28th day which might play an important role in plant resistance and defense system activation. In fact, it has been shown that exogenous SA treatment prior to inoculation provided increased *F. oxysporum* resistance as evidenced by reduced foliar necrosis and plant death in *Arabidopsis* [78]. Furthermore, exogenous SA stimulated the systemic resistance and significantly reduced Fusarium wilt severity in chickpea [31]. In the same sense, pre-treatment of asparagus roots with SA primed plants for a potentiated defence response toward *F. oxysporum* f. sp. *asparagi* associated with increased levels of peroxidases, phenylalanine ammonia-lyase and lignifications [79]. External application of SA to *Arabidopsis* and tobacco boosts endogenous SA signal production, and induces systemic acquired resistance responses against pathogens including expression of the pathogenesis related genes that are implicated in disease resistance [80-82].

Recently Jendoubi, et al. [26] found that SA, applied at 200 μM directly to the root system, had activated systemically resistance against FORL in hydroponically grown tomato plants. In the same sense, [83] demonstrated the induced resistance to VD toxin in cotton achieved using exogenous SA which was coupled with an increased in β-1,3-glucanase production while cellular integrity was maintained and damage to cell wall and plasma membrane was avoided. These results are also in line with those of [84] who showed that chickpea seed treatment with SA reduced Fusarium wilt disease by 40%. In this regards, [27] mentioned that in field trials, treatment of tomato plants with SA at 100 mM resulted in reduction of root rot incidence and disease severity caused by *F. solani*, *R. solani* and *S. rolfsii*. [85] found that tomato plants sprayed with SA, as inducer, for 48 h before root inoculation with FOL provided induced resistance in plant against this pathogen and resulted in disease incidence decline, 15 days post inoculation.

In the present study, tomato plants treated with 10 mM SA and 4 mg/ml chitosan enhanced all tomato growth parameters tested (height, aerial part and root fresh weights) compared to pathogen-inoculated control. In fact, SA-based-treatment had significantly increased plant height, root and aerial part fresh weights by 17.94, 52.17 and 33.33%, by 23.01, 55.40% and 29.72%, and by 17.72, 50 and 46.84%, respectively, while compared to VD-, FOL- and FORL-inoculated and untreated plants. Furthermore, a significant improve, by 20.18%, in plant height was also recorded in SA- treated and VD-inoculated plants compared to uninoculated and untreated control. In fact, growth-stimulating effects of SA have been reported in many plants such as soybean, wheat, maize, and chamomile [86]. Furthermore, Abo-Hamed, et al. [87] reported that soil drench with salicylate led to an increase in the fresh and dry weight of shoot and at lower concentration appeared to enhance plant height and leaf area of wheat plant. Rivas-San, et al. [86] mentioned that the effect of exogenous SA supply on growth depends on plant species, developmental stage, and SA concentrations tested. In the same sense, [27] have also reported that the application of SA was among the most efficient RIs tested for increase of growth parameters (plant height and number of branches), yield and quality of tomato fruits under field conditions during two cropping seasons. Abd El-Gawad and Bondok [88] found that foliar application of SA (2 mM/l) and chitosan (0.1%) significantly improves tomato vegetative growth.

In the current study, the growth-stimulating effect of chitosan was observed mainly on FORL-inoculated plants. In fact, a significant improve of tomato height by 13.81% was noted on chitosan-treated plants, compared to FORL-inoculated and untreated control ones. Furthermore, an enhanced root fresh weight by about 62.16 and 40.54% was also recorded, compared to FORL-inoculated and non-controls. The aerial part fresh weight recorded on chitosan-treated plants was significantly increased by 38.97 and 40% compared to FORL- and VD-inoculated plants, respectively. The positive effect of chitosan on the growth of roots, shoots and leaves of various plants have been previously reported [89-91]. In fact, Algam, et al. [92] reported that chitosan had successfully controlled Ralstonia wilt in tomato as well as promoted tomato growth. El-Mougy, et al. [93] also demonstrated that tomato root rot pathogens were successfully controlled using chitosan, and recorded 66.7% increased yield using this compound. Abdel-Mawgoud, et al. [90] indicated that chitosan application improved strawberry plant height, number of leaves, fresh and dry weights of the leaves and yield components. Also, Sheikha and Al-Malki [91] indicated that chitosan works as a positive factor in enhancing bean shoot and root length, fresh and dry weights of shoots and roots as well as leaves area. In this regard, Farouk [89] mentioned that a positive effect of chitosan was observed on the growth of roots, shoots and leaves of various plants including cucumber. In the same sense, El-Tantawy [94] found that spraying tomato plants with chitosan increased all vegetative growth parameters expressed in plant height, number of branches, number of leaves and plant fresh and dry weight. Recently, Algam and Elwagia [95] found that tomato growth parameters were significantly increased using chitosan-based treatments compared to control. In fact, the application of chitosan at 5 mg/ml as foliar spray combined with 5 mg/ml as seed treatment had increased plant height and fresh and dry weight by 16, 36, 24%, respectively.

Conclusion

In this study, it was observed that RIs tested possess variable antifungal activity *in vitro* depending on fungal tomato pathogens tested and concentrations used. They were shown to have direct antifungal potential and also indirect effect through by decreasing tomato wilt incited by FOL, FORL and VD and by enhancing growth. Based on these findingss, it could be concluded that, when applied as soil drench, chitosan used at 4 mg/ml and SA applied at 10 Mm may markedly suppress Verticillium wilt, Fusarium wilt, and Fusarium Crown and Root Rot severities and enhance tomato plant growth. Thus, SA and chitosan may be used as potential inducer of systemic acquired resistance against the devastating soil-borne vascular wilt pathogens of tomato. Further studies are needed to more elucidate the exact mechanism deployed by these inducers via different modes and timings of application. The current study indicates that inducing plant's own defense mechanisms using these RIs can be integrated in plant disease management together with other control environmentally safe alternatives such as biocontrol, grafting, and solarisation.

Acknowledgement

This work was carried out during a Collaborative Project between Tunisia and Egypt, titled Integrated management for controlling tomato fungal diseases under Egyptian and Tunisian conditions, funded by the Ministry of Scientific Research, in Egypt (Grand no.4/10/4) and the Ministry of Higher Education and Scientific Research of Tunisia.

References

1. Olaniyi JO, Akanbi WB, Adejumo TA, Akande OG (2010) Growth, fruit yield and nutritional quality of tomato varieties. Afr J Food Sci 4: 398-402.
2. Rakha M, Scott J, Hutton S, Smith H (2011) Identification of trichomes, loci and chemical compounds derived from *Solanum habrochaites* accession LA1777 that are associated with resistance to the sweet potato whitefly, *Bemisia tabaci* in tomato, *S. lycopersicum*. 43rd Tomato Breeders Meeting. University of Florida.
3. Ananymous (2015) Groupement Interprofessionnel des légumes, filières des legumes.
4. Gajanana TM, Krishna Moorthy PN, Anupama HL, Raghunatha R, Kumar GTP (2006) Integrated pest and disease management in tomato: an economic analysis. Agric Econ Res Rev 19: 269-280.
5. Hibar K, Daami-Remadi M, El-Mahjoub M (2007) Induction of resistance in tomato plants against *Fusarium oysporum* f. sp. radicis-*lycopersici* by *Trichoderma* spp. Tunisian J Plant Prot 2: 47-58.
6. Jabnoun-Khiareddine H, Daami-Remadi M, Ayed F, El Mahjoub M (2009) Biocontrol of tomato Verticillium wilt by using indigenous *Gliocladium* spp. and *Penicillium* sp. isolates. Dynamic Soil Dynamic Plant 3: 70-79.
7. Jabnoun-Khiareddine H, Daami-Remadi M, Platt HW, Ayed F, El Mahjoub M (2010) Variation in aggressiveness of Tunisian *Verticillium dahliae* races 1 and 2 isolates and response of differential tomato cultivars to Verticillium wilt. International Journal of Plant Breeding 4: 63-70.
8. Hassine M, Aydi-Ben Abdallah R, Jabnoun-Khiareddine H, Ben Jannet H, Daami-Remadi M (2013) Effet des températures d'incubation et des méthodes de confrontation sur le pouvoir inhibiteur exercé par *Penicillium* sp. et *Gliocladium* spp. sur *Botrytis cinerea*. Tunis J Med Plants Nat Prod 9: 41-51.
9. Hassine M, Aydi-Ben Abdallah R, Jabnoun-Khiareddine H, Daami-Remadi M (2014) Pouvoir antifongique des *Penicillium* sp. et des *Gliocladium* spp. Contre Alternaria solani in vitro et sur fruits de tomate. Tunis J Med Plants Nat Prod 12: 9-28.
10. Batish DR, Singh HP, Kohli RK, Kaur S (2008) Eucalyptus essential oil as a natural pesticide. For Ecol Manage 256: 2166-2174.
11. Nicholls CI, Altieri MA (2013) Plant biodiversity enhances bees and other insect pollinators in agroecosystems. Agron Sustain Dev 33: 257-274.
12. Kamble A, Koopmann B, von Tiedemann A (2013) Induced resistance to *Verticillium longisporum* in *Brassica napus* by b-aminobutyric acid. Plant Pathol 62: 552-561.

13. Kessmann H, Staub T, Hofmann C, Maetzke T, Herzog J, et al. (1994) Induction of systemic acquired disease resistance in plants by chemicals. Annu Rev Phytopathol 32: 439-459.

14. Kuc J (2001) Concepts and direction of induced systemic resistance in plants and its application. Eur J Plant Pathol 107: 7-12.

15. Edreva A (2004) A novel strategy for plant protection: Induced resistance. J Cell Mol Biol 3: 61-69.

16. Lyon GD, Reglinski T, Newton AC (1995) Novel disease control compounds: the potential to 'immunize' plants against infection. Plant Pathol 44: 407-427.

17. Schneider M, Schweizer P, Meuwly P, Métraux JP (1996) Systemic acquired resistance in plants. Int. Rev. Cytology 168: 303-340.

18. Benhamou N, Picard K (1999) La résistance induite: une nouvelle stratégie de défense des plantes contre les agents pathogènes. Phytoprotection 80: 137-168.

19. Cohen Y (2001) The BABA story of induced resistance. Phytoparasitica 29: 375-378.

20. Benhamou N (1996) Elicitor-induced plant defence pathways. Trends Plant Sci 1: 233-240.

21. Bautista-Baños S, Hernández-Lauzardo AN, Velázquez-del Valle MG, Hernández-López M, Ait Barkab E, et al. (2006) Chitosan as a potential natural compound to control pre and postharvest diseases of horticultural commodities. Crop Prot 25: 108-118.

22. Bhattacharya A (2013) Fungicidal potential of chitosan against phytopathogenic *Fusarium Solani*. J Exp Biol & Agr Sci 1: 259-263.

23. Hammerschmidt R, Smith-Becker JA (2000) The role of salicylic acid in disease resistance in mechanisms of resistance to plant diseases. Eds: Slusarenko A, Fraser RSS, Van Loon LC, Kluwer Academic Publisher.

24. Zgnen H, Mehmet B, Erkili A (2001) The Effect of salicylic acid and endomycorrhizal fungus *Glomus etunicatum* on plant development of tomatoes and Fusarium wilt caused by *Fusarium oxysporum* f. sp *lycopersici*. Turk J Agric 25: 25-29.

25. El-Khallal SM (2007) Induction and modulation of resistance in tomato plants against Fusarium wilt disease by bioagent fungi (*arbuscular mycorrhiza*) and/ or hormonal elicitors (jasmonic acid and salicylic acid): 1- changes in growth, some metabolic activities and endogenous hormones related to defence mechanism. Australian J Basic Appl Sci 1: 691-705.

26. Jendoubi W, Harbaoui K, Hamada W (2015) Salicylic acid-induced resistance against *Fusarium oxysporum* f.s.p *radicis-lycopercisi* in hydroponic grown tomato plants. Journal of New Sciences 21: 985-995.

27. El-Mohamedy RSR, Jabnoun-Khiareddine H, Daami-Remadi M (2014) Control of root rot diseases of tomato plants caused by *Fusarium solani, Rhizoctonia solani* and *Sclerotium rolfsii* using different chemical plant resistance inducers. Tunisian J Plant Prot 9: 45-55.

28. Abdou ES, Abd-Alla HM, Galal AA (2001) Survey of sesame root rot/wilt disease in Minia and their possible control by ascorbic and salicylic acids. Assuit J of Agric Sci 32: 135-152.

29. Mahesh HM, Sharada MS (2014) Role of salicylic acid in induction of resistance in Brinjal (*Solanum melongena L.*) against *Verticillium dahliae*. Int J Recent Sci Res 5: 1865-1870.

30. He CY, Wolyn DJ (2005) Potential role for salicylic acid in induced resistance of asparagus roots to *Fusarium oxysporum* f.sp. *asparagi*. Plant Pathol 54: 227-232.

31. Saikia S, Singh T, Kumar R, Srivastava J, Srivastava AK, et al. (2003) Role of salicylic acid in systemic resistance induced by *Pseudomonas fluorescens* against *Fusarium oxysporum* f. sp. *ciceri* in chickpea. Microbiol Res 158: 203-213.

32. Pharand B, Carisse O, Benhamou N (2002) Cytological aspects of compost-mediated induced resistance against fusarium crown and root rot in tomato. Phytopathology 92: 424-438.

33. El Ghaouth A, Arul J, Ponnampalam R (1991) Use of chitosan coating to reduce water loss and maintain quality of cucumbers and bell pepper fruits. J Food Process Preservation 15: 359-368.

34. Tiru M, Muleta D, Bercha G, Adugna G (2013) Antagonistic effect of rhizobacteria against coffee wilt disease caused by Gibberella xylarioides. Asian J Plant Pathol 7: 109-122.

35. Romanazzi G, Nigro F, Ippolito A (2001) Chitosan in the control of postharvest decay of some Mediterranean fruits. In: Muzarelli, R.A.A. (Ed.), Chitin Enzymology, Atec, Italy.

36. El Ghaouth A, Arul J, Wilson C, Benhamou N (1997) Biochemical and cytochemical aspects of the interactions of chitosan and *Botrytis cinerea* in bell pepper fruit. Postharvest Biol Technol 12: 183-194.

37. Bhaskara Reddy MV, Arul J, Angers P, Couture L (1999) Chitosan treatment of wheat seeds induces resistance to Fusarium graminearum and improves seed quality. J Agric Food Chem 47: 1208-1216.

38. Rhoades J, Roller S (2000) Antimicrobial actions of degraded and native chitosan against spoilage organisms in laboratory media and foods. Appl Environ Microbiol 66: 80-86.

39. El Ghaouth A, Ponnampalam R, Castaigne F, Arul J (1992) Chitosan coating to extend the storage life of tomatoes. HortScience 27: 1016-1018.

40. Al-Najada AR, Gherbawy YA (2015) Molecular identification of spoilage fungi isolated from fruit and vegetables and their control with chitosan. Food Biotechnol 29: 166-184.

41. Wade HE, Lamondia JA (1994) Chitosan inhibits *Rhizoctonia fragariae* but not strawberry black root rot. Adv Strawberry Res 13: 26-31.

42. Bell AA, Hubbard JC, Liu L, Davis RM, Subbarao KV (1998) Effects of chitin and chitosan on the incidence and severity of Fusarium yellows in celery. Plant Dis 82: 322-328.

43. Trotel-Aziz P, Couderchet M, Vernet G, Aziz A (2006) Chitosan stimulates defense reactions in grapevine leaves and inhibits development of *Botrytis cinerea*. Eur J plant pathol 114: 405-13.

44. Liu J, Tian S, Meng X, Xu Y (2007) Effects of chitosan on control of postharvest diseases and physiological responses of tomato fruit. Postharvest Biol Technol 44: 300-306.

45. Badawy MEI, Rabea EI (2011) A biopolymer chitosan and its derivatives as promising antimicrobial agents against plant pathogens and their applications in crop protection. Int J Carbohyd Chem.

46. El-Ghaouth A, Arul J, Grenier J, Benhamou N, Asselin A, et al. (1994) Effect of chitosan on cucumber plants: suppression of *Pythium aphanidermatum* and induction of defence reactions. Phytopathology 84: 313-20.

47. Bautista-Baños S, Hernández-López M, Bosquez-Molina E, Wilson CL (2003) Effects of chitosan and plant extracts on growth of *Colletotrichum gloeosporioides*, anthracnose levels and quality of papaya fruit. Crop Prot 22: 1087-1092.

48. Bautista-Baños S, Hernández-López M, Bosquez-Molina E (2004) Growth inhibition of selected fungi by chitosan and plant extracts. Mexican Journal of Phytopathology 22: 178-186.

49. Palma-Guerrero J, Jansson HB, Salinas J, Lopez-Llorca LV (2008) Effect of chitosan on hyphal growth and spore germination of plant pathogenic and biocontrol fungi. J Appl Microbiol 104: 541-553.

50. Xu J, Zhao X, Han X, Du Y (2007) Antifungal activity of oligochitosan against *Phytophtora capsici* and other pathogenic fungi in vitro. Pest Biochem Physiol 87: 220-228.

51. Guerra-Sánchez MG, Vega-Pérez J, Velázquez-del Valle MG, Hernández-Lauzardo AN (2009) Antifungal activity and release of compounds on *Rhizopus stolonifer* (Ehrenb.:Fr.) Vuill. by effect of chitosan with different molecular weights. Pesti Biochem Physiol 93: 18-22.

52. Song Y, Babiker EE, Usui M, Saito A, Kato A (2002) Emulsifying properties and bactericidal action of chitosan-lysozyme conjugates. Food Res Int 35: 459-466.

53. Hirano S, Nagao N (1989) Effect of chitosan, pectic acid, lysozyme and chitinase on growth of several phytopathogens. Agric Biol Chem 53: 3065-3066.

54. Ait Barka E, Eullaffroy P, Clément C, Vernet G (2004) Chitosan improves development, and protects *Vitis vinifera* L. against *Botrytis cinerea*. Plant Cell Rep 22: 608-614.

55. Abdel-Monaim MF, Abdel-Gaid MAW, Armanious HAH (2012) Effect of chemical inducers on root rot and wilt diseases, yield and quality of tomato. Int J Agr Sci 2: 210-220.

56. Wu HS, Raza W, Fan JQ, Sun YG, Bao W, et al. (2008) Antibiotic effect of exogenously applied salicylic acid on in vitro soilborne pathogen, *Fusarium oxysporum* f.sp.niveum. Chemosphere 74: 45-50.

57. Yao HJ, Tian SP (2005) Effects of pre- and post-harvest application of salicylic acid or methyl jasmonate on inducing disease resistance of sweet cherry fruit in storage. Posthar Biol Tech 35: 253-262.

58. Amborabé BE, Fleurat-Lessard P, Chollet JF, Roblin G (2002) Antifungal effects of salicylic acid and other benzoic acid derivatives towards Eutypa lata: structure-activity relationship. Plant Physiol Biochem 40: 1051-1060.

59. Mandal S, Mallick N, Mitra A (2009) Salicylic acid-induced resistance to *Fusarium oxysporum* f. sp. *lycopersici* in tomato. Plant Physiol Biochem 47: 642-649.

60. Panahirad S, Zaare-Nahandi F, Safaralizadeh R, Alizadeh-Salteh S (2012) Postharvest control of *Rhizopus stolonifer* in peach (Prunus persica L. Batsch) fruits using salicylic acid. J of Food Safety 32: 502-507.

61. El-Mohamedy SR, Abdel-Kader MM, Abd-El-Kareem F, El-Mougy NS (2013) Essential oils, inorganic acids and potassium salts as control measures against the growth of tomato root rot pathogens in vitro. International Journal of Agricultural Technology 9: 1507-1520.

62. da Rocha Neto AC, Maraschin M, Di Piero RM (2015) Antifungal activity of salicylic acid against *Penicillium expansum* and its possible mechanisms of action. Int J Food Microbiol 215: 64-70.

63. Thulke O, Conrath U (1998) Salicylic acid has a dual role in the activation of defence-related genes in parsley. Plant J 14: 35-42.

64. Benhamou N, Belanger RR (1998) Benzothiadiazole-mediated induced resistance to *fusarium oxysporum* f. sp. *radicis-lycopersici* in tomato Plant Physiol 118: 1203-1212.

65. Benhamou N, Bélanger RR (1998) Induction of systemic resistance to Pythium damping-off in cucumber plants by benzothiadiazole: ultrastructure and cytochemistry of the host response. Plant J 14: 13-21.

66. Guzmán-Téllez E, Montenegro DD, Benavides-Mendoza A (2014) Concentration of salicylic acid in tomato leaves after foliar aspersions of this compound. Am J Plant Sci 5: 2048-2056.

67. El Hadrami A, Adam LR, El Hadrami I, Daayf F (2010) Chitosan in plant protection. Mar Drugs 8: 968-987.

68. Rabea EI, Badawy ME, Stevens CV, Smagghe G, Steurbaut W (2003) Chitosan as antimicrobial agent: applications and mode of action. Biomacromolecules 4: 1457-1465.

69. Benhamou N, Kloepper JW, Tuzun S (1994) Induction of systemic resistance to Fusarium crown rot and root rot in tomato plants by seed treatment with chitosan. Phytopathology 84: 1432-1444.

70. Benhamou N, Lafontaine PJ (1994) Ultrastructural and cytochemical characterization of elicitor-induced structural responses in tomato root tissues infected by *Fusarium oxysporum* f. sp. radicis-*lycopersici*. Planta 197: 89-102.

71. Cretoiu MS, Korthals GW, Visser JHM, Elsas JD (2013) Chitin Amendment Increases Soil Suppressiveness toward Plant Pathogens and Modulates the Actinobacterial and Oxalobacteraceal Communities in an Experimental Agricultural Field. Appl Environ Microbiol 79: 5291-5301.

72. Amini J (2015) Induced resistance in potato plants against Verticillium wilt invoked by chitosan and Acibenzolar-S-methyl. Aust J Crop Sci 9: 570-576.

73. El Ghaouth A, Smilanick JL, Wilson CL (2000) Enhancement of the performance of *Candida saitoana* by the addition of glycolchitosan for the control of postharvest decay of apple and citrus fruit. Postharvest Biol Technol 19: 103-110.

74. Cavalcanti FR, Resende MLV, Carvalho CPS, Silveira JAG, Oliveira JTA (2007) An aqueous suspension of *Crinipellis perniciosa* mycelium activates tomato defence responses against *Xanthomonas vesicatoria*. Crop Prot 26: 729-738.

75. Spletzer ME, Enyedi AJ (1999) Salicylic Acid Induces Resistance to Alternaria solani in Hydroponically Grown Tomato. Phytopathology 89: 722-727.

76. Abdel-Monaim MF (2012) Induced Systemic Resistance in Tomato Plants Against Fusarium Wilt Disease. International Research Journal of Microbiology 3: 14-23.

77. Ojha S, Chatterjee NC (2012) Induction of resistance in tomato plants against *Fusarium oxysporum* f. sp. *lycopersici* mediated through salicylic acid and *Trichoderma harzianum*. J Plant Prot Res 52: 220-225.

78. Edgar CI, McGrath KC, Dombrecht B, Manners JM, Maclean DC, et al. (2006) Salicylic acid mediates resistance to the vascular wilt pathogen *Fusarium oxysporum* in the model host *Arabidopsis thaliana*. Aust Plant Pathol 35: 581-591.

79. He YL, Liu YL, Chen Q, Bian AH (2002) Thermotolerance related to antioxidation induced by salicylic acid and heat hardening in tall fescue seedlings. J Plant Physiol Mol 28: 89-95.

80. Delaney TP, Uknes S, Vernooij B, Friedrich L, Weymann K, et al. (1994) A central role of salicylic Acid in plant disease resistance. Science 266: 1247-1250.

81. Summermatter K, Sticher L, Metraux JP (1995) Systemic Responses in Arabidopsis thaliana Infected and Challenged with *Pseudomonas syringae* pv syringae. Plant Physiol 108: 1379-1385.

82. Métraux JP, Nawrath C, Genoud T (2002) Systemic acquired resistance. Euphytica 124: 237-243.

83. Zhen XH, Li YZ (2004) Ultrastructural changes and location of ß-1,3-glucanase in resistant and susceptible cotton callus cells in response to treatment with toxin of *Verticillium dahliae* and salicylic acid. J Plant Physiol 161: 1367-1377.

84. Sarwar N, Zahid MHC, Haq I (2010) Seed treatments induced systemic resistance in chickpea against Fusarium wilt in wilt sick field. Pak J Bot 42: 3323-3326.

85. Biswas SK, Pandey NK, Mohd R (2012) Inductions of defense response in tomato against Fusarium wilt through inorganic chemicals as inducers. J Plant Pathol Microb 3: 1-7.

86. Rivas-San Vicente M, Plasencia J (2011) Salicylic acid beyond defence: its role in plant growth and development. J Exp Bot 62: 3321-3338.

87. Abo-Hamed SA, Mansour FA, Aldesuquy HS (1987) Shoot growth and morphological characteristics of wheat as influenced by sodium salicylate, alar, asulam and kinetin. Mansoura Sci Bull 14: 203-221.

88. Abd El-Gawad HG, Bondok AM (2015) Response of Tomato Plants to Salicylic Acid and Chitosan under Infection with Tomato mosaic virus. American-Eurasian J Agric & Environ Sci 15: 1520-1529.

89. Farouk S, Ghoneem KM, Ali AA (2008) Induction and expression of systematic resistance to downy mildew disease in cucumber plant by elicitors. Egypt J Phytopathol 36: 95-111.

90. Abdel-Mawgoud AM, Tantawy AS, El-Nemr MA, Sassine YN (2010) Growth and yield responses of strawberry plants to chitosan application. Eur J Sci Res 39: 161-168.

91. Sheikha SA, Al-Malki FM (2011) Growth and chlorophyll responses of bean plants to chitosan applications. Eur J Sci Res 50: 124-134.

92. Algam SAE, Xie GL, Li B, Yu SH, Su T, et al. (2010) Effects of Paenibacillus strains and chitosan on plant growth promotion and control of Ralstonia wilt in tomato. J Plant Pathol 92: 593-600.

93. El-Mougy NS, El-Gamal NG, Fotouh YO, Abd-El-Kareem F (2006) Tomato root rot disease under greenhouse and field conditions. Res J Agric and Biol Sci 2: 190-195.

94. El-Tantawy EM (2009) Behavior of tomato plants as affected by spraying with chitosan and aminofort as natural stimulator substances under application of soil organic amendments. Pak J Biol Sci 12: 1164-1173.

95. Algam SAE, Elwagia MEA (2015) Evaluation of chitosan efficacy on tomato growth and control of early blight disease. Jordan Journal of Agricultural Sciences 11: 27-36.

PERMISSIONS

All chapters in this book were first published in JPPM, by OMICS International; hereby published with permission under the Creative Commons Attribution License or equivalent. Every chapter published in this book has been scrutinized by our experts. Their significance has been extensively debated. The topics covered herein carry significant findings which will fuel the growth of the discipline. They may even be implemented as practical applications or may be referred to as a beginning point for another development.

The contributors of this book come from diverse backgrounds, making this book a truly international effort. This book will bring forth new frontiers with its revolutionizing research information and detailed analysis of the nascent developments around the world.

We would like to thank all the contributing authors for lending their expertise to make the book truly unique. They have played a crucial role in the development of this book. Without their invaluable contributions this book wouldn't have been possible. They have made vital efforts to compile up to date information on the varied aspects of this subject to make this book a valuable addition to the collection of many professionals and students.

This book was conceptualized with the vision of imparting up-to-date information and advanced data in this field. To ensure the same, a matchless editorial board was set up. Every individual on the board went through rigorous rounds of assessment to prove their worth. After which they invested a large part of their time researching and compiling the most relevant data for our readers.

The editorial board has been involved in producing this book since its inception. They have spent rigorous hours researching and exploring the diverse topics which have resulted in the successful publishing of this book. They have passed on their knowledge of decades through this book. To expedite this challenging task, the publisher supported the team at every step. A small team of assistant editors was also appointed to further simplify the editing procedure and attain best results for the readers.

Apart from the editorial board, the designing team has also invested a significant amount of their time in understanding the subject and creating the most relevant covers. They scrutinized every image to scout for the most suitable representation of the subject and create an appropriate cover for the book.

The publishing team has been an ardent support to the editorial, designing and production team. Their endless efforts to recruit the best for this project, has resulted in the accomplishment of this book. They are a veteran in the field of academics and their pool of knowledge is as vast as their experience in printing. Their expertise and guidance has proved useful at every step. Their uncompromising quality standards have made this book an exceptional effort. Their encouragement from time to time has been an inspiration for everyone.

The publisher and the editorial board hope that this book will prove to be a valuable piece of knowledge for researchers, students, practitioners and scholars across the globe.

LIST OF CONTRIBUTORS

Bimal S. Amaradasa
Department of Plant Pathology, University of Nebraska-Lincoln, Lincoln, NE 68583, USA

Dilip Lakshman
Floral and Nursery Plants Research Unit and the Sustainable Agricultural Systems Lab, Beltsville Agricultural Research Center-West, Beltsville, MD 20705, USA

Keenan Amundsen
Department of Agronomy and Horticulture, University of Nebraska-Lincoln, Lincoln, NE 68583 USA

Michael E Foley, Münevver Doğramacı and William R Underwood
USDA-Agricultural Research Service, Sunflower and Plant Biology Research Unit, Fargo, ND 58102-2765, USA

Mark West
USDA-Agricultural Research Service, 2150 Centre Ave., Suite 300 Fort Collins, CO 80526, USA

Teshome E and Tegegn A
Sinana Agricultural Research Center, Bale-Robe, Ethiopia

Habtamu Terefe and Chemeda Fininsa
School of Plant Sciences, Haramaya University, P.O.Box 138, Dire Dawa, Ethiopia

Samuel Sahile
Natural and Computational Science, University of Gondar, Ethiopia

Kindie Tesfaye
The International Maize and Wheat Improvement Center (CIMMYT), Addis Ababa, Ethiopia

Nada Ouhaibi-Ben Abdeljalil
Higher Agronomic Institute of Chott-Mariem, Sousse University, 4042-Chott Mariem, Tunisia
UR13AGR09-Integrated Horticultural Production in the Tunisian Centre-East, Regional Centre of Research on Horticulture and Organic Agriculture, University of Sousse, 4042, Chott-Mariem, Tunisia

Mejda Daami-Remadi
UR13AGR09-Integrated Horticultural Production in the Tunisian Centre-East, Regional Centre of Research on Horticulture and Organic Agriculture, University of Sousse, 4042, Chott-Mariem, Tunisia

Jessica Vallance, Emilie Bruez and Patrice Rey
INRA, UMR1065 Santé et Agroécologie du Vignoble (SAVE), ISVV, F-33140 Villenave d'Ornon, France
Université de Bordeaux, Bordeaux Sciences Agro, ISVV, UMR1065 SAVE, F-33140 Villenave d'Ornon, France

Jonathan Gerbore
BIOVITIS, 15400 Saint Etienne de Chomeil, France

Guilherme Martins
USC Oenologie-INRA, Université Bordeaux Segalen, Bordeaux Sciences Agro, ISVV, Villenave d'Ornon, France

Eshetu Belete
Department of Plant Sciences, P.O. Box 1145, Wollo University, Dessie, Ethiopia

Amare Ayalew
School of Plant Sciences, P.O. Box 241, Haramaya University, Ethiopia

Seid Ahmed
International Center for Agricultural Research in the Dry Areas, P O Box 5689, Addis Ababa, Ethiopia

Harnet Abrha
Tigray Agricultural Research Institute, Alamata Agricultural Research Center, P.O.Box-56 Alamata, Ethiopia

Alem Gebretsadik
Capacity Building for Scaling up of Evidence-Based Best Practices in Agricultural Production in Ethiopia (CASCAPE) project, Mekelle, Ethiopia

Girmay Tesfay
Department of Land Resources Management and Environmental Protection, Mekelle University, Ethiopia

Girmay Gebresamuel
Department of Natural Resources Economics and Management, Mekelle University, Ethiopia

Hemant Pathak, Saurabh Maru, Satya HN and Silawat SC
Forest Research and Extension Circle, Indore, Madhya Pradesh, India

Bekriwala TH and Chaudhary DA
Department of Plant Pathology, N. M. College of Agriculture, Navsari Agricultural University, Navsari, India

Kedar Nath
Regional Rice Research Station, Navsari, Agricultural University, Vyara, India

Dagnew Bitew
Department of Microbial, Cellular and Molecular Biology, College of Natural Science, Addis Ababa University, Ethiopia

Ilondu EM
Department of Botany, Faculty of Science, Delta State University, Abraka, Nigeria

Bosah BO
Department of Agronomy, Faculty of Agriculture, Delta State University, Asaba Campus, Nigeria

Al-Hmoud G and Al-Momany A
University of Jordan, Faculty of agriculture, Department of Plant Protection, Amman, Jordan

Narayan Chandra Talukdar
Institute of Bioresources and Sustainable Development, Takyelpat, Imphal, Manipur 795001, India

Louis Bengyella and Sayanika Devi Waikhom
Institute of Bioresources and Sustainable Development, Takyelpat, Imphal, Manipur 795001, India
Department of Biotechnology, Haldia Institute of Technology, Haldia 721657, West Bengal, India
Centre of Advanced Study in Life Sciences, Manipur University, Imphal, Manipur 795003, India

Pranab Roy
Department of Biotechnology, Haldia Institute of Technology, Haldia-721657, West Bengal, India

Tadele S and Emana G
Addis Ababa University, College of Natural and Computational Sciences, Department of Zoological Sciences, Ethiopia

Thulasi G Pillai
Department of Forest Pathology, Kerala Forest Research Institute, Peechi, Thrissur-680751, India

Jayaraj R
Divisions of Forest Ecology and Biodiversity Conservation, Kerala Forest Research Institute, Peechi, Thrissur-680751, India

Montaser F Abdel-Monaim, Marwa AM Atwa and Kadry M Morsy
Plant Pathology Research Institute, Agriculture Research Center, Giza 12619, Egypt

Sarmad Mansoor, Asmlam Khan M and Nasir Ahmed Khan
Department of Plant Pathology, University of Agriculture, Faisalabad, Pakistan

Mareeswaran J, Nepolean P, Jayanthi R, Premkumar Samuel Asir R and Radhakrishnan B
Plant Pathology Division, UPASI Tea Research Institute, Valparai-642127 Coimbatore District, Tamil Nadu, India

Firas Ali Ahmed, Brent S Sipes and Anne M Alvarez
Department of Plant and Environmental Protection Sciences, 3190 Maile Way University of Hawaii at Manoa, Honolulu, HI, 96822, USA

Abdullah S, Sehgal SK and Ali S
Department of Agronomy, Horticulture, and Plant Science (AHPS), South Dakota State University, Brookings, USA

Belay Habtegebriel
Ethiopian Institute of Agricultural Research, Plant Protection Research Center, P.O. Box 37, Ambo, Ethiopia

Anteneh Boydom
Ethiopian Institute of Agricultural Research, Holleta Agricultural Research Center, Holleta, Ethiopia

Rania Aydi Ben Abdallah
National Agronomic Institute of Tunisia, 1082, Tunis, University of Carthage, Tunisia
UR13AGR09- Integrated Horticultural Production in the Tunisian Centre-East, Regional Center of Research on Horticulture and Organic Agriculture, University of Sousse, 4042, Chott-Mariem, Tunisia

Hayfa Jabnoun-Khiareddine and Mejda Daami-Remadi
UR13AGR09- Integrated Horticultural Production in the Tunisian Centre-East, Regional Center of Research on Horticulture and Organic Agriculture, University of Sousse, 4042, Chott-Mariem, Tunisia

Boutheina Mejdoub-Trabelsi
UR13AGR09- Integrated Horticultural Production in the Tunisian Centre-East, Regional Center of Research on Horticulture and Organic Agriculture, University of Sousse, 4042, Chott-Mariem, Tunisia
Higher Agronomic Institute of Chott-Mariem, University of Sousse, 4042, Chott-Mariem, Tunisia

Muhammad Ali, Muhammad Adnan and Mehra Azam
Institute of Agricultural Sciences,University of the Punjab, Lahore-54590, Pakistan

Adelene SM Auyong
Faculty of Science, The University of Melbourne, Parkville 3010 VIC, Australia

Rebecca Ford
School of Environment, Griffith University, Southport 4222 QLD, Australia

Paul WJ Taylor
Faculty of Veterinary and Agricultural Sciences, The University of Melbourne, Parkville 3010 VIC, Australia

Rajendra Kumar Seth and Shah Alam and Shukla DN
Department of Botany, Bharagawa Agricultural Botany Laboratory, University of Allahabad, Allahabad, U.P., India

Hardik N Lakhani and Dinesh N Vakharia
Biotechnology Department, Junagadh Agricultural University, India

Abeer H Makhlouf
Agricultural Botany Department, Menoufiya University, Egypt

Ragaa A Eissa
Genetics Department, Menoufiya University, Egypt

Mohamed M Hassan
Genetics Department, Menoufiya University, Egypt
Scientific Research Center, Biotechnology and Genetic Engineering Unit, Taif University, KSA

Manica Tomar
Directorate of extension education, Dr. Y.S.P Univ. of Hort. and forestry, Nauni, Solan H.P, 173230, India

Harender Raj
Department of Plant Pathology, Dr. Y.S.P Univ. of Hort. and forestry, Nauni, Solan H.P, 173230, India

Kedar Nath and Solanky KU
Department of Plant Pathology, N.M. College of Agriculture, Navsari Agricultural University, Navsari-396450,India

Madhu Bala
Department of Plant Breeding and Genetics, College of Agriculture, Junagadh Agricultural University, Junagadh-362001, India

Chemeda Dilbo and Tariku Hunduma
Ethiopian Institute of Agricultural Research, Ambo Plant Protection Research Center, P.O. Box 37, Ambo, Ethiopia

Melaku Alemu
Ethiopian Institute of Agricultural Research, National Agricultural Biotechnology Research Laboratory, P.O. Box 31, Holetta, Ethiopia

Alemu Lencho
College of Agriculture and Veterinary Sciences, Department of Plant science, Ambo University, P.O. Box 19, Ambo, Ethiopia

Solanky KU
Department of Plant Pathology, N. M. College of Agriculture, Navsari Agricultural University, Navsari-645039, India

Kedar Nath
Department of Plant Pathology, N. M. College of Agriculture, Navsari Agricultural University, Navsari-645039, India
Regional Rice Research Station, Navsari Agricultural University, Vyara 39 4650, India

Madhubala
Regional Rice Research Station, Navsari Agricultural University, Vyara 39 4650, India

Rakesh M Swami
Department of Plant Molecular Biology and Biotechnology, Navsari Agricultural University, Navsari 39 450, India

Mahatma MK
Department of Plant Molecular Biology and Biotechnology, Navsari Agricultural University, Navsari 39 450, India
Directorate of Groundnut Research, Indian Council of Agricultural Research, P. B. No.5, Junagadh 362 001, India

Boutheina Mejdoub-Trabelsi
Institut Supérieur Agronomique de Chott-Mariem, Université de Sousse, 4042, Chott-Mariem, Tunisia
UR13AGR09, Production Horticole Intégrée au Centre Est Tunisien, Centre Régional des Recherches en Horticulture et Agriculture Biologique de Chott-Mariem, Université de Sousse, 4042, Chott-Mariem, Tunisia

Hayfa Jabnoun-Khiareddine and Mejda Daami-Remadi
UR13AGR09, Production Horticole Intégrée au Centre Est Tunisien, Centre Régional des Recherches en Horticulture et Agriculture Biologique de Chott-Mariem, Université de Sousse, 4042, Chott-Mariem, Tunisia

Mouna Gueddes-Chahed, Hayfa Jabnoun-Khiareddine and Mejda Daami-Remadi
UR13AGR09-Integrated Horticultural Production in The Tunisian Centre-East, Regional Center of Research on Horticulture and Organic Agriculture, University of Sousse, 4042, Chott-Mariem, Tunisia

Riad SR El-Mohamedy and Farid Abdel-Kareem
Plant Pathology Department, National Research Center, Dokki, Giza, Egypt

Rania Aydi Ben Abdallah
UR13AGR09-Integrated Horticultural Production in The Tunisian Centre-East, Regional Center of Research on Horticulture and Organic Agriculture, University of Sousse, 4042, Chott-Mariem, Tunisia
National Agronomic Institute of Tunisia, 1082 Tunis Mahrajène, University of Carthage, Tunisia

Index

www.ingramcontent.com/pod-product-compliance
Lightning Source LLC
Chambersburg PA
CBHW080623200326
41458CB00013B/4480